Nano/Micro Science and Technology in Biorheology

Rio Kita · Toshiaki Dobashi
Editors

Nano/Micro Science and Technology in Biorheology

Principles, Methods, and Applications

Springer

Editors
Rio Kita
Department of Physics
Tokai University
Hiratsuka, Kanagawa, Japan

Toshiaki Dobashi
Division of Molecular Science,
Faculty of Science and Technology
Gunma University
Kiryu, Gunma, Japan

ISBN 978-4-431-54885-0 ISBN 978-4-431-54886-7 (eBook)
DOI 10.1007/978-4-431-54886-7

Library of Congress Control Number: 2015942247

Springer Tokyo Heidelberg NewYork Dordrecht London

Printed on acid-free paper

Springer Japan KK is part of Springer Science+Business Media (www.springer.com)

Preface

The Japanese Society of Biorheology was established in 1977 by scientists and experts in a wide variety of fields who were interested in deformation and flow in biological systems and their components, with Syoten Oka, a pioneer in this field, being the first president. Since then, the Society has been offering researchers the opportunity to have common information and recognition in this interesting interdisciplinary field that relates life science, clinical medicine, medical engineering, food science, physics, chemistry, and other fields. It is evident that recent developments in biorheology have not only expanded the range of fields that it covers but also deepened each field by integrating new methods, technologies, and theories. Thus, to provide current leading information, the Society intends to publish a series of books on recent topics in biorheology. This book, the first one of the series, deals with nano/micro science and technology in biorheology. Measurement techniques on the nano- and microscale developed over the past few decades are now indispensable for analyzing complex phenomena in biological systems from the aspect of biorheology, because the main targets of biorheology are generally inhomogeneous at these scales. The topics raised in this book should be of particular interest to readers working at research facilities in both industry and academia dealing with cutting-edge bio-related subjects, and it is hoped that the book will provide suggestions helping them to solve problems they encounter in their work.

Hiratsuka, Kanagawa, Japan — Rio Kita
Kiryu, Gunma, Japan — Toshiaki Dobashi

Contents

Introduction of Nano/Micro Science and Technology in Biorheology

1

Rio Kita and Toshiaki Dobashi

Rheology is the science of the deformation and flow responses of viscoelastic materials resulting from stimulation, mainly by inducing mechanical stress. The stimulus-response relationship or response function is a characteristic of materials, and thus rheology provides general theoretical and experimental approaches that can be applied to the study of an object. This enables us to understand the macroscopic elasticity, viscosity, and viscoelasticity of materials on a molecular basis.

Over the past few decades, knowledge of the fundamental science on which rheology is based has increased considerably. For example, the phase transition of soft matter such as polymers and liquid crystals has been explained in a unified manner [1]. A description of irreversible thermodynamics in relation to rheology has been attempted. Critical phenomena associated with the anomalous growth of fluctuations near the critical point indicate the universality of such phenomena independent of materials [2]. The direct application of these progresses in physics and the modification of the rheological concept, however, seem to be limited. On the other hand, considerable progress in measurement techniques and newly developed methods such as mass spectroscopy [3], nuclear magnetic resonance (NMR) [4, 5], femtosecond spectroscopy [6], the use of imaging CCD sensors [7], and scanning probe microscopy [8] has markedly affected the study of rheology. The method of particle tracking using a microscope known as microrheology is carried out by measuring the motion of colloidal particles to determine the mechanical

R. Kita
Department of Physics, Tokai University, 4-1-1 Kitakaname,
Hiratsuka, Kanagawa 259-1292, Japan
e-mail: rkita@keyaki.cc.u-tokai.ac.jp

T. Dobashi (✉)
Division of Molecular Science, Faculty of Science and Technology,
Gunma University, 1-5-1 Tenjin, 376-8515 Kiryu, Gunma, Japan
e-mail: dobashi@gunma-u.ac.jp

R. Kita, T. Dobashi (eds.), *Nano/Micro Science and Technology in Biorheology*,
DOI 10.1007/978-4-431-54886-7_1

properties of the surrounding medium [9]. In active microrheology, the response function is determined by measuring the displacement of colloidal particles in response to an external force. In passive microrheology, the response function is determined by measuring the fluctuation density of colloidal particles using the fluctuation-dissipation theorem in the equilibrium state. Both methods have been developed to lead to a response function that is also in a nonequilibrium state [10]. Improvements of scanning probe microscopy (SPM) such as atomic force microscopy (AFM) have enabled us to perform mechanical measurements at the nanoscale [11]. By combining such methods with laser trapping or the use of optical tweezers to fix the position of molecules or to shift the position of their ends, we can now perform the mechanical characterization of a single molecule directly. These methods are called of nanorheology [12, 13]. In addition to these methods, a wide variety of other techniques are available such as local area measurement using surface plasmons [14], measurement of the dynamics of a fluorescent probe molecule from the autocorrelation function of fluctuations in fluorescent intensity [15], and measurement of the distribution of the elastic modulus in real space determined by nuclear magnetic resonance [16], which is obtained using an inverse program as in X-ray computer tomography (CT). Thus, measurement in nano- or micro-domains by a complementary approach has helped deepen our understanding of macroscopic phenomena at the molecular level. This type of rheology that treats responses at the molecular level induced by various types of physical stimulation is generally called molecular rheology. Thus, rheology can be applied to the study of various soft substances such as polymers, colloids, blood, foods, and cosmetics.

It is reasonable that these new technologies have been actively applied to the study of biological structures, such as cells, tissues, and organs, since they are inhomogeneous in elasticity and viscosity even at the nano- and microscales and the corresponding time scale, generally having multiple characteristic length and time scales, as well as hierarchical structures considered to play an essential role in vivo. The main goal of classical biorheology seems to be to macroscopically understand complex phenomena by simple modeling, called phenomenological rheology [17]. The recent remarkable development and improvement of analytical methodologies, apparatus, and theories, especially at the nano- and microscales, have promoted the development of additional methods of rheological analysis at these scales. Furthermore, rheological methodologies have been used to detect anomalies in the human body in clinical medicine, which is closely related to biorheology. Therefore, the establishment of nano-/micro-science and technology in biorheology is a natural extension of molecular rheology.

At the nano- and microscales, it is necessary to investigate various vibration and rotation modes, the segmental motion of macromolecules, and inter- and intramolecular interactions, since the behavior of a system at these scales is related to diffusion, collision, and molecular aggregation resulting from thermal motion. Such molecular mechanisms cannot always be clarified only by studies based on mechanical stimulation because of the limited experimental range in the time window of rheological experiments. However, combinations of complementary methods utilizing electric, magnetic, and thermal stimulus-response relationships [18, 19]

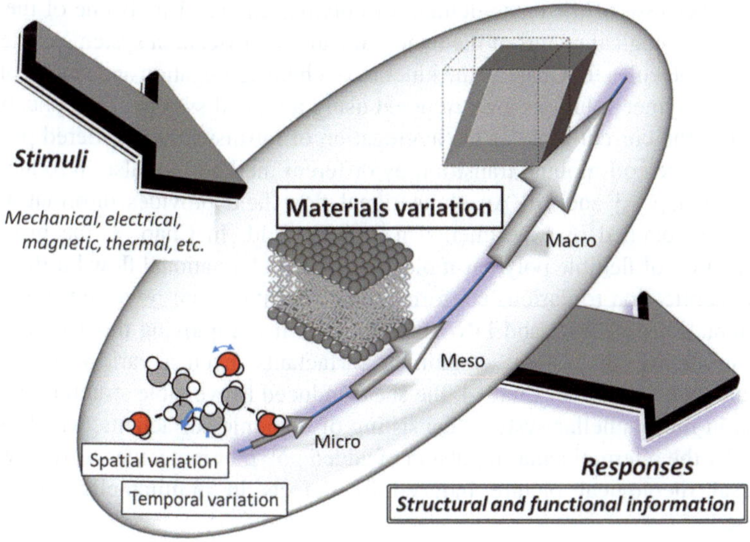

Fig. 1.1 Image of nano-/micro-science and technology in biorheology regarding variation in materials, time and space, and stimulus-response function

can cover a broad experimental range in time and space, which enables us to understand the interdependence of molecular mechanisms and ultimately of macroscopic properties. This aspect is even more important in biorheology, the main targets of which are usually inhomogeneous in both time and space as described above. In this book, not only mechanical stimuli but also electric and magnetic stimuli are addressed comprehensively, as illustrated in Fig. 1.1. Such stimuli cover the nano- and microscale ranges, in which the rheological aspects of various phenomena should be clarified in terms of molecular dynamics, which should cover a broad time scale. For this purpose, specific techniques should be employed. In writing this book, we did not restrict ourselves to the research fields conventionally called microrheology and nanorheology, and we went as far as to introduce a variety of research results that can be integrated into biorheology. We, however, limited our discussion of nanorheology and microrheology, which have already been published in reviews and books, to the introduction of references and a brief description.

In this book, recent developments in methodologies and novel results or information obtained are described by experts in each field, together with the biological and medical aspects of their applications. Here we summarize the content of the subsequent chapters, and their biomedical applications are appended as follows:

When a dilute polymer solution is quenched below the consolute point, the polymer chains collapse and chain aggregation occurs competitively owing to the intra- and intermolecular interactions between polymer segments. At very low concentrations, polymer chains collapse first (the coil-globule transition) and the collapsed chains aggregate, whereas at high concentrations, it is even more complex because of the entanglement of polymer chains. This is one of the most essential rheological events at the nanoscale and microscale in systems containing polymer molecules. In Chap. 2, the kinetics of chain aggregation and chain collapse in dilute polymer solutions are discussed using a typical system. The concepts are applied to the current topic of the aggregation of intrinsically disordered proteins. Studies on the coil-globule transition by different methods are also introduced for DNA in Chaps. 3 and 14. An elongational flow field provides information that cannot be obtained in experiments in a shear field. In Chap. 3, the history of investigation of flexible polymer molecules by the elongational flow birefringence and its application to various conformational transitions are reviewed. Studies on the interactions of DNA and DNA-binding proteins employing this technique are also introduced. In systems containing surfactants, even a variety of complex structures are formed. In Chap. 4, the shear-induced irreversible structure transformation in multilamellar systems consisting of nonionic surfactants, which is controlled by the intermolecular repulsion of added polymer molecules, is reviewed. A universal rheological picture induced by defects in multilamellar systems is proposed.

Although it is well known that the collective diffusion and the thermal diffusion are coupled to affect each other in accordance with Onsager's reciprocal theorem, its application to biopolymer solutions has been achieved only recently in a newly developed methodology and apparatus. The Soret coefficient, which characterizes the coupling of two types of current determined from measurements, is a promising tool for microfluidic separation and thermophoresis in biomolecular solutions. In Chap. 5, recent developments in apparatus and experimental results for studying the thermal diffusion using laser holography are reviewed.

In Chaps. 6 and 7, recent progress in the biorheological application of nuclear magnetic resonance is introduced. The development of nucleic magnetic resonance has afforded many measurement modes. In pulsed gradient diffusion experiments called PGSE, the magnetization is excited with a 90° radiofrequency pulse and then dispersed, and a spin echo is the refocusing of spin magnetization by a pulse of resonant electromagnetic radiation. Using this method, the high mobility of water in biological systems can be observed. In Chap. 6, the diffusion of water in typical biological systems is reviewed. By taking an image of the wave intensity distribution in a tissue, induced externally using magnetic resonance imaging, magnetic resonance elastography (MRE) can be performed noninvasively. In Chap. 7, the theoretical basis of this technique is surveyed, the current state of medical applications is reviewed, and a physical and mathematical basis is given.

In Chaps. 8 and 9, recent topics related to the progress of dielectric spectroscopy are introduced. The improvement of electronic devices enables measurements to be performed with high precision and in situ broadband dielectric relaxation spectra to be obtained. In particular, the various states of water in biological systems can now be identified. In Chap. 8, the basis of this technique and the dynamics of water and biomacromolecules, along with their application in vivo to tissues such as skin, are described. Chapter 9 focuses on the glass transition of water in various biomaterials such as foods.

Nano-/microspheres and nano-/microcapsules are now used in a variety of bio-related engineering applications such as in pharmaceutical, medical, and food engineering. The rheological method is important in the process of the preparation procedure and in analyzing the resulting characteristic structure, the physicochemical properties of these small particles, and their functions in biological systems. In Chap. 10, the essential properties of microcapsules with a gel wall membrane are summarized focusing on the rheological aspects. Some typical systematic studies using conventional and novel preparation methods of preparing microcapsules having isotropic or anisotropic wall membranes are introduced. In Chap. 11, biological and medical uses of microparticles are reviewed. Most tissues in biological systems are anisotropic and sometimes their physical quantities have a gradient; thus, such properties are also required for artificial tissues such as artificial vessels. Anisotropy or a gradient in scaffolds often affects the cell viability and growth rate. In Chap. 12, the design and control methods of scaffolds for bone cells are reviewed.

In the functioning of biological molecules, various parameters such as the charge, density, and refractive index as well as viscoelastic properties change over time. The electrostatic detection of biomolecular recognition is possible using a biologically coupled field-effect transistor (bio-FET) in accordance with a principle based on the potentiometric detection of a change in charge density induced at a gate insulator/solution interface by specific biomolecular recognition. Such FET devices have also been used in combination with pulsatile living cells to study extracellular potential changes. This technique is also applied for drug transport analysis. In Chap. 13, noninvasive monitoring of the uptake kinetics of substrates mediated by membrane-bound transporters and the discrimination of transporting ability among the genotypes of transporters using an oocyte-based FET are reviewed. Recent progress of the fluorescence probe imaging of single molecules has led to the clarification of the mechanism of the conformation transition of various biomacromolecules. In Chap. 14, the all-or-none transition and a pathway through intrachain phase-separated structures by the imaging of DNA in real time and space are reviewed.

Applications to gene delivery are also introduced. The rheological properties of living cells play important roles in regulating their various biological functions. AFM is a useful technique for single-cell diagnosis because the rheological properties of adherent cells at any region on a surface can be obtained without any modifications. Therefore, by combining AFM and rheological techniques, we can obtain information of nanoscale and microscale image. In Chap. 15, AFM techniques for examining single-cell rheology in the frequency and time domains are reviewed. The specificity and universality of cell rheology are demonstrated using AFM.

To facilitate understanding of the recent advances for beginners in this field, we briefly introduce fundamental explanations of the methodologies developed and their physical background in Supplement (Chap. 16). In Supplement I, we describe the structure and motion of the various materials discussed in this book. In Supplement II, we survey the principles of the physical and methodological aspects of nano-/micro-science and technology in biorheology, in relation to external forces and related phenomena, such as the transport behavior, diffusion, aggregation, and relaxation of molecules. In Supplement III, principles of experimental method and related physical property are summarized.

Finally, we would like to express our thanks to Dr. Makoto Kaibara and Dr. Kazuo Tanishita for suggesting the idea of the concept of this book.

References

1. de Gennes PG (1979) Scaling concepts in polymer physics. Cornell University Press, Ithaca
2. Domb C, Green M, Lebowitz JL (1991) Phase transitions and critical phenomena. Academic, London/New York
3. Schlag EW (2012) Time of flight mass spectroscopy and its application. Elsevier, Amsterdam
4. James TL (ed) (2005) Nuclear magnetic resonance of biological macromolecules, Part C, vol 394, Methods in enzymology. Elsevier Academic Press, New York
5. Michel M, Modo J, Bulte JWM (eds) (2007) Molecular and cellular MR imaging. CRC Press, Boca Raton
6. Hannaford P (2005) Femtosecond laser spectroscopy. Springer, New York
7. Flora L, Nathan A (2010) CCD image sensors in deep-ultraviolet: degradation behavior and damage mechanisms (microtechnology and MEMS). Springer, Berlin/New York
8. Bhushan B, Fuchs H, Tomitori M (2008) Scanning probe microscopy techniques. Springer, Berlin/New York
9. MacKintosh FC, et al. (1999) Microrheology. Curr Opin Colloid Interface Sci 4:300; Capsi A, et al. (2000) Enhanced diffusion in active intracellular transport. Phys Rev Lett 85:5655
10. Wilson LG, Poon WCK (2011) Small-world rheology: an introduction to probe-based active microrheology. PCCP 22:10617; Lee H, Ferrer JM, Nakamura F, Lang MJ, Kamm RD (2010) Passive and active microrheology for cross-linked F-actin networks in vitro. Acta Biomater 6 (4):1207–1218
11. Haga H, et al. (2000) Elasticity mapping of living fibroblasts by AFM and immunofluorescence observation of the cytoskeleton. Ultramicroscopy 82:253; Bhanu P, Heinrich J, Hoerber JK (eds) Atomic force microscopy. Academic Press, San Diego
12. Matthew MJ, Fordyce PM, Engh AM, Neuman KC, Block SM (2004) Simultaneous, coincident optical trapping and single-molecule fluorescence. Nat Methods 1:133–139
13. Neuman KC, Nagy A (2008) Single-molecule force spectroscopy: optical tweezers, magnetic tweezers and atomic force microscopy. Nat Methods 5:491–505
14. Willets KA, Van Duyne RP (2007) Localized surface plasmon resonance spectroscopy and sensing. Annu Rev Phys Chem 58:267–297
15. Masuda A, Ushida K, Okamoto T (2005) Direct observation of spatiotemporal dependence of anomalous diffusion in inhomogeneous fluid by sample-volume-controlled fluorescence correlation spectroscopy. Phys Rev E 72:060101R
16. Muthupillai R, Lomas DJ, Rossman PJ, Greenleaf JF, Manduca A, Ehman RL (1990) Magnetic resonance elastography by direct visualization of propagating acoustic strain waves. Science 269:1854
17. Oka S (1981) Cardiovascular hemorheology. Cambridge University Press, Cambridge
18. Kremer F, Schonhals A (2003) Broadband dielectric spectroscopy. Springer, Berlin/New York
19. Callaghan PT (2006) Rheo-NMR and velocity imaging. Curr Opin Colloid Interface Sci 11:13

Rheological Aspects of Conformational Change and Molecular Aggregation of Macromolecules

2

Yasuyuki Maki

Abstract

When the temperature of a dilute polymer solution is changed gradually far below the Θ point, the conformation of individual polymer chains changes from a random coil to a compact globular state (coil-globule transition), followed by intermolecular aggregation owing to phase separation. In this chapter, the coil-globule transition and the aggregation behavior in dilute solutions of typical neutral synthetic homopolymers, poly(*N*-isopropylacrylamide) (PNIPAM) and poly(methyl methacrylate) (PMMA), are reviewed, and some rheological aspects of the kinetics of chain collapse and chain aggregation are demonstrated. The applicability of the rheological concepts derived from the aggregation behavior of synthetic homopolymers to that of intrinsically disordered proteins (IDPs) is considered.

Keywords

Coil-globule transition • Aggregation • Mesoglobule • Phase separation

2.1 Introduction

The conformation of a linear, flexible polymer in a dilute solution depends on the interactions between monomers of polymer and solvent molecules. Such interactions can be represented by the effective interactions of monomers through a solvent, which are commonly referred to as excluded-volume interactions. The effective interactions of monomers are renormalized by the presence of the solvent and, therefore, should depend on the type of solvent species as well as on

Y. Maki (✉)
Division of Molecular Science, Faculty of Science and Technology,
Gunma University, Tenjin-cho 1-5-1, Kiryu 376-8515, Japan
e-mail: maki@gunma-u.ac.jp

R. Kita, T. Dobashi (eds.), *Nano/Micro Science and Technology in Biorheology*,
DOI 10.1007/978-4-431-54886-7_2

temperature [1]. According to Flory, the swelling of a single neutral linear chain in a given solvent is determined by two main factors: the excluded-volume interactions and entropic elastic forces [2]. When the attractive and repulsive forces between monomers due to excluded-volume interactions balance each other, i.e., at the Θ point, the polymer chain assumes a Gaussian coil conformation. In the case that the repulsive forces between monomers prevail over attractive ones, i.e., in a thermodynamically good solvent, a polymer chain swells and adopts the conformation of an expanded coil. Conversely, if the attractive interactions between monomers become dominant, i.e., in a thermodynamically poor solvent, a polymer chain shrinks. Theoretically, a polymer coil collapses abruptly below the Θ point and assumes a compact globular conformation (coil-globule transition) [1]. If there are many chains in the solutions, a decrease in solvent thermodynamic quality tends to enhance the attraction between monomers in different polymer chains, resulting in intermolecular aggregation and the subsequent phase separation.

Polymers could show thermosensitive behaviors because the excluded-volume interactions between monomers depend on temperature: their chain conformation and phase behavior can be controlled by adjusting temperature. If polymer chains are cross-linked, the resultant polymer network shows a volume phase transition in response to a change in temperature [3]. The volume phase transition of the gel can be regarded as the coil-globule transition of a polymer of infinitely large molecular weight [1]. The thermosensitive behavior of polymer solutions and gels seems to have potential in biomedical and biotechnological applications such as drug delivery devices and intelligent materials [4–6].

Conformational change and aggregation phenomena of macromolecules may also be important in understanding biological systems. Proteins are biopolymers with a globular structure in aqueous solutions. Protein aggregation is one of the crucial problems in molecular biology and biotechnology, while the aggregation of globular proteins denatured by heat or pressure can be a useful property for developing food texture and has been extensively investigated owing to its importance in food application [7]. Moreover, protein aggregation is well represented in human etiology. More than 20 different diseases are caused at least partially by abnormal protein aggregation. Some of these diseases are characterized by the formation of protein fibrils, such as Alzheimer's disease. Achieving better understanding of protein aggregation is critical not only in various biopharmaceutical processes but also in finding a cure for those devastating diseases [8].

Rheology is defined as the science of flow and deformation of matter. Because even a single polymer chain has a large number of degrees of freedom in its conformation, the dynamics of polymer chains in a solvent would show some aspects of rheological processes, such as relaxation, dissipation, and diffusion. The scope of this chapter is to show some rheological aspects in the kinetics of the coil-globule transition of each polymer and that of intermolecular aggregation when the temperature of a dilute polymer solution under good solvent condition is changed rapidly to cross the Θ point. In the case of the coil-globule transition, the dimensions of a single polymer chain decrease with time to reach an equilibrium value corresponding to a globular state. This relaxation process would be relevant to viscous dissipation due to the interactions between polymer chains and solvent molecules or due to self-entanglement in polymer chains. In the case of the intermolecular aggregation, two polymer chains or clusters of many chains will collide and merge into one large aggregate by means of interdiffusion.

There have been numerous studies of conformational changes and intermolecular aggregation under poor solvent conditions for various polymer-solvent systems [9, 10]. This chapter, however, is not intended to encompass all of these polymer-solvent systems, but to describe only some representative polymer-solvent systems in order to demonstrate the abovementioned rheological aspects of these processes. In the next section, the phase behaviors observed in typical polymer-solvent systems are described. In the subsequent sections, the coil-globule transition and the kinetics of chain collapse and chain aggregation observed in dilute solutions of poly(*N*-isopropylacrylamide) (PNIPAM) and poly(methyl methacrylate) (PMMA), which are model systems showing typical phase behaviors, are reviewed, and some rheological aspects in the kinetics of chain collapse and chain aggregation are demonstrated. In the last part of this chapter, the applicability of the rheological concepts derived from the aggregation behavior of synthetic homopolymers to that of intrinsically disordered proteins (IDPs) [11] is considered.

2.2 Typical Phase Behaviors of Polymer Solutions

For solutions of neutral polymers in organic solvents, the interaction energy $u(r)$ between monomers is as shown schematically in Fig. 2.1 [1]. At shorter distances, repulsion is predominant because of the geometric volume of monomers, while at longer distances, the monomers are attracted by van der Waals forces. At high temperatures ($kT >> \varepsilon$), the attractive (negative) part of $u(r)$ is negligible, but as temperature decreases, the attractive part becomes more predominant. Thus, the excluded-volume interaction between monomers is repulsive at higher

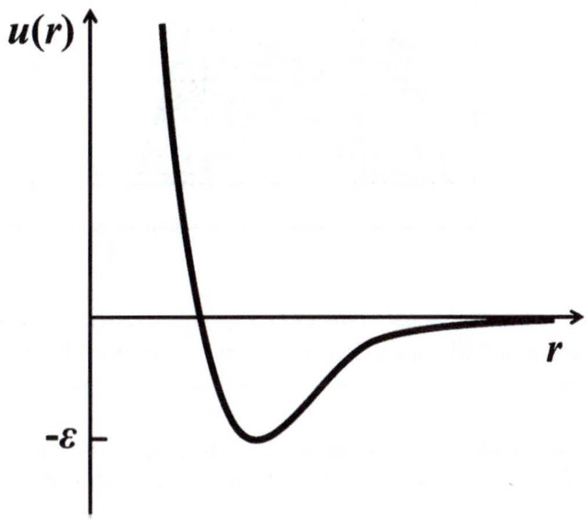

Fig. 2.1 Schematic illustration of effective interaction energy $u(r)$ between monomers

temperatures and attractive at lower temperatures for these polymer-solvent systems; polymer chains collapse on cooling, followed by chain aggregation due to phase separation. The temperature at which phase separation takes place depends on the polymer concentration and molecular weight. The phase behavior of a polymer-solvent system as a function of temperature is conveniently represented by a phase diagram where the phase-separation boundary, or binodal, indicates the temperature at which a given polymer-solvent mixture passes from a one-phase system to a two-phase system consisting of a polymer-rich phase and a polymer-poor phase. The maximum of the binodal is the upper critical solution temperature (UCST) (Fig. 2.2a) [1].

There is another type of phase behavior observed in some polymer-solvent systems. Solutions of nonionic water-soluble polymers in water can be described by, in many cases, phase diagrams with a lower critical solution temperature (LCST) (Fig. 2.2b).[1] For polymer-solvent systems showing LCST behavior, polymer chains collapse on heating, followed by phase separation. In water-soluble polymers, the hydrophilicity of a polymer and its ability to mix with water arise from specific interactions between its functional groups and water molecules. The solubility of polymers can be traced to the formation of hydrogen bonds between hydrophilic groups and water molecules. Hydrophilic groups, which are capable of forming hydrogen bonds with water molecules, are usually attached to side chains, whereas the hydrocarbon backbone itself is insoluble in water. The entire polymer is solubilized in aqueous media if the number of hydrophilic groups along the polymer chain is large enough. Water molecules, which are incapable of forming

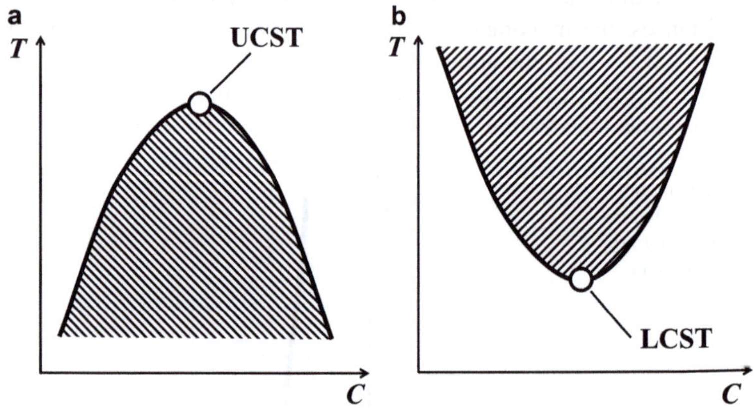

Fig. 2.2 Schematic illustrations of phase diagram for a UCST-type polymer-solvent system (**a**) and an LCST-type polymer-solvent system (**b**). The shaded areas represent two-phase regions

[1] An aqueous solution of water-soluble polymers shows an LCST-type behavior in many cases. However, a group of water-soluble polymers showing a UCST-type behavior has attracted much attention in recent years [30].

hydrogen bonds with nonpolar groups, reorient themselves around these nonpolar groups, forming regions of structured water and thus decreasing entropy upon mixing (the hydrophobic effect) [12]. Hydrophobic groups have a tendency to associate in order to decrease their overall surface area of interaction with water and release structured water molecules in the bulk. At higher temperatures, the entropy term predominates over the exothermic enthalpy of hydrogen-bond formation. Therefore, the free-energy change upon mixing becomes positive, and phase separation occurs.

The LCST-type phase behaviors of thermosensitive polymers in aqueous media are important because of the possible biopharmaceutical and biotechnological applications of these polymers. However, the mechanism of LCST is rather complex, and thus the quantitative comparison between theory and experiment may be difficult. Among various water-soluble homopolymers showing the LCST behavior, poly(*N*-isopropylacrylamide) (PNIPAM), which has a Θ temperature of approximately 31 °C, has been studied of its coil-globule transition and intermolecular aggregation extensively [13]. To study the cases of UCST-type homopolymer solutions, dilute solutions of poly(methyl methacrylate) (PMMA) have been used for the quantitative investigation of the coil-globule transition and interchain aggregation [14–29]. Thus, in the subsequent sections, studies of dilute solutions of PNIPAM and PMMA as model systems of LCST-type solutions and UCST-type solutions[2,3], respectively, are reviewed to demonstrate the rheological aspects in chain collapse and chain aggregation.

2.3 Coil-Globule Transition of a Polymer Chain

2.3.1 Quantitative Theories of Coil-Globule Transition

Since Stockmayer [32] predicted collapse of the single polymer chain below the Θ point, many theoretical investigations on coil-globule transition have been conducted [9]. Birshtein and Pryamitsyn [33] developed a theory based on the simple scheme of Flory [2] and derived an equation for the quantitative comparison between theory and experiment. In the Flory-type theory, the expansion or shrinkage of a polymer chain under different conditions is represented by the expansion factor $\alpha = R/R_0$, where R is a characteristic polymer dimension and R_0 is R in the ideal state (at the Θ point). Experimentally, the characteristic size of a polymer chain depends on the measurement technique. For example, it is the mean-square radius of gyration $<S^2>$ for the static light scattering (SLS) experiment but the hydrodynamic radius

[2] Note that phase diagrams of some polymer-solvent systems show both UCST- and LCST-type behaviors [31].

[3] Although this chapter focuses mostly on homopolymers, introducing comonomers and the resultant copolymers or heteropolymers is also of great importance from the viewpoint of applications [10, 13].

<R_h> for the dynamic light scattering (DLS) experiment. According to the theoretical approach of Flory, polymer free energy can be written as

$$F = F_{el} + F_{int} \tag{2.1}$$

where F_{el} and F_{int} are the contributions of the deviation of the conformational entropy from the ideal chain and of the excluded-volume interactions, respectively. The latter is calculated by mean-field approximation with virial expansion:

$$F_{int} = VkT\left[v\tau n^2 + wn^3 + \cdots\right] \tag{2.2}$$

Here, V, n, $v\tau = v(1 - \Theta/T)$, and w are the volume occupied by a polymer chain, the segment density in the volume V, and the second and third virial coefficients, respectively. With respect to the entropy term, $F_{el} \sim kT\alpha^2$ is expected for the good solvent condition. For the poor solvent condition, Birshtein et al. [33] derived an equation for F_{el} as

$$F_{el} = \gamma kT\left[\alpha_S^{-2} + 2\ln \alpha_S\right] \tag{2.3}$$

where γ is a constant and α_S is the expansion factor defined as $\alpha_S{}^2 = \langle S^2\rangle/\langle S^2\rangle_0$. By using Eqs. (2.1), (2.2), and (2.3), an equation of the expansion factor α_S is derived in the form of

$$\alpha_S^3 - \alpha_S - C\left(\alpha_S^{-3} - 1\right) = B\tau M^{1/2} \tag{2.4}$$

where M is the molecular weight of the polymer. The parameters B and C in Eq. (2.4) are related explicitly to v and w in Eq. (2.2), respectively. Thus, experimental data can be readily compared with Eq. (2.4) to obtain the parameters of the second ($v\tau$) and third (w) virial coefficients of segment interactions.

Among the experimentally observable characteristic of chain dimensions, only the radius of gyration was considered in the Flory-type theory of Birshtein and Pryamitsyn. Grosberg and Kuznetsov [34–36] developed a theory (GK theory) of the coil-globule transition according to a more rigorous and complex scheme derived by Lifshitz et al. [37]. In the GK theory, both the radius of gyration $\langle S^2\rangle^{1/2}$ and the hydrodynamic radius <R_h> have been considered, and a comparison between the theoretical and experimental results can be made quantitatively.

2.3.2 Coil-Globule Transition of PNIPAM

There have been only a few experimental studies of the coil-globule transition of a polymer chain in solution. This is because there are numerous practical problems to overcome, particularly those associated with intermolecular aggregation and the subsequent macroscopic phase separation, which may occur, in most cases, before chain collapse finishes and obscures coil-globule transition. To minimize the effects of aggregation, most experiments have been carried out with highly diluted polymer solutions. For example, light scattering experiments have been performed on polystyrene (PS) in cyclohexane, which is a typical UCST-type system, at extremely low concentrations ($<10^{-7}$ g/cm^3), and coil-globule transition was observed [38]. Because of difficulties in the experiments such as the low signal-to-noise ratio due to the very low concentrations, data for the coil-globule transition of PS in cyclohexane obtained by different groups seem contradictory [39].

Several groups [40–42] have observed the coil-globule transition of PNIPAM in water, which shows an LCST-type phase behavior, by means of static light scattering (SLS) and dynamic light scattering (DLS) techniques. In their studies, SLS and DLS measurements were carried out on dilute solutions of PNIPAM in water below and above the Θ temperature (30.6 °C), and the temperature dependences of the radius of gyration $<S^2>^{1/2}$ and the hydrodynamic radius $<R_h>$ were determined.

Kubota et al. [40] carried out experiments on PNIPAM samples with weight-average molecular weights M_w ranging from 1.6×10^6 to 2.5×10^7 at concentrations from 1×10^{-5} to 2×10^{-4}. The dimensions of the polymer chains showed a sharp decrease just above the Θ temperature. The data obtained under the poor solvent condition were limited because at temperatures 1–2 K higher than the Θ temperature, the solution became metastable where scattered light intensity increased gradually, suggesting that chain aggregation was in progress.

Wu and Zhou [41] also observed chain collapse above the Θ temperature in dilute solutions of narrowly distributed ($M_w/M_n < 1.05$), high molecular weight PNIPAM with $M_w = 1.1 \times 10^7$ and 1.2×10^7 at a concentration of 5×10^{-6} g/cm^3. They observed intermolecular aggregation above ca. 32 °C, where they noted that individual PNIPAM chains first collapsed to densely packed single-chain globules that were stable for a limited period (~ 10^2 min), which were then thermodynamically driven to aggregate. In this kinetically stable two-phase region, the dimensions of the polymer chain were measured up to ca. 33 °C.

Wang et al. [42] studied the coil-globule transition of PNIPAM using an extremely dilute (7×10^{-7} g/cm^3) solution of narrowly distributed ($M_w/M_n < 1.05$) PNIPAM with $M_w = 1.3 \times 10^7$. Because the solution was thermodynamically stable up to ca. 39 °C, the dimensions of the polymer chain in the highly collapsed state were measured at temperatures far above Θ.

From Eq. (2.4), it is conjectured that the coil-globule transition of PNIPAM is represented by the chain expansion factor α_S as a function of $\tau M_w^{1/2}$, independent of the molecular weight M_w. Figure 2.3 shows the temperature dependence of the chain dimensions of PNIPAM in water in a plot of α_S versus $\tau M_w^{1/2}$ made from data cited from different works [40–42]. The sharp decrease in α_S above the Θ temperature indicates the coil-globule transition of PNIPAM. In Fig. 2.3, the coil-globule transition curves observed by the different groups are almost the same, independent of the molecular weight of the sample. Figure 2.4 shows the same data as in Fig. 2.3, plotted as $\alpha_S^3 \tau M^{1/2}$ versus $\tau M^{1/2}$ for closer inspection. Figure 2.4 shows that the data obtained by the different groups are in rather good agreement except that measured by Kubota et al. [40] for the sample with the highest molecular weight $M_w = 2.5 \times 10^7$. They estimated the polydispersity of this sample as $M_w/M_n = 1.3$

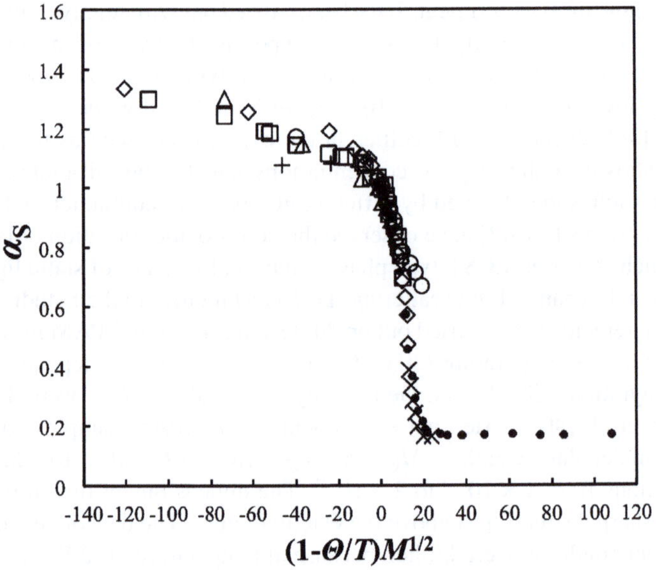

Fig. 2.3 Coil-globule transition curves for PNIPAM in water. Data obtained from studies by Kubota et al. [40] (○: $M_w = 25 \times 10^6$, □: $M_w = 9.1 \times 10^6$, △: $M_w = 4.1 \times 10^6$, +: $M_w = 1.6 \times 10^6$), Wu et al. [41] (◇: thermodynamically stable one-phase region, ×: kinetically stable two-phase region), and Wang et al. [42] (●)

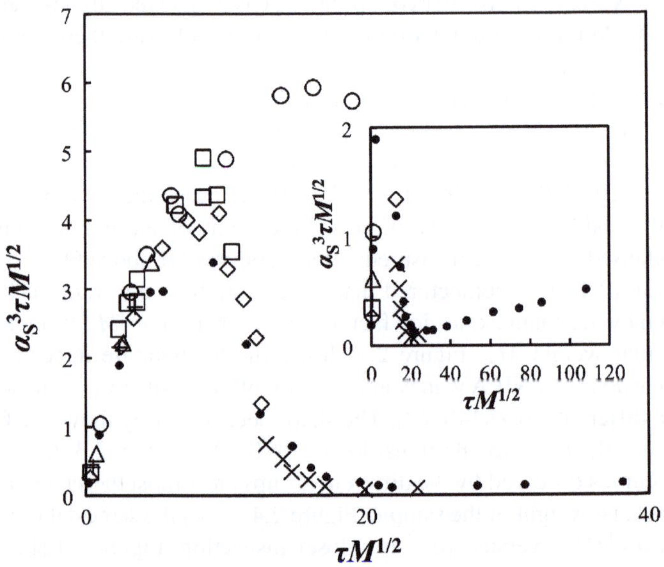

Fig. 2.4 Coil-globule transition curves for PNIPAM in water obtained using the plots of $\alpha_S{}^3\tau M^{1/2}$ against $\tau M^{1/2}$. The symbols are the same as those in Fig. 2.3

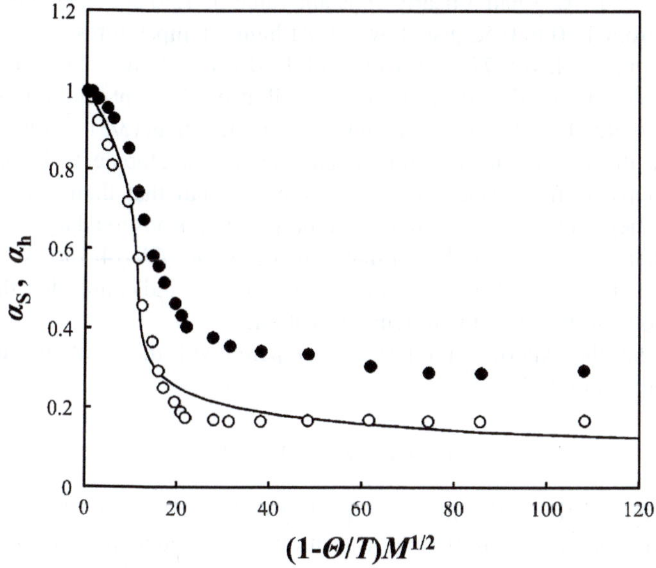

Fig. 2.5 Coil-globule transition curves for PNIPAM in water obtained using the plots of α_S (○) and α_h (●) against $\tau M^{1/2}$. The solid curve represents the data calculated from Eq. (2.5)

from $\mu_2/<\Gamma>^2$ determined from the correlation function of DLS data; the discrepancy of the data of this sample from the others would be due to effects of the molecular weight distribution.

In Figs. 2.3 and 2.4, the data for the kinetically stable two-phase region obtained by Wu and Zhou [41] are shown with the data for the thermodynamically stable phase. The data of the kinetically stable region are almost in accord with those obtained by Wang et al. [42] with an extremely dilute solution in the thermodynamically stable region, suggesting that the equilibrium size of a globule can be estimated if the measurement is carried out when chain collapse has finished before the intermolecular aggregation starts, even in a kinetically stable two-phase region.

Figure 2.5 shows the data obtained by Wang et al. [42] in the plots of α_S and α_h ($=<R_h>/<R_h>_0$) as a function of $\tau M_w^{1/2}$, showing that α_h also decreases sharply above the Θ temperature owing to the coil-globule transition. In Fig. 2.5, the chain dimensions in terms of α_S decrease by approximately sixfold; the decrease in α_h is much less in this temperature range, i.e., ~3.5 times. This difference between the changes in the radius of gyration $<S^2>^{1/2}$ and the hydrodynamic radius $<R_h>$ is understandable because they are defined in different ways. $<S^2>^{1/2}$ is related to the actual space occupied by the polymer chain, whereas $<R_h>$ is the radius of an equivalent hard sphere that has an identical diffusion coefficient as the polymer chain in the solution. When a polymer chain is in an extended coil form under good solvent conditions, the polymer chain is nearly free draining, so that $<R_h>$ is much smaller than $<S^2>^{1/2}$. At the Θ point, the polymer chain becomes nondraining so that the ratio $<S^2>^{1/2}/<R_h>$ becomes smaller. According to the data of Wang et al., $<S^2>^{1/2}/<R_h>$ had a nearly constant value of 1.50 at $T < \Theta$ and decreased markedly from 1.50 to 0.56 just above Θ. At higher temperatures, $<S^2>^{1/2}/<R_h>$ increased and reached 0.77. The values of 1.50 and 0.77 are close to 1.504 and $(3/5)^{1/2}$, which are predicted for a random coil at the Θ point and a uniform hard sphere, respectively [43]. The value of 0.56 being less than $(3/5)^{1/2}$ was interpreted to indicate the formation of a molten globule, i.e., a globule with many small stressed loops on the surface. The loops are so small that their contribution to $<S^2>^{1/2}$ is negligible, but they are not draining, which leads to a larger $<R_h>$. The average chain density ρ can be estimated using $\rho = M_w/[N_A(4/3)\pi<R_h>^3]$, and ρ was estimated to be 0.34 g/cm^3 in the globule region, indicating that the globule contains 66 % water in its hydrodynamic volume.

In Fig. 2.5, the experimental data are compared with those calculated from the following equation [42]:

$$\alpha_S - \alpha_S^3 + C'\left(\alpha_S^{-3} - 1\right) = B'\tau M^{1/2} \tag{2.5}$$

which is obtained by replacing τ in Eq. (2.4) with $(-\tau)$ in order to reproduce the LCST behavior. The coil-globule transition curve calculated using Eq. (2.5) coincides with the experimental data near the Θ temperature, but there is a deviation between the calculated and experimental results in the globular region. The deviation would be attributed to the complex mechanism of LCST; Equation (2.5) based on the simple excluded-volume theory may not be applicable to LCST-type polymers. A more elaborate theory in which cooperativity in hydration is taken into account has recently been developed [44, 45] for representing coil-globule transition of LCST-type polymers. According to Eq. (2.5), the plot of $\alpha_S{}^3\tau M^{1/2}$ approaches an asymptotic value of (C′/B′) in the poor solvent limit ($\tau M^{1/2} \to \infty$). However, Fig. 2.4 shows that after reaching the minimum, $\alpha_S{}^3\tau M^{1/2}$ starts to increase with $\tau M^{1/2}$. This behavior can be explained by a finite compressibility of a collapsed globule [46]. Although the globule contains 66 % water inside, the observed $<S^2>^{1/2}$ in the globule state is independent of temperature, which causes the increase in $\alpha_S{}^3\tau M^{1/2}$ in a high-temperature limit.

2.3.3 Coil-Globule Transition of PMMA

Nakata [14] has studied chain collapse below the Θ temperature using PMMA solutions of UCST-type phase behaviors. He carried out SLS measurements with moderately dilute solutions (~10^{-4} g/cm^3). Although the concentrations were higher than those used in the studies of PS in cyclohexane [38, 39], he observed coil-globule transition to be not affected by intermolecular aggregation. This is because chain aggregation occurs very slowly in PMMA solutions. For example, the rate of chain aggregation in a dilute solution of PMMA in acetonitrile (AcN) is more than one hundred times lower than in that of PS in cyclohexane [15]. When the temperature of a dilute solution of PMMA in AcN is changed from the Θ temperature to a lower temperature, the mean-square radius of gyration $<S^2>$ of a single PMMA chain in the dilute solution decreases with time and reaches equilibrium before the polymer chains start to aggregate. The coil-globule transition curve is obtained by plotting the equilibrium values of $<S^2>$ against temperature [15].

Figure 2.6(a) shows the coil-globule transition curves for PMMA in AcN [15], isoamyl acetate (IAA) [16], and a mixed solvent of t-butyl alcohol (TBA) + water (2.5 vol%) [17], in the plots of $\alpha_S{}^2 = <S^2>/<S^2>_0$ versus $\tau M^{1/2}$. The experiments employed PMMA with the molecular weights M_w of 6.4×10^6 and 1.1×10^7 in AcN [15], 2.4×10^6 in IAA [16], and 1.6×10^6, 2.8×10^6, 4.0×10^6, and 1.2×10^7 in TBA + water [17]. The data points for different molecular weights construct a single curve for each polymer-solvent system. The coil-globule transition is sharp in TBA + water and broad in AcN. For PMMA in TBA + water, $\alpha_S{}^2$ in the globular region reaches 0.055, from which ρ is estimated to be 0.24 g/cm^3.

The coil-globule transition curves were well represented by Eq. (2.4). The solid curves in Fig. 2.6 were depicted with the values of the parameters B and C shown in Table 2.1. The values of B and C of the other systems are also shown in Table 2.1 [15, 18]. Figure 2.6b shows the same data as those in Fig. 2.6a plotted as $-\alpha^3\tau M^{1/2}$ versus $-\tau M^{1/2}$. As is predicted in Eq. (2.4), the plot of $-\alpha^3\tau M^{1/2}$ approaches an asymptotic value of (C/B) in the poor solvent limit ($\tau M^{1/2} \to -\infty$).

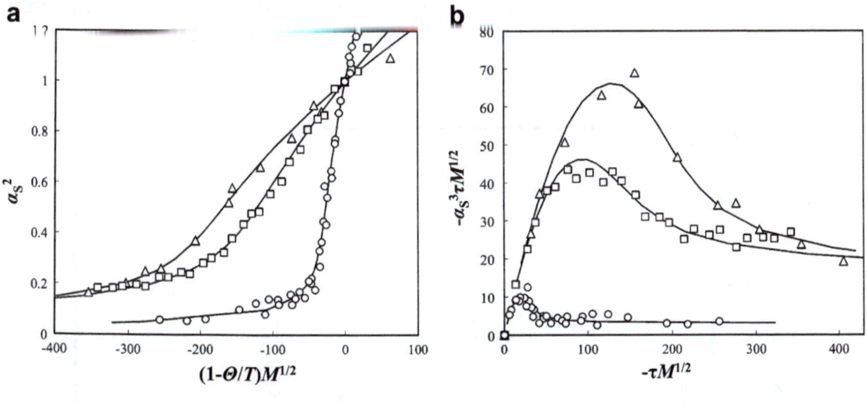

Fig. 2.6 Coil-globule transition curves for PMMA in AcN (△), IAA (□), and the mixed solvent of TBA + water (2.5 vol.%) (○) obtained using the plots of $\alpha_S{}^2$ against $\tau M^{1/2}$ and $-\alpha_S{}^3\tau M^{1/2}$ against $-\tau M^{1/2}$. The *solid curve* represents the data calculated from Eq. (2.4)

Table 2.1 The parameters B and C and the second (v) and third (w) virial coefficients determined by the analyses of coil-globule transition curves using Eq. (2.4) for solutions of PMMA in various solvents

Solvent	B	C	C/B	v (10^{-23} cm^3)	w (10^{-45} cm^6)
AcN	0.0025	0.041	16.4	0.42	0.39
IAA	0.0039	0.067	17.2	0.66	0.64
TBA	0.018	0.057	3.1	3.1	0.55
TBA + water (2.5 vol.%)	0.017	0.043	2.6	2.8	0.41
AcN + water (10 vol.%)	0.0052	0.045	8.7	8.8	0.43

In Eq. (2.4), B and C stem from the second ($v\tau$) and third (w) virial coefficients among monomers, respectively. According to Grosberg and Kuznetsov [34], $B = 1.09v(N/M)^{1/2}/a^3$ and $C = 3.55w/a^6$, where a and N are the monomer size and number, respectively. Using the equations, v and w were estimated and their values are given in Table 2.1. The virial coefficients, $v\tau$ and w, in the free-energy expansion in Eq. (2.2) of monomer interactions can be related to the binary β_2 ($=\beta_0\tau$) and the ternary cluster integral β_3 among monomers through $v\tau = \beta_2/2$ and $w = \beta_3/6$, respectively. For the solutions of PMMA in AcN, β_0 and β_3 have been estimated from the experimental values of the virial coefficients A_2 and A_3 for dilute solutions as $\beta_0 = 3.5 \times 10^{-23}$ cm^3 and $\beta_3 = 4.8 \times 10^{-45}$ cm^6, respectively [47, 48], and v and w corresponding to β_0 and β_3 are 1.75×10^{-23} cm^3 and 0.8×10^{-45} cm^6, respectively, which are comparable to those in Table 2.1. The coil-globule transition curve for PMMA in AcN has been analyzed on the basis of a more rigorous theory formulated by Grosberg and Kuznetsov (GK theory) [34], and v and w are estimated to be 0.88×10^{-23} cm^3 and 0.83×10^{-45} cm^6, respectively, which are closer to those due to A_2 and A_3 [15]. Thus, the shape of the coil-globule transition curves for the different polymer-solvent systems in Fig. 2.6 can be explained in terms of the excluded-volume interactions between monomers, which are characterized by v and w.

In the GK theory, both the radius of gyration $\langle S^2 \rangle$ and the hydrodynamic radius $\langle R_h \rangle$ are considered [34, 35]. Figure 2.7 shows the coil-globule transition curve in

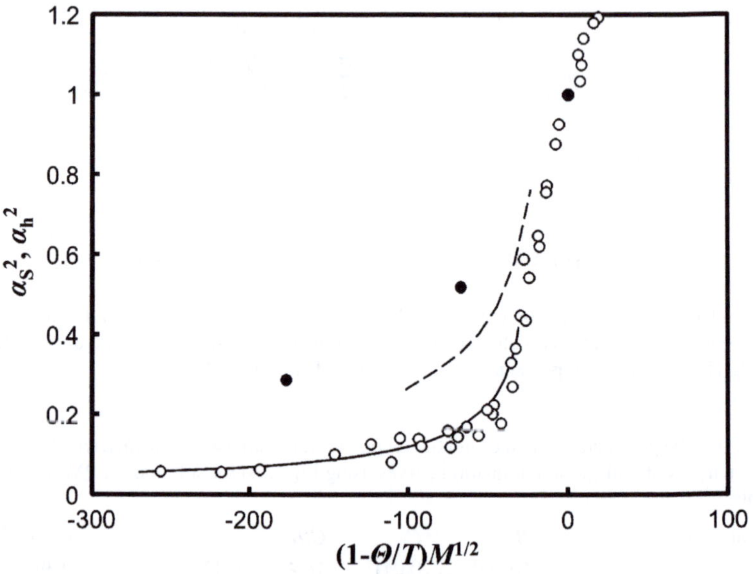

Fig. 2.7 Coil-globule transition curves for PMMA in TBA + water (2.5 vol.%) obtained using the plots of α_S (○) and α_h (●) against $\tau M^{1/2}$. The solid and dashed lines are theoretical curves calculated according to the theories of Grosberg and Kuznetsov for radius of gyration and hydrodynamic radius, respectively

the plot of α_S^2 against $\tau M_w^{1/2}$ for PMMA in the mixed solvent of TBA + water (2.5 vol%) together with data of α_h^2 obtained by a preliminary DLS experiment. The solid and dashed curves are obtained for α_S^2 and α_h^2 according to the GK theory [34, 35]. The theoretical values of α_h^2 are smaller than the experimental data, which would be attributed to the breakdown of the supposition that all monomers are equally bathed by the solvent even in the dense globular state [36]. The coil-globule transition of PMMA studied by viscometry [49, 50] and DLS [51, 52] has been reported by several groups.

2.4 Kinetics of Chain Collapse

2.4.1 Phenomenological Models of Chain Collapse Kinetics

The previous section focused on the equilibrium chain dimensions as a function of temperature. When a polymer solution quenched from the Θ temperature to temperatures of poor solvent conditions, the chain dimensions will decrease with time until reaching their equilibrium value. The kinetics of chain collapse has been studied theoretically by means of two main approaches: phenomenological models [53–57] and computer simulations [58–64]. Phenomenological models show the dependence of the characteristic time of chain collapse on chain length, which can be verified by experimental observations.

A phenomenological theory of the kinetics of chain collapse was first considered by de Gennes [53]. A chain was assumed to occupy a sausage-shaped domain, and its chain collapse was modeled as a lateral swelling and a longitudinal contraction of the domain. In phenomenological models, the balance of driving and dissipative forces is considered to give scaling laws. In de Gennes's theory, driving free energy is related to interfacial free energy, and the dissipation is attributed to hydrodynamic flow. The characteristic time τ_{col} of the chain collapse is predicted to show the dependence of the number N of monomers as $\tau_{col} \sim N^2$.

The topological constraints due to the nonphantomness of the chain and possible self-entanglement formation were not considered in de Gennes's model. On the basis of these concepts, Grosberg et al. [54] developed a two-stage chain collapse model: a fast crumpling of the unknotted polymer chain to a crumpled globule and the subsequent slow knotting due to the reptation-like chain motion of the crumpled globule to a more compact globule. The characteristic times τ_{crum} and τ_{eq} for the first and second stages are written as $\tau_{crum} \sim N^2$ and $\tau_{eq} \sim N^3$, respectively.

A different view was later proposed by Buguin et al. [55]. In their model, a polymer chain is assumed to form a shape consisting of globular clusters ("pearls") of a dense phase connected by strings immediately after the quench ("pearl-necklace" model), as demonstrated by computer simulations [58, 59]. The collapse occurs in two stages: the coalescence of "pearls" to form a semi-compact state and the subsequent Ostwald ripening of the solvent "bubbles" remaining inside the semi-compact state. Two scenarios were considered for the first stage. In one, the driving energy is free energy of transfer of monomers in the strings into the "pearls" of the dense phase (transfer free energy), and the associated dissipation is due to hydrodynamic flow, yielding the characteristic time $\tau_{col} \sim N^{1/2}$. In another, the chain is envisioned as two pearls connected by a string of monomers, and the relation $\tau_{col} \sim N^{4/3}$ is obtained based on the assumption that the driving energy and the associated dissipation mechanism are the transfer free energy and the Stokes drag force on the "pearls," respectively. For the second stage, the characteristic time $\tau_{comp} \sim N$ is obtained.

Several authors developed phenomenological models based on the "pearl-necklace" picture but obtained different scaling relations of collapse time depending on their assumptions. Klushin [56] described a self-similar coalescence stage in the collapse of a necklace of clusters. In this model, the driving free energy is the transfer free energy, and the dissipation is due to the Stokes drag force; the scaling relation $\tau_{col} \sim N^{0.93}$ is obtained in the nondraining limit. Halperin and Goldbart [57] considered the early stages of the chain collapse in detail. In the first stage, nascent droplets of a dense phase are formed, with little effect on the configuration of the bridges between the droplets. The nascent droplets then grow by accreting monomers from the bridges, thus causing the bridges to stretch. During these two stages, the overall dimensions of the chain decrease only slightly. In the third stage, the collapse of the "pearl-necklace" occurs, which causes the shrinkage of the overall dimensions of the chain. The characteristic times of the three stages scale as N^0, $N^{1/5}$, and $N^{6/5}$, respectively.

2.4.2 Chain Collapse Kinetics of PNIPAM

Experimentally, the first attempt to measure chain collapse kinetics was made by Chu et al. for PS in cyclohexane [65]. They observed two stages in chain collapse kinetics with characteristic times of $\tau_{crum} = 357$ s and $\tau_{eq} = 323$ s, using the dynamic light scattering of PS solution in a capillary tube with a thin wall (0.01 mm). They attributed the observed two steps to the fast crumpling and the slow knotting proposed by Grosberg et al. [54], respectively. The observed characteristic times are much longer than those predicted on the basis of the de Gennes's theory [53]. In their experiments, although such a thin wall scattering cell was used, it still took at least 120 s for the temperature jump to reach its equilibrium, which is comparable to the observed chain collapse time. Moreover, interchain association occurred within 100 s after the temperature jump, so that collapsed single-chain globules coexisted with multichain aggregates in a solution.

Wu and Zhou [41] attempted to observe the chain collapse kinetics of PNIPAM using DLS after temperature jumps from 30.6 to 31.8 °C (thermodynamically stable region) and from 30.6 to 33.0 °C (kinetically stable two-phase region). In their experiments, the temperature jump took ~400 s. In the case of the temperature jump to 33.0 °C, the collapse kinetics was too fast to be monitored. The two-stage chain collapse kinetics was observed for the temperature jump to 31.8 °C, with the time scales for the two stages estimated to be ~50 and ~300 s, respectively.

Xu et al. [66] used a stopped-flow device to study the transition kinetics of a narrowly distributed PNIPAM chain randomly labeled with pyrene (PNIPAM-Py). It has been shown that PNIPAM chains exhibit coil-globule transition in a mixture of methanol and water at a specific composition, although both pure methanol and pure water are good solvents for PNIPAM [67]. The kinetics of coil-globule transition after mixing equal volumes of PNIPAM-Py in methanol and water was observed by recording the time dependence of the ratio of excimer emission intensity to monomer emission intensity (I_E/I_M). The time dependence of I_E/I_M was represented by a double exponential function with two distinct characteristic times ($\tau_{fast} \sim 12$ ms and τ_{slow} ~270 ms). The observed τ_{fast} was in reasonable agreement with the value estimated by de Gennes's theory [53].

Ye et al. [68] studied the collapse kinetics of linear PNIPAM homopolymer chains in dilute aqueous solutions using a fast infrared laser heating pulse and water-soluble 8-anilino-1-naphthalenesulfonic acid ammonium salt (ANS) in solution as a fluorescence probe owing to its sensitivity to its hydrophobic/hydrophilic surroundings. ANS fluorescence intensity increased rapidly after the temperature jump caused by the heating pulse. The time dependence of fluorescence intensity was represented by a double exponential function with two characteristic times $\tau_{fast} \sim 0.1$ ms and $\tau_{slow} \sim 0.8$ ms. In the range of the molecular weight between $M_w = 1.2 \times 10^6$ and 2.3×10^7, τ_{fast} is independent of chain length, and τ_{slow} shows a weak dependence on chain length. The first and second stages characterized by τ_{fast} and τ_{slow} could be attributed to the formation of droplets of the dense phase on the chain (~N^0) and the growth of the dense phase by accreting monomers from strings (~$N^{1/5}$), respectively, as proposed in the theory of Halperin and Goldbart [57].

2.4.3 Chain Collapse Kinetics of PMMA

In the DLS experiments on PS in cyclohexane [65] and PNIPAM in water [41], the characteristic times of chain collapse are short compared with the time required for the thermal equilibration; therefore, it seems difficult to analyze the collapse process in detail. Fluorescence experiments [66, 68] have successfully been used to monitor the fast process during the chain collapse in the aqueous PNIPAM solution. However, fluorescence intensity probes only the microscopic environment around the polymer chain, and the size of the polymer chain itself is not measured directly.

In contrast, Nakata and coworkers [15, 17–20] have reported that PMMA chains in dilute solutions collapse very slowly after the quench depending on molecular weight, temperature, and the type of solvent species. In their SLS studies, the temperature of the PMMA solution in an optical cell was changed by transferring the optical cell preserved at the Θ temperature into a thermostatted cylindrical cell at the center of the photometer. The measurement of scattered light was started 30 min after the setup of the cell for thermal equilibration. Figure 2.8 shows Zimm plots of the light scattering data for PMMA of $M_w = 6.4 \times 10^6$ in AcN at concentrations $c = 0.75$, 1.3, 1.9, and 2.6×10^{-4} g/cm^3. The plots a (circles) and

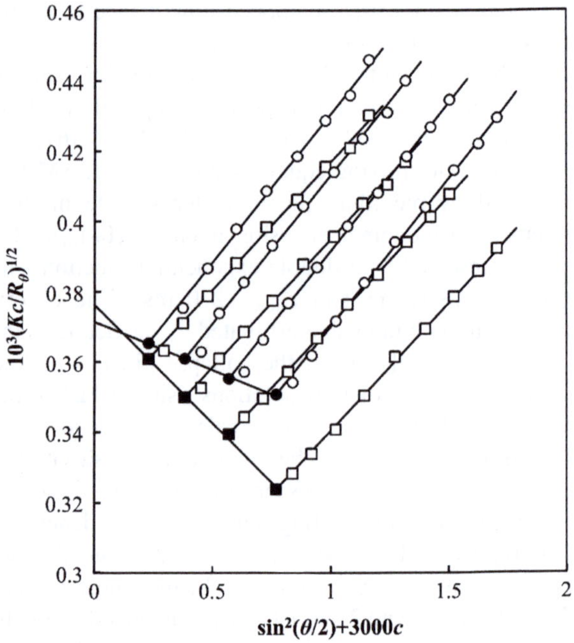

Fig. 2.8 Zimm plot of $(Kc/R_\Theta)^{1/2}$ as a function of $\sin^2(\theta/2)$ and c (g/cm^3) for SLS data of PMMA with $M_w = 6.4 \times 10^6$ in AcN, obtained at 30 min (○) and 600 min (□) after quenching at 20.0 °C, respectively

b (squares) were obtained 30 and 600 min after the quench from the Θ-temperature (44.0 °C) to 20.0 °C, respectively. The intercepts at $c = 0$ of the straight lines seem to coincide within experimental error. Thus, the determination of $<S^2>$ was carried out without being affected by the intermolecular aggregation. Figure 2.9 shows the chain collapse kinetics expressed by plotting the transient expansion factor $\alpha_S{}^2(t) = <S^2>(t)/<S^2>_0$ against ln t, where t is the time in minutes after the quench. It is clear that $\alpha_S{}^2(t)$ decreases rapidly during the first 30 min after the quench and slowly approaches an equilibrium value in roughly 2,000 min. Chain collapse kinetics has been represented by the stretched exponential function [15, 17, 18]:

$$\alpha_S^2(t) = \alpha_{S,eq}^2 + \left(1 - \alpha_{S,eq}^2\right)\exp\left[-\left(t/\tau^*\right)^\beta\right] \tag{2.6}$$

where β and τ^* are constants independent of time t. The solid curve in Fig. 2.9 is given by Eq. (2.6).

Chain collapse kinetics has been studied by SLS for PMMA in IAA [19], AcN [15], TBA [20], and the mixed solvents of AcN + water (10 vol%) [18] and TBA + water (2.5 vol%) [17]. The collapse of PMMA with $M_w = 1.2 \times 10^7$ to equilibrium globules occurred very slowly in TBA + water (2.5 vol.%) but rapidly in pure TBA: chain collapse in the mixed solvent required days to weeks depending on quench depth, while that in pure TBA was completed within 90 min after the quench. PMMA chains with $M_w = 1.6 \times 10^6$ in the mixed solvent of TBA + water

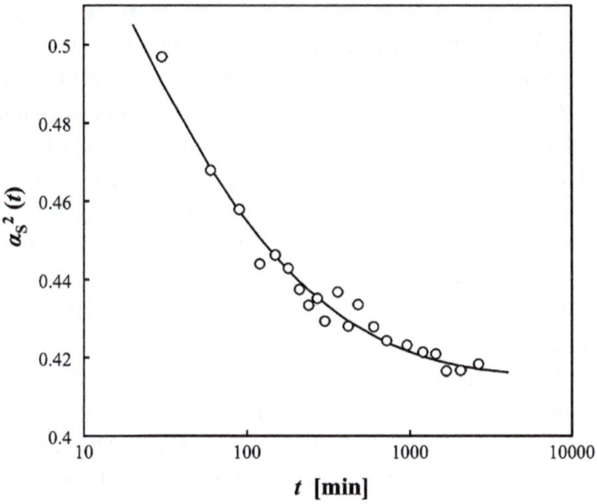

Fig. 2.9 Chain collapse kinetics of PMMA ($M_w = 6.4 \times 10^6$) in AcN at 20.0 °C

(2.5 vol.%) collapsed to equilibrium globules within the first 30 min after the quench. The chain collapse of PMMA with $M_w = 1.2 \times 10^7$ in IAA and that of PMMA with $M_w = 1.1 \times 10^7$ in AcN were very slow and required 2–7 days depending on temperature. Maki et al. [18] studied the molecular weight and solvent dependences of the characteristic time of chain collapse for PMMA with $M_w = 6.4 \times 10^6$ and 1.1×10^7 in pure AcN and the mixed solvent of AcN + water (10 vol.%). Chain collapse in the mixed solvent was found to occur much faster than that in pure AcN, although the former collapse process required 2–30 h depending on the molecular weight and temperature. The characteristic time τ_c of chain collapse was conjectured to be $\tau_c \sim \tau_0 N^z$, where τ_0 reflects the nature of the solvent species. The ratio of τ_0 for PMMA in pure AcN to that in the mixed solvent, $\tau_{0,\text{pure}}/\tau_{0,\text{mix}}$, was estimated to be 9.5 for $M_w = 6.4 \times 10^6$ and 12.0 for 1.1×10^7. The observed exponent value was $z = 2.4 \pm 0.7$ for AcN + water and 3.0 ± 0.7 for pure AcN, which are compatible with the value $z = 3$ for slow knotting via the reptation-like motion proposed in the two-stage phenomenological model by Grosberg et al. [54]. In the phenomenological model, the coefficient τ_0 is a function of solvent viscosity and the second virial coefficient $v\tau$ [54]. However, the large difference in τ_0 between the two systems cannot be explained theoretically. Moreover, the observed characteristic times are longer by several orders of magnitude than those of the theoretical prediction (~100 ms). The large deviation between the experimental and theoretical time scales would be attributed to the effective viscosity induced by the interaction between the polymer chain and the solvent [18].

Kayaman et al. [52] attempted to observe the collapse kinetics of PMMA samples with $M_w = 3.3 \times 10^6$ and 6.5×10^6 in IAA by the use of DLS. Chain collapse process was observed to take place within almost 35 s. In their experiment, however, they found that the unperturbed dimension was not recovered for the sample after the measurement even when the sample cell was transferred to the oven at the Θ temperature and kept there for 3 days. In contrast, Nakata and Nakagawa [19] reported that, when the solutions of PMMA with $M_w = 1.3 \times 10^7$ in IAA were suddenly heated to the Θ temperature (61.0 °C) after the preservation at 30.0 °C for 10,080 min, the size of the polymer chain recovered its unperturbed value within 60 min. Recently, Maki and Dobashi [21] observed the kinetics of chain collapse by using SLS and DLS for dilute solutions of PMMA with $M_w = 1.1 \times 10^7$ in a mixed solvent of TBA + water (2.5 vol.%). Both $<S^2>$ and $<R_h>$ continued to decrease for more than 24 h, but $<R_h>$ seemed to reach its equilibrium value faster than $<S^2>$. The ratio $<S^2>^{1/2}/<R_h>$ decreased from 1.4 to 0.78, indicating a conformational change from a random coil to a globular state [43].

The molecular weight dependence of the collapse kinetics of PMMA suggested knot formation in the collapsed globule. Recently, a dynamic Monte Carlo simulation of the collapse transition has been presented, which reveals the occurrence of knotting during chain collapse [69]. If a globule has knots inside, its rapid chain expansion would be constrained by the knots, resulting in an arrested partially swollen state. The formation of arrested swollen globules has been studied theoretically [70–72] but not observed experimentally [19, 41, 73]. More recently, Nakata and coworkers [22, 23] measured the chain expansion from a globule to a coil for PMMA in the mixed solvent of TBA + water (2.5 vol.%) by means of SLS. The chain size increased monotonically with time to its unperturbed value after the temperature rise, and the expansion kinetics depends on molecular weight and the aging time and temperature of the globular state, which supported knot formation in the globule. The characteristic time of chain expansion depended on chain length as $N^{2.7}$, of which the exponent close to 3 demonstrated a disentanglement via self-reptation [23].

2.5 Rheological Aspects in Polymer Chain Aggregation

2.5.1 Colloidal Stability of PNIPAM Mesoglobules

Wu and Zhou [41] observed that, when a dilute solution of PNIPAM in water is heated above the Θ temperature up to the kinetically stable two-phase region, a conformational change of individual polymers owing to coil-globule transition occurs and then polymer chains start to aggregate, which can be detected as an increase in the intensity of scattered light. For solutions at moderate concentrations, interchain aggregation and the subsequent phase separation can be observed visually as a decrease in the clarity of the solutions. Thus, turbidimetry is suitable for detecting a cloud point, the temperature at which a transparent polymer solution turns opaque. The cloud point as a function of polymer concentration is represented by a cloud point curve that is, in many cases, similar to the phase diagram in Fig. 2.2.

Many studies have been carried out on the LCST-type phase behavior of aqueous solutions of PNIPAM by turbidimetry. The first detailed study was reported by

Heskins and Guillet [74], who determined the cloud point as a function of polymer concentration by visible observation. Schild and Tirrell [75] showed that the cloud point for PNIPAM in water depends on polymer molecular weight and concentration. From the optical transmittance measurement, Fujishige et al. [76] showed that the cloud point curves for PNIPAM samples ranging in molecular weight from 5.0×10^4 to 8.4×10^6 were superimposable. The cloud points for the dilute solutions of the highest molecular weight sample were almost independent of polymer concentrations, in the range from 0.033 to 1.0 %. A phase behavior that shows no molecular weight dependence was also reported in the calorimetric study of PNIPAM solutions [77]. Tong et al. [78] measured the cloud point for PNIPAM samples with $M_w = 4.9 \times 10^4$ and 1.0×10^5 in the concentration range from 0.58 to 70 wt%. The observed cloud point curves depended on polymer concentration, and the cloud points for $M_w = 1.0 \times 10^5$ were higher than those for $M_w = 4.9 \times 10^4$, which seems counterintuitive behavior for LCST-type polymer-solvent systems. In contrast, Xia et al. [79] used narrowly distributed PNIPAM samples obtained by atom transfer radical polymerization (ATRP) and reported that the cloud points measured for 1 wt% solutions decreased with M_n in the range from 2.8×10^3 to 2.7×10^4. Kawaguchi et al. [80] showed that the cloud point curves for PNIPAM in water depended on solvents used for the radical polymerization of PNIPAM owing to the different branching structures of PNIPAM. Kawaguchi et al. [81] measured the optical transmittance at a temperature slightly higher than the cloud point for a linear, narrowly distributed PNIPAM sample prepared by living anionic polymerization and found that the cloud point curve does not correspond to the binodal but arises from the formation of large aggregates. Recently, it has been revealed that phase behavior also strongly depends on the tacticity of PNIPAM [82–84].

As discussed above, the phase behaviors of aqueous PNIPAM solutions previously reported by different groups seem confusing. However, it is commonly accepted that the cloud point is observed at approximately 32 °C for dilute solutions of high molecular weight PNIPAM. Above the cloud point, the solution looks opaque owing to microscopic phase separation caused by intermolecular aggregation. The microscopically separated phase is expected to be metastable, and the polymers should gradually precipitate. However, dilute PNIPAM solutions at concentrations ranging from 10^{-5} to 10^{-3} g/g yield colloidally stable dispersions of small (ca. 100 nm) PNIPAM aggregates above the cloud point; the solutions do not phase separate at the macroscopic level during the conventional experimental time scales [85]. The formation of colloidally stable aggregates of PNIPAM was also reported by Gorelov et al. [86]. They performed DLS experiments on aqueous solutions of PNIPAM samples obtained by persulfate-initiated radical polymerization with molecular weights ranging from 2.0×10^6 to 9.3×10^6 at concentrations between 10^{-5} and 10^{-4} g/cm^3. They observed small aggregates above the cloud point, the size of which is independent of temperature when solutions were heated to 65 °C. Their size depended on polymer concentration, i.e., increasing with increasing initial concentration, as well as on heating rate, i.e., decreasing with increasing heating rate. For solutions maintained well above the transition temperature, the size of the particles remained constant over several days and was stable even when diluted. The electron micrographs showed that the aggregates were spherical particles with a narrow size distribution and a mean radius in agreement with $<R_h>$ obtained by DLS. They attributed the uniform size distribution of PNIPAM particles to the spinodal decomposition of polymer solutions. In this picture, when crossing the spinodal line with moderately rapid heating, fluctuations in concentration with a certain correlation length appear. Then, if the temperature is increased further, these fluctuations are kinetically frozen owing to the collapse and aggregation of polymer chains, resulting in globular particles consisting of many chains with a size reflecting the effective correlation length. This picture seems to be supported by the concentration and heating rate dependence of the resulting particle size. According to Timoshenko and Kuznetsov [87], such PNIPAM particles are examples of mesoglobules, of which the definition is "equally sized spherical aggregates consisting of more than one and less than all polymer chains that are colloidally stable in solution." Aseyev et al. [88] carried out SLS and DLS measurements on dilute (10^{-5}–10^{-4} g/cm^3) aqueous solutions of PNIPAM samples synthesized by ATRP ($M_w = 2.7 \times 10^4$) and persulfate-initiated free-radical polymerization ($M_w = 1.6 \times 10^5$) and observed the formation of mesoglobules for both PNIPAM samples. The density ρ of the particles was calculated using the equation $\rho = M_{w,agg}/[N_A(4/3)\pi <R_h>_{agg}{}^3]$, where $M_{w,agg}$ and $<R_h>_{agg}$ are the weight-average molecular weight and hydrodynamic radius of aggregates, respectively, and ρ was estimated to be 0.35–0.4 g/cm^3, which is comparable to the density of a single PNIPAM chain in the globular state. The ratio of the radius of gyration to the hydrodynamic radius of aggregates, $<S^2>_{agg}{}^{1/2}/<R_h>_{agg}$, is approximately 0.77, which is consistent with the value for the uniform hard sphere. The scaling of the molecular weight of PNIPAM mesoglobules was shown as $M_{w,agg} \propto (<S^2>_{agg}{}^{1/2})^{2.7}$, the exponent of which is slightly smaller than 3 for the hard sphere. Various thermally responsive polymers form mesoglobules upon heating their aqueous solutions. A compilation of the types of polymer reported exhibiting this behavior has been presented in recent reviews [10, 13].

The colloidal stability of intermolecular aggregates formed by thermally responsive polymers under thermodynamically poor condition can be achieved by modifying the surface of the particles either sterically or electrostatically. However, both stabilizing mechanisms seem not applicable to mesoglobules of PNIPAM homopolymers. Several mechanisms underlying the colloidal stability of mesoglobules have been discussed. One is related to the fact that the mesoglobular phase has been reported for rather dilute PNIPAM solutions. As Gorelov et al. [86] pointed out, the low probability of Brownian collisions in dilute solutions will delay the aggregation of particles, resulting in colloidal stability. The effect of concentration on the characteristic time of chain collision can be estimated using the theory of diffusion-controlled aggregation. According to Smoluchowski [89, 90], the characteristic time for the diffusion-limited cluster aggregation (DLCA) is given by

$$\tau_{\mathrm{DLCA}} = \frac{3\eta M}{4N_{\mathrm{A}}kTc} \tag{2.7}$$

where c is the concentration of polymers. Under the conditions of $\eta \sim 1$ mPa s, $M \sim 10^6$, and $c \sim 10^{-5}$ g/cm^3, the characteristic time is roughly estimated to be 0.1 s, which is several orders of magnitude smaller than the experimental time scale. Thus, some additional mechanisms are needed to explain the colloidal stability of the mesoglobules. Chan et al. [91] considered the effect of the end-group structure of PNIPAM. In many studies, NIPAM samples have been synthesized using ammonium persulfate as an initiator. They studied the effect of salt addition on the mesoglobule formation and argued that, despite the small number of ionic end groups of PNIPAM, these charges may be situated on the surfaces of PNIPAM mesoglobules, thus significantly contributing to the stability. However, Aseyev et al. [88] observed mesoglobule formation for PNIPAM homopolymers bearing a neutral end group. Another mechanism of the colloidal stability is related to the translation entropy of polymer chains [88]. A polymer in a poor solvent optimizes its interaction energy via the intermolecular association among polymer chains, which results in a decrease in the translational entropy of the chains. When these two energy terms are compensated, the system reaches a metastable state characterized by the formation of polymeric mesoglobules.

Another and possibly more important cause of colloidal stability is the visco-elastic effect, recently discussed in the literature [92, 93]. Tanaka [92] observed an unusual behavior of small droplets in the phase separation of poly(vinyl methyl ether) in water under a microscope: droplets did not undergo coalescence and appeared to be stable. For dilute polymer solutions, the coalescence of two droplets was considered to be brought about by the chain migration between droplets forming a bridge during contact. This idea was substantiated by introducing two characteristic times, that is, the chain diffusion time τ_{ent} and the contact time τ_{cont}. For $\tau_{\mathrm{ent}}/\tau_{\mathrm{cont}} << 1$, the coalescence of two droplets would occur in each collision by the Brownian motion, and the aggregation would occur via DLCA. For $\tau_{\mathrm{ent}}/\tau_{\mathrm{cont}} >> 1$, chains could hardly bridge two droplets during contact, and the collision is ineffective. Tanaka roughly estimated the ratio $\tau_{\mathrm{ent}}/\tau_{\mathrm{cont}}$ using τ_{ent} for the time of reptation of a chain in a dense polymer matrix and τ_{cont} for the diffusion time of a droplet defined as

$$\tau_{\mathrm{ent}} \approx \frac{a^2 N^3 \phi^{3/2}}{D_1}, \quad \tau_{\mathrm{cont}} \approx \frac{r_0^2}{D_{\mathrm{R}}} \tag{2.8}$$

respectively, where r_0 is the range of interaction, ϕ is the segment volume fraction, and D_1 and D_{R} are the diffusion constants of a monomer and a droplet, respectively. This concept of the coalescence of two droplets would be applicable to the merging of mesoglobules consisting of a limited number of polymer chains or globules of a single chain. If mesoglobules or globules are quite dense and compact, the characteristic time of the collision is less than the time required to establish a permanent entanglement between two colliding mesoglobules/globules via chain reptation;

consequently, the merging of two mesoglobules/globules into a single, larger one may take a long time. This approach not only accounts for the stability of mesoglobules, but also explains why mesoglobules prepared by quenching are smaller than those prepared with a slow temperature change. Quenching results in quick intrachain contraction, preventing extensive interchain association. Chuang et al. [94] argued the viscoelastic effect of the coalescence of two globules in a different way. They demonstrated the existence of entanglement force acting between two globules by molecular dynamics simulation at the prereptational time scale. When globules approaching each other are dense, topological restrictions induce repulsive forces on the globules on a time scale shorter than the time of reptation.

Recently, Kujawa et al. [95] have studied the kinetics of merging and/or chain exchange between mesoglobules in dilute ($\sim 10^{-4}$ g/cm^3) aqueous solutions of fluorescently labeled PNIPAM via nonradiative energy transfer (NRET) using polymers carrying a trace of either naphthyl (Np, energy donor) or pyrene (Py, energy acceptor). Their measurement of the time course of the efficiency of NRET immediately upon mixing aqueous solutions of PNIPAM-Np and PNIPAM-Py incubated above the cloud point allowed them to monitor the kinetics of interparticle chain exchange. The results show that solutions heated within the temperature range between 31 and 36 °C consist of fluidlike particles that can merge and grow in size. At higher temperatures, the mesoglobules act as rigid spheres unable to merge upon collision. They also observed the effect of thermal history on chain aggregation: the merging of particles occurred less frequently for mesoglobules prepared at higher temperatures. These results confirmed the contribution of the viscoelastic effect on the mechanism of the stability of PNIPAM mesoglobular phases.

The formation of mesoglobules above the cloud point is observed typically for dilute PNIPAM solutions. In contrast, a different behavior has been observed for solutions of higher concentrations. Pamies et al. [96] measured DLS for solutions of PNIPAM prepared via ATRP in water at 1 wt%. The correlation functions of DLS obtained at temperatures below the cloud point were described by the sum of a single exponential function at shorter time scales and a stretched exponential function at longer time scales, which means the coexistence of large aggregates. The formation of large aggregates below the cloud point was also shown by Kawaguchi et al. [97], which was attributed to hydrophobic end groups.

2.5.2 Slow Aggregation Kinetics of PMMA

When a dilute solution is quenched below the phase-separation temperature, chain collapse and chain aggregation would occur competitively in a manner that depends on polymer concentration. Chain aggregation kinetics becomes slower with decreasing concentration, while chain collapse kinetics hardly depends on the concentration. Accordingly, at very low concentrations, polymer chains would collapse first, and intermolecular aggregation would start after the chain collapse has finished; at higher concentrations, chain collapse and aggregation occur

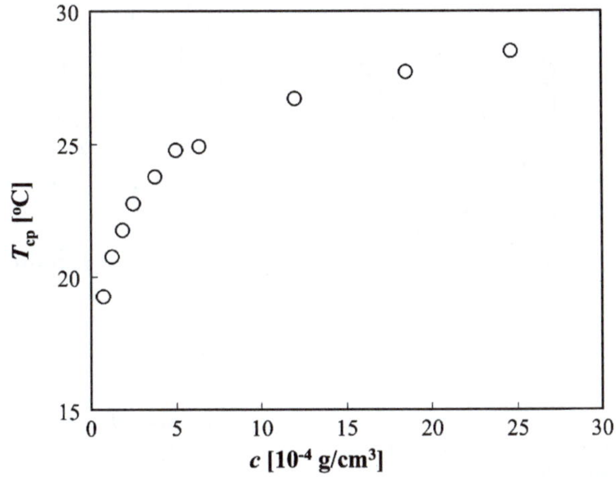

Fig. 2.10 Cloud points plotted against concentration of PMMA ($M_w = 6.4 \times 10^6$) in AcN

simultaneously, and the size of each chain comprising intermolecular aggregates would decrease with time.

Dilute solutions of PMMA in some solvents show a very slow phase separation even far below the cloud point. It has been shown that the characteristic time of the intermolecular aggregation of PMMA in AcN is more than a hundred times larger than that of PS in cyclohexane [15]. Thus, for dilute solutions, the coil-globule transition of PMMA in various solvents has been studied quantitatively by SLS [14–20] and DLS [21], without being affected by intermolecular aggregation. On the other hand, taking advantage of the very slow phase separation of PMMA solutions, intermolecular chain aggregation processes have been studied by SLS in detail for dilute solutions of PMMA in IAA [24, 25], AcN [26], the mixed solvent of AcN + water (10 vol.%) [26], and the mixed solvent of TBA + water (2.5 vol.%) [27–29].

Figure 2.10 shows the cloud point curve for PMMA with $M_w = 6.4 \times 10^6$ in AcN in the concentration range from 0.7 to 24.5 (10^{-4} g/cm^3). When the solutions are quenched from the Θ temperature (44.0 °C) to temperatures below the cloud point T_{cp}, chain collapse and aggregation occur, which can be detected by SLS measurement. From the SLS data, the weight-average molecular weight $M_{w,agg}$ and the mean-square radius of gyration $\langle S^2 \rangle_{agg}$ (nm^2) for clusters of PMMA are estimated. The plots in Figs. 2.11 and 2.12, respectively, show the time evolutions of ln $M_{w,agg}$ and ln $\langle S^2 \rangle_{agg}$ after the quench of the solutions to 20.0 °C. Different symbols are used for concentrations of c (10^{-4} g/cm^3) = 24.5 (open diamonds), 18.3 (open triangles), 12.4 (open squares), 6.4 (open circles), 5.2 (filled diamonds), 3.9 (filled triangles), 2.6 (filled squares), and 1.9 (filled circles). The rate of chain aggregation increased with increasing concentration. The measurement was stopped before the effect of multiple scattering appeared. The values of ln $M_{w,agg} \sim 19$ and ln$\langle S^2 \rangle_{agg} \sim 9.5$ in the last measurements correspond to the average number of chains in a cluster, $N_{agg} = (M_{w,agg}/M_w) \sim 30$ and $\langle S^2 \rangle_{agg}{}^{1/2} \sim 100$ nm, where the average segment density ρ of clusters is estimated to be 0.02 g/cm^3 using the equation $\rho = M_{w,agg}/[N_A(4/3)\pi[(5/3)\langle S^2 \rangle_{agg}]^{3/2}]$.

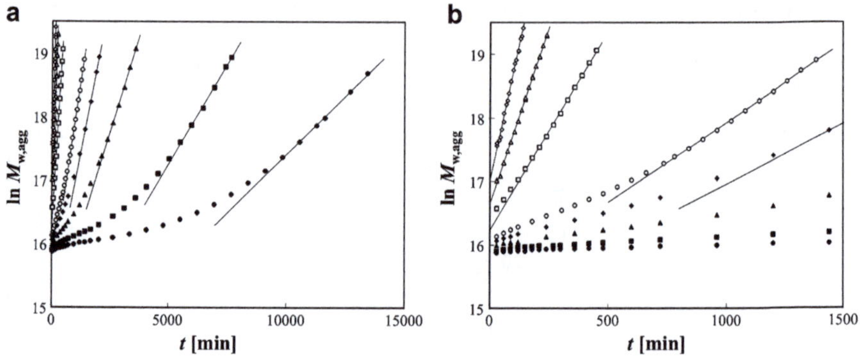

Fig. 2.11 Time evolution of ln $M_{w,agg}$ for PMMA ($M_w = 6.4 \times 10^6$) in AcN at 20.0 °C obtained at the concentrations c (10^{-4} g/cm^3) = 24.5 (◇), 18.3 (△), 12.4 (□), 6.4 (○), 5.2 (◆), 3.9 (▲), 2.6 (■), and 1.9 (●)

The aggregation processes were so slow that it took hours to weeks for clusters containing 30 chains on average to be formed. The low aggregation rate would be explained by the viscoelastic effect. As shown in Sect. 2.5.1, the viscoelastic effect was represented by the ratio of the two characteristic times, τ_{ent}/τ_{cont}, defined in Eq. (2.8). Here, the interaction range l_0 is assumed to be equal to $<S^2>_{agg}^{1/2}$ [78], although l_0 would be much smaller than $<S^2>_{agg}^{1/2}$. Then, the ratio $\tau_{ent}/\tau_{cont} \sim (a^3N^3\phi^{3/2})/(<S^2>_{agg}^{3/2})$ is roughly estimated to be 10^5. The result $\tau_{ent}/\tau_{cont} >> 1$ indicates the contribution of the viscoelastic effect to the slow aggregation of PMMA.

As shown in Figs. 2.11b and 2.12b, plots for higher concentrations can be represented by straight lines, indicating the exponential growth of clusters: $M_{w,agg} \sim e^{Gt}$ and $<S^2>_{agg} \sim e^{Ht}$, where G and H are coefficients independent of time. In Figs. 2.11a and 2.12a, plots for lower concentrations can be represented by straight lines in the later stage of chain aggregation. The exponential growth of clusters is characteristic of the reaction-limited cluster aggregation (RLCA) [98]. As mentioned in Section 2.5.1, in the diffusion-limited cluster aggregation (DLCA), each collision of two clusters would give rise to their coalescence, and the characteristic time of such aggregation is given by τ_{DLCA} in Eq. (2.7). In contrast, in RLCA, only a small fraction of collisions between two clusters induce the coalescence of clusters. In the case of $\eta \sim 1$ mPa s, $M \sim 6 \times 10^6$, $T \sim 300$ K, and $c = 1$–10×10^{-4} g/cm^3, τ_{DLCA} is estimated to be 1–10 ms, which is much shorter than the experimental time scales. Because of the slow exponential growth of clusters, it is clear that the present aggregation kinetics is controlled by RLCA. Chain aggregation processes controlled by RLCA have been reported for solutions of atactic [24–29] and syndiotactic PMMA [99]. The deviation from the exponential growth for solutions at lower concentrations in the initial stage, as shown in Figs. 2.11 and 2.12, was also observed previously [24, 26, 28].

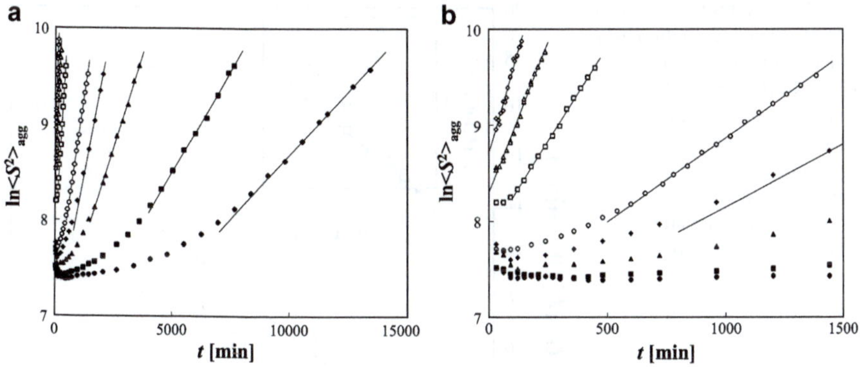

Fig. 2.12 Time evolution of ln$<S^2>_{agg}$ for PMMA ($M_w = 6.4 \times 10^6$) in AcN at 20.0 °C. The symbols are the same as those in Fig. 2.11

Figures 2.11 and 2.12 show that the behavior of $M_{w,agg}$ and $<S^2>_{agg}$ in the later stage of chain aggregation can be written as

$$lnM_{w,\,agg} = lnM_0 + Gt, \quad ln\langle S^2\rangle_{agg} = lnS_0^2 + Ht \tag{2.9}$$

where M_0 and S_0 are constants and G and H are coefficients representing the rate of chain aggregation. In Fig. 2.13, the values of G and H estimated from the data in Figs. 2.11 and 2.12 are plotted against c. The inset shows that the data points of G and H for low concentrations seem to fit the straight lines given by $G = g(c–c_0)$ and $H = h(c–c_0)$, where g and h represent the intrinsic rates independent of c although the straight lines deviated from the experimental data at higher concentrations. In Fig. 2.13, a filled triangle represents the concentration c_{cp} of the phase separation onset estimated from the cloud point curve in Fig. 2.10, at which the rate of aggregation should be zero. The common intercept c_0 coincides with c_{cp}. In previous studies of solutions of PMMA with $M_w = 6.4 \times 10^6$ in AcN at 15.0 °C [26] and solutions of PMMA in other solvents [24–26, 28, 29], the plots of G and H against c were fitted to a straight line passing through the origin. The absence of an intercept on the concentration axis would be due to the deeper quench depth than that used in the present study. In the case of a deep quench far below T_{cp}, c_{cp} becomes very small and c_0 can be approximated to be zero.

The dependence of chain aggregation rate on molecular weight or type of solvent species can be examined by comparing the intrinsic aggregation rates g and h. Nakata et al. [29] analyzed the aggregation processes for PMMA in IAA and TBA + water (2.5 vol.%) and estimated the molecular weight dependence of aggregation rate as $g \sim M_w^{-3}$ independent of the type of solvent species. Maki et al. [26] studied the effect of the type of solvent species on aggregation rate. Under the conditions of comparable quench depths, the ratio of the intrinsic rate in AcN + water (10 vol.%) to that in pure AcN, i.e., g_{mix}/g_{pure} or h_{mix}/h_{pure}, was roughly 9. They also pointed out the correlation between the rates of chain aggregation and

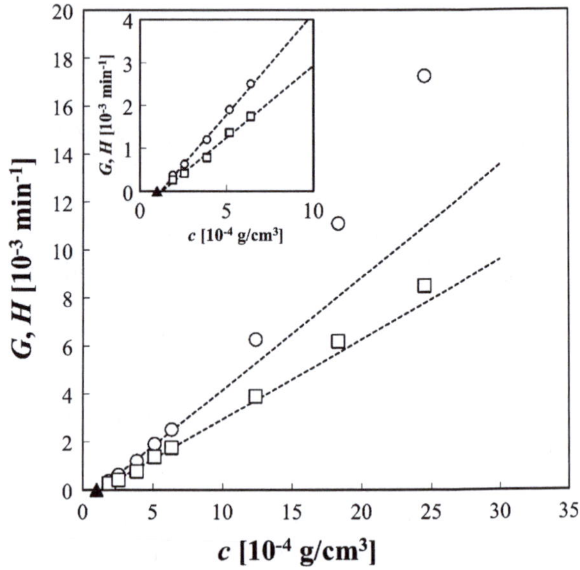

Fig. 2.13 Rates G (○) and H (□) of chain aggregation as a function of concentration of PMMA ($M_w = 6.4 \times 10^6$) in AcN at 20.0 °C. A filled triangle represents the concentration of the onset of the phase separation estimated from the cloud point curve (Fig. 2.10)

chain collapse. The ratio of the characteristic collapse time for PMMA in pure AcN to that in AcN + water (10 vol.%) has been roughly estimated to be 11 [18], which was close to the ratio of the aggregation rates. The ratio of the collapse times was unexpectedly large considering the solvent viscosity and the quench depth and was attributed to the effective viscosity. The effective viscosity would be induced by interactions between the polymer and the solvent, which are identical irrespective of whether the segments belong to the same chain or to different chains. Thus, they argued that the close ratios indicated the solvent nature affecting the rates of chain collapse and chain aggregation through a similar mechanism such as through the effective viscosity.

The rate of chain aggregation, G, depends on temperature significantly. Figure 2.14 shows the plot of G/c against the temperature T for PMMA with $M_w = 6.4 \times 10^6$ in AcN at $c = 4.5 \times 10^{-4}$ g/cm^3. The aggregation rate G increased with increasing quench depth for 15 °C $< T < T_{cp}$ but decreased with quench depth for $T <$ 15 °C. The slower chain aggregation in UCST-type PMMA solutions for a deeper quench at lower temperatures would be relevant to the formation of smaller mesoglobules in LCST-type PNIPAM solutions for faster heating [86, 88].

Figure 2.15 shows double-logarithmic plots of $M_{w,\mathrm{agg}}$ vs $\langle S^2 \rangle_{\mathrm{agg}}$ (nm^2) for the data in Figs. 2.11 and 2.12, with the same symbols. Figure 2.15a shows the data for concentrations ranging between 1.9×10^{-4} g/cm^3 and 5.2×10^{-4} g/cm^3. The data except those at the smallest $M_{w,\mathrm{agg}}$ construct a single straight line for c = 2.6, 3.9, and 5.2×10^{-4} g/cm^3, which suggests that clusters with similar structures are formed in the solutions of different concentrations [22–29]. The dashed straight line in Fig. 2.15a was depicted to fit those data. The observed linear relationship between ln $M_{w,\mathrm{agg}}$ and ln$\langle S^2 \rangle_{\mathrm{agg}}$ can be expressed by the power law as

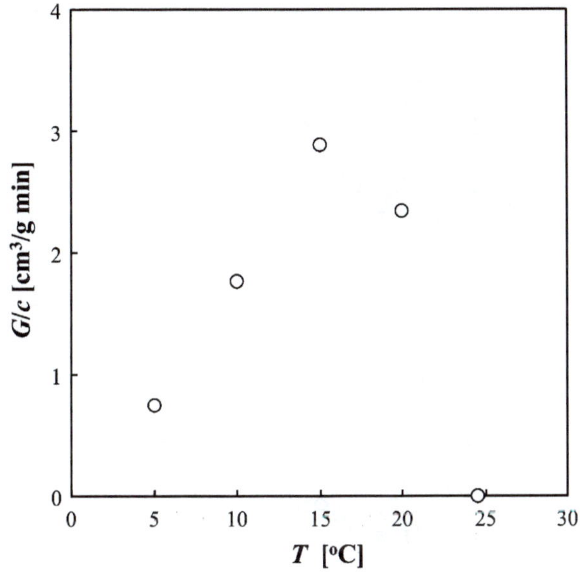

Fig. 2.14 Temperature dependence of the chain aggregation rate for PMMA ($M_w = 6.4 \times 10^6$) in AcN

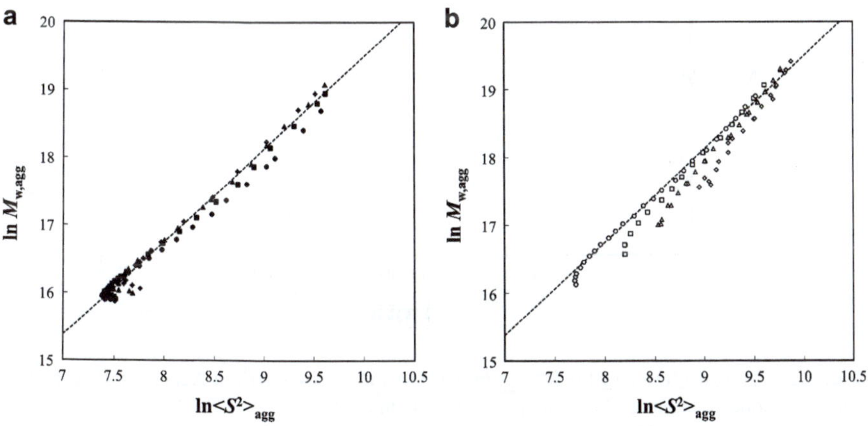

Fig. 2.15 Double-logarithmic plots of $M_{w,\text{agg}}$ vs $\langle S^2 \rangle_{\text{agg}}$ for PMMA ($M_w = 6.4 \times 10^6$) in AcN at 20.0 °C. The symbols are the same as those in Fig. 2.11

$$M_{w,\text{agg}} = A\langle S^2 \rangle_{\text{agg}}^{D/2} \tag{2.10}$$

where the exponent D is dependent on the structure and size distribution of clusters. The observed $D = 2.7$ is slightly smaller than that for a uniform hard sphere. The data for $c = 1.9 \times 10^{-4}$ g/cm^3 show similar behavior to those for higher concentrations but slightly deviate downwards, which would be due to the small quench depth. The deviation from the straight line observed in the initial stage of

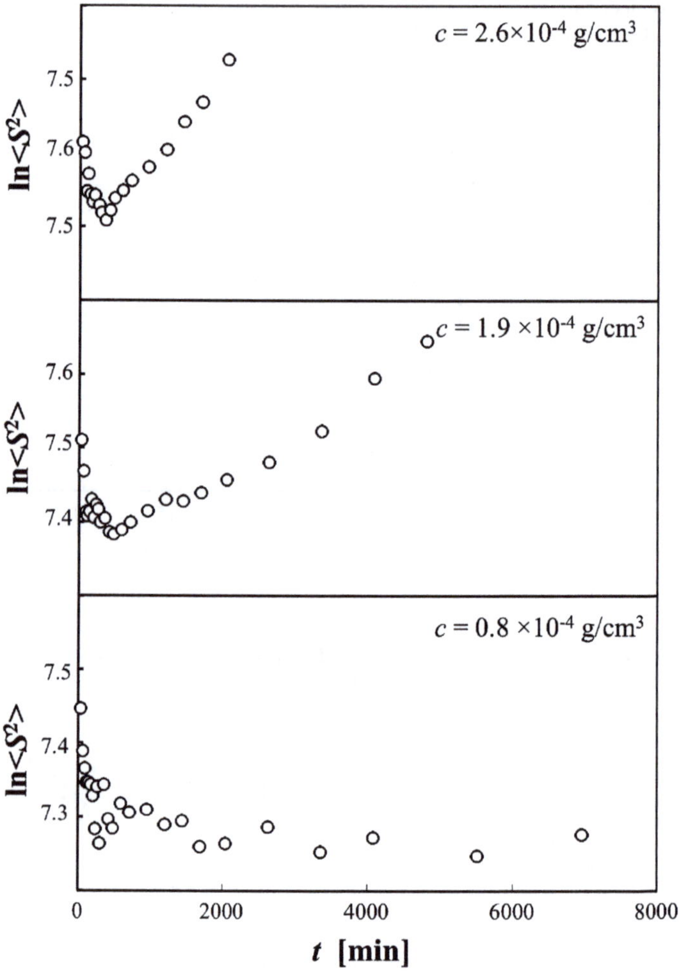

Fig. 2.16 Time evolution of ln$<S^2>_{agg}$ for PMMA ($M_w = 6.4 \times 10^6$) in AcN at 20.0 °C at short time t for concentrations c (10^{-4} g/cm^3) = 2.6, 1.9, and 0.8

Fig. 2.15a is attributed to chain collapse. In order to clarify the effect of chain collapse kinetics, Fig. 2.16 shows the time evolution of the radius of gyration, $<S^2>$, obtained in the initial stage for c (10^{-4} g/cm^3) = 0.8, 1.9, and 2.6. Because the solution of 0.8×10^{-4} g/cm^3 is thermodynamically stable at 20.0 °C, chain aggregation does not occur and chain collapse kinetics is measured. For the solutions of higher concentrations, $<S^2>$ decreases similarly owing to chain collapse, while $<S^2>$ increases gradually with time owing to chain aggregation. The decrease in $<S^2>$ in the initial stage where the chain aggregation is still not significant (Fig. 2.16) causes the deviation from the straight line at the smallest $M_{w,agg}$ in Fig. 2.15a. Figure 2.15b shows the data for concentrations ranging between 6.4×10^{-4} and 24.5×10^{-4} g/cm^3. The dashed straight line is the same as that shown in Fig. 2.15a. The behavior of the data for $c = 6.4 \times 10^{-4}$ g/cm^3 is almost the same as those for $c = 5.2 \times 10^{-4}$ g/cm^3 in Fig. 2.15(a). For the data for higher concentrations, the effect of chain collapse at the smallest $M_{w,\mathrm{agg}}$ shown in Fig. 2.15a becomes unclear, and the data approach the dashed straight line asymptotically with an increase in $M_{w,\mathrm{agg}}$. This behavior can be explained by assuming that the size of each polymer chain in clusters decreases with time. Thus, Fig. 2.15a and b demonstrates that chain collapse and chain aggregation occur competitively after the quench. At lower concentrations, chain collapse occurs first, and the collapsed chain aggregates to form clusters, resulting in the behavior shown in Fig. 2.15a. In contrast, chain collapse and chain aggregation occur simultaneously at higher concentrations, resulting in the behavior shown in Fig. 2.15b. The transitional behavior from chain collapse to chain aggregation shown in Fig. 2.15a has recently been studied by Nakata et al. [29].

2.6 Applicability of Rheological Concepts to Slow Aggregation of Polyglutamine

Proteins often have a relatively rigid and compact globular structure in aqueous solution owing to the well-defined folding of the polypeptide chain. Folding to a specific globular conformation is essential for their biological functions. Most globular proteins are regarded as amphiphilic copolymers that have both hydrophilic and hydrophobic units in one chain. It is accepted that, in globular proteins, hydrophilic ionic units mainly cover the surface of the globule, thus preventing interprotein association by electrostatic repulsion, while hydrophobic units mainly form the core of the globule. In addition, protein may experience a short-range attractive interaction, the origin of which is not yet fully understood. Van der Waals, opposite charge, hydrophobic, and hydrogen interactions may be involved [100].

Electrostatic repulsion between proteins may be modified by adding salt or reducing protein charge density, i.e., by adjusting pH closer to the isoelectric point (pI). If the degree of repulsion is decreased, the dominant short-range attraction between proteins can cause phase separation into a low-density phase and a high-density liquid or crystalline phase. If the attraction is strong, the bonds between proteins become long-lived so that it takes a very long time before the system reaches its equilibrium state. In this case, proteins aggregate, which may lead to the formation of a nonequilibrium gel or the precipitation of large amorphous flocs [100].

A traditional view of protein aggregation is that it is the association of the unfolded states of proteins. However, there is overwhelming evidence that partially unfolded states are much more prone to aggregation than both completely folded and unfolded conformations. In the completely folded or unfolded states, hydrophobic side chains are either mostly buried not in contact with water or randomly scattered. In contrast, partially unfolded states have large patches of contiguous surface hydrophobicity, which initiate aggregation effectively [8].

If protein aggregation occurs in living organisms, it can lead to a wide range of diseases. Much of the current work on protein aggregation is aimed at elucidating the structure and formation of amyloid fibrils. Although amyloid formation has been shown for a wide range of different proteins, it is clearly a very specific association process leading to the so-called cross-β structure [101].

Because of obvious differences in the primary structure and origin of stability in globular states, it seems that there is little relevance between the aggregation behaviors of homopolymers and proteins. However, the concepts derived from the chain aggregation of homopolymers would be applicable to the aggregation of a group of proteins referred to as "intrinsically disordered proteins (IDPs)" [11]. In this section, studies of the conformation and intermolecular aggregation of a polyglutamine homopolymer, which is an archetypal sequence of IDPs, are reviewed.

Intrinsically disordered proteins (IDPs) are functional proteins that do not fold into well-defined, ordered tertiary structures under physiological conditions [11]. These proteins are termed intrinsically disordered because disorder prevails under nondenaturing conditions and their amino acid sequence encodes the propensity to be disordered. Generic IDP sequences have a combination of low overall hydrophobicity and low sequence complexity. For example, exon 1 of huntingtin, the N-domain of the yeast prion protein Sup25, and α-synuclein are IDP sequences.

Among the above IDP sequences, the simplest IDP sequences are homopolymers such as polyglutamine. The expansion of polyglutamine tracts in several proteins is a genetic indicator of neurological diseases. Nine neurodegenerative diseases, including Huntington's disease, are associated with proteins containing the genetic expansion of polyglutamine tracts above a normal threshold length of 35 glutamine residues [102]. The expanded polyglutamine region in proteins is also associated with their enhanced deposition in intracellular inclusions in vivo and the propensity for amyloid-like β-sheet-rich fibrillar structures in vitro. The age of onset of a disease shows a nonlinear, inverse correlation with the length of polyglutamine expansion. The majority of the proposed mechanisms centers on the aggressive, length-dependent ability of polyglutamine to form ordered intermolecular aggregates [103, 104]. The conformation and aggregation behaviors and their dependences on the repeat length (molecular weight, in other words) of polyglutamine have been studied intensively.

Chen et al. [105] obtained the circular dichroism (CD) spectra of polyglutamine peptides with repeat lengths of 5, 15, 28, and 44 residues that were nearly identical and consistent with a high degree of random coil structure, suggesting that the length dependence of a disease is not related to the conformational change in the single chain of expanded polyglutamine sequences. They also studied the aggregation kinetics of a series of polyglutamines with different repeat lengths at a concentration of ~10 μM (~10^{-6} g/cm^3) by monitoring scattered light intensity. The aggregation occurred very slowly after lag times in the range of 10–100 h. In contrast to the CD results, there is a significant increase in aggregation rate for polyglutamine with a repeat length greater than 37 residues. In their subsequent study, Chen et al. [106] studied the in vitro aggregation kinetics of polyglutamine by means of CD, thioflavin T (ThT) fluorescence, high-performance liquid chromatography (HPLC), and light scattering methods. They found that the kinetics of conformation transition from a random coil to a β-sheet monitored by CD and ThT fluorescence is superimposable on that of chain aggregation monitored by HPLC and light scattering, suggesting the absence of soluble, β-sheet-rich intermediates in aggregation kinetics. Details of the time course of aggregate growth confirm that polyglutamine aggregation occurs by nucleated growth polymerization. However, in contrast to situations of conventional models of nucleated growth polymerization of proteins, the aggregation nucleus was a single polyglutamine chain. That is, the nucleation of polyglutamine aggregation corresponds to an energetically unfavorable protein-folding reaction.

Recently, Walters and Murphy [107] have measured DLS for polyglutamine with repeat lengths of 8, 12, 16, 20, and 24 residues at a concentration of ~10 μM (~10^{-6} g/cm^3). The average hydrodynamic radius $\langle R_h \rangle$ for the repeat lengths of 16, 20, and 24 continuously increased and reached several hundreds of nanometers within 4 h, indicating aggregate formation. For the repeat lengths of 8 and 12 residues, no aggregation was observed. They also studied the kinetics of the growth of insoluble aggregates by sedimentation assay [107]. Sedimentable aggregates are detected for the repeat lengths of 20 and 24 residues after lag phases of approximately 5 days and 3 days, respectively. Furthermore, a slow internal conformational rearrangement in the aggregates was hypothesized from the results of TEM image analysis. The results suggested that polyglutamine aggregation is unlikely to follow a homogeneous nucleation mechanism with the monomer as the critical nucleus, as shown by Chen et al. [106]. Instead, the results supported the scenario that disordered, non-β-sheet-like soluble molten oligomers form as early intermediates, followed by a conformational rearrangement. A kinetic model of aggregation that includes oligomers of different sizes has been developed by Vitalis and Pappu [108].

CD [105] and NMR [109] data have indicated that a single polyglutamine chain prefers a random coil state under physiological conditions. However, a globular conformation of a polyglutamine chain has been suggested recently. Crick et al. [102] employed fluorescence correlation spectroscopy (FCS) for polyglutamine solutions of 50 nM (~10^{-8} g/cm^3) in order to quantify the hydrodynamic size of polyglutamine chains as a function of chain length N in the range of 15–53. The observed translational diffusion time τ_D scaled with N as $\tau_D \sim N^{0.32}$, the exponent of which is comparable to 1/3 for the polymer globules. On the basis of these observations, they concluded that water is a poor solvent for polyglutamine. Previous results of CD and NMR studies [105, 109] indicating an intrinsically disordered structure of polyglutamine combined with the FCS data [102] suggested that an ensemble of polyglutamine in aqueous solutions is made up of a heterogeneous collection of collapsed structures. Digambaranath et al. [103] performed fluorescence resonance energy transfer (FRET) experiments on polyglutamine of $N = 16$ in water at 1 μM. They compared their experimental results with the results of molecular dynamics simulation using different theoretical models and observed that the models based on random coils and extended β-strands did not reproduce the FRET data well but that another model predicting compact conformation for $N > 16$ reproduced the FRET results. Using FRET, Walters and Murphy [107] observed an increasing propensity for collapsed conformations as the number of glutamine residues is increased; polyglutamine was relatively extended in solution for $N = 8$ or 12 and was relatively collapsed for $N = 20$ or 24.

The molecular conformation of polyglutamine has also been studied by means of computer simulations. Khare et al. [110] used a coarse-grain model and showed that although polyglutamine peptides of length $N < 37$ are disordered, chains with $N > 37$ are likely to form marginally stable β-helices. On the other hand, Vitalis et al. [111] analyzed data from molecular simulation of 20-residue polyglutamine peptide and concluded that water is a poor solvent for polyglutamine, despite its polar residues. They suggested that intrabackbone interactions provide at least part of the driving force for the collapse of polyglutamine in water. In their subsequent study [112], they studied the characteristics of the dimerization of two polyglutamine globules as a function of N and temperature by molecular simulation. They showed that polyglutamine molecules form disordered collapsed globules for a chain length $N > 15$ and that the approaching two globular chains associate with each other, indicating that similar driving forces govern both the collapse and a spontaneous dimerization of polyglutamine in aqueous milieus. This result supports the formation of soluble oligomers lacking a regular secondary structure as early intermediates [107].

The heterogeneous, disordered, and compact nature of the collapsed structure of polyglutamine demonstrated by experiments [102] and computer simulation [111] is common to globules of homopolymers below the Θ point, and therefore, mechanisms for slow aggregation of synthetic homopolymers below the cloud point could be applicable to the slow aggregation of polyglutamine. Although an extremely dilute solution (~10^{-6} g/cm^3) was used, the slow aggregation and long lag time observed by Chen et al. [105] cannot be explained by the low probability of Brownian collision, because the characteristic time of DLCA (Eq. 2.7) is estimated to be $\tau_{\mathrm{DLCA}} \sim 1$ ms. Therefore, some other mechanisms such as the nucleation growth polymerization proposed by Chen et al. [106] are surely needed. In the homogeneous nucleation mechanism [106] with a single polyglutamine chain as the critical nucleus, the slow aggregation is attributed to the very slow conversion from globules to an energetically unfavorable conformation, which is assumed to be rich in the β-structure. On the other hand, in the mechanism of an amorphous oligomeric intermediate formation followed by the conformational rearrangement [107], the slow aggregation could be due to the viscoelastic effect of polyglutamine globules. A molecular simulation study [112] suggested that two globular polyglutamines in contact have a propensity to form a dimer. In the case of synthetic homopolymers in a dilute solution, although the free energy of an equilibrium dimer consisting of two polymer chains is significantly lower than that of two separate globules in a poor solvent [113], the formation of an equilibrium dimer could be kinetically less favorable owing to the suppression of interdiffusion motion required for merging two globules, i.e., the viscoelastic effect, as shown by Tanaka [92, 93]. The viscoelastic effect has been observed in globules of PNIPAM and PMMA with a monomer density $\phi > 0.1$. The volume fraction of polyglutamine globules can be estimated to be $\phi \sim 0.1$ from the hydrodynamic radius $R_{\mathrm{h}} \sim 2.5$ nm for a chain length

$N \sim 50$ obtained by the FCS study [102]. Unlike in the case of PNIPAM or PMMA, the viscoelastic effect in polyglutamine is irrelevant to the reptational motion of chains because of the short chain length $N < 100$. Alternatively, a high segment density in globules may lead to a higher effective viscosity or partial vitrification [77] of polymers, which would make the chain diffusion time much longer than the contact time, causing the viscoelastic effect. In order to verify the possible contribution of the viscoelastic effect to the slow aggregation of polyglutamine, more information on the dynamics of the structural relaxation of a single polyglutamine chain is needed.

References

1. Grosberg AY, Khokhlov AR (1994) Statistical physics of macromolecules. AIP Press, New York
2. Flory PJ (1952) Principles of polymer chemistry. Cornell University Press, Ithaca
3. Hirokawa Y, Tanaka T (1984) J Chem Phys 81:6379
4. Gil ES, Hudson SM (2004) Prog Polym Sci 29:1173–1222
5. Hoffman AS, Stayton PS (2004) Macromol Symp 207:139–151
6. Kikuchi A, Okano T (2002) Adv Drug Deliv Rev 54:53–77
7. Clark AH (1998) Gelation of globular proteins. In: Hill SE, Ledward DA, Mitchell JR (eds) Functional properties of food macromolecules. Aspen Publishers, Gaithersburg
8. Wang W (2005) Int J Pharm 289:1–30
9. Baysal BM, Karasz FE (2003) Macromol Theory Simul 12:627
10. Aseyev VO, Tenhu H, Winnik FM (2006) Adv Polym Sci 196:1–85
11. Pappu RV, Wang X, Vitalis A et al (2008) Arch Biochem Biophys 469:132–141
12. Tanford C (1966) Physical chemistry of macromolecules. Wiley, New York
13. Zhang G, Wu C (2005) Adv Polym Sci 195:1
14. Nakata M (1995) Phys Rev E 51:5770
15. Maki Y, Sasaki N, Nakata M (2004) Macromolecules 37:5703–5709
16. Nakata M, Nakagawa T (1997) Phys Rev E 56:3338
17. Nakamura Y, Sasaki N, Nakata M (2001) Macromolecules 34:5992–6002
18. Maki Y, Dobashi T, Nakata M (2007) J Chem Phys 126:134901
19. Nakata M, Nakagawa T (1999) J Chem Phys 110:2703
20. Nakamura Y, Sasaki N, Nakata M (2003) J Chem Phys 118:3861
21. Maki Y, Dobashi T (2012) Kobunshi Ronbunshu 69:373–381
22. Nakata M, Nakamura Y, Maki Y (2004) Macromolecules 37:4917
23. Nakata M, Nakamura Y, Sasaki N et al (2012) Phys Rev E 85:021802
24. Nakata M, Nakagawa T, Nakamura Y et al (1999) J Chem Phys 110:2711
25. Nakagawa T, Nakamura Y, Sasaki N et al (2001) Phys Rev E 63:031803
26. Maki Y, Dobashi T, Nakata M (2008) Phys Rev E 78:041802
27. Nakamura Y, Nakagawa T, Sasaki N et al (2001) Macromolecules 34:5984–5991
28. Nakamura Y, Sasaki N, Nakata M (2002) Macromolecules 35:1365–1372
29. Nakata M, Nakamura Y, Sasaki N (2007) Phys Rev E 76:041805
30. Seuring J, Agarwal S (2012) Macromol Rapid Commun 33:1898–1920
31. Koningsveld R, Stockmayer WH, Nies E (2001) Polymer phase diagrams, a text book. Oxford University Press, Oxford
32. Stockmayer WH (1960) Makromol Chem 35:54
33. Birshtein TM, Pryamitsyn VA (1991) Macromolecules 24:1554–1560
34. Grosberg AY, Kuznetsov DV (1992) Macromolecules 25:1970–1979
35. Grosberg AY, Kuznetsov DV (1992) Macromolecules 25:1996–2003

36. Grosberg AY, Kuznetsov DV (1992) Macromolecules 25:1980–1990
37. Lifshitz IM, Grosberg AY, Khokhlov AR (1978) Rev Mod Phys 50:683–713
38. Sun ST, Nishio I, Swislow G et al (1980) J Chem Phys 73:5971–5975
39. Park IH, Wang QW, Chu B (1987) Macromolecules 20:1965–1975
40. Kubota K, Fujishige S, Ando I (1990) J Phys Chem 94:5154–5158
41. Wu C, Zhou S (1995) Macromolecules 28:8381–8387
42. Wang X, Qiu X, Wu C (1998) Macromolecules 31:2972–2976
43. Burchard W (1996) Combined static and dynamic light scattering. In: Brown W (ed) Light scattering: principles and development. Clarendon, New York
44. Okada Y, Tanaka F (2005) Macromolecules 38:4465–4471
45. Tanaka F, Koga T, Kojima H et al (2009) Macromolecules 42:1321–1330
46. Witelski TP, Grosberg AY, Tanaka T (1998) J Chem Phys 108:9144–9149
47. Abe F, Einaga Y, Yamakawa H (1995) Macromolecules 28:694
48. Yamakawa H, Abe F, Einaga Y (1994) Macromolecules 27:3272
49. Baysal BM, Kayaman N (1998) J Chem Phys 109:8701–8707
50. Dogan M, Kuntman A (2000) Polym Int 49:1648–1652
51. Kayaman N, Gürel EE, Baysal BM et al (2000) Polymer 41:1461–1468
52. Kayaman N, Gürel EE, Baysal M et al (1999) Macromolecules 32:8399–8403
53. de Gennes PG (1985) J Phys Lett 46:L-639
54. Grosberg AY, Nechaev SK, Shakhnovich EI (1988) J Phys (France) 49:2095
55. Buguin A, Brochart-Wyart F, de Gennes PG (1996) C R Acad Sci Ser IIb Mec Phys Chim Astron 322:741
56. Klushin LI (1998) J Chem Phys 108:7917
57. Halperin PM, Goldbart PM (2000) Phys Rev E 61:565
58. Byrne A, Kiernan P, Green D et al (1995) J Chem Phys 102:573
59. Kuznetsov YA, Timoshenko EG, Dawson KA (1995) J Chem Phys 103:4807
60. Kuznetsov YA, Timoshenko EG, Dawson KA (1996) J Chem Phys 104:3338
61. Abrams CF, Lee N-K, Obukhov SP (2002) Europhys Lett 59:391–397
62. Frisch T, Verga A (2002) Phys Rev E 66:041807
63. Kikuchi N, Ryder JF, Poooley CM et al (2005) Phys Rev E 71:061804
64. Kamata K, Araki T, Tanaka H (2009) Phys Rev Lett 102:108303
65. Chu B, Ying Q, Grosberg AY (1995) Macromolecules 28:180–189
66. Xu J, Zhu X, Luo S et al (2006) Phys Rev Lett 96:027802
67. Zhang GZ, Wu C (2001) Phys Rev Lett 86:822
68. Ye X, Lu Y, Shen L et al (2007) Macromolecules 40:4750–4752
69. Mansfield ML (2007) J Chem Phys 127:244902
70. Rabin Y, Grosberg AY, Tanaka T (1995) Europhys Lett 32:505
71. Lee N-K, Abrams CF, Johner A et al (2003) Phys Rev Lett 90:225504
72. Lee N-K, Abrams CF, Johner A et al (2004) Macromolecules 37:4917
73. Chu B, Ying Q (1996) Macromolecules 29:1824–1826
74. Heskins M, Guillet JE (1968) J Macromol Sci Chem A2:1441
75. Schild HG, Tirrell DA (1990) J Phys Chem 94:4352
76. Fujishige S, Kubota K, Ando I (1989) J Phys Chem 93:3311–3313
77. Van Durme K, Van Assche G, Van Mele B (2004) Macromolecules 37:9596
78. Tong Z, Zeng F, Zheng X et al (1999) Macromolecules 32:4488–4490
79. Xia Y, Yin X, Burke NAD, Stöver HDH (2005) Macromolecules 38:5937–5943
80. Kawaguchi T, Kojima Y, Osa M et al (2008) Polym J 40:455
81. Kawaguchi T, Kojima Y, Osa M et al (2008) Polym J 40:528
82. Katsumoto Y, Kubosaki N (2008) Macromolecules 41:5955–5956
83. Katsumoto Y, Kubosaki N, Miyata T (2010) J Chem Phys B 114:1044–1045
84. Nishi K, Hiroi T, Hashimoto K et al (2013) Macromolecules 46:6225–6232
85. McPhee W, Tam KC, Pelton R (1993) J Colloid Interface Sci 156:24
86. Gorelov AV, Du Chesne A, Dawson KA (1997) Phys A 240:443–452

87. Timoshenko EG, Kuznetsov YA (2001) Europhys Lett 53:322–327
88. Aseyev V, Hietala S, Laukkanen A et al (2005) Polymer 46:7118–7131
89. von Smoluchowski M (1916) Phys Z 17:585
90. von Smoluchowski M (1917) Z Phys Chem Stoechiom Verwandtschafts 92:129
91. Chan K, Pelton R, Zhang J (1999) Langmuir 15:4018–4020
92. Tanaka H (1992) Macromolecules 25:6377–6380
93. Tanaka H (2000) J Phys Condens Matter 12:R207–R264
94. Chuang J, Grosberg AY, Tanaka T (2000) J Chem Phys 112:6434
95. Kujawa P, Aseyev V, Tenhu H et al (2006) Macromolecules 39:7686–9693
96. Pamies R, Zhu K, Kjøniksen A-L et al (2009) Polym Bull 62:487–502
97. Kawaguchi T, Kobayashi K, Osa M et al (2009) J Phys Chem B 113:5440
98. Vicsek T (1989) Fractal growth phenomena. World Scientific, Singapore
99. Chen W, Zhao Y, Jiang Y et al (2004) Chem Phys Chem 5:1745–1749
100. Nicolai T, Durand D (2007) Curr Opin Colloid Interface Sci 12:23–28
101. Uversky VN, Fink AL (2004) Biochim Biophys Acta 1698:131–153
102. Crick SL, Jayaraman M, Frieden C et al (2006) Proc Natl Acad Sci 103:16764–16769
103. Digambaranath JL, Campbell TV, Chung A et al (2010) Proteins 79:1427–1440
104. Nekooki-Machida Y, Kurosawa M, Nukina N et al (2009) Proc Natl Acad Sci 24:9679–9684
105. Chen S, Berthelier V, Yang W et al (2001) J Mol Biol 311:173–182
106. Chen S, Ferrone FA, Wetzel R (2002) Proc Natl Acad Sci 99:11884–11889
107. Walters RH, Murphy RM (2009) J Mol Biol 393:978–992
108. Vitalis A, Paapu RV (2011) Biophys Chem 159:14–23
109. Masino L, Kelly G, Leonard K et al (2002) FEBS Lett 513:267–272
110. Khare SD, Ding F, Gwanmesia KN et al (2005) PLoS Comput Biol 1:230–235
111. Vitalis A, Wang X, Pappu RV (2007) Biophys J 93:1923–1937
112. Vitalis A, Wang X, Pappu RV (2008) J Mol Biol 384:279–297
113. Grosberg AY, Kuznetsov DV (1992) J Phys II 2:1327

Elongational Flow Birefringence Investigation of Dynamics of DNA Molecules

3

Naoki Sasaki

Abstract

Elongational flow studies on the dynamics of phage DNA molecules are reviewed in this chapter, where the elongational flow field was used as a perturbation to obtain a flow-induced birefringence signal from a dilute sample solution. First, techniques and apparatus for generating elongational flow are discussed. Then, the response of biopolymers in dilute solutions to an elongational flow field is discussed, as well as the use of a birefringence technique to detect that response to flow. Historically, it was unclear whether flexible polymer molecules in an elongational flow field are actually elongated. As a first topic in this review, I will report the evidence that polymer molecules in an elongational flow field are in reality elongated, using observations with a flow birefringence method. Dynamical properties of DNA molecules, such as helix–coil transitions, scission of molecules caused by the flow field, and coil–globule transitions, as observed by flow birefringence and fluorescent microscopic methods, are also presented. Finally, studies on the interaction of DNA molecules with the DNA-binding protein HU are shown. As HU's function was regarded as bending and folding of DNA molecules, the interaction has been expected to be similar to a coil–globule transition. However, this similarity was disproved through elongational flow birefringence experiments and fluorescent microscopy. In general, the elongational flow birefringence technique has proven to be useful for studying dynamic properties of polymer molecules.

Keywords

Elongational flow field • DNA molecules • Flow birefringence • Helix–coil transition • Scission of molecules • Coil–globule transition

N. Sasaki (✉)
Faculty of Advanced Life Science, Hokkaido University, Kita-ku, Sapporo 060-0810, Japan
e-mail: nasa5131@sci.hokudai.ac.jp

R. Kita, T. Dobashi (eds.), *Nano/Micro Science and Technology in Biorheology*,
DOI 10.1007/978-4-431-54886-7_3

3.1 Introduction

According to J. C. Kendrew, biopolymers are "threads of life" [26]. A number of reaction sites are located on the threads, which assign biological functions to the threads of life. Because of the remarkable functions of biopolymers, most studies have focused on their chemical structure and biochemistry, while the physical characteristics of the threads themselves have been given less weight. But considering biopolymers interacting with an external field, a flow field, for example, we have to have knowledge about their threadlike nature. In the case of a macroscopic thread, much information would be obtained by pulling both ends of the thread material and elongating it. In this article, I will review methods for elongating biopolymers and observing the outcomes of their dynamic transitions. In particular, the use of a flow field to create extension of biopolymers will be presented.

Most of the important characteristics of polymeric materials in solution, intrinsic viscosity, for example, are determined by using transport phenomena in a shear flow. In a shear flow, however, though deformation of a polymer chain would be possible, its elongation is difficult to realize because of unavoidable rotation of the chain. In order to elongate the polymer chain, an extensional or elongational flow field is needed. Frank [15] expected that in an elongational flow field, a flexible polymer chain would be extended by the tensile force exerted on the polymer molecule by the flow field. The extension was considered to occur as a transition, with the extensional force overcoming an inherent entropic contraction force within the polymer molecule. This expectation by Frank was theoretically formulated for dilute flexible polymer solutions by P. G. de Gennes [9] as a coil–stretch transition, using a dumbbell model proposed by A. Peterlin [39]. The coil–stretch transition was regarded as a runaway process; once the flexible polymer molecule starts deforming above a critical strain rate, the tensile force exerted also starts increasing. Experiments confirming the Frank expectation and the de Gennes formulation were performed extensively by the Bristol group [25]. Since then, analysis using elongational flow-induced birefringence has been established as a new tool for investigating molecular properties of biopolymers that are not addressed by the traditional shear flow field method [14, 19–21, 23, 40, 43].

3.2 Elongational Flow Generation

In Figs. 3.1, 3.2, 3.3, and 3.4, several methods for generating an elongational flow field are shown. A common feature in all the methods shown is a stagnation point in the flow. From the stagnation point, elongational flow exists in a direction toward the outlet. In the opposed jets (OJ) (Fig. 3.1) [16, 36] and the cross slots (CS) (Fig. 3.2) [6, 44], the elongational flow field continues while fluid is translated from reservoir 1 to reservoir 2, requiring a sufficiently large amount of sample solution for long observation periods. Both OJ and CS are used for a strain rate $\dot{\varepsilon}$ up to 10^3–10^4 s^{-1}. Miniaturized versions have been developed for microflow experiments [17, 19]. Figure 3.3 shows a scheme for another elongational flow

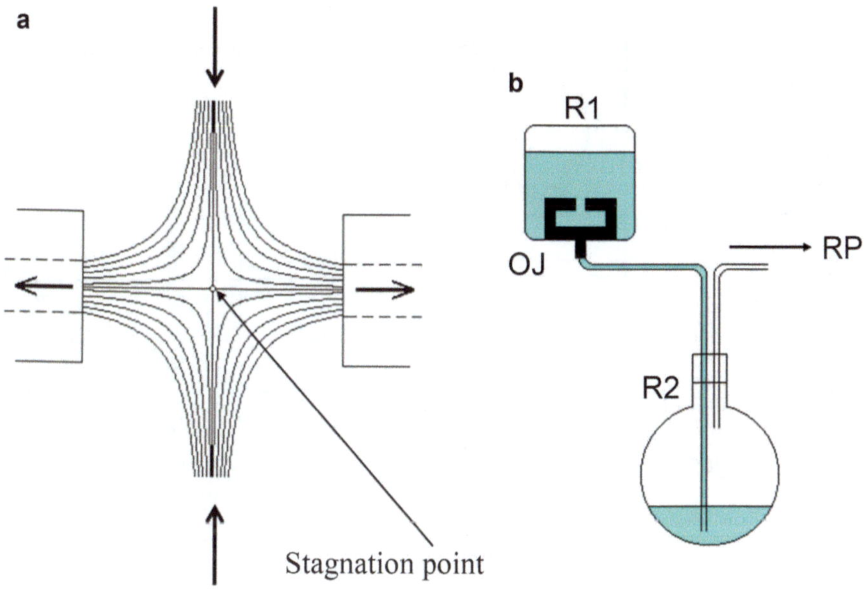

Fig. 3.1 (**a**) Opposed jets and flow lines of the resulting flow field. A small circle in the center of the field represents the stagnation point. From the stagnation point to both exit nozzles, there is a one-dimensional elongational flow field along the exit symmetry axis. Both the distance between nozzles and the orifice dimension are usually in the order of 1 mm. *Arrows* indicate flow directions. (**b**) Total flow circuit used with opposed jets. The opposed jet (OJ) apparatus is immersed in a reservoir R1. By evacuating a reservoir R2 with a rotary pump (RP), polymer solution in R1 is sucked into nozzles of opposed jets causing flow along the symmetry axis

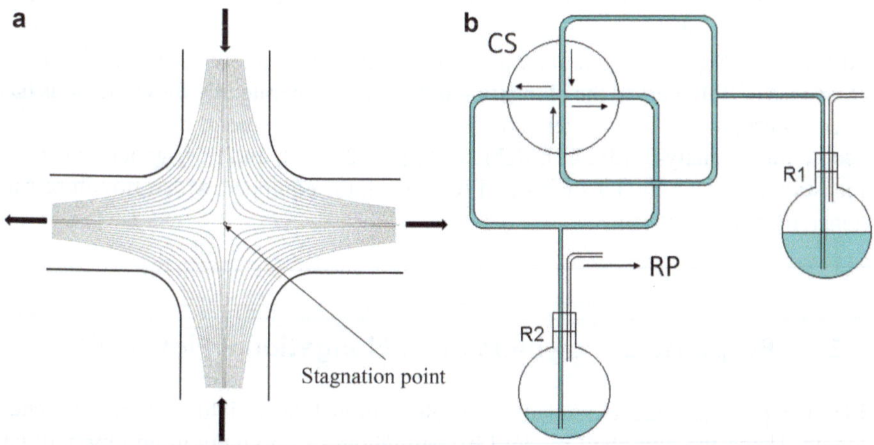

Fig. 3.2 (**a**) Cross slots and flow lines of generated flow field. At the center of the cross slots, there is a stagnation point. *Arrows* indicate flow directions. (**b**) A total flow circuit used with cross slots. Solution in the reservoir R1 is drawn into the cross slots (CS) when the reservoir R2 is evacuated by a rotary pump (RP). *Arrows* indicate flow directions

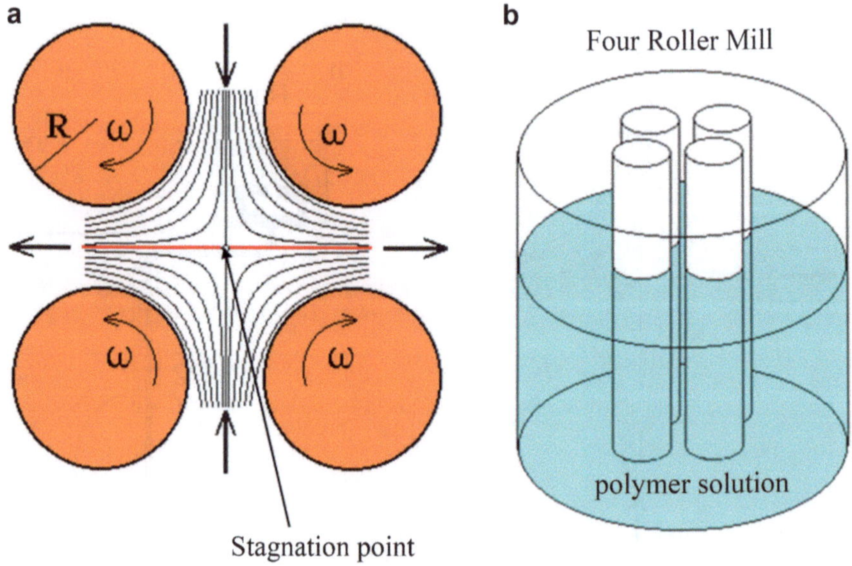

Fig. 3.3 (**a**) A four-roller mill and generated flow field. Top view along rollers. By rotating rollers at angular velocity $\omega(s^{-1})$ to the indicated directions, a planar elongational flow field parallel to the rollers is generated in the space among the rollers. *Arrows* indicate flow directions. A small circle in the center of the field represents the stagnation point. *Arrows* indicate flow directions. (**b**) Four rollers of the same size are set parallel to one another. The rollers are immersed in polymer solution

field generator, the four-roller mill (FRM) [8, 52]. Flow lines in the four-roller mill near the surface of rollers are parabolic curves, while the roller surface is circle. This difference of the curvatures leads to the separation of the flow from the roller surface [53]. Strain rate generally achieved in an FRM is a few 10^2 s^{-1} at the most. Response of polymers to the elongational flow field generated by these methods has been observed by a flow-generated birefringence measurement, fluorescent microscopic image analysis [38, 47], light scattering [27, 29], and flow-generated stress measurements. Figure 3.4 shows an FRM apparatus equipped with a flow birefringence measuring system.

3.3 Response of Polymers to an Elongational Flow Field

Flexible polymers are extended in an elongational flow field to their stretched length. Using the dumbbell model [39], stretching of polymers in an elongational flow field is well described. The force exerting on a flexible random coil polymer molecule is

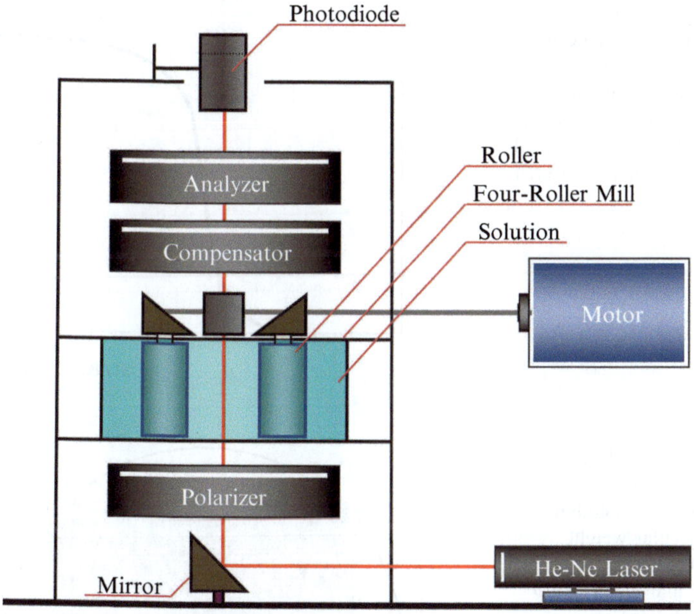

Fig. 3.4 A setup of the four-roller mill in an optical polarization measuring system. Optical signal is detected by a photodiode

$$\Delta F(l) = \frac{1}{2}\zeta\dot{\varepsilon}l - F_{\text{cont}}(l) \tag{3.1}$$

where l is polymer molecular size, ζ is friction coefficient, $\dot{\varepsilon}$ is elongational strain rate, and F_{cont} is the entropic contraction force of the polymer molecule. The first term is the drag force by the flow field. On increasing $\dot{\varepsilon}$, the sign of ΔF changes from − to + on passing through the point $\Delta F = 0$. The strain rate $\dot{\varepsilon}_{\text{c}}$ at $\Delta F = 0$ is termed the critical strain rate for the coil–stretch transition. In a flow birefringence experiment, for $\dot{\varepsilon} < \dot{\varepsilon}_c$, no birefringence is observed. As described above, the coil–stretch transition is considered to be a runaway process. When the drag force exceeds the entropic contraction force, the polymer chains are deformed along the flow direction. Once the polymer chain starts to become deformed, the drag force increases proportionally to the deformed length along the flow direction. Thus, for a strain rate slightly larger than $\dot{\varepsilon}_c$, birefringence starts to appear. Figure 3.5 shows a schema of empirically obtainable birefringence intensity, Δn, resulting from the coil–stretch transition. From the strain rate dependence of Δn, for $\dot{\varepsilon} < \dot{\varepsilon}_c$, after the coil–stretch transition occurs, localized birefringence appears in the pure elongational flow field. Figure 3.6 shows the localized birefringence in the central region of FRM, where the strain rate is larger than the critical value.

The critical strain rate $\dot{\varepsilon}_{\text{c}}$ relates to the longest relaxation time τ_L of the molecule through the inverse relation

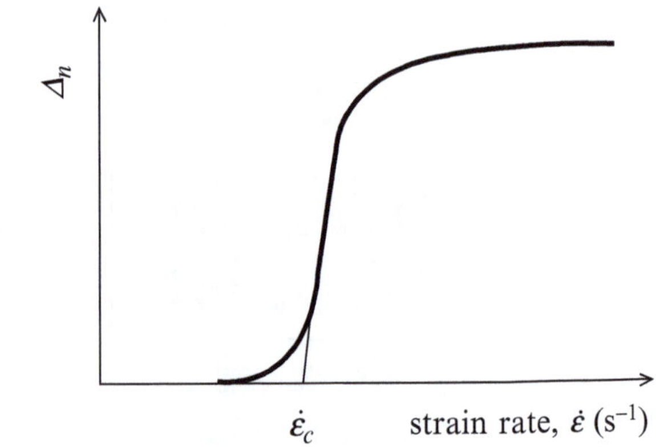

Fig. 3.5 Schematic of elongational flow-induced birefringence intensity of flexible polymers in the sample solution as a function of strain rate $\dot{\varepsilon}$. $\dot{\varepsilon}_c$ is a critical strain rate for the coil–stretch transition

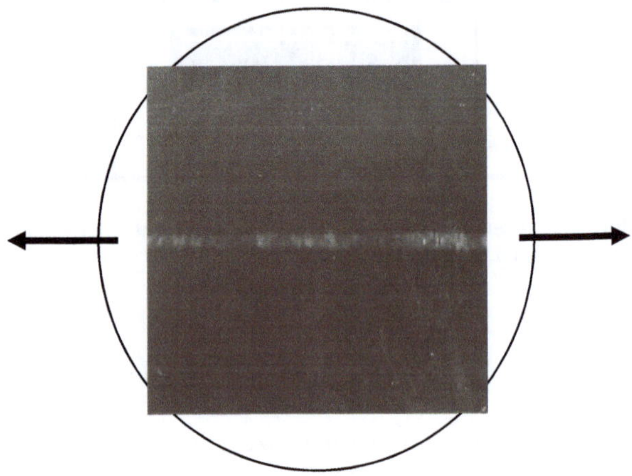

Fig. 3.6 Flow-induced birefringence pattern in a flexible polymer solution (high-molecular-weight polyethylene oxide (PEO) in low-molecular-weight PEO aqueous solution). *Arrows* indicate flow directions, and a *circle* represents optical microscopic field (From Sasaki et al. [43]. With permission from Elsevier)

$$\dot{\varepsilon}_c \tau_L \sim C, \tag{3.2}$$

where C is a numerical constant of the order of unity [9, 20, 31, 38].

When the polymer molecules in a solution are rigid rodlike, α-helical polypeptides, for example, such molecules respond to the flow field by rotating to orient in a definite direction [25]. The solution shows almost homogeneous birefringence throughout the irradiating field, where the orientation direction of each rodlike molecule is considered to be almost identical. Figure 3.7 shows the non-localized birefringence response of a solution containing rigid rodlike molecules as solute. By observing color information of the birefringence, Odell et al. determined that the orientation direction of rodlike molecules was parallel to the outlet flow direction [34]. In this case, birefringence starts to increase from $\dot{\varepsilon} \geq 0$ and there is no critical strain rate for the appearance of birefringence. Figure 3.8 shows the

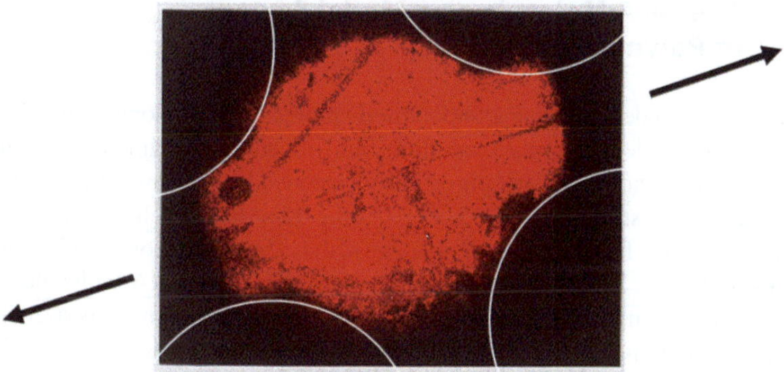

Fig. 3.7 Flow-induced birefringence pattern of a rigid rodlike polymer solution (poly(γ-methyl D-glutamate) in chloroform). *White circles* indicate four rollers. *Arrows* show flow directions (From Sasaki et al. [43]. With permission from Elsevier)

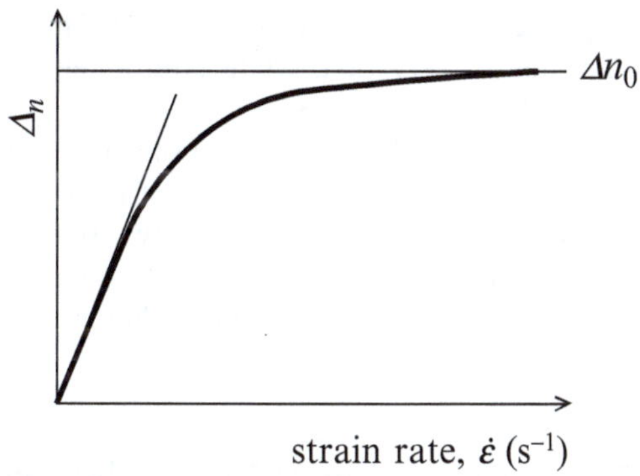

Fig. 3.8 Schematic drawing of elongational flow-induced birefringence intensity of rigid rodlike molecules in the sample solution as a function of strain rate $\dot{\varepsilon}$

birefringence intensity, Δn, plotted against strain rate $\dot{\varepsilon}$. The shape of the curve is described by the strain rate dependence of the orientational order parameter S,

$$S = \frac{2}{3}\left[\frac{1}{4} - \frac{3}{2\xi} + \left(\frac{9}{16} - \frac{3}{4\xi} + \frac{9}{4\xi^2}\right)^{\frac{1}{2}}\right] \tag{3.3}$$

where $\xi = \dot{\varepsilon}/D_r$ and D_r is the rotational diffusion coefficient [11]. The D_r value for a rodlike molecule can be determined from its birefringence data [34]. One method to determine the rotational diffusion coefficient of rigid rodlike molecules is to observe the decay of Δn after a sudden cessation of flow. The birefringence decay of a molecule having rotational diffusion coefficient D_r is described by [3],

$$\Delta n = \Delta n_0 \exp(-6D_r t). \tag{3.4}$$

The decay results from disorientation processes affecting the molecule after flow ceases.

3.4 The DNA Molecule as a Model System of Polymer Dynamics

There had been a debate whether a polymeric molecule in an elongational flow field is actually extended. Elongational flow birefringence experiments indicate that flexible polymer molecules are almost fully extended by the flow field in the regime above the critical strain rate [6, 35, 41]. On the other hand, light scattering studies on the elongating flexible polymers suggest that the elongation should not be so remarkable [7, 27, 29]. Using fluorescent microscopy and DNA molecules from a bacteriophage, Chu et al. showed that the DNA elongated to almost its stretched-out state, using an elongational flow field beyond the critical strain rate [38, 47]. This method of single-molecules visualization looks at each molecule in the solution, while flow birefringence measurements are macroscopic, and the result is averaged over all molecular in the solution. Correspondence between these two methods has been investigated, and the coil–stretch transition was observed to be sharper with single-molecular visualization than with birefringence. This discrepancy likely originates from the selection of molecules in a steady extension in the single-molecular observation, while birefringence averages over molecules in a broad range of extension states.

DNA molecules are often used as a model system for visualizing flexible polymer dynamics because each of them after staining is easily visualized separately by fluorescent microscopy. Before and after the elongational flow experiments, molecular weight was confirmed to be unchanged for the DNA molecules examined in the studies below.

3.5 Elongation of DNA Molecules in the Flow Field

As a model system, DNA was used to demonstrate that a flexible polymer chain could be stretched almost to its limit by an elongational flow. Figure 3.9 shows elongational flow-induced birefringence plotted against elongational strain rate for λ-phage DNA 0.2 M NaCl aqueous solutions of three different DNA concentrations [41]. The elongational flow field was generated by an FRM. Though at small values, there is a critical strain rate $\dot{\varepsilon}_c$ seen in each profile. After a small non-birefringent range of $\dot{\varepsilon}$, Δn increased rapidly at first and then gradually. The observed birefringence pattern in the irradiated region of the FRM was broadly localized around the elongational flow field, the broadness originated from the semiflexible nature of DNA chains. Before and after the experiments that covered a range up to $\dot{\varepsilon}{\sim}160\ s^{-1}$, no evidence for DNA molecular scission by the elongational flow field was observed; this was also confirmed by an agarose gel electrophoresis assessment. Figure 3.10 shows a set of birefringence profiles along the inlet line containing the stagnation point in the FRM at indicated strain rates. The pattern and these profiles document the molecular elongation and underlying flow field evolution from quiescent state; (1) an almost non-localized birefringence appears first,

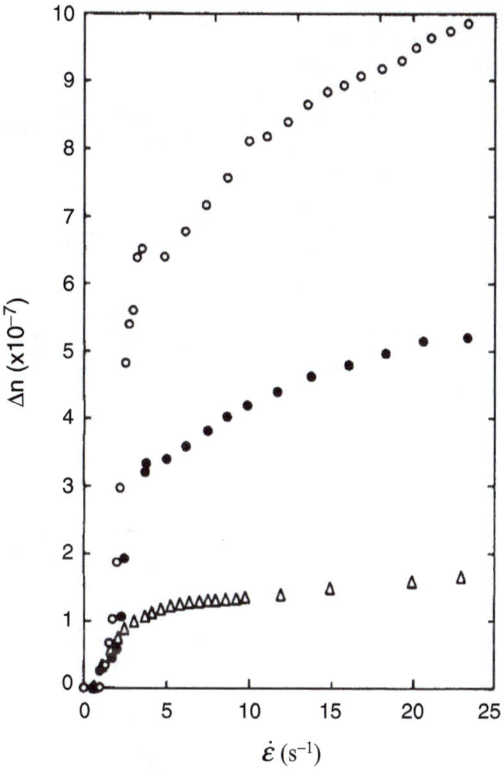

Fig. 3.9 Elongational flow-induced birefringence intensity, Δn, plotted against strain rate, $\dot{\varepsilon}$, for (○) 10 μg/ml, (●) 7 μg/ml, and (△) 5 μg/ml of λ-phage DNA solutions (From Sasaki et al. [41]. With permission from John Wiley & Sons, Inc.)

(2) a sharp localized birefringence line then appears, and (3) the localized line thickens. The second process in the profile evolution corresponds to the rapid increasing process of Δn in the Δn vs. $\dot{\varepsilon}$ plot (Fig. 3.9). Odell and Taylor [33] regarded the criticality at $\dot{\varepsilon}_c$ as the manifestation of the coil–stretch transition of DNA coils, where the chain hydrodynamic analogy changes from non-free draining to free draining. Figure 3.11 shows the decay in Δn after sudden cessation of the flow at $\dot{\varepsilon} = 24\ \ \mathrm{s}^{-1}$ for 10 mg/ml of DNA solution. For all solutions measured, there were two distinct processes: an initial rapid relaxation (stage 1) and a slow relaxation (stage 2). This measurement was performed using a photodiode, and the observed birefringence was integrated over the entire microscopic field inside the FRM. Figure 3.12 shows the time evolution of the similar profile as Fig. 3.10 after the sudden cessation of the flow at $\dot{\varepsilon} = 24\ \ \mathrm{s}^{-1}$. It is clear that the birefringence at the off-symmetrical plane or foot region in the profile decreases faster than that near the stagnation point. Figure 3.13 shows Δn decay at indicated points in the inside area of FRM after the sudden cessation of the flow at $\dot{\varepsilon} = 24\ \ \mathrm{s}^{-1}$. Δn is plotted against the frame number of the video still, which is proportional to time after stopping the mill.

In this stopped flow experiment for semiflexible DNA molecules, the Δn relaxation is considered to contain both a disorientation process of deformed molecules

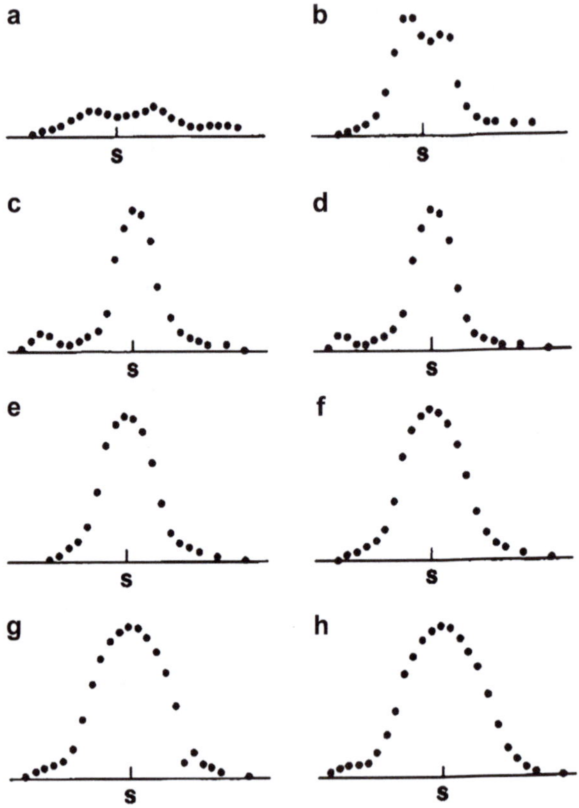

Fig. 3.10 Birefringence profile along the inlet symmetry axis including the stagnation point in an FRM for 10 μg/ml of λ-phage DNA solution at eight stages from $\dot{\varepsilon} = 2$ to 24 s^{-1}: at (**a**) $\dot{\varepsilon} = 2\ s^{-1}$, (**b**) 3 s^{-1}, (**c**) 4 s^{-1}, (**d**) 5 s^{-1}, (**e**) 7 s^{-1}, (**f**) 10 s^{-1}, (**g**) 16 s^{-1}, (**h**) 24 s^{-1}. s indicates the stagnation point (From Sasaki et al. [41]. With permission from John Wiley & Sons, Inc.]

oriented by the flow and the recovery process from a deformed internal conformation. Because both the stage 1 and the stage 2 relaxations are described by simple exponential decay, each of them is regarded as a single mechanism. The relaxation time for stage 1 was 0.52 s and for stage 2, 6.7 s. According to a direct imaging studies, the relaxation time for the extended λ-phage DNA to contract has been reported to be 3 s [37] for molecules of 12.8 μm long. From these facts, the main contribution to stage 2 was concluded to be the contraction process of extended DNA molecules and for stage 1, the disorientation process of deformed DNA molecules.

As the classification of stages was made by observing the Δn decay after flow stopped, processes in Fig. 3.13 should be categorized as stage 1. Figure 3.14 shows the birefringence relaxation time τ at each point in the FRM, estimated from Fig. 3.13. All relaxation times plotted are in stage 1. From Eq. (3.4), the orientation relaxation time is related to the rotational diffusion coefficient,

$$\tau^{-1} = 6D_{\mathrm{r}}. \tag{3.5}$$

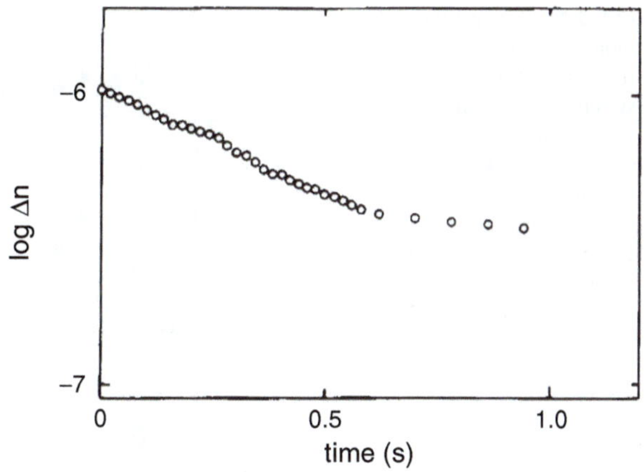

Fig. 3.11 A typical decay curve of flow-induced birefringence intensity, Δn, after sudden cessation of flow at strain rate $\dot{\varepsilon} = 24\ \mathrm{s}^{-1}$, for 10 μg/ml of DNA solution (From Sasaki et al. [41]. With permission from John Wiley & Sons, Inc.)

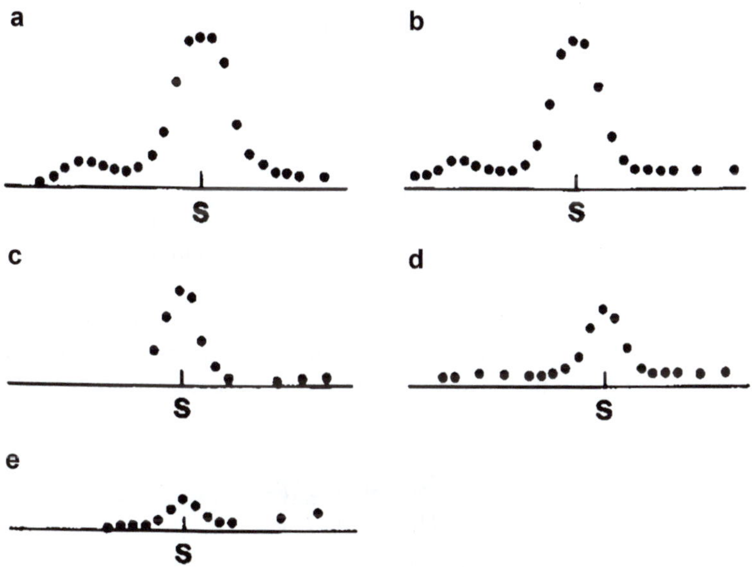

Fig. 3.12 Change in the birefringence profile along the inlet symmetry axis for 10 μg/ml DNA solution at (**a**) $t = 0.2$ s, (**b**) 0.3 s, (**c**) 0.5 s, (**d**) 0.8 s, and (**e**) 1 s, after sudden cessation of flow (From Sasaki et al. [41]. With permission from John Wiley & Sons, Inc.)

At the same time, D_r is a function of an aspect ratio ($p = b/a \leq 1$) when the molecules are assumed to be a prolate spheroid with longer radius a and shorter radius b,

$$D_r = \frac{3kT}{16\pi a b^2 \eta_s}\left[\frac{p^2}{1-p^4}\right]\left\{\left[\frac{2-p^2}{2\sqrt{1-p^2}}\right]\ln\left(\frac{1+[1-p^2]^{\frac{1}{2}}}{1-[1-p^2]^{\frac{1}{2}}}\right)-1\right\} \tag{3.6}$$

where k is the Boltzmann constant, T is the absolute temperature, and η_s is the viscosity of the solvent [11]. In order to compare the shape of molecules just

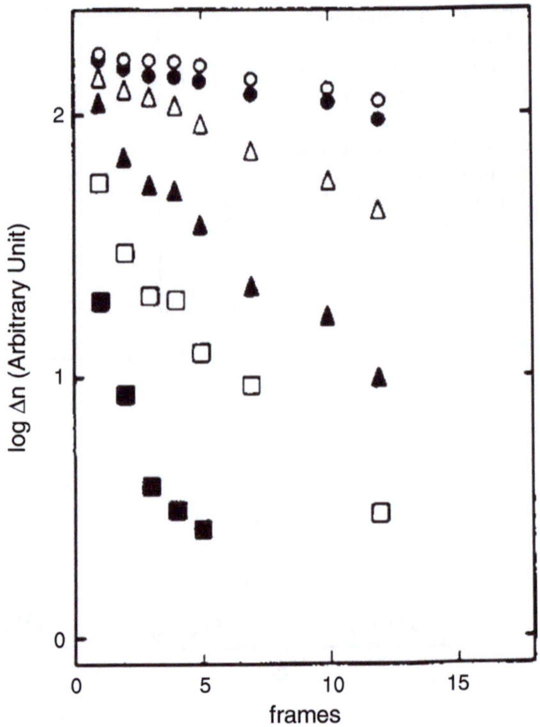

Fig. 3.13 Logarithm of Δn plotted against time (in video frame units) on the inlet symmetry plane at (○) the stagnation point, (●) 0.5 mm, (△) 1.0 mm, (▲) 1.5 mm, (□) 2.0 mm, and (■) 205 mm from the stagnation point for 10 μg/ml DNA solution (From Sasaki et al. [41]. With permission from John Wiley & Sons, Inc.)

entering the inner region of the FRM to those around the stagnation point, the ratio of D_r values at each point is used. The diffusion coefficient at the entrance was defined as D_e and that at the stagnation point was D_s. The ratio of D_e against D_s was defined as α,

$$\alpha = \frac{D_e}{D_s} = \frac{a_s^3 p_s^2}{a_e^3 p_e^2} \frac{f(p_e)}{f(p_s)} \tag{3.7}$$

where

$$f(p) = \left[\frac{p^2}{1-p^4}\right]\left\{\left[\frac{2-p^2}{2\sqrt{1-p^2}}\right] \ln\left(\frac{1+[1-p^2]^{\frac{1}{2}}}{1-[1-p^2]^{\frac{1}{2}}}\right) - 1\right\}. \tag{3.8}$$

For the DNA random coil, assuming incompressible deformation

$$\frac{4\pi a_s^3 p_s^2}{3} = \frac{4\pi a_e^3 p_e^2}{3}, \tag{3.9}$$

hence,

$$\alpha = \frac{f(p_s)}{f(p_e)}. \tag{3.10}$$

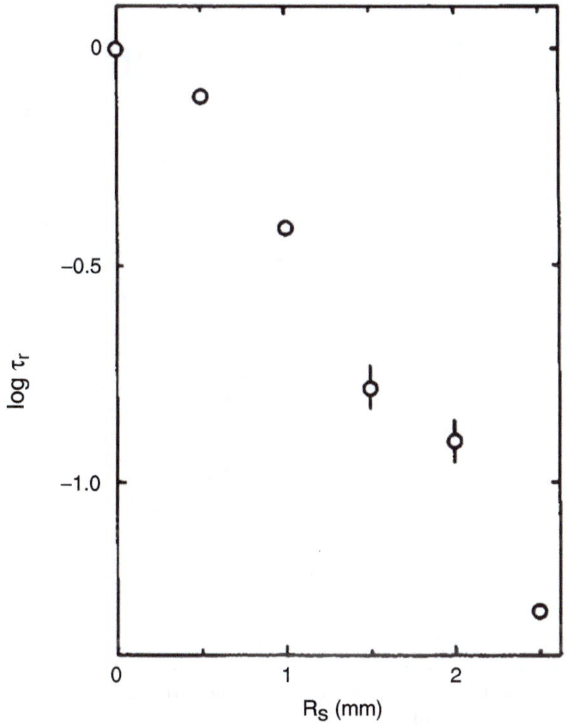

Fig. 3.14 Relative relaxation time, τ_r, plotted against distance from the stagnation point, R_s. The relaxation time at each point was normalized by the value at the stagnation point (From Sasaki et al. [41]. With permission from John Wiley & Sons, Inc.)

From Fig. 3.14, α was found to be 20. A DNA molecule just entering the mill space must have an aspect ratio slightly smaller than 1 because weak flow-induced birefringence was observed. If we assume $p_e \sim 0.95$, p_s would be around 0.08; $p_s/p_e \sim 1/12$. Menasveta and Hoagland reported that the polystyrene (PS) molecule with $M_w = 2 \times 10^7$ Da shows a coil–stretch transition in an elongational flow field in toluene, where R_g in the stretched state was ~650 nm and that in the coil state was ~350 nm [29]. The stretched state was observed in an opposed jet apparatus at $\dot{\varepsilon}$~2×10^4 s^{-1}. This difference in R_g between both states of PS molecules corresponds to a p_s/p_e value of 1/6. This value is significantly different from our DNA value observed by the stopped flow experiments at $\dot{\varepsilon} = 24$ s^{-1}. The difference is attributed to the semiflexible nature of DNA molecules, in contrast to the ideal flexible, non-free draining nature of PS in toluene. DNA molecules at $\dot{\varepsilon} = 24 s^{-1}$ are regarded as being stretched to their extension limit.

3.6 Helix–Coil Transition of DNA Molecules

Though double-stranded DNA molecules can be a model system for flexible or semiflexible polymer chains, at the same time, they show a helix–coil transition at a certain condition. In this section, I will discuss the helix–coil transition of DNA

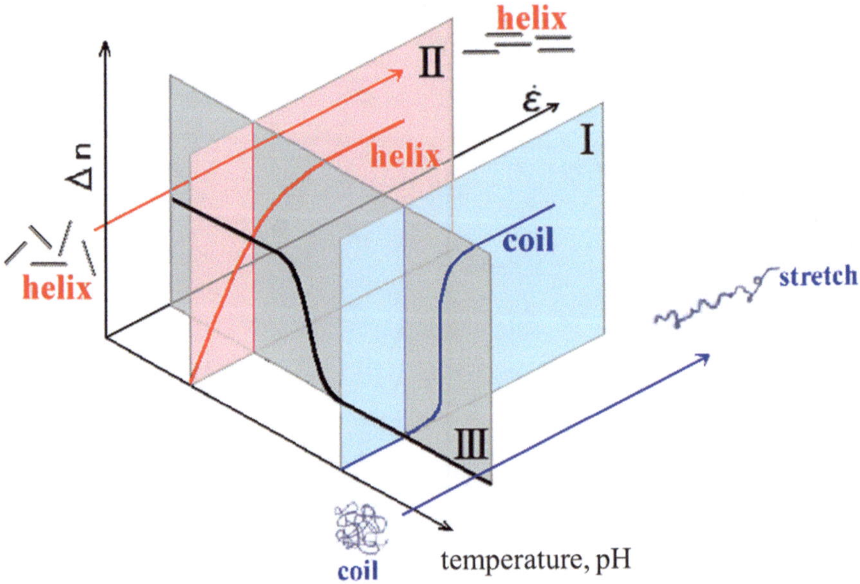

Fig. 3.15 A three-dimensional schematic picture of Δn against temperature and/or pH. On plane (I), Δn is plotted against $\dot{\varepsilon}$ for coil state of the molecules. On plane (II), Δn is plotted against $\dot{\varepsilon}$ for helix state of the molecules. On plane (III), Δn at a particular strain rate is plotted against temperature or pH

molecules observed by elongational flow birefringence studies. In Fig. 3.15, a schema for detecting the helix–coil transition of α-helical peptide molecules by an elongational flow birefringence method is shown [22]. Peptides are regarded as rigid rods in their helix conformation and flexible in their coil conformation. The difference in conformations is expected to be detectable by the difference in Δn. Because a double-stranded DNA molecule is semiflexible, the dynamics of its helix–coil transition are thought to be hydrodynamically different from those of an α-helical polypeptide chain. A heat-induced helix–coil transition in DNA has been studied: Figure 3.16 shows Δn vs. $\dot{\varepsilon}$ plot for T4-phage DNA at different temperatures from 25 to 65 °C. Up to 53 °C, flow-induced birefringence Δn was observed, while at 55 and 65 °C, Δn was not detected [42]. In each curve, there was a critical strain rate, $\dot{\varepsilon}_c$. The birefringence pattern was localized at the elongational flow field containing the stagnation point. These observations suggest an occurrence of the coil–stretch transition of DNA molecules, induced by an elongational flow field. Figure 3.17 shows the plateau value of Δn for each isothermal birefringence profile, plotted against temperature. With increasing temperature, Δn decreases gradually up to 40 °C and rapidly over 50 °C. Above 55 °C, no birefringence was detected. Figure 3.18 shows the temperature dependence of the critical strain rate, $\dot{\varepsilon}_c$, for the coil–stretch transition. Over 50 °C, $\dot{\varepsilon}_c$ increases rapidly with temperature. Both the decrease in Δn and the increase in $\dot{\varepsilon}_c$ are considered to include the effect of the decrease in solvent viscosity and the conformational transition of DNA molecules. Figure 3.19 is the Arrhenius plot for $\dot{\varepsilon}_c$. From 25 to

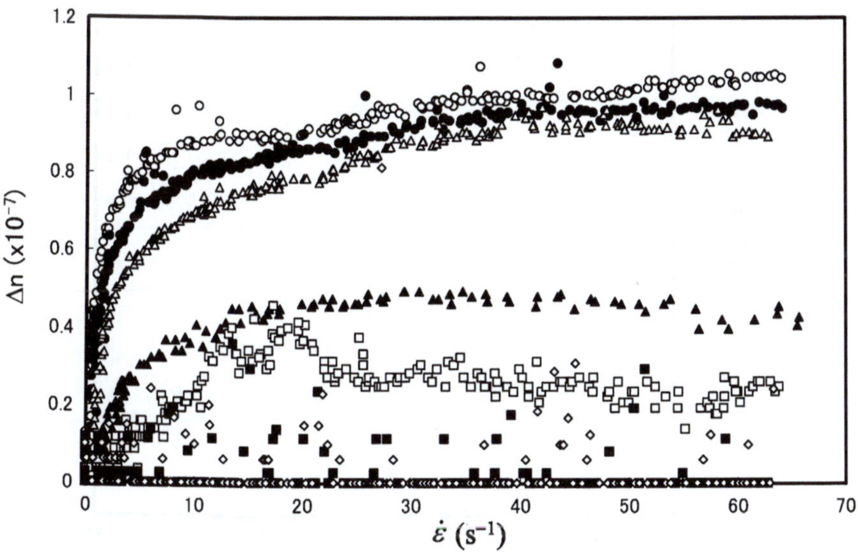

Fig. 3.16 Flow-induced birefringence intensity, Δn, plotted against $\dot{\varepsilon}$ at (○) 25.0 °C, (●) 30.0 °C, (△) 40.0 °C, (▲) 50.0 °C, (□) 53.0 °C, (■) 55.0 °C, and (◇) 65 °C for 5 μg/ml T4-phage DNA solution (From Sasaki et al. [42]. With permission from John Wiley & Sons, Inc.)

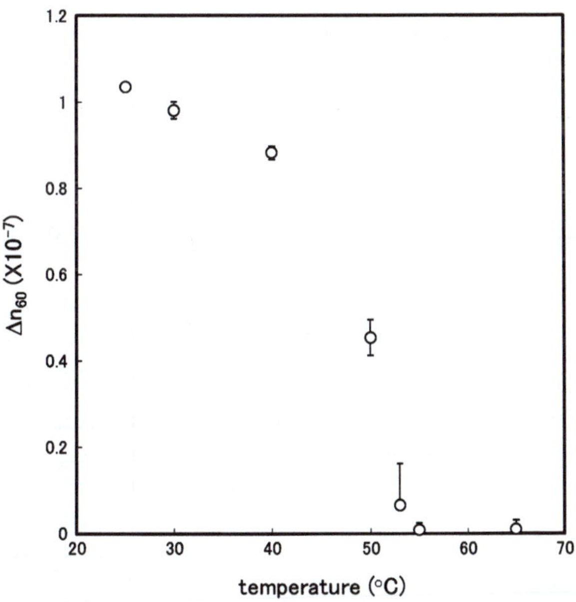

Fig. 3.17 Δn at $\dot{\varepsilon} = 60$ s^{-1} plotted against temperature (From Sasaki et al. [42]. With permission from John Wiley & Sons, Inc.)

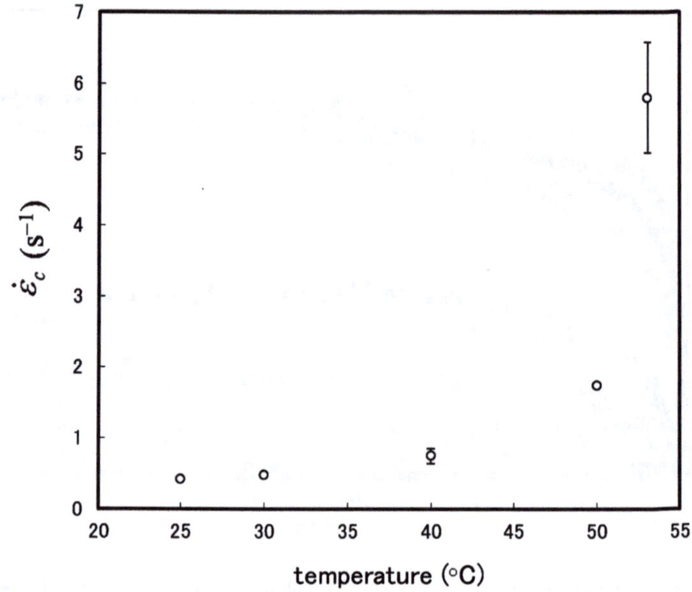

Fig. 3.18 Critical strain rate for the coil–stretch transition, $\dot{\varepsilon}_c$, plotted against temperature (From Sasaki et al. [42]. With permission from John Wiley & Sons, Inc.)

Fig. 3.19 The Arrhenius plot for $\dot{\varepsilon}_c$, for the data in Fig. 3.18 (From Sasaki et al. [42]. With permission from John Wiley & Sons, Inc.)

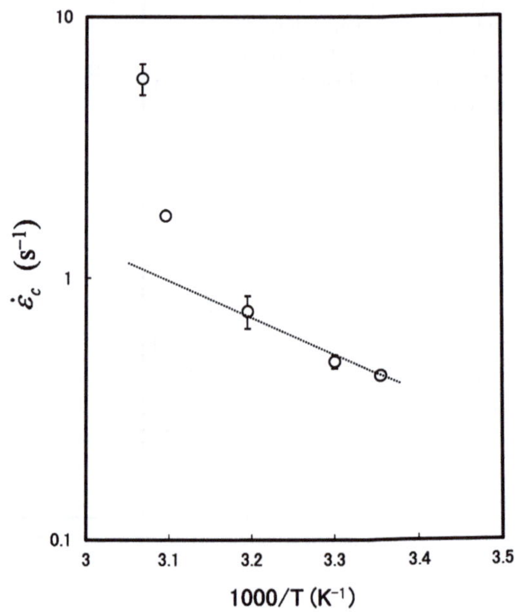

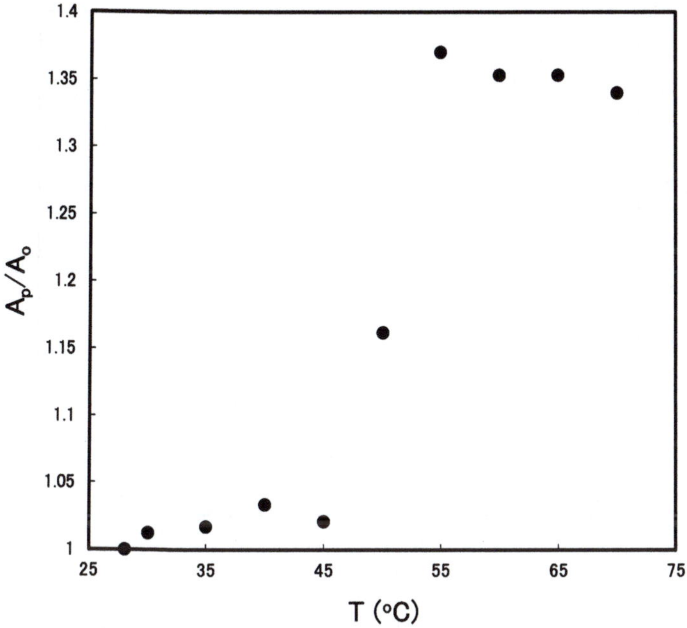

Fig. 3.20 UV absorption at 260 nm of T4 DNA sample solution plotted against temperature. Relative absorption value, A_p/A_0, was plotted, where A_0 is the absorption value for native DNA and A_p is that at indicated temperature (From Sasaki et al. [42]. With permission from John Wiley & Sons, Inc.)

40 °C, the plot is linear but becomes nonlinear over 50 °C. It is expected that in the linear section the activation energy for the coil–stretch transition does not change, indicating that the hydrodynamic shape of a DNA chain remains unchanged in this temperature range. Figure 3.20 shows UV absorption at 260 nm as a function of temperature for the same DNA solution as that used for the elongational flow experiments. From these results, the conformational change in a DNA solution expected over 50 °C is regarded as a change from a double-stranded coil to an untwined one. In a partly untwined DNA molecule, an untwined part and a double-stranded part coexist along the chain. The untwined part is not birefringent, causing the remarkable reduction in Δn in temperatures over 50 °C (Fig. 3.17). At the same time, the untwined part is more flexible than a double-stranded chain. The increased flexibility produces a larger entropic contraction force in the DNA chain. Thus, the rapid increase in $\dot{\varepsilon}_c$ is also explained by untwining (Fig. 3.18). Figure 3.21 shows the similar Δn-temperature plot as Fig. 3.17 but with Δn values (filled circle) remeasured for those samples after cooling for 30 min at room temperature. The latter values for 50 and 53 °C recover the room-temperature values, although at these temperatures the chain shows untwining. The Δn value at 55 °C is 0, but after cooling, the value reaches about half of the room-temperature Δn. Over 60 °C, even after cooling, flow birefringence was not observed. At both 55 and 60 °C and at

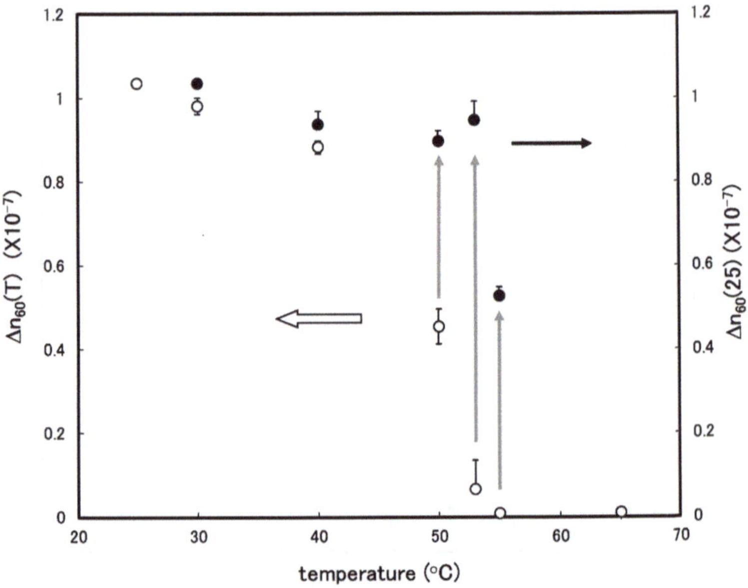

Fig. 3.21 (○) Δn at $\dot{\varepsilon} = 60\ \mathrm{s}^{-1}$ plotted against temperature (identical with Fig. 3.17). Δn at $\dot{\varepsilon} = 60\ \mathrm{s}^{-1}$ measured 25 °C after the first flow experiments at each temperature (From Sasaki et al. [42]. With permission from John Wiley & Sons, Inc.)

Table 3.1 Molecular weight of DNA after isothermal flow experiments at indicated temperatures

Temperature (°C)	Molecular weight (Da)
25	1.1×10^8
55	6.0×10^7
65	4.7×10^6

higher temperatures, DNA molecules are considered almost completely untwined. The partial recovery in Δn at 55 °C could be due to incorrect and/or incomplete repairing of base pairs among untwined DNA chains. Another explanation for this could be scission of DNA molecules at the untwined region, as well as untwining by the flow field, explaining the absence of flow birefringence for 60 °C and higher temperature solutions. Table 3.1 lists molecular weight values of DNA molecules after isothermal flow birefringence measurements at the temperatures indicated. The value at 25 °C is the molecular weight of intact T4-phage DNA. At 55 °C, the molecular weight of DNA is reduced to 1/2 of the intact value, and at 65 °C, it is only 1/20. These values confirm the validity of the scission mechanism of DNA molecules at these temperatures in an elongational flow field. In a previous section, I stated that DNA molecules did not reduced in molecular weight after elongational flow experiments. Observed results here seem to be a contradiction to this.

It is well known that scission of polymers can occur in flow fields where the mechanical forces associated with the flow field are sufficient to rupture covalent bonds along the chain. A number of fracture studies have been performed, and the results have been rationalized within a theoretical framework known as the thermally activated barrier to scission (TABS) model [32, 35]. According to the TABS model, fracture proceeds in an elongational flow field as a two-stage process. In the first stage, molecules are stretched by the elongational flow field into an extended conformation and are aligned with their molecular axes parallel to the flow lines. In the second stage, the extended conformation is then subjected to fracture. Furthermore, scission of an extended chain always occurs near the midpoint, where stress in the molecule reaches a maximum [24]. The result listed in Table 3.1 for 55 °C should be investigated on the basis of the TABS model. According to the TABS model, the critical strain rate for fracture $\dot{\varepsilon}_{\mathrm{f}}$ is related to contour length L and to molecular weight M_{w} of the polymer molecule through the relationship

$$\dot{\varepsilon}_{\mathrm{f}} \propto 1/L^2 \sim 1/{M_{\mathrm{w}}}^2. \tag{3.11}$$

Using this relation, $\dot{\varepsilon}_{\mathrm{f}}$ for T4-phage DNA ($M_{\mathrm{w}} \sim 110$ MDa) was estimated in comparison with that for T7-phage DNA ($M_{\mathrm{w}} \sim 25$ MDa). The minimum of $\dot{\varepsilon}_{\mathrm{f}}$ for T4 phage DNA was found to be 620 s^{-1} ($\dot{\varepsilon}_{\mathrm{f}} > 620\ \mathrm{s}^{-1}$) [2]. This value is larger than the uppermost strain rate of this experiment, 70 s^{-1}. The DNA molecules in the estimates above are double stranded. On the other hand, according to the UV absorption data, DNA molecules in 55 °C and higher temperature solutions are almost completely untwined. The extensional force required for breaking chain molecules would be smaller for single-stranded DNA than that for double-stranded DNA. According to the TABS model, the scission rate, $K(T)$, is described as

$$K(T) = A\exp[-u_0(\dot{\varepsilon})/kT] \tag{3.12}$$

and

$$u_0 = U_{\mathrm{d}} - \left(Ca\eta_{\mathrm{s}} l^2 N^2 \dot{\varepsilon}/8\right) \tag{3.13}$$

where U_{d} is the dissociation energy of polymer chain, C is a correction factor determined by molecular shape, a is the stretched bond length, η_{s} is the solvent viscosity, l is the monomer length, and N is the degree of polymerization. The pre-exponential factor A is regarded as the attempt frequency for fracture, modifying the scission rate for $\dot{\varepsilon} = \dot{\varepsilon}_{\mathrm{f}}$. If experiments are performed with $\dot{\varepsilon} \le \dot{\varepsilon}_{\mathrm{f}}$, $u_0(\dot{\varepsilon}) > 0$, then $K(T)$ would change with temperature according to the usual Arrhenius law, increasing with temperature. Figure 3.22 shows *K(T)* schematically plotted against strain rate at different temperatures. For a strain rate smaller than $\dot{\varepsilon}_f$, even for *K(T)* ~ 0 at T_0, then at $T > T_0$, *K(T)* could have a definite value. The developed foot (slightly below $\dot{\varepsilon}_{\mathrm{f}}$) of the *K(T)* versus $\dot{\varepsilon}$ plot can explain the difference with increasing temperature in flow behavior at 55 and 65 °C of untwined DNA molecules. The difference suggests that $\dot{\varepsilon}_{\mathrm{f}}$ for untwined T4 DNA molecules is larger than 70 s^{-1}.

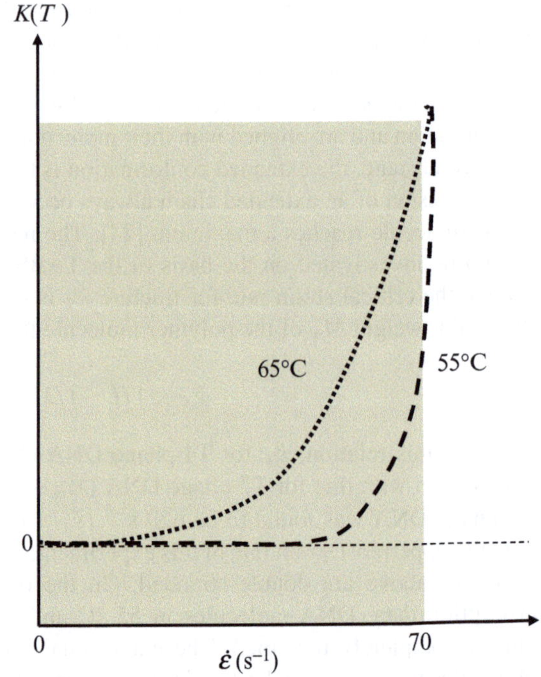

Fig. 3.22 Schematic drawing of the temperature effect on the flow-induced fracture rate, $K(T)$. At a strain rate $\dot{\varepsilon}$ slightly smaller than $\dot{\varepsilon}_f$, the fracture rate increases as a result of broadening of the foot in $K(T)$ vs. $\dot{\varepsilon}$ plot with increase in temperature. $\dot{\varepsilon}$ used in this study covers the gray-colored region (From Sasaki et al. [42]. With permission from John Wiley & Sons, Inc.)

3.7 Observation of Coil–Globule Transitions by the Elongational Flow Birefringence Method

In the previous section we treated the helix–coil transition of DNA molecular chains. Another characteristic conformational transition is the coil–globule transition. This transition is considered biologically important; the size of T4-phage, a double-stranded DNA molecule, in a good solvent has been reported to be several tens of μm, while the radius in the phage capsid is several tens of nm [46, 48]. The latter is considered to be similar to the globular conformation DNA molecules take in their poor solvent. The change in conformation from an extended random coil to a globule structure is regarded as thermodynamically a first-order phase transition. The coil–globule transition of polymer molecular chains has been studied by measuring their size [49]. In the case of DNA molecules, a fluorescent microscopic method has been utilized [45, 54]. In this section, I will pursue a possibility of using an elongational flow birefringence method for observing dynamics of the coil–globule transition of DNA molecules [23, 56].

Coil–globule transitions of DNA are known to be inducible by adding polyethylene glycol molecules [58], multivalence metal ions [1], and/or surfactant molecules into an aqueous DNA solution [43, 45]. Here I will discuss the results for the coil–globule transition of T4-phage DNA induced by polyethylene glycol

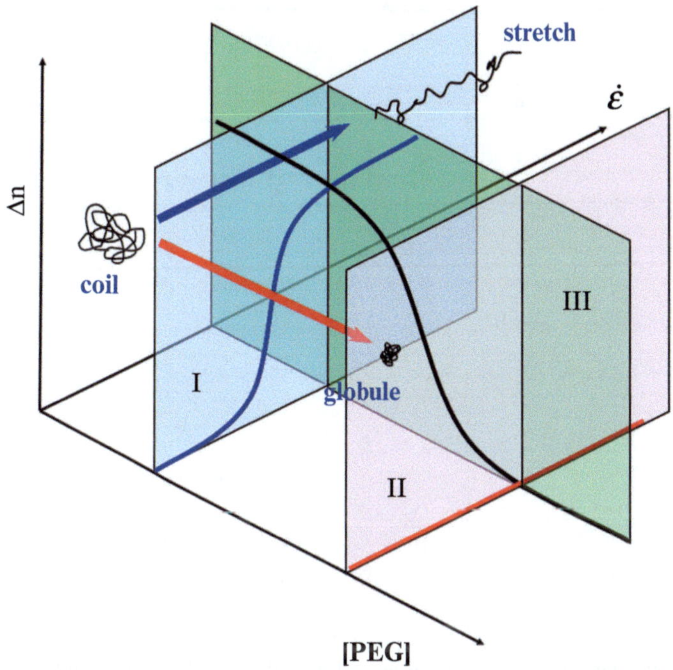

Fig. 3.23 A three-dimensional schematic of birefringence Δn against polyethylene glycol concentration [PEG], in monomer units. On the plane (I), Δn is plotted against $\dot{\varepsilon}$ at a certain [PEG] for coil state of the molecules, where $\dot{\varepsilon} > \dot{\varepsilon}_c$. On plane (II), Δn is plotted against $\dot{\varepsilon}$ at a certain [PEG] for globule state of the molecules. On plane (III), Δn at a particular strain rate $\dot{\varepsilon} > \dot{\varepsilon}_c$ is plotted against [PEG]

(PEG). Figure 3.23 shows the expected flow birefringence observations of the coil–globule transition induced by the PEG addition. At a fixed PEG concentration, denoted [PEG] in monomer units, a coil–stretch transition observation would be performed. The contraction force in a deformed polymer molecule is expected to be larger for a globular state than for an extended random coil structure, because in a globular structure a polymer chain is too tightly and densely packed to move. $\dot{\varepsilon}_c$ for a globular polymer would be larger than for a polymer in a good solvent. It is expected that at a small value of [PEG], a DNA molecule takes on a relatively extended structure and the $\dot{\varepsilon}_c$ value is similar to that of a DNA solution without PEG. Increasing the [PEG] value above the coil–globule transition, $\dot{\varepsilon}_c$ will be infinitely large because a globular DNA molecule will not be deformed. On the other hand, the addition of [PEG] increases solvent viscosity; the critical strain rate is inversely proportional to the solvent viscosity. Then, for [PEG] up to the transition value, $\dot{\varepsilon}_c$ will decrease. The birefringence Δn at a constant $\dot{\varepsilon}$ would

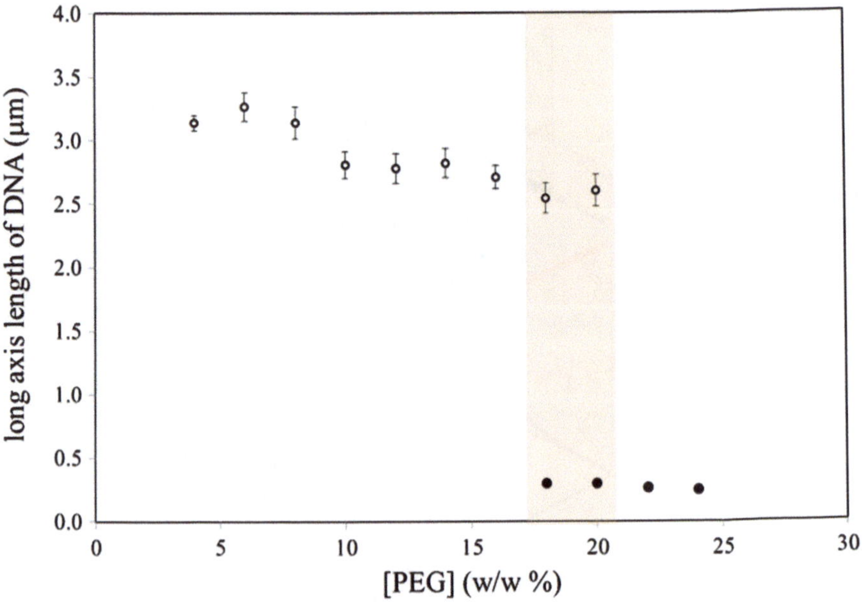

Fig. 3.24 The long-axis length of T4-phage DNA molecular fluorescent images in aqueous solution containing 0.1 M NaCl plotted against polyethylene glycol concentration in monomer unit, [PEG]. In the shaded region, the distribution has two distinct peaks. The points indicate the maximum of the distribution (From Wakabayashi et al. [56]. With permission from John Wiley & Sons, Inc.)

suddenly decrease at the critical [PEG] for the coil–globule transition, the experimental curve thereby documenting the transition.

As the purpose of this section is to examine the applicability of elongational flow birefringence to detecting coil–globule transitions in DNA molecules, it is necessary to confirm the transition by a conventional method such as fluorescent microscopy. Figure 3.24 shows the length of the long axis of DNA molecules l as a function of PEG concentration in weight, [PEG], for a solution containing 0.10 M NaCl. At [PEG] 18 % (weight % of monomer unit), l suddenly decreased, indicating that the coil–globule transition occurred at this PEG concentration. In the shaded region in Fig. 3.24, both coiled and globular DNA molecules were observed in the same field of the microscope. The bimodal distribution of l around the transition point is caused by spatial fluctuation of the PEG concentration in the solution [28, 30, 55, 57].

Figure 3.25 shows Δn plotted against $\dot{\varepsilon}$ for DNA molecules under similar conditions to Fig. 3.24, using a flow birefringence technique. PEG weight concentrations ranged from 7.3 to 24.3 %. For PEG concentrations from 7.3 to 20.6 %, a birefringence signal was detected, but at 24.3 %, no birefringence was observed. The observed birefringence was localized on the outlet center line containing the stagnation point. In these observations, flow-induced birefringence

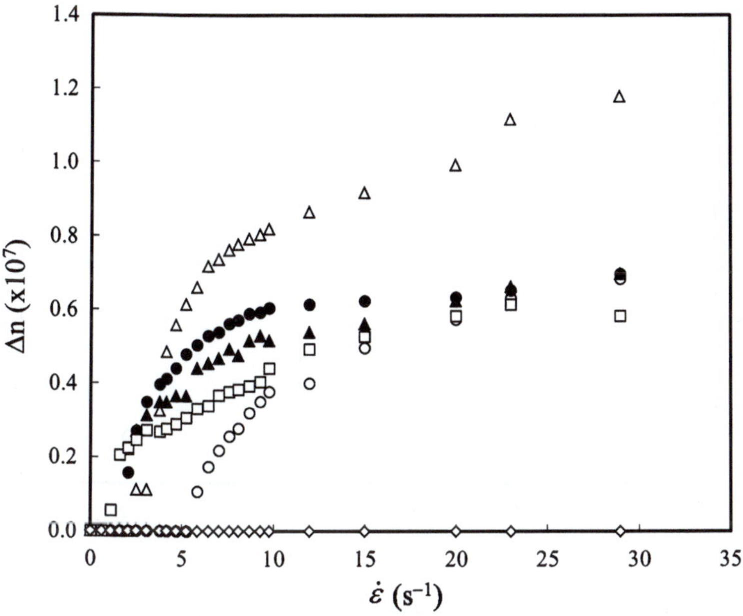

Fig. 3.25 Δn plotted against $\dot{\varepsilon}$ for a T4 DNA aqueous solution containing 0.1 M NaCl at PEG weight concentrations in monomer unit [PEG] of 7.3 % (○), 10.6 % (●), 13.7 % (△), 15.3 % (▲), 20.6 % (□), and 24.3 % (◇) (From Wakabayashi et al. [56]. With permission from John Wiley & Sons, Inc.)

originates in the coil–stretch transition of DNA molecules in the sample solutions. Figure 3.26 shows $\dot{\varepsilon}_c$ plotted against [PEG]. As expected, $\dot{\varepsilon}_c$ decreases with [PEG]. Multiplying solvent viscosity by $\dot{\varepsilon}_c$ at each [PEG] value, the product is roughly independent of PEG concentration. Figure 3.27 shows the Δn value at $\dot{\varepsilon} = 29\ \text{s}^{-1}$, Δn_{29}, plotted against [PEG]. Up to [PEG] = 15 %, the Δn_{29} value increases almost linearly, while above this concentration it decreases. From the fluorescent microscopy observations, the coil–globule transition is expected to occur above [PEG] = 18 %. The change in Δn_{29} against [PEG] at around 15 % corresponds to the coil–globule transition of DNA molecules in the system. On the other hand, the initial increase in Δn_{29} against [PEG] is considered to result from the increase in solvent viscosity by adding PEG to the DNA aqueous solution containing 0.1 M NaCl. In Fig. 3.28, viscosity dependence of flow-induced birefringence at $\dot{\varepsilon} = 29\ \text{s}^{-1}$ is shown. The corresponding data for DNA–glycerol NaCl solution is also shown for reference. Δn_{29} for a DNA–glycerol system increases almost linearly when plotted against the logarithm of solvent viscosity, log η_s. The results for the DNA–PEG system are similar to DNA–glycerol in the lower viscosity range, but depart from the linear relation at higher viscosity. The departure indicates that near the deflection point, a conformation of DNA changes to a form with smaller hydrodynamic friction than the random coil. This result is explained by the occurrence of a coil–globule transition at a PEG concentration of about log $\eta_s \sim 1$ (cp), that is, [PEG]

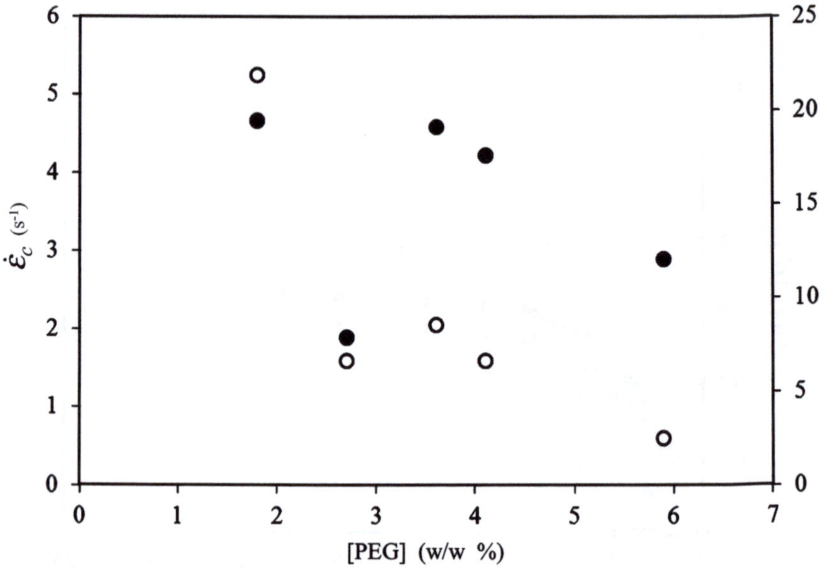

Fig. 3.26 Critical strain rate for the coil–stretch transition, $\dot{\varepsilon}_c$(○), and the product with the solvent viscosity value, $\dot{\varepsilon}_c\eta_s$, (●) plotted as functions of [PEG]

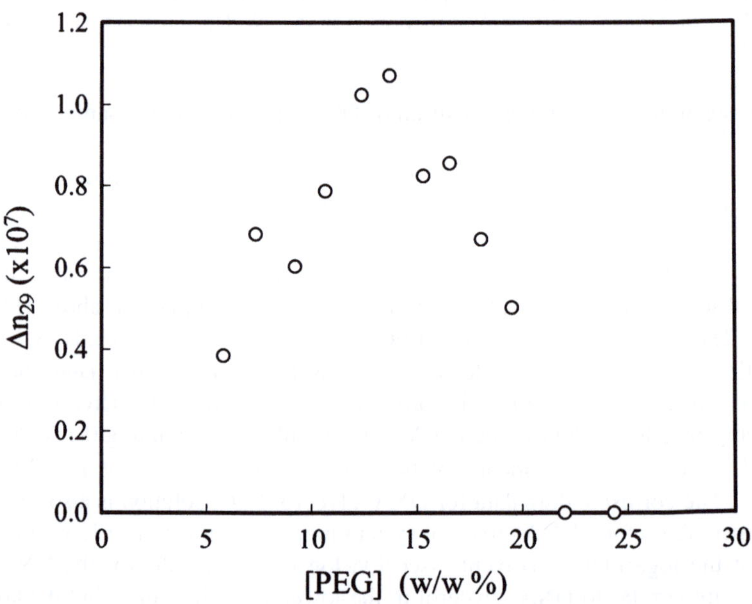

Fig. 3.27 Δn at $\dot{\varepsilon} = 29\ \mathrm{s}^{-1}$ of T4-phage DNA molecules in aqueous solution containing 0.1 M NaCl plotted against [PEG] (From Wakabayashi et al. [56]. With permission from John Wiley & Sons, Inc.)

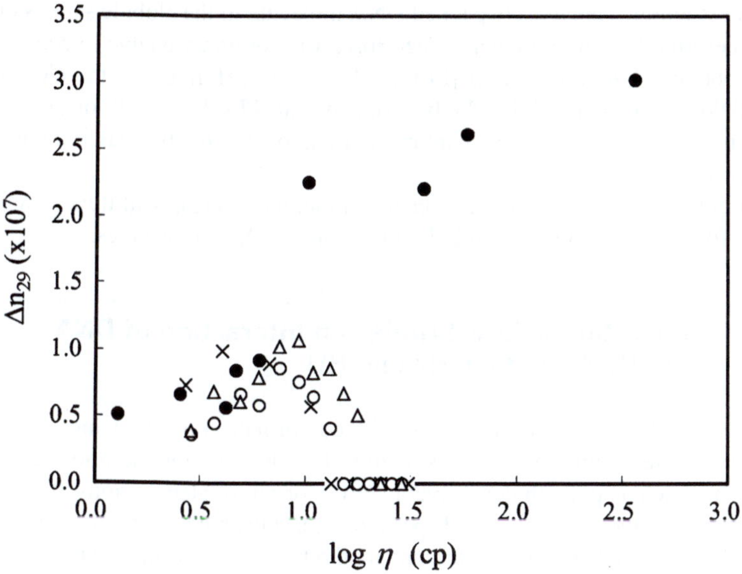

Fig. 3.28 Δn at $\dot{\varepsilon} = 29\ s^{-1}$ of T4-phage DNA molecules in aqueous solution containing PEG and 0.2 M NaCl (X), 0.15 M NaCl (○), and 0.1 M NaCl (△) plotted against logarithm of solvent viscosity. The results for T4 DNA aqueous solution containing glycerol and 0.2 M NaCl (●) were also plotted (From Wakabayashi et al. [56]. With permission from John Wiley & Sons, Inc.)

~ 15 %. Detailed comparison of the fluorescent results (Fig. 3.24) to flow birefringence (Fig. 3.27) reveals that a DNA molecule in DNA–PEG solution at [PEG] = 20.6 % should be a birefringent globule. The following explanation is offered for this unexpected observation.

The conformational change in DNA molecules between coil and globule states, with the addition of PEG, is induced by a decrease in the solvent's affinity to polymer segments. If coexistence of molecules in both coil and globule states occurs at this PEG concentration, there is the possibility of observing birefringence from the few remaining DNA coils. However, no molecules were observed in the coil state in fluorescent micrograph images for the solution of [PEG] =19.5 %. Thus, coexistence of DNA coils and globular DNA molecules cannot explain the Δn_{29} observed at [PEG] = 19.5 %.

Another possible explanation for the observed Δn_{29} at [PEG] = 19.5 % is the deformation of DNA molecules in the globule state by the elongational flow field. The deformation of the polyelectrolyte globular chain in an elongational flow field was investigated by Borisov et al. [5]. They considered that the force required for the deformation of a polymer depends on the solvent strength. In a DNA globule chain, segments are closely packed at high PEG concentrations. In a DNA globule chain near the transition concentration, however, segments are thought to be packed loosely, compared with high PEG concentrations, because of the relatively low solvent strength. This loosely packed DNA molecule in the globule state is thought to be deformed by the flow field. Therefore, it is not unreasonable to consider that the globular DNA molecule in [PEG] = 19.5 % is deformed by the elongational flow field. It is concluded that the birefringence at [PEG] = 19.5 % originates from deformed globule chains, which are responding to a relatively weak intramolecular condensation force.

From these observations, it is concluded that the elongational flow technique is useful for investigating coil–globule transitions of DNA molecules.

3.8 Elongational Flow Studies on Interaction of DNA with DNA-Binding Protein HU

As mentioned in the previous section, the conformation of DNA molecules in an aqueous buffer solution becomes compact following the addition of neutral polymers, such as polyethylene glycol (PEG), or multivalent cations (MVC), such as hexamine cobalt, spermine, and spermidine. Using neutral polymers degrades the solvent quality for DNA molecules, and methods using MVC make DNA molecules mechanically compact. In the previous section, I discussed the globulization of DNA by adding PEG into aqueous solution. As mentioned, the process of compacting DNA molecules is biologically important. In the case of eukaryotes, histone, a DNA-binding protein, behaves as a spindle of the double-stranded DNA molecule to construct a nucleosome and then a chromosome, which are compact structures of DNA to be capsulized in the nucleus. In the case of prokaryotes, nucleosome-like structures of DNA threads have also been reported [18]. HU and HU-like proteins, a group of histone-like DNA-binding proteins, are thought to be involved in the formation of those nucleosome-like structures [12]. Molecular mechanical models of the HU–DNA interaction based on results from X-ray crystallographic and NMR spectroscopic studies have been proposed [4, 50, 51]. An HU dimer unit grips a double-stranded DNA molecule by its “arms,” adjacent units are combined together by a hydrophobic interaction, and the DNA molecule is then bent at the interaction region. From this model, it is expected that a conformation of DNA could result from interaction with a certain amount of HU, which is as tightly packed as bacterial virus capsids. However, there has been no evidence of HU-induced collapse of DNA molecules. In this section, I will discuss the compacting process of T2-phage DNA by HU [14].

HU protein was obtained from *Bacillus stearothermophilus* (HUBst) and purified. The methods of protein isolation and purification are as described by Dijk et al. [10]. In this study, coliphage T2 DNA was used as a DNA sample. The molecular weight of T2 DNA is 1.10×10^8 Da, corresponding to 160 kbp, and the contour length is about 50 μm. In Figs. 3.29 and 3.30, fluorescent microscopic images are shown for solutions, one without HUBst and the other containing 9.3×10^{-7} M, twice the stoichiometric concentration of HU occupying available binding sites on DNA molecules in the solution [13]. A change from random coil conformations without HU to globular ones with HU was confirmed. Figure 3.31

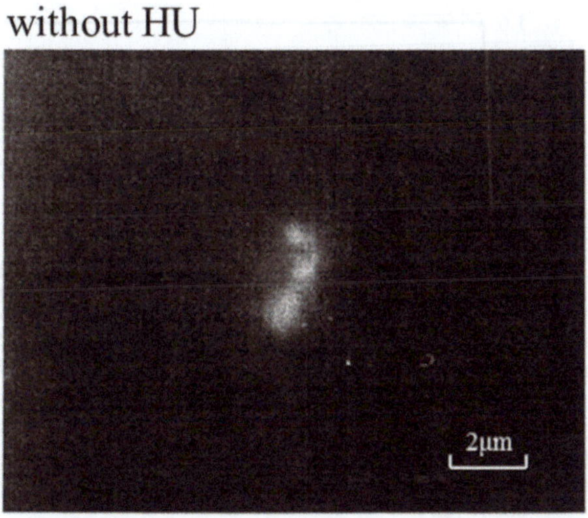

Fig. 3.29 Fluorescent microscopic image of T2 DNA in 0.2 M NaCl aqueous solution

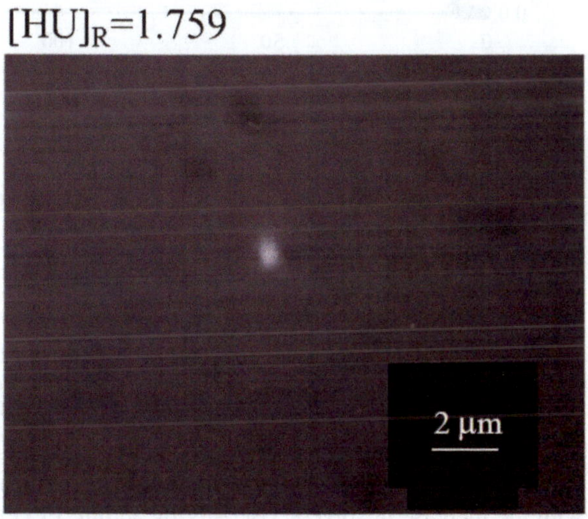

Fig. 3.30 Fluorescent microscopic image of T2 DNA in 0.2 M NaCl aqueous solution containing $[HU]_R$ times of stoichiometric HU concentration

shows flow-induced birefringence, Δn, plotted against elongational strain rate, $\dot{\varepsilon}$, at HUBst dimer concentrations from 0 to 9.34×10^{-7} M [13]. Figure 3.32 is a magnification of the lower-left corner of Fig. 3.29. The general features of birefringence in the solutions at each HUBst dimer concentration, $[HUBst]_D$, are as follows: (1) birefringence was localized along the pure elongational flow field including stagnation point in the four-roller mill; (2) in the strain rate dependence, there is a critical strain rate, $\dot{\varepsilon}_c$, for the appearance of flow-induced birefringence, and beyond $\dot{\varepsilon}_c$ Δn increased rapidly at first and then gradually with increasing $\dot{\varepsilon}$, finally reaching a plateau-like region. These features indicate that all the

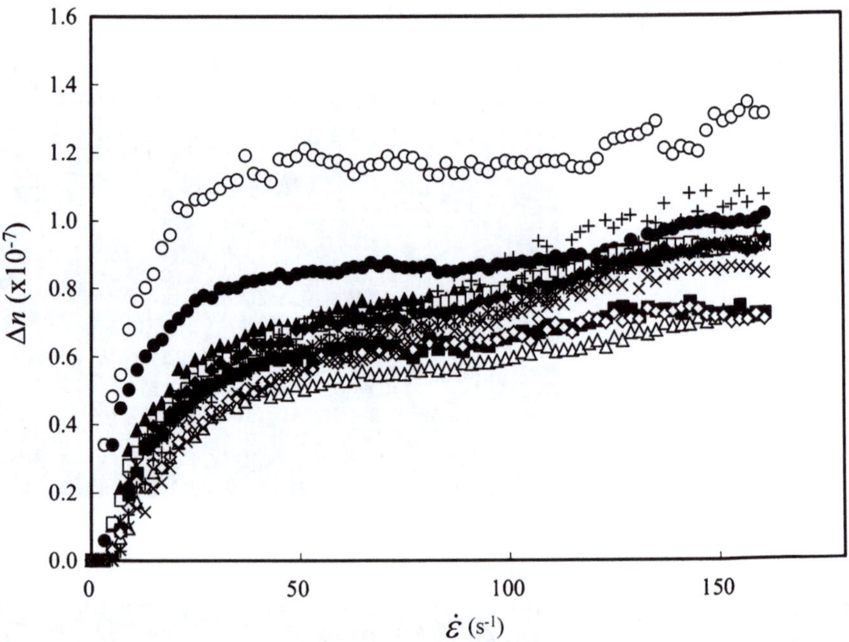

Fig. 3.31 Elongational flow-induced birefringence of DNA-HU mixed solutions as a function of strain rate for HU concentrations of 0 M (○), 0.934×10^{-7} M (●), 1.87×10^{-7} M (Δ), 2.80×10^{-7} M (▲), 3.74×10^{-7} M (□), 4.67×10^{-7} M (■), 5.60×10^{-7} M (♦), 6.54×10^{-7} M (◊), 7.47×10^{-7} M (+), 8.41×10^{-7} M (×), and 9.34×10^{-7} M (*) (From Endoh et al. [14], with permission)

birefringence responses observed have characteristics suggesting a coil–stretch transition of DNA molecules even under the interaction with HUBst dimer units.

Figure 3.33 shows the critical strain rate $\dot{\varepsilon}_c$ for the coil–stretch transition plotted against $[\text{HUBst}]_D$; $\dot{\varepsilon}_c$ increases with $[\text{HUBst}]_D$. To see how $\dot{\varepsilon}_c$ is expected to change for a typical coil–globule transition, $\dot{\varepsilon}_c$ values plotted against [PEG] are shown in Fig. 3.26. In the typical coil–globule transition case, $\dot{\varepsilon}_c$ could, however, decrease with the increase in solvent viscosity by adding PEG to the solution. In Fig. 3.26, the product of $\dot{\varepsilon}_c$ and solvent viscosity is also plotted, and the value is roughly independent of [PEG]. As the response to a typical coil–globule transition, $\dot{\varepsilon}_c$ is constant when the molecules involved are random coils but infinitely large for globular conformations. From these data, the compacting of DNA molecules by HU looks different from a coil–globule transition. Figure 3.34 shows histograms of DNA molecular size for different concentrations $[\text{HUBst}]_D$, and the average size is plotted against $[\text{HUBst}]_D$. There is no critical value of $[\text{HUBst}]_D$ evident in the plots of both $\dot{\varepsilon}_c$ and DNA size.

It is notable that in Fig. 3.33 the slope, $d\dot{\varepsilon}_c/d[HUBst]_D$, changed at $[\text{HUBst}]_D$ of ~2×10^{-7} M. This change in slope indicates that the decrement in the friction coefficient (Eq. 3.1), ζ, at HU concentrations less than 2×10^{-7} M is larger than

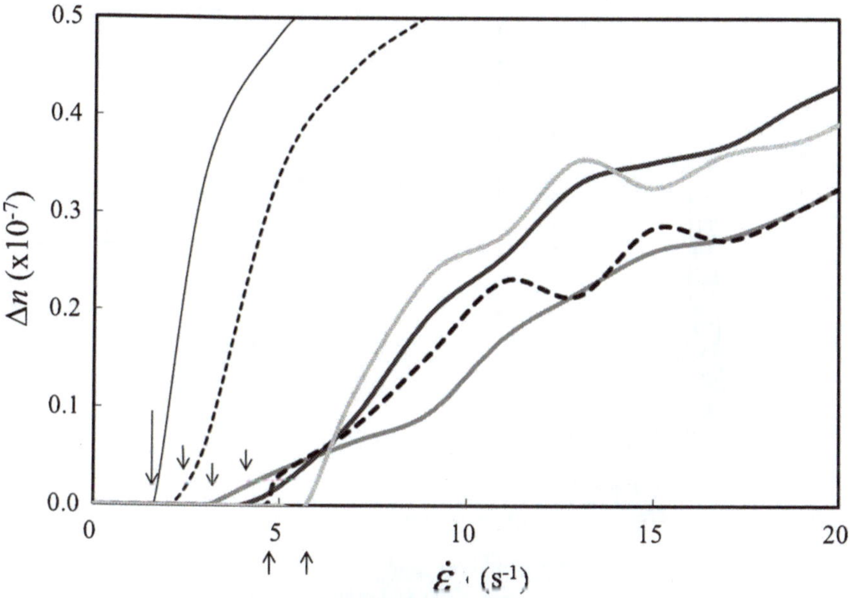

Fig. 3.32 Magnification of Fig. 3.31 at the *lower-left* corner. For the sake of clarity, plots are shown by curves for some $[\mathrm{HUBst}]_{\mathrm{D}}$s: 0 M (*solid line*) , 0.934×10^{-7} M (*short dashed line*), 1.87×10^{-7} M (*medium-shaded line*), 4.67×10^{-7} M (*dark-shaded line*), 6.54×10^{-7} M (*long dashed line*), and 9.34×10^{-7} M (*light-shaded line*). $\dot{\varepsilon}_c$ value at each HU concentration is shown by *small arrow* (From Endoh et al. [14]. With permission from John Wiley & Sons, Inc.)

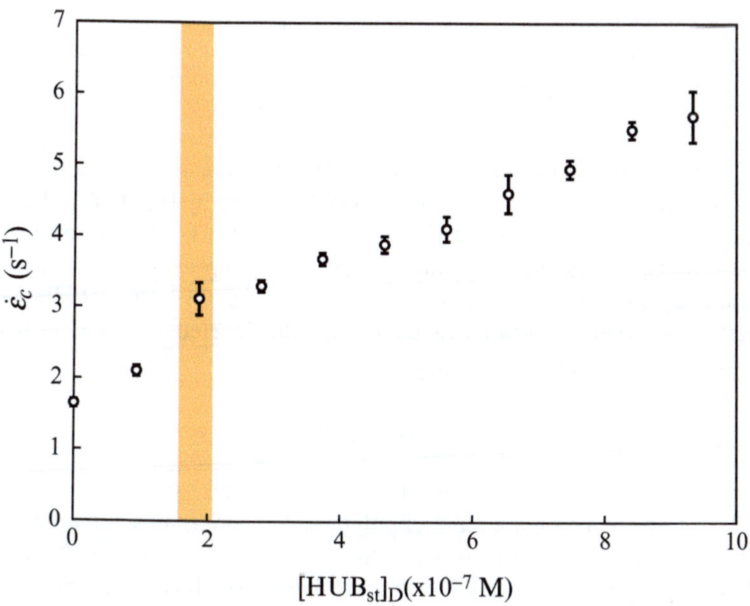

Fig. 3.33 Critical strain rate, $\dot{\varepsilon}_c$, for the coil–stretch transition plotted against HUBst dimer concentration, $[\mathrm{HUBst}]_{\mathrm{D}}$ (From Endoh et al. [14]. With permission from John Wiley & Sons, Inc.)

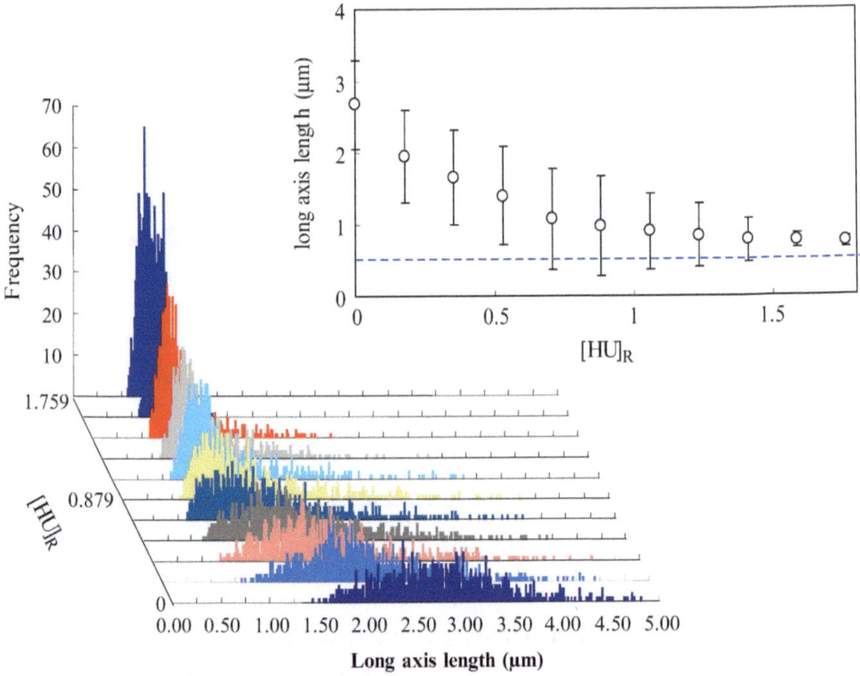

Fig. 3.34 Histograms at different $[\mathrm{HUBst}]_D$ and the average size of DNA–HU complex as a function of $[\mathrm{HUBst}]_D$ (*inset*). In this figure, all of the histograms measured at several HU concentrations are plotted in the same scale. It can be seen that by adding HU proteins the broad distribution gradually becomes sharp with reducing the average value. This graph shows the average size of DNA image as a function of HU concentration. A *horizontal line* indicates the size of globular DNA induced by polyethylene glycol. You can see that the average size of DNA gradually decreases with HU to the size comparable to that of globular DNA by polyethylene glycol

that at HU concentrations of more than 2×10^{-7} M. This result suggests that there are at least two stages in the contraction of DNA induced by HU. In Fig. 3.35, $\dot{\varepsilon}_c$ and the birefringence value in the plateau Δn_p at $\dot{\varepsilon}/\dot{\varepsilon}_c = 10$ of DNA solution are plotted against $[\mathrm{HUBst}]_D$. For smaller values of $[\mathrm{HUBst}]_D$, up to ~2×10^{-7} M, a rapid increase in $\dot{\varepsilon}_c$ and rapid decrease in Δn_p were observed. Over this $[\mathrm{HUBst}]_D$ value, the increase in $\dot{\varepsilon}_c$ became gradual, and the Δn_p value leveled off. As a birefringence intensity Δn is generally described as

$$\Delta n \sim \rho^{\frac{1}{2}} S \delta A \tag{3.14}$$

where ρ is the solvent density, S is the order parameter, and δA is the refractive index anisotropy in a monomer unit of polymer. As changes in ρ and δA are usually negligible, the change in Δn_p as a function of $[\mathrm{HUBst}]_D$ indicates the change in S with increase in HU. At first, S decreased with increase in HU and then leveled off at HU concentrations of more than about 2×10^{-7} M. On the other hand, during the two stages of change with HU in the contracting process indicated by $\dot{\varepsilon}_c$ and Δn_p,

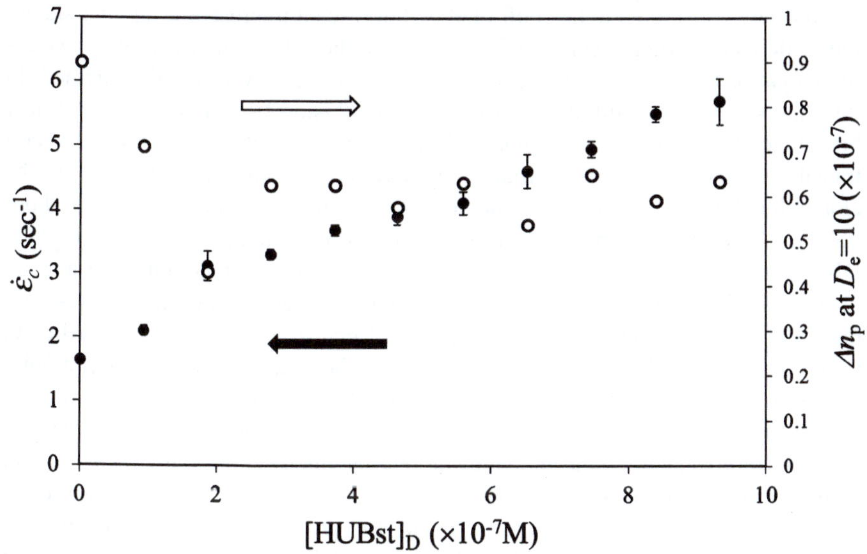

Fig. 3.35 Critical strain rate for the coil–stretch transition and the plateau birefringence value plotted against [HUBst]$_D$

DNA size continuously decreased with an increase in [HUBst]$_D$ as shown in Fig. 3.35 [13]. Both the decrease in S and the increase in $\dot{\varepsilon}_c$ with HU concentration are understandable considering that many partially tangled structures are constructed on a long phage DNA. In the first stage of the DNA–HU interaction, the number of partially tangled regions is increased on the DNA chain like the beads of a rosary, where each tangled region is independent of each other. When [HUBst]$_D$ is increased, in the second stage, tangled blobs are weakly interacting with one another, and the whole DNA molecule takes on a spherulite shape comprised of the blobs. Extending the spherulite does not require much greater energy. It is still not certain why [HUBst]$_D \sim 2 \times 10^{-7}$ M is a critical value. This value is about 1/3 of the stoichiometric concentration for HU binding to the T2-phage DNA molecule. The observed result that the Δn_p value over [HUBst]$_D \sim 2 \times 10^{-7}$ M is 2/3 of the initial value (Fig. 3.35) is believed to correspond to the expected scheme for the two stages of collapsing DNA with HU. These arguments infer that the structural change of DNA molecules induced by HU proteins is not the coil–globule transition discussed in the previous section.

3.9 Concluding Remarks

In this chapter the applicability of the elongational flow birefringence method to study dynamics of polymer molecules was examined. As DNA molecules have been previously used for visualization studies of single molecules, the birefringence results were considered to be confirmed by fluorescent microscopy, another single-molecule visualization technique. The debate on the extension of flexible polymers in elongational flow was concluded by a single-molecule visualization method. In this chapter, we have further confirmed this conclusion from an elongational flow birefringence method. Conformational transitions of DNA such as helix–coil transitions and coil–globule transitions have also been investigated by the elongational flow birefringence method. The results accorded well with those obtained by traditional methods. In the helix–coil transition, it was found that the DNA chain was easily degraded after the transition at high temperatures, as a result of the drag force exerted by the flow field being concentrated on the single-chain region of untwined DNA. As the contraction force in a globular DNA chain had been considered to be much larger than that in the random coil structure, a coil–stretch transition was not expected for the globular DNA in this study. We did not expect the birefringence originated from a coil–stretch transition. Comparison of the flow birefringence data with the fluorescent microscopic ones revealed that there is evidence of DNA deformation even in the globular state. DNA-binding protein HU functions as a bending agent for prokaryote DNA molecules. A similarity of the interaction between DNA and HU with the transition from coil to globule in DNA was examined by the flow-induced birefringence method and fluorescent microscopy. With increase in HU concentration, the size of DNA was reduced. However, the reduction did not have a criticality reminiscent of the coil–globule transition dependence on HU concentration, indicating it is not similar to that transition. Instead it was found that the folding process of DNA by HU proceeded in two distinct stages. The elongational flow birefringence technique in conjunction with fluorescent microscopy has been proven to be a useful method for studying dynamic properties of biopolymers.

References

1. Arscott PG, Li A-Z, Bloomfield VA (1990) Condensation of DNA by trivalent cations. 1. Effects of DNA length and topology on the size and shape of condensed particles. Bioploymers 30:619–630
2. Atkins EDT, Taylor MA (1992) Elongational flow studies on DNA in aqueous solution and stress-induced scission of double helix. Biopolymers 32:911–923
3. Bird RB, Hasager O, Armstrong RG, Curtiss CF (1977) Dynamics of polymeric liquids, vol II. Wiley, New York
4. Boelens R, Vis H, Vorgias CE, Wilson KS, Kaptein R (1997) Structure and dynamics of the DNA binding protein HU from Bacillus stearothermophilus by NMR spectroscopy. Biopolymers 40:553–559
5. Borisov OV, Darinskii AA, Zhulina EB (1995) Stretching of polyelectrolyte coils and globules in an elongational flow. Macromolecules 28:7180–7187
6. Carrington SP, Tatham JP, Odell JA, Sáez AE (1997) Macromolecular dynamics in extensional flows: 1. Birefringence and viscometry. Polymer 38(16):4151–4164
7. Cooper SD, Batchelder DN, Ramalingam P (1998) The effect of extensional flow on the Raman spectra of dilute polymer solutions. Polymer 39(14):3017–3027
8. Crowley DG, Frank FC, Mackley MR, Stephenson RG (1976) Localized flow birefringence of polyethylene oxide solution in a four roll mill. J Polym Sci Polym Phys Ed 14:1111–1119

9. De Gennes PG (1974) Coil-stretch transition of dilute flexible polymers under ultrahigh velocity gradient. J Chem Phys 60:5030–5042
10. Dijk J, White SW, Wilson KS, Appelt K (1983) On the DNA binding protein II from bacillus stearothermophilus. J Biol Chem 258(6):4003–4006
11. Doi M, Edwards SF (1986) The theory of polymer dynamics. Clarendon, Oxford
12. Drlica K, Rouviere-Yaniv J (1987) Histonelike protein of bacteria. Microbiol Rev 51 (3):301–319
13. Endo T, Sasaki N, Tanaka I, Nakata M (2002) Compact form of DNA induced by DNA-binding protein HU. Biochem Biophys Res Commun 290:546–551
14. Endoh T, Iyaguchi D, Sasaki N, Tanaka I, Nakata M (2003) Elongational flow studies on conformational change in DNA induced by DNA-binding protein HU. Biopolymers 68:486–496
15. Frank FC (1970) The strength and stiffness of polymers. Proc R Soc A 319:127–136
16. Frank FC, Keller A, Mackley MR (1971) Polymer chain extension produced by impinging jets and its effect on polyethylene solution. Polymer 12:467–473
17. Galindo-Rosales FJ, Alves MA, Oliveira MSN (2013) Microdevices for extensional rheometry of low viscosity elastic liquids: a review. Microfluid Nanofluid 14:1–9
18. Griffith JD (1976) Visualization of prokaryotic DNA in a regularly condensed chromatin-like fiber. Proc Natl Acad Sci U S A 73(2):563–567
19. Haward SJ (2014) Characterization of hyaluronic acid and synovial fluid in stagnation point elongational flow. Biopolymers 101(3):287–305
20. Haward SJ, Sharma V, Odell JA (2011) Extensional opto-rheometry with biofluids and ultra-dilute polymer solutions. Soft Matter 7:9908–9921
21. Haward SJ, Jaishankar A, Oliveira MSN, Alves MA, Mckinley GH (2013) Extensional flow of hyaluronic acid solutions in an optimized microfluidic cross-slot device. Biomicrofluid 7:1–14. 044108
22. Hayakawa I, Sasaki N, Hikichi K (1995) Elongational flow field as a tool for investigating helix-coil transition: Observation of helix-coil transition in poly(L-glutamic acid) induced by pH change. J Appl Polym Sci 56:661–665
23. Hayakawa I, Sasaki N, Hikichi K (1998) The response of DNA macromolecules to an elongational flow field and coil-globule transition of DNA. Polymer 39(6–7):1393–1397
24. Henyey FS, Rabin Y (1985) Rotational isomeric model of the coil stretching transition of semiflexible polymers in elongational flow. J Chem Phys 82(9):4362–4368
25. Keller A, Odell JA (1985) The extensibility of macromolecules in solution; a new focus for macromolecular science. Colloid Polym Sci 263:181–201
26. Kendrew JC (1966) The thread of life: an introduction to molecular biology. Bell & Hyman, London
27. Lee EC, Muller SJ (1999) Flow light scattering studies of polymer coil conformation in solutions in extensional flow. Macromolecules 32:3295–3305
28. Mel'nikov SM, Sergeyev VG, Yoshikawa K (1995) Discrete coil-globule transition of large DNA induced by cationic surfactant. J. Am Chem Soc 117:2401–2408
29. Menasveta MJ, Hoagland DA (1991) Light scattering from dilute poly(styrene) solution in uniaxial elongational flow. Macromolecules 24:3427–3433
30. Minagawa K, Matsuzawa Y, Yoshikawa K, Khokhlov AR, Doi M (1994) Direct observation of the coil-globule transition in DNA molecules. Biopolymers 34:555–558
31. Odell JA, Carrington SP (2006) Extensional flow oscillatory rheometry. J Non-Newtonian Fluid Mech 137:110–120
32. Odell JA, Keller A (1986) Flow-induced chain fracture of isolated linear macromolecules in solution. J Polym Sci Part B Polym Phys 24:1889–1916
33. Odell JA, Taylor M (1994) Dynamics and thermomechanical stability of DNA solution. Biopolymers 34:1483–1493
34. Odell JA, Keller A, Atkins EDT (1985) Diffusion and orientability of rigid-rodlike molecules in solution. Macromolecules 18:1443–1453

35. Odell JA, Keller A, Rabin Y (1988) Flow-induced scission of isolated macromolecules. J Chem Phys 88(6):4022–4028
36. Odell JA, Müller AJ, Narh KA, Keller A (1990) Degradation of polymer solutions in extensional flows. Macromolecules 23:3092–3103
37. Perkins TT, Quake SR, Smith DE, Chu S (1994) Relaxation of a single DNA molecule observed by optical microscopy. Science 246:822–826
38. Perkins TT, Smith DE, Chu S (1997) Single polymer dynamics in an elongational flow. Science 276:2016–2021
39. Peterlin A (1966) Hydrodynamics of linear macromolecules. Pure Appl Chem 12:563–586
40. Sasaki N, Atkins EDT, Fulton S (1991) Elongational flow studies on the molecular properties of collagen and its thermal denaturation. J Appl Polym Sci 42:2975–2985
41. Sasaki N, Hayakawa I, Hikichi K, Atkins EDT (1996) Deformation of λ-phage DNA molecules in an elongational flow field. J Appl Polym Sci 59:1389–1394
42. Sasaki N, Maki Y, Nakata M (2002) Elongational flow studies of DNA as a function of temperature. J Appl Polym Sci 83:1357–1365
43. Sasaki N, Ashitaka H, Ohtomo K, Fukui A (2007) Hydrodynamic properties of DNA and DNA-lipid complex in an elongational flow field. Int J Biol Macromol 40:327–335
44. Scrivener O, Berner C, Cressely R, Hocquart R, Sellen R, Vlachos NS (1979) Dynamic behavior of drag-reducing polymer solutions. J Non-Newtonian Fluid Mech 5:475–495
45. Sergeyev VG, Mikhailenko SV, Pyshkina OA, Yaminsky IV, Yoshikawa K (1999) How does alcohol dissolve the complex of DNA with a cationic surfactant? J Am Chem Soc 121:1780–1785
46. Simpson AA, Tao Y, Lieman PG, Badasso MO, He Y, Jardine PJ, Olson NH, Morais MC, Grimes S, Anderson DL, Baker TS, Rossmann MG (2000) Structure of the bacteriophage phi-29 DNA packaging motor. Nature 408:745–750
47. Smith DE, Chu S (1998) Response of flexible polymers to a sudden elongational flow. Science 281:1335–1340
48. Smith DE, Tans SJ, Smith SB, Grimes S, Anderson DL, Bustamante C (2001) The bacteriophage phi-29 portal motor can package DNA against a large internal force. Nature 413:748–752
49. Sun S-T, Nishio I, Swislow G, Tanaka T (1980) The coil-globule transition: Radius of gyration of polystyrene in cyclohexane. J Chem Phys 73:5971–5975
50. Tanaka I, Appelt K, Dijk J, White SW, Wilson KS (1984) 3-A resolution structure of a protein with histone-like properties in prokaryotes. Nature 310:376–381
51. Tanaka H, Yasuzawa K, Kohno K, Imamoto F (1995) Role of HU proteins in folding and constructing supercoils of chromosomal DNA in Escherichia coli. Mol Gen Genet 248:518–526
52. Tayler GI (1934) The formation of emulsions in definable fields of flow. Proc R Soc Lond Ser A 146:501–523
53. Torza SJ (1975) Shear-induced crystallization of polymers. I The four-roller apparatus. J Polym Sci Polym Phys Ed 13:43–57
54. Ueda M, Yoshikawa K (1996) Phase transition and phase segregation in a single double-stranded DNA molecule. Phys Rev Lett 77:2133–2136
55. Vasilevskaya VV, Khokhlov AR, Matsuzawa Y, Yoshikawa K (1995) Collapse of single DNA molecule in poly(ethylene glycol) solutions. J Chem Phys 102(16):6595–6602
56. Wakabayashi K, Sasaki N, Hikichi K (2000) Elongational flow studies of conformation of DNA molecules in the globular state. J Appl Polym Sci 76:1351–1358
57. Yoshikawa K, Matsuzawa Y (1995) Discrete phase transition of giant DNA dynamics of globule formation from a single molecular chain. Phys D 84:220–227
58. Yoshikawa K, Matsuzawa Y (1996) Nucleation and growth in single DNA molecules. J Am Chem Soc 118:929–930

Nonequilibrium Structure Formation of Complex Bilayer Membrane Lamellar Phase Under Shear

4

Shuji Fujii

Abstract

We present recent developments on the shear-induced onion phase (multilamellar vesicle) formation of the lyotropic lamellar phase from the viewpoint of the defect-mediated rheology and the polymer-reinforced membrane rigidity. We found that the increase in the defect (focal conic domain) density is essential as a pretransition of the onion phase formation. Critical shear stress for the onion formation is well scaled by taking account of the effective increment of the bending rigidity of complex bilayers. Shear-induced onion phase formation will be achieved by a realignment of bilayers from focal conic domain with negative Gaussian curvature to the positive one. Bending rigidity of bilayers preventing the realignment will determine the critical shear stress.

Keywords

Shear-induced phase • Lamellar phase • Multilamellar vesicle • Defect • Rheology

4.1 Introduction

Shearing lamellar phases often induces instability, which leads to nonequilibrium structure formation such as a multilamellar vesicle phase, so-called onion phase, and layer orientation transition. Although these structural transitions induced by shear have been fascinating topics in soft matter physics, rheological properties associated with the structural transition have not intensively studied. While the

S. Fujii (✉)
Department of Materials Science and Technology, Nagaoka University of Technology, 1603-1, Nagaoka, Niigata 940-2188, Japan
e-mail: sfujii@mst.nagaokaut.ac.jp

R. Kita, T. Dobashi (eds.), *Nano/Micro Science and Technology in Biorheology*,
DOI 10.1007/978-4-431-54886-7_4

shear diagram of the structural transition under shear has been well documented on the basis of rheophysical techniques such as scattering ones and NMR, the structural origin of the rheological behavior originated from the transition is not clear at all. In this chapter, we will present recent development on the shear-induced structural transition, especially shear-induced onion phase formation. We should mention that it is also not clear what controls the nonequilibrium structural transition and how the specific nonequilibrium phase is chosen between these two alternatives.

First, we briefly mention about some findings on the shear-induced onion phase formation and layer orientation transition that might be related to the onion phase formation. Then, we will give the information on the model for the shear-induced instability of the lamellar phase. We will also mention about the defect-mediated rheology as one of the key ideas in order to explain the rheological behavior associated with the onion phase formation in Sect. 4.1. In Sect. 4.2, we will present our recent achievements on the shear-induced onion phase formation of lamellar phase composed of complex bilayer membranes. And finally, we will present new insights on the onion phase formation in Sect. 4.3.

4.1.1 Some Findings on the Shear-Induced Onion Formation

The onion of the lamellar phase has a unique meso-structure in which a number of bilayer membranes stacked concentrically form a vesicle. Shear-induced onion phase formation in the lyotropic lamellar phase was presented for ionic surfactant SDS system by Roux et al. for the first time [1, 2]. They built up a shear diagram as a function of shear rate range and surfactant concentration. At low shear rate, the lamellae are slightly oriented with defects. In the intermediate shear rate, the lamellar phase shows the structural transition into the onion phase. Size of the onion, R, decreases with the shear rate in a manner of a power law, $R\sim\dot{\gamma}^{-0.5}$. At high shear rate, the orientation is similar to that at low shear rate but defects disappear. They also gave simple explanation on the power law behavior of the onion size by considering balance between shear stress and elastic stress of the bilayer membranes. Although they could identify the structure of the onion phase, detailed transition mechanism was not mentioned.

In order to understand the onion formation mechanism from planar lamellar phase, Nettesheim et al. traced the transition process by using small-angle neutron scattering under shear (Rheo-SANS) and found existence of intermediate structure as schematically shown in Fig. 4.1 [3]. They identified the intermediate structure with multilamellar cylinder structure or coherently buckled lamellae. Medronho et al. recently presented the experimental results supporting the multilamellar cylinder structure on the basis of the anisotropic diffusion of water measured by Rheo-NMR [4]. On the other hand, Kato et al. very recently presented that the coherently buckled lamellae are plausible compared to the cylinder structure by observing a small-angle X-ray scattering profiles under shear (Rheo-SAXS) [5]. Although new experimental results have been steadily accumulated, the comprehensive studies on the other physical quantities such as critical condition and membrane flexibility necessary for the onion phase formation are still lacking.

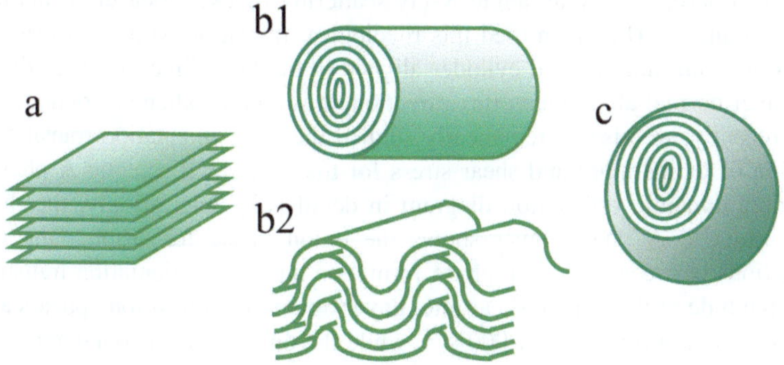

Fig. 4.1 Schematic diagram of (**a**) planar lamellar phase, (**b1**) multilamellar cylinder structure and (**b2**) coherently buckled lamellae, and (**c**) onion phase

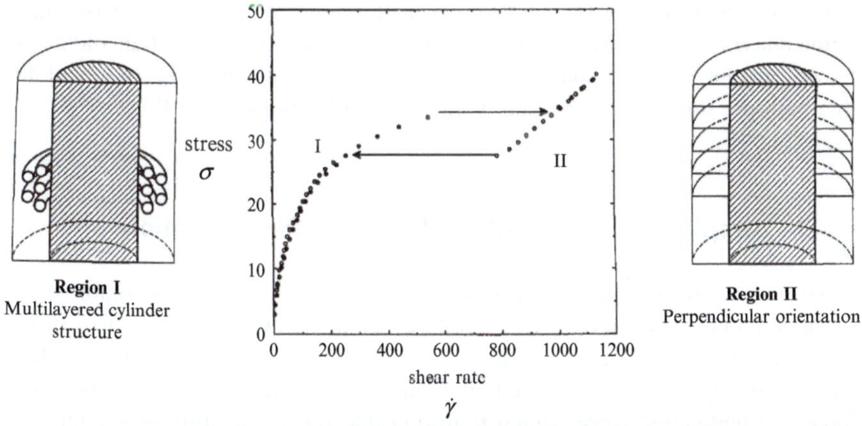

Fig. 4.2 Stress vs. shear rate for thermotropic 8CB smectic phase and schematic diagrams of multilayered cylinder structure and perpendicular orientation expected for shear rate regions I and II, respectively

4.1.2 Other Transitions of Layered System Under Shear

Shear-induced structural transition is an unsolved topic in a soft matter science. It has been widely reported that many different surfactant lamellar systems show the shear-induced onion phase formation [1–10]. In addition to the lyotropic lamellar systems, thermotropic smectic A phase, more studied layered system rather than lyotropic system, also shows a unique shear-induced transition as presented by Safinya et al. and Panizza et al. [11, 12].

Figure 4.2 presents a stress as a function of shear rate performed for one component material 8CB, one of the most known thermotropic smectic phases. The flow curve shows a characteristic feature of the shear-induced transition between shear-thinning region (described as I) and Newtonian region (II). Panizza et al. also performed small-angle X-ray scattering (SAXS) measurements in two regions I and II. They identified this rheological transition with layer orientation transition from multilayered cylinder aligned in the flow direction (I) to the layer with layer normal along the vortex direction, thus a perpendicular orientation (II). Very recently, we also systematically studied the layer orientation transition as a function of temperature and shear stress for the same 8CB smectic A phase and presented dynamic orientation diagram in detail [13]. Interestingly, the thermotropic smectic A phase never shows the onion phase formation. One of the distinctions between the onion phase formation and layer orientation transition is the magnitude of the critical shear rate. Layer orientation transition appears at high shear rate region roughly at 1,000 s^{-1}, while the onion phase is obtained at rather low shear rate such as 1 s^{-1}. Salient difference in the critical shear rate suggests that the instability triggered by shear flow has different physical origin. However, it has not been discussed from what the difference in these transitions comes. It has not been clear whether these transitions have the same physical origin or different physical background. In this chapter, we will mainly focus on the shear-induced onion phase formation. We will not discuss on the layer orientation transition here. However, one can find some articles on the layer orientation transition of the thermotropic smectic A phase and lyotropic lamellar phase [11–13].

4.1.3 Simple Explanation of Models on Onion Formation

Two models are generally suggested to explain the origin of the onion phase formation. One explanation is described by Zilman and Granek [14]. Their model considers the coupling of the short wavelength undulation of the lamellar phase with the shear flow. Short wavelength undulations produce a long-range repulsive force, the undulation force, which is responsible for the stability of the lamellar phase. Reduction of the excess area and suppression of membrane undulations due to shear flow thus generate an effective dilatation. As the dilatation becomes higher than the critical shear rate, the coherent buckling of the lamellar membranes appears, and the onion phase formation is achieved. This proposed mechanism that attracted many scientists engaging in the shear-induced onion phase formation. However, one finds mismatch in the critical shear rate. The critical shear rate $\dot{\gamma}_{c,ZG}$ ~1,000s^{-1} calculated on the basis of typical parameters such as a repeating distance of the lamellar membranes of 10 nm, a bending modulus of $\kappa \sim k_B T$ and solvent viscosity, is much higher than that of experimentally obtained one $\dot{\gamma}_{c,\exp} \sim$ 1-10 s^{-1}. Although good agreement can be obtained by using the effective viscosity of the lamellar phase, the origin of the high viscosity is unclear. One of the reasons for such differences might be a contribution of defects.

The other mechanism mainly based on the work of Oswald considers the motion of defects [15, 16]. In layered systems, nonuniform gap in the sample cell leads to the formation of defects. In the presence of defects, the number of layers shows local inhomogeneity. This inhomogeneity can lead a dilatational and compressive strain normal to the layers. Dislocations in the lamellar phase are able to move with flow by permeation at low shear rate. Here, permeation is the diffusion of surfactant monomer and solvent across the membranes. However, the permeation process is too low to allow dislocations to flow at high shear rates. Therefore, a delay of dislocations against the shear flow generates a dilative and compressive strain. This nonuniform shear flow becomes the origin of the undulation instability and results in the onion phase formation. Qualitative explanation argued by Roux et al. is also similar to this idea [1, 2]. Interestingly, common origin of the instability in the layered systems is a dilative strain in spite of the inherent mechanism of the shear-induced onion phase formation. However, we should note that the dilative strain should also produce defects before the onion phase is formed. Eventually, we will also mention about the role of defects on the onion phase formation.

4.1.4 Defects and Dislocations

Lamellar phase always includes locally unregulated alignment of layers, i.e., defects. As typical defects in layered systems, two dislocations, edge and screw dislocations, and a texture with meso-scale size, focal conic domain, are known [17, 18]. Edge and screw dislocations are linear defects. Since these dislocations never possess terminal, a pair of edge dislocations and screw dislocations with opposite signs usually form a dislocation loop as shown in Fig. 4.3.

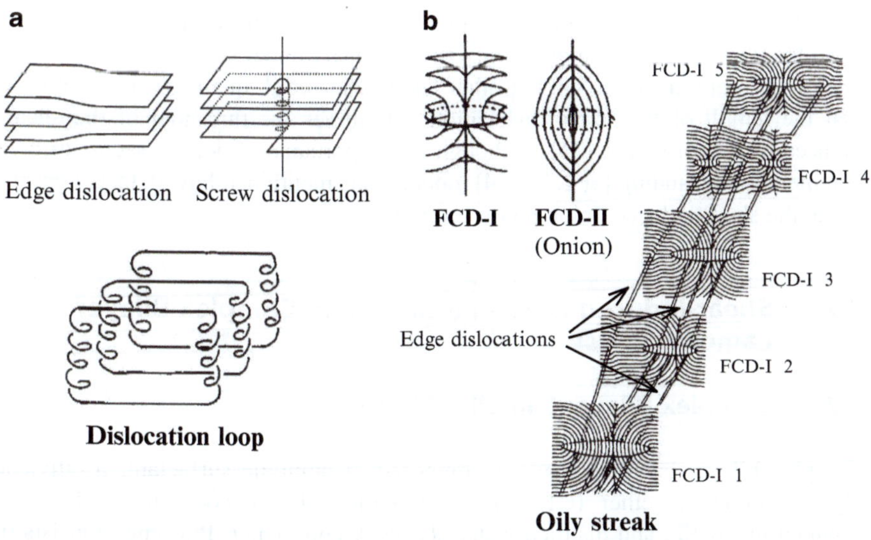

Fig. 4.3 Schematic diagram of (**a**) edge and screw dislocations and dislocation loop in which two dislocations are paired and (**b**) focal conic domain type I (FCD-I), type II (FCD-II), and oily streak defects composed of FCD-I connected by edge dislocations

Formation of the dislocation loop accompanies unfavorable local compression deformation of the lamellar membranes. Thus, the increase in the dislocation loop density allows the strain energy to accumulate. Such accumulation of the strain energy is relaxed down by creating new texture, focal conic domains (FCDs). FCD can be easily obtained because of the constant layer repeating distance inside it, and this texture is actually often observed in lyotropic lamellar systems. Depending on the sign of the curvature of membranes, FCD is further classified into two species: FCD-I with torus-like deformation of membranes and FCD-II with onion-like structure. Because of the similarity of the structure, the onion phase itself is sometimes referred to as one kind of defect. One should also note that several FCD-Is are often connected in series by edge dislocations. The connected FCD-I is referred to as oily streak defects [17].

4.1.5 Smectic Rheology

Here, we propose a mesoscopic structure-mediated rheology in soft matter systems, called "structural rheology." This is a new concept in the soft matter physics. One of such examples of the structural rheology is a smectic rheology in which the defects in the meso-scale structure significantly affect the rheology.

It is well recognized that the defects govern the rheological properties of the lamellar phase [19–22]. Oily streaks and dislocation loops mentioned above dominate the elasticity and the shear-thinning behavior of the lamellar phase, respectively. It has also been considered that defects are deeply correlated with the onion formation. Medronho et al. actually presented that the increase in the defect density is necessary as a pretransition of the onion phase formation by measuring the shear modulus [23]. The increase in the oily streak density under shear might be considered to be a pretransition. On the other hand, Dhez et al. qualitatively mentioned that the length of the screw dislocation determines the threshold of the shear-induced onion formation [24]. Although the importance of defects seems to be a common understanding [1, 2, 23, 24], nobody has mentioned how defects contribute to the shear-induced onion phase formation.

4.2 Shear-Induced Onion Formation in Complex Bilayer Lamellar Phase

4.2.1 Complex Bilayer Lamellar Phases

Complex bilayer lamellar phase is composed of nonionic surfactant, triethylene glycol monodecyl ether (C10E3), amphiphilic triblock copolymers, Pluronic supplied by BASF, and distilled water. Triblock copolymer, Pluronic, consists of two hydrophilic ethylene oxide (EO) blocks bounded to a central hydrophobic propylene oxide (PO) block. Hydrophilic EO chains at both sides have an identical chain length. In complex bilayer membranes, the hydrophobic PO block

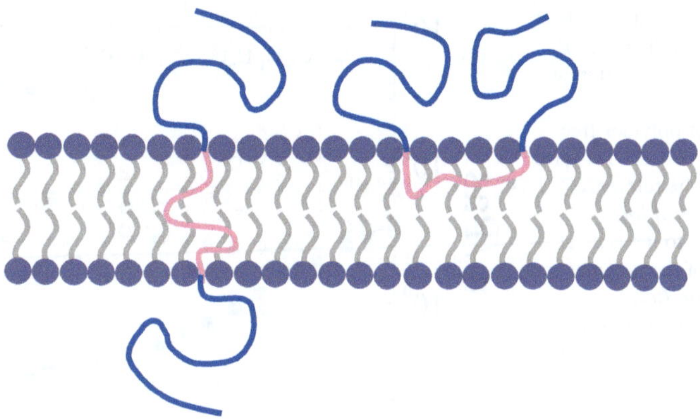

Fig. 4.4 Schematic diagram of complex bilayer lamellar phase composed of surfactant bilayer and grafted triblock copolymer chains

interpenetrates or stacks inside the hydrophobic region of the surfactant bilayers, and EO blocks decorate the membranes as presented by Masui et al. for nonionic surfactant C12E5/Pluronic/D2O systems [25]. Schematic diagram of the complex bilayer membrane is shown in Fig. 4.4.

In the triblock copolymer-decorated lamellar phase, on the one hand, several triblock copolymers having different degrees of polymerization of EO chains ($N_{EO} = 17$, 27, and 37) are used, while the degree of polymerization of PO block is fixed at $N_{PO} = 60$.

In the triblock copolymer-embedded lamellar phase, on the other hand, the degree of polymerization of PO chains is varied as $N_{PO} = 16$, 30, 42, 59, and 69, while the degree of polymerization of EO block is fixed at $N_{EO} = 4$, which is almost the same as the hydrophilic chain length of the surfactant.

The mole fraction of triblock copolymer in the surfactant bilayer X_P is calculated by the following relation: $X_P = n_P/n_S + n_P$. Here, n_P and n_S are, respectively, the mole number of the triblock copolymer and surfactant. Total mole number $n_S + n_P$ is fixed at 8.01 mmol, which corresponds to the weight fraction of the surfactant of 40 wt% in $C_{10}E_3/H_2O$ binary system.

The bending modulus of such polymer-grafted membranes depends on the size and a number density of the hydrophilic polymer chain as theoretically predicted by Hiergeist and Lipowsky [25, 26]. Varying these quantities, effects of the defect density and the bending rigidity on the shear-induced onion phase formation can be systematically studied [27].

4.2.2 Shear Viscosity

4.2.2.1 Triblock Copolymer-Decorated System

Figure 4.5 shows the shear rate dependence of the viscosity for C10E3/P-EO(37)/H2O systems with different mole fractions of $X_P = 0.2$, 0.8, 1.0, and 1.2 %. Here,

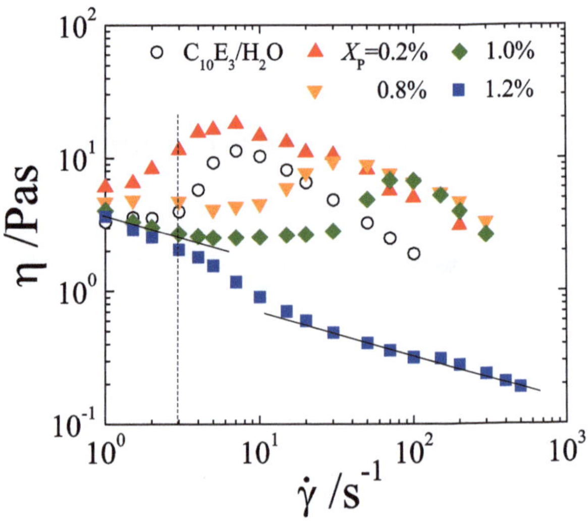

Fig. 4.5 Viscosity vs. shear rate for C10E3/P-EO(37)/H2O system with different mole fractions of X_P. Viscosity of polymer-free system (C10E3/H2O) is also plotted for the reference. *Vertical dashed line* corresponds to the critical shear rate of the polymer-free system. *Solid lines* corresponds to a slope of −1/3 [27]. Reproduced by permission of The Royal Society of Chemistry

P-EO(37) expresses Pluronic with EO blocks of $N_{EO} = 37$. For the comparison, the viscosity of the polymer-free (C10E3/H2O) system is also shown as open circle. The shear-induced onion phase formation of the polymer-free system accompanies a shear-thickening behavior [10]. Appearance of the shear thickening can be identified as the initiation of the onion phase formation for nonionic surfactant lamellar system [3, 10]. The shear-thickening region, however, corresponds to a transition region where a coexistence of the planar lamellar and the onion phases is expected, and the onion phase formation is completed in the shear-thinning region at the higher shear rate. A critical shear rate of the shear thickening in the polymer-free system is $\dot{\gamma}_c = 3\ s^{-1}$. In the polymer-grafted system, anomalous dependence of the critical shear rate $\dot{\gamma}_c$ on X_P is observed.

At low $X_P(=0.2\ \%)$, $\dot{\gamma}_c$ shifts to lower shear rate indicating that the onion formation was easily achieved. On the contrary, remarkable shift of $\dot{\gamma}_c$ into high shear rate region is observed at higher $X_P(=1.0\ \%)$. For such systems, the first shear thinning at low shear rate followed by the shear thickening and the second shear thinning at high shear rate can be observed. The first shear thinning follows a power law relation of $\eta{\sim}\dot{\gamma}^{-1/3}$. Once X_P exceeded the overlap concentration of EO chains in the water layer $C^* = 1.1\ \%$, the shear thickening is never observed as can be seen for $X_P = 1.2\ \%$. In this system, the same power law exponent −1/3 as the first shear thinning can be observed in a wide shear rate region. Qualitatively, similar behavior on the X_P dependence of the flow curve is observed for other systems with different EO chain lengths [27]. The shear-induced onion formation is favored at lower mole fraction, while $\dot{\gamma}_c$ remarkably shifts into higher shear rate region at higher mole fraction. The EO block-decorating membranes have strong impact on the critical shear rate.

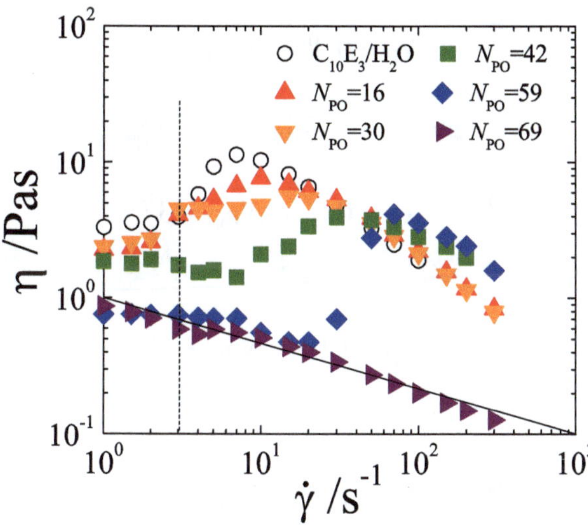

Fig. 4.6 Viscosity vs. shear rate for C10E3/P-PO(N_{PO})/H2O system with different mole fractions of X_P

4.2.2.2 Triblock Copolymer-Embedded System

Figure 4.6 shows a typical shear rate dependence of the viscosity for triblock copolymer-embedded systems with different degrees of polymerization of $N_{PO} = 16$, 30, 42, 59, and 69 at $X_P = 1$ mol% instead of fixed EO block of $N_{EO} = 3$. At low N_{PO} (=16 and 30), $\dot{\gamma}_c$ is not remarkably influenced by embedded hydrophobic chain. As N_{PO} is increased, on the contrary, significant shift of the critical shear rate into high shear rate region has appeared. Finally, at N_{PO} =69, the shear thickening as the signature of the onion formation disappears. Similar tendency has been observed when the hydrophilic chain length N_{EO} is increased [27]. In X_P dependence, increasing X_P makes the critical shear rate higher (for more information, please see [38]).

4.2.3 Structural Observations

Figures 4.7 summarizes typical small-angle light-scattering (2D-SALS) patterns of different systems at several shear rates. The data shown in Fig. 4.7 corresponds to results in Fig. 4.5. SALS pattern of the polymer-free system is markedly allowed to develop with the shear rate. In the beginning of the shear thinning (at $\dot{\gamma} = 10\mathrm{s}^{-1}$), a four-leaves pattern, a typical feature of the shear-induced onion phase, is observed. As the shear rate is increased, a refinement of the onion structure yields the Bragg peak in addition to the four leaves. The Bragg peak is a representative of the shear-induced onion phase with monodispersed size distribution. In general, such pattern is always observed in the shear-thinning region [3, 28]. Kato et al. attributed the four leaves with the Bragg peak to closed-pack onions in FCC packing.[6].

Grafting polymer chains onto the membrane remarkably affects the scattering profile as well as the flow curve depends on X_P, N_{EO}, and N_{PO}. Increasing X_P at fixed

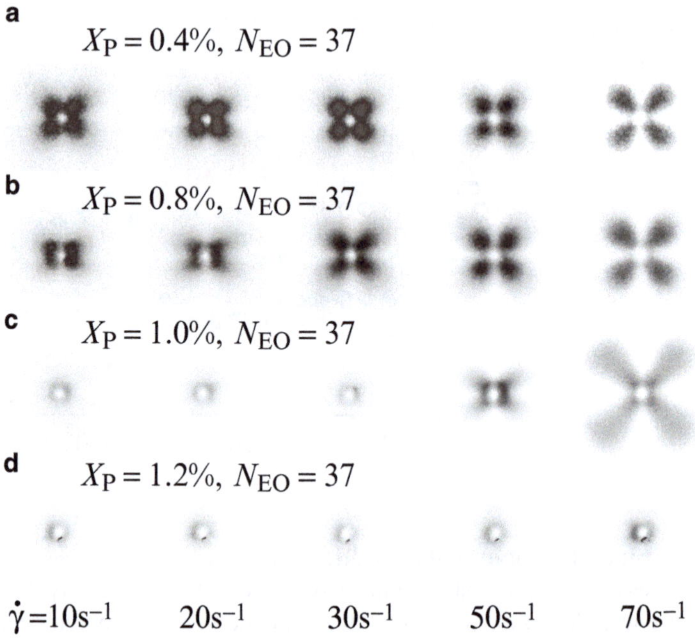

Fig. 4.7 SALS pattern obtained for (**a**) polymer-free system and for several systems with different EO chain length N_{EO} at fixed mole fraction of $X_P = 0.2$ % under several shear rates [27]. Reproduced by permission of The Royal Society of Chemistry

N_{EO} (or N_{PO}), or vice versa, disturbs the development of the four-leaves pattern and allows disappearance of the Bragg peak even in the shear-thinning region. Shear rate dependence of the scattering pattern for a system with $N_{EO} = 37$ at $X_P = 0.4$ % indicates no significant development of the onion structure. Weak shear effect on the onion radius was also observed for the hydrophobically modified polymer-grafted lamellar systems by Yang et al. [29]. As X_P is increased, the four-leaves pattern eventually vanishes at $X_P = 1.2$ %. The shear-thinning behavior at $X_P = 1.2$ % in Fig. 4.5 is thus attributed to the lamellar phase but not to the onion phase.

Figure 4.8 shows a sequence of the microscope images as a function of the shear rate for polymer-free system and polymer-grafted systems with P-EO(37) at different X_P. Each image was taken after the cessation of shear. Please note that the system with $X_P = 1.0$ % shows the onion formation, while no shear-induced onion formation is observed at $X_P = 1.2$ %. Before shearing, an oily streak network can be seen for every system. In the polymer-free system in Fig. 4.8(a), applying shear flow generates a number of thinner oily streaks. As the shear rate is increased, the oily streak density is increased, and the shear-induced onion phase is achieved at $\dot{\gamma} = 3\ s^{-1}$. Further increase in $\dot{\gamma}$ yields the fine texture indicating the monodispersed onion phase formation. In the polymer-grafted system with $X_P = 1.0$ % in Fig. 4.8 (b), the sequence of the meso-scale structure shows a significant orientation of the oily streak along the flow direction below $\dot{\gamma}_c$. Aligned oily streaks thin and the

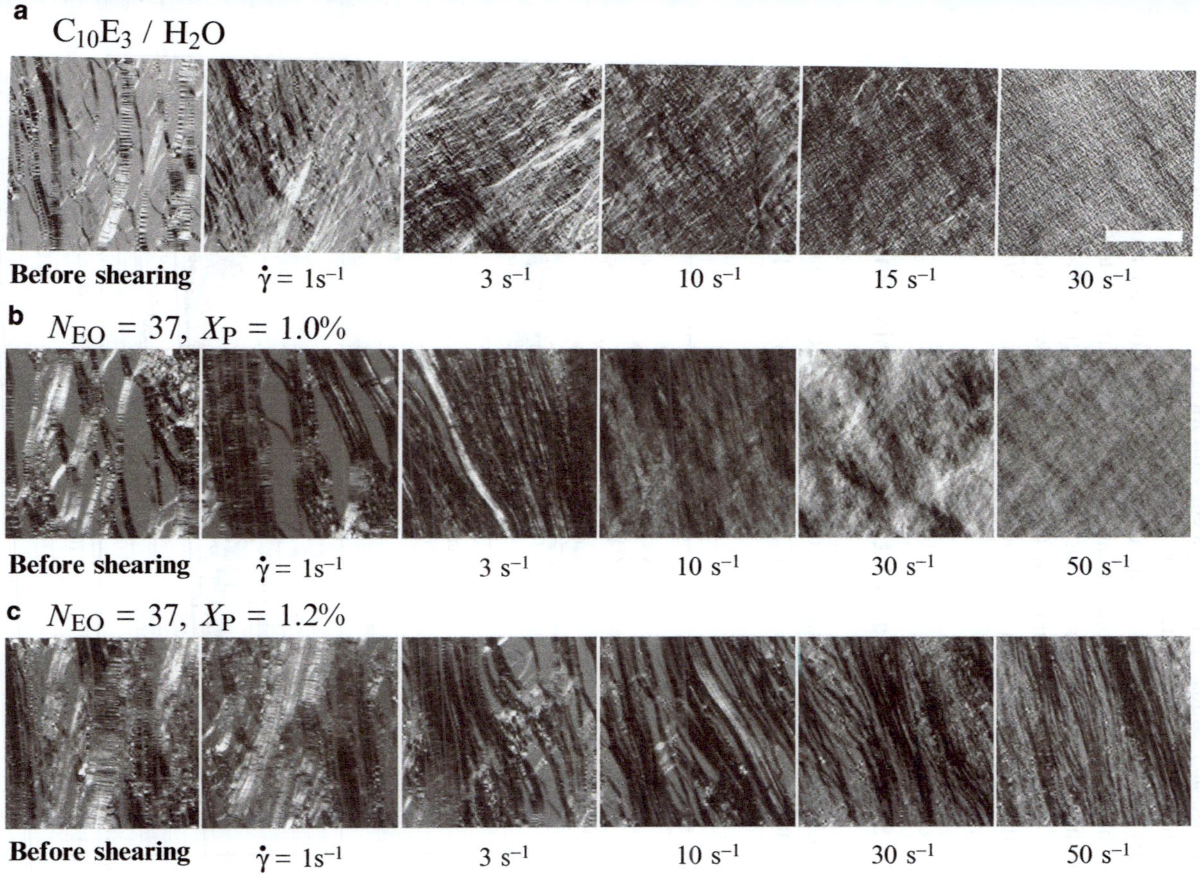

Fig. 4.8 Depolarized microscope images of (**a**) polymer-free system and C10E3/P-EO(37)/H2O system with the mole fraction of (**b**) $X_P = 1.0$ % and (**c**) $X_P = 1.2$ % at several shear rates. Flow direction corresponds to longitudinal one. Scale bar corresponds to 150 μm [27]. Reproduced by permission of The Royal Society of Chemistry

density is increased with $\dot{\gamma}$. At $\dot{\gamma}$ $=10\text{s}^{-1}$, the development of the oily streak network makes it difficult to recognize each single oily streak under an optical microscope. Significantly, increased oily streaks eventually transform into the onion phase at $\dot{\gamma}$ $=30\text{s}^{-1}$. In the system with $X_P = 1.2$ % in Fig. 4.8(c), the oily streaks thin at low shear rate in the same way as $X_P = 1.0$ %. And its density is gradually increased. However, the oily streak does not thin remarkably, and a rate of the increase in the oily streak density is quite moderate compared to the system showing the onion formation. One can still recognize the oily streak even at $\dot{\gamma}$ $=50\text{s}^{-1}$. It thus seems like that the difference between the system with and without the onion formation is attributed to the structural development of the oily streak density under shear.

4.2.4 Shear Modulus of the Lamellar and Onion Phases

Figure 4.9 shows typical examples of the frequency dependence of the dynamic storage and loss moduli G′ and G″ measured after applying preshear $\dot{\gamma}_p$ to the

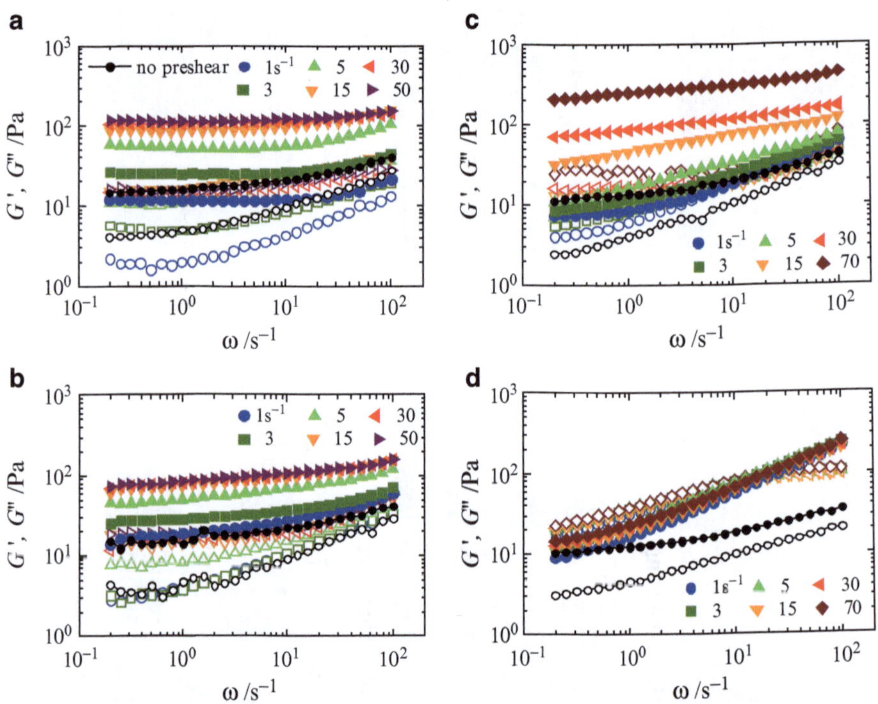

Fig. 4.9 G′ and G″ as a function of angular frequency for (**a**) the polymer-free system and the polymer-grafted lamellar phase with P-EO(37) (**b**) $X_P = 0.2$ %, (**c**) $X_P = 1.0$ %, and (**d**) $X_P = 1.2$ %, respectively. Each symbol corresponds to preshear rate applied before dynamic measurements. For the sake of comparison, dynamic moduli before preshearing are also shown as symbols connected with lines. Filled and open symbols are G′ and G″, respectively [27]. Reproduced by permission of The Royal Society of Chemistry

polymer-free and polymer-grafted systems with P-EO(37) at different X_P. For each system showing the onion formation, the value of G′ develops higher as the preshear rate $\dot{\gamma}_p$ increases. The lamellar and onion phases show a solid-like behavior since no terminal region can be seen in measured frequency region. As shown by Basappa et al. [21], the extremely slow viscoelastic relaxation at low frequency regime is dominated by the oily streak networks. Shear modulus mediated by defects has been also reported by Ramos et al. [22]. $\dot{\gamma}_p$ dependence of G′ will thus represent the development of the oily streak networks with $\dot{\gamma}_p$.

Comparing $\dot{\gamma}_p$ dependence of G′ with the flow curve in Fig. 4.5 makes it obvious that G′ increases in the first shear-thinning and shear-thickening regions. The microscope observation indicates that the enhancement of the shear modulus originates from the significant increase in the oily streak density. In some systems, the shear moduli measured before preshearing are higher than after applying preshear. Low shear modulus after preshear might be attributed to a degree of the orientation of the oily streak. Other polymer-grafted systems with a different X_P or N_{EO}, except for systems with no shear-induced onion phase formation, also showed qualitatively similar behavior. Contrary to the system showing the onion formation, G′ and G″ for the system exhibiting no onion formation do not show $\dot{\gamma}_p$ dependence. Although the solid-like behavior is observed before preshearing, G′ after the preshear shows no plateau. Orientation and the weak $\dot{\gamma}_p$ dependence of the oily streak density seem to cause the fluid-like behavior.

In addition to dislocation loops, it is known that the shear flow causes a formation of oily streak networks composed of focal conic domains (FCDs) connected in series via edge dislocations [17]. Such increase of the oily streak network density can be monitored by investigating the preshear rate $\dot{\gamma}_p$ dependence of the shear modulus G′ through a defect-mediated elasticity [21, 22]. Basappa et al. estimated G′ of the oily streak networks with spacing l and line tension Γ to be

$$G' \approx \frac{\Gamma}{l^2}. \tag{4.1}$$

Γ depends on the internal structure of oily streaks. The line energy of the oily streak is described by two terms. One is the contribution from a mean curvature energy, and the other is from a layer compression within a narrow region connecting the homotropic region with the boundary of FCD [17].

Since the curvature term is negligibly small, the compression modulus B originating from the layer compression acts as a dominant contribution. The effect of the polymer chain will be thus included into the shear modulus G′ through the line tension Γ of the oily streak which is mainly dominated by the compression modulus. Although the shear modulus depends on the degree of orientation of the oily streaks, by reducing G′ with polymer effect on Γ, the shear rate dependence of l will be extracted as a measure of the oily streak density.

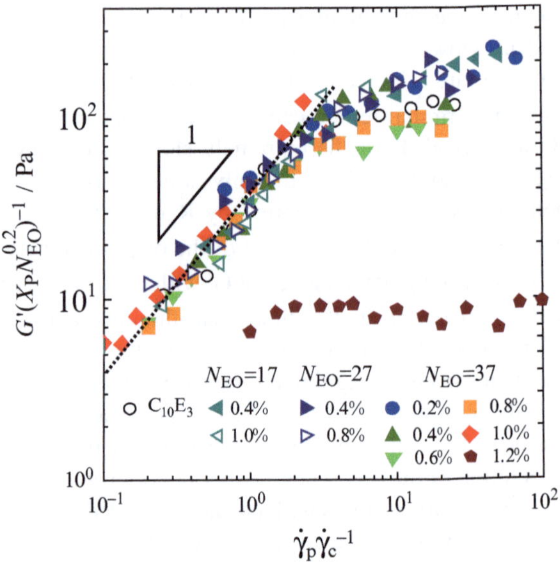

Fig. 4.10 Scaled shear modulus as a function of the reduced shear rate. Shear modulus was reduced by the scaling factor $X_P N_{EO}^{0.2}$, and the shear rate was reduced by the critical shear rate. For the comparison, G′ for the polymer-free system is also plotted [27]. Reproduced by permission of The Royal Society of Chemistry

Figure 4.10 shows a scaling plot on $\dot{\gamma}_p$ dependence of G′. G′ and $\dot{\gamma}_p$ were, respectively, reduced by a scaling factor and the critical shear rate of the shear-thickening $\dot{\gamma}_c$. In order to obtain the scaling plot, we first determined the power law relation on the X_P dependence of G′. We found that G′ at $\dot{\gamma}_c$ follows a linear relation of $G' \sim X_P$ in the range of $0.2 < X_P < 1.0$. We should note that G′ deviates from the linear relation as X_P is lowered in $X_P < 0.2$ and approaches to a finite value of the polymer-free system.

The final scaling behavior $X_P N_{EO}^{0.2}$ including N_{EO} dependence was determined so that all data points fall on the same line. In the same panel, G′ of the polymer-free system, which has no scaling factor, is also plotted. The data collapse is satisfactory, and the scaled G′ almost coincides with G′ of the polymer-free system except for the system with $X_P = 1.2$ % at $N_{EO} = 37$, in which the horizontal axis corresponds to the preshear rate but not the reduced shear rate, because of no critical shear rate. At low shear rates corresponding to the first shear-thinning and the shear-thickening regions, the scaled G′ linearly increases with $\dot{\gamma}_p$. On the other hand, G′ in the onion phase shows a weak $\dot{\gamma}_p$ dependence and is not able to scale by $X_P N_{EO}^{0.2}$. Crossover between two slopes has appeared at $\dot{\gamma}_p/\dot{\gamma}_c \cong 2$ which corresponds to the beginning of the second shear thinning.

Shear rate dependence of the defect size a was theoretically predicted to be $a \propto 1/\sqrt{\dot{\gamma}}$ by Larson et al. [30]. If the defect size a is assumed to be identical to the defect spacing l, Eq. (4.1) is rewritten to be $G' \sim \dot{\gamma}_p$, which is the same as experimentally found relation for the scaled G′. Linear increase of scaled G′ in the defective lamellar phase is thus mainly attributed to the increase of the oily streak density. At high X_P where the shear-induced onion formation is prohibited, no $\dot{\gamma}_p$ dependence of G′ indicates that the development of the oily streak density is not remarkable as shown in the microscope observation. These results evidently suggest that the onion formation necessitates the increase of the oily streak network density.

Influence of the grafted polymers on the elastic properties has been carefully investigated by some researchers [25, 28–37]. It is known that grafting polymers enhances the bending modulus and the repulsive interaction between membranes. Castro-Roman et al. explained the enhancement of the repulsive Helfrich interaction by renormalizing the thickness of polymer layer into an effective thickness of membranes [35]. Furthermore, Masui et al. successfully explained the interlamellar interaction by taking account of the compression of the polymer chains in the renormalized Helfrich interaction. These elastic properties of the polymer-grafted membrane depend on the polymer density on the membrane surface and the radius of gyration. Collapse of the scaled G′ with the polymer-free system may suggest that the scaling factor $X_P {N_{EO}}^{0.2}$ is a contribution from the grafted polymers on the line tension of the oily streak. Although we have no explanation how the polymer affects line energy, the scaling factor $X_P {N_{EO}}^{0.2}$ seems to reflect the influence of the grafted polymers on the line energy of the oily streak.

G′ of the onion phase is originated from the effective surface tension of the membrane, $G' \cong \sqrt{KB}/R$, as Panizza et al. suggested [12]. Scattering of the scaled G′ data in the onion phase may be attributed to the lack of the consideration of the effective surface tension $\sqrt{KB}$.

4.2.5 Critical Shear Stress

4.2.5.1 Triblock Copolymer-Decorated Membranes

The oily streak is composed of focal conic domains (FCDs), which have negative Gaussian curvature, while the onion has a spherical form with positive Gaussian curvature. As mentioned in Fig. 4.3, both structures are recognized as the FCDs and referred to as FCD-I and FCD-II (onion), respectively. The difference is the opposite sign of the Gaussian curvature. For the onion phase formation to occur, therefore, the Gaussian curvature of the membrane has to be reorganized from negative to positive by shear force. For the reorganization of the layer orientation, external energy is necessary to overcome the bending energy of the membrane.

The effect of the grafted-polymer chain on the bending modulus κ is described as an effective bending modulus:

$$\kappa_{\text{eff}} = \kappa_b + Ck_BT\sigma_P R_g^2, \tag{4.2}$$

where κ_b is the bare bending modulus of the membrane, C is the numerical constant, σ_P is the number density of the polymer chain, and R_g is the radius of gyration of the polymer chain [25, 26]. Using the mole fraction of the polymer X_P and degree of polymerization N, the effective increment of the bending modulus due to the polymer $\Delta\kappa$, the second term, can be represented as $\Delta\kappa \sim X_P\, N^{2\nu}$. Here, σ_P was

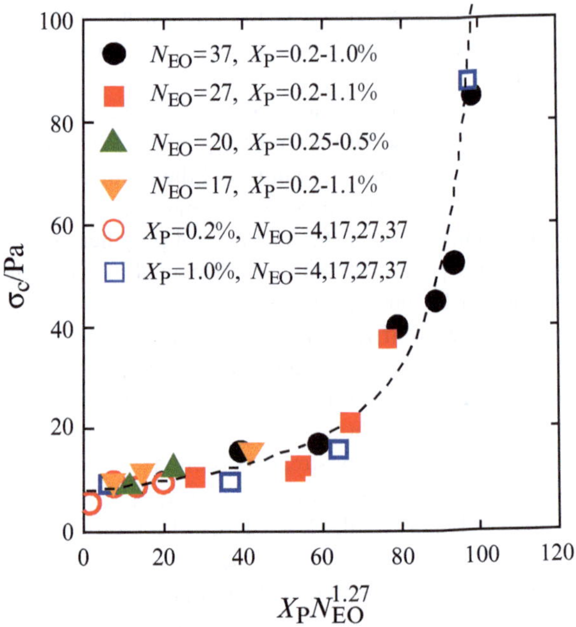

Fig. 4.11 Critical shear stress obtained for different complex lamellar phases with different NEO and XP as a function of the increment of bending rigidity $X_p N_{EO}^{1.27}$. Dashed curve is a best fit to $\sigma_c \sim (\Delta\kappa_c - \Delta\kappa)^{-1}$. Here, $\Delta\kappa_c$ is a critical value of $\Delta\kappa$. [27]. Reproduced by permission of The Royal Society of Chemistry

replaced by X_P. The physical origin of the enhanced membrane rigidity due to grafted-polymer chain is attributed to the configuration entropy of polymers on the membrane. For the hydrophilic polymer-grafted lamellar phase, σ_c monotonically increases with X_P and N.

In Fig. 4.11, the critical shear stress σ_c obtained for various combinations of X_P and N_{EO} was compared with the effective increment term $\Delta\kappa$. σ_c obtained for several systems with different X_P and N_{EO} are well scaled onto the same curve. σ_c monotonically increases with X_P and N_{EO} although the critical shear rate shows the promotion and hindrance of the onion formation depending on X_P. This suggests that the shear-induced onion phase formation is mechanically suppressed by grafting polymers. Scaled σ_c was followed by $\sigma_c \sim (\Delta\kappa_c - \Delta\kappa)^{-1}$, where $X_P {N_{EO}}^{1.27}$ was rewritten by $\Delta\kappa$ as the effective incremental term of the bending modulus. $\Delta\kappa_c$ represents a critical value where the shear-induced onion formation is hindered. As $\Delta\kappa_c$ is approached, σ_c shows approximately diverging behavior. This behavior suggests the existence of the critical membrane rigidity for the shear-induced onion formation. Estimated critical value $\Delta\kappa_c \approx 107$ corresponds to $X_P = 1.09$ mol% which almost coincides with $C^* = 1.1$ mol%. The correspondence of $\Delta\kappa_c$ and C^* suggests that a repulsive interaction between polymers irons out the undulation fluctuation, which results in the hardening of the membranes. Scaling behavior of σ_c via $\Delta\kappa$ suggests that the shear stress drives the shear-induced onion phase formation when it overcomes the bending energy of the membranes. Inverse proportionality between σ_c and $\Delta\kappa_c - \Delta\kappa$ might be regarded as a power law behavior. The shear-induced onion phase formation is a stress-induced mechanical phenomenon.

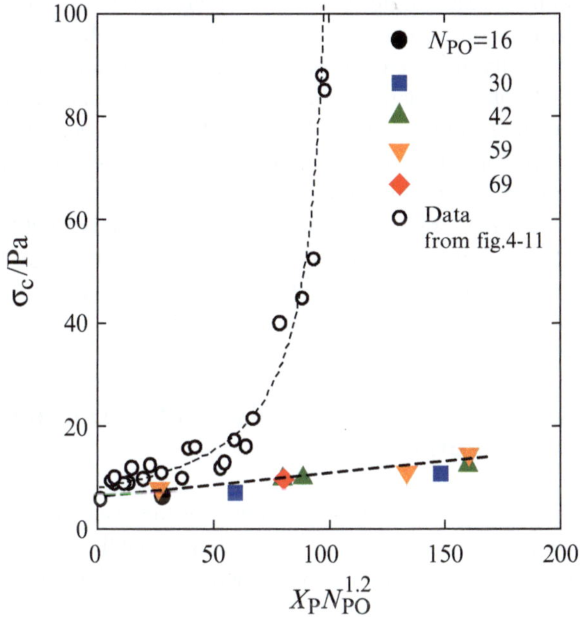

Fig. 4.12 Critical shear stress obtained for several systems with different PO chain length N_{PO} and mole fraction X_P. Critical shear stress is plotted as a function of the incremental term of bending modulus for the polymer-grafted membrane. For the comparison, the critical shear stress of the hydrophilic polymer-grafted lamellar systems is also shown as *open circle*

4.2.5.2 Triblock Copolymer-Embedded Membranes

We also would like to discuss the effect of the hydrophobic chains on the critical shear stress [38]. Although it is not clear how the bending rigidity of the hydrophobic polymer-embedded membrane can be described, the same scaling procedure is attempted in Fig. 4.12. Assuming that the radius of gyration of the hydrophobic propylene oxide (PO) chain behaves as a good solvent in the bilayer, $R_g \sim N_{PO}^{0.6}$, σ_c obtained for several systems with different combinations of X_p and N_{PO} are almost superimposed. The same scaling behavior of σ_c via $\Delta\kappa$ as that for the hydrophilic chain seems to be applicable for the hydrophobic polymer-embedded lamellar phase. Validity for using Eq. (4.2) may be appropriate if it is assumed that the hydrophobic layer in the membrane is decorated by PO chain. Embedding the hydrophobic chain seems to enhance the bending modulus by following Eq. (4.2). The data in Fig. 4.11 is also shown in the same figure. The contribution from the embedded hydrophobic chain on the critical shear stress, thus on the bending modulus of the membrane, is rather weak than that from the decorated hydrophilic chain. Such difference of the polymer contribution on the bending modulus will be influenced by the efficiency of polymer chains, which depends on how they are attached on the membrane. Reminding the structure of the amphiphilic triblock copolymers, Pluronic, two hydrophilic chains are grafted on both sides of the membrane, while single hydrophobic chain is embedded inside. Twice higher number density of the hydrophilic chain will more efficiently act to enhance the bending modulus. Additionally, the bending deformation of the membrane sandwiched by hydrophilic chains seems to be highly disturbed more than single hydrophobic chain-embedded membrane does. Effective enhancement of the bending rigidity via the grafted-hydrophilic chain compared to the embedded hydrophobic chain was actually implied by Masui et al. Increase in the membrane viscosity and the decrease in the lamellar repeating distance due to the embedded hydrophobic chain might be one of the reasons for the slight enhancement of the bending modulus. In order to control the critical condition of the shear-induced onion formation, it is effective to modify the bending modulus of the membrane by grafting hydrophilic polymer chain.

Biomembrane dynamics is one of the active fields in the soft matter physics. Our findings on the effect of the embedded component in the bilayer membrane will be commonly understood with the structural formation of such soft matter complex system such as multicomponent vesicles.

4.3 Possible Mechanism of Shear-Induced Onion Phase Formation

We propose a possible mechanism of the shear-induced onion phase formation by summarizing our findings. According to the theory proposed by Lu et al., the shear-thinning mechanism is governed by the dynamic balance of the production-annihilation processes of the dislocation loops [19]. Increase in the dislocation loop density with the shear rate yields the shear-thinning behavior. On the other hand, the shear modulus is linearly increased with preshear rate at the same time as the first shear thinning and the shear thickening in the low shear rate region. The linear increase in the shear modulus is attributed to the increase in the density of the oily streak network composed of the focal conic domains. It is known that accumulated dislocations constitute the focal conic domains [39]. Increase in the dislocation loops results in FCD formation, i.e., the oily streak network. However, FCDs have unfavorable energy contribution originating from the layer compression because of the inhomogeneous layer spacing in narrow space [17]. Excess increase in the oily streak network density will trigger the realignment of the layer orientation from FCD with the negative Gaussian curvature to new state so that the inhomogeneous layer spacing is eliminated. The realignment might induce the buckling of layers and results in the onion phase with positive Gaussian curvature and homogeneous layer spacing.

Kato et al. recently observed a significant enhancement of the lamellar orientation with the layer normal along the velocity gradient direction [6]. This significant lamellar orientation prior to the onion formation may correspond to the realignment of the membrane from FCDs with negative curvature to the onion with positive one. Unfortunately, our results give no information on the exact structure of the intermediate state, which has been mentioned by several researchers [3, 6, 10]. There still remain open questions. In the above discussion, we focused on the critical shear stress of the shear thickening. However, the physical origin of the shear thickening is still unknown. Linear increase in G' in the first shear-thinning and the shear-thickening regions may indicate that the onion phase formation is not completed in the shear-thickening region. Scaling behavior of the critical shear stress, on the other hand, seems to show that the layer realignment occurs at the same time as the shear thickening. We still need other experimental studies in order to elucidate the origin of the shear thickening.

4.4 Conclusion

Lyotropic lamellar phase is one of the simplest phases in the soft matter systems. A variety of the nonequilibrium behavior under shear in such simple phase structure suggests how difficult it is to find out the universal behavior in the rheology of the soft matter systems. In this chapter, we have shown that the elastic properties of the bilayer membrane dominate the structural transition in the lyotropic lamellar phase. Understanding microscopic origin of the shear-induced structural transition will be useful for many applications, including a biorheological problem such as a rheology of a thrombocyte in a blood flow and rheology of a red blood cell. We have also found that the structural transition leads a significant variation in the rheological features, such as shear viscosity and modulus. Scaling laws we found on the shear modulus from the viewpoint of defect-mediated rheology well correlates the elasticity to the structural development. Small success for understanding the shear-induced onion phase formation in terms of the defect-mediated rheology will shed the light on not only the nonequilibrium structural transition of the lamellar phase under shear but also on all sorts of rheological phenomena in soft matter systems. Defect-mediated rheology is one of the possibilities to unify the rheology of soft matter systems and thus the topics of biorheology. Rheology of soft matter systems mediated by mesoscale structure such as defects is referred to as structural rheology. Structural rheology as a new concept of the soft matter physics will help us to find out keys in order to understand complicated phenomena in the biorheology.

References

1. Roux D, Nallet F, Diat O (1993) Rheology of lyotropic lamellar phases. Europhys Lett 24:53–58
2. Diat O, Roux D, Nallet F (1993) Effect of shear on a lyotropic lamellar phase. J Phys II France 3:1427–1452
3. Nettesheim F, Zipfel J, Olsson U, Renth F, Lindner P, Richtering W (2003) Pathway of the shear induced transition between planar lamellae and multilamellar vesicles as studied by time resolve scattering techniques. Langmuir 19:3603–3618
4. Medronho B, Brown J, Miguel MG, Schmidt C, Olsson U, Galvosas P (2011) Planar lamellae and onions: a spatially resolved rheo-NMR approach to the shear-induced structural transformations in a surfactant model system. Soft Matter 7:4938–4947
5. Ito M, Kosaka Y, Kawabata Y, Kato T (2011) Transition processes from lamellar to onion state with increasing temperature under shear flow in a nonionic surfactant/water system studied by rheo-SAXS. Langmuir 27:7400–7409
6. Kosaka Y, Ito M, Kawabata Y, Kato T (2010) Lamellar-to-onion transition with increasing temperature under shear flow in a nonionic surfactant/water system. Langmuir 26:3835–3842

7. Bergenholtz J, Wagner NJ (1996) Formation of AOT/Brine Multilamellar Vesicles. Langmuir 12:3122–3126
8. Courbin L, Delville JP, Rouch J, Panizza P (2002) Instability of a lamellar phase under shear flow: formation of multilamellar vesicles. Phys Rev Lett 89:148305-1–148305-4
9. Courbin L, Panizza P (2004) Shear-induced formation of vesicles in membrane phases: kinetics and size selection mechanisms, elasticity versus surface tension. Phys Rev E 69:021504-1–021504-12
10. Oliviero C, Coppola L, Gianferri R, Nicotera I, Olsson U (2003) Dynamic phase diagram and onion formation in the system C10E3/D2O. Colloids Surf A 228:85–90
11. Safinya CR, Sirota EB, Plano RJ (1991) Nematic to smectic-A phase transition under shear flow: a nonequilibrium synchrotron x-ray study. Phys Rev Lett 66:1986–1989
12. Panizza P, Archambault P, Roux D (1995) Effects of shear on the smectic A phase of thermotropic liquid crystals. J Phys II France 5:303–311
13. Fujii S, Ishii Y, Komura S, Lu C-YD (2010) Smectic rheology close to the smectic-nematic transition. Europhys Lett 90(64001):1–6
14. Zilman AG, Granek R (1999) Undulation instability of lamellar phases under shear: a mechanism for onion formation? Eur Phys J B 11:593–608
15. Oswald P, Ben-Abraham SI (1982) Undulation instability under shear in smectic A liquid crystals. J Phys France 43:1193–1197
16. Oswald P, Kleman M (1982) Lubrication theory of smectic A phases. J Phys Lett 43:L411–L415
17. Boltenhagen P, Lavrentovich O, Kleman M (1991) Oily streaks and focal conic domains in La lyotropic liquid crystals. J Phys II France 1:1233–1252
18. Kleman M, Lavrentovich OD (2003) Soft matter physics: an introduction. Springer, New York
19. Lu C-YD, Chen P, Ishii Y, Komura S, Kato T (2008) Non-linear rheology of lamellar liquid crystals. Eur Phys J E 25:91–101
20. Meyer C, Asnacios S, Kleman M (2001) Universal properties of lamellar systems under weak shear. Eur Phys J E 6:245–253
21. Basappa G, Sneel-Kumaran V, Nott PR, Remaswamy S, Naik VM, Rout D (1999) Structure and rheology of the defect-gel states of pure and particle-dispersed lyotropic lamellar phases. Eur Phys J B 12:269–276
22. Ramos L, Zapotocky M, Lubensly TC, Weitz DA (2002) Rheology of defect networks in cholesteric liquid crystals. Phys Rev E 66(031711):1–10
23. Medronho B, Rodrigues M, Miguel MG, Olsson U, Schmidt C (2010) Shear-induced defect formation in a nonionic lamellar phase. Langmuir 26:11304–11313
24. Dhez O, Nallet F, Diat O (2001) Influence of screw dislocations on the orientation of a sheared lamellar phase. Europhys Lett 55:821–826
25. Masui T, Imai M, Nakaya K, Taniguchi T (2006) Effects of grafted polymer chains on lamellar membranes. J Chem Phys 124:074904, 12 pp
26. Hiergeist C, Lipowsky R (1996) Elastic properties of polymer-decorated membranes. J Phys France 6:1465–1481
27. Fujii S, Mitusmasu D, Isono Y, Richtering W (2012) Shear-induced onion formation of polymer-grafted lamellar phase. Soft Matter 8:5381–5390
28. Bergenhausen J, Zipfel J, Lindner P, Richtering W (2001) Influence of water-soluble polymers on the shear-induced structure formation in lyotropic lamellar phases. J Phys Chem B 15:11081–11088
29. Yang BS, Russel WB, Prudhomme RK (2005) Effect of hydrophobically modified polymers on shear-induced multilamellar vesicles. Langmuir 21:10038–10045
30. Larson RG, Mead DW (1992) Development of orientation and texture during shearing of liquid-crystalline polymers. Liq Cryst 12:751–768
31. Yang BS, Lal J, Kohn J, Huang JS, Russel WB, Prud'homme RK (2001) Interaction of surfactant lamellar phase and a strictly alternating comb-graft amphiphilic polymer based on PEG. Langmuir 17:6692–6698

32. Yang BS, Lal J, Mihailescu M, Monkenbusch M, Richter D, Huang JS, Kohn J, Russel WB, Prud'homme RK (2002) Neutron spin-echo study of dynamics of hydrophobically modified polymer-doped surfactant bilayers. Langmuir 18:6–13
33. Nettesheim F, Zipfel J, Lindner P, Richtering W (2001) Influence of sodium dodecyl sulfate on the structure and rheology of aqueous solution of the nonionic surfactant tetraethylene glycol-monododecyl ether (C12E4). Colloids Surf A 183–185:563–574
34. Ligoure C, Bouglet G, Porte G (1993) Polymer induced phase separation in lyotropic smectics. Phys Rev Lett 71:3600–3603
35. Castro-Roman F, Porte G, Ligoure C (2001) Smectic phase of fluid membranes decorated by amphiphilic copolymers. Langmuir 17:5045–5058
36. Warriner HE, Keller SL, Idziak SH, Slack NL, Davidson P, Zasadzinski JA, Safinya CR (1998) The influence of polymer molecular weight in lamellar gels based on PEG-lipids. Biophys J 75:272–293
37. Imai M, Mawatari R, Nakaya K, Komura S (2004) Inter-lamellar interactions modulated by addition of guest components. Eur Phys J E 13:391–400
38. Fujii S, Mitsumasu D, Isono Y (2013) Shear-induced onion formation of triblock copolymer-embedded surfactant lamellar phase. J Soc Rheol Jpn 41:29–34
39. Williams CE, Kleman M (1975) Dislocations, grain boundaries and focal conics in smectic A. J Phys Colloq 36:C1-315–C1-320

Diffusion and Thermal Diffusion by Means of Dynamic Light Scattering and Laser Holography

5

Rio Kita

Abstract

Diffusion processes of biological molecules in a medium are important subjects for Nano/Micro Biorheology. Developments of the laser as a light source have made a considerable impact in the field of spectroscopy, owing to its high intensity and coherence where the diffusion processes have been studied extensively with a high precision by many researchers. In this chapter, two types of diffusion are reviewed. Firstly, self-diffusion under a homogeneous temperature condition, known as Fick's diffusion, is shown where the diffusion coefficient of biomaterials is measured by dynamic light scattering (DLS). Second, thermal diffusion, also called the Ludwig-Soret effect, is followed-up to study biological molecules and related water-soluble polymers. Thermal diffusion occurs when the system is in nonequilibrium thermodynamically, where a temperature gradient induces a mass diffusion in the system. The experiments for studying the thermal diffusion mentioned in this chapter involve a holographic method of laser interferometry. Characterizations of these diffusion processes are related to insights on the molecular interactions between biological molecules and solvents and are utilized to understand their unique behaviors.

Keywords

Dynamic light scattering • Diffusion • Temperature gradient • Ludwig-Soret effect

R. Kita (✉)
Department of Physics, Tokai University, 4-1-1 Kitakaname,
Hiratsuka, Kanagawa 259-1292, Japan
e-mail: rkita@keyaki.cc.u-tokai.ac.jp

R. Kita, T. Dobashi (eds.), *Nano/Micro Science and Technology in Biorheology*,
DOI 10.1007/978-4-431-54886-7_5

5.1 Introduction

Dynamic light scattering (DLS) is one of the standard methods of studying the dynamics of polymers in solvents. In practice, it is often utilized to study particles with size in the nanometer range. In addition to simple spherical particles, DLS is a powerful method for studying complex systems, e.g., slow dynamics of a network structure of gels, characterizations of the anisotropic structure of aggregates, and critical concentration fluctuations. There are numerous papers and books regarding DLS [1–4] owing to the wide range of applications of this method. Thus, in this chapter, the fundamental aspects of DLS experiments and applications of DLS to studying the gel formation process are reviewed as in the first part. The gel reviewed in this chapter is the fibrin gel, which is known as the primary structural component of blood clots.

In the second part, transport phenomena of solute molecules in a solution under a stable temperature gradient condition over the system will be reviewed, where the mixture is assumed to be confined to closed reservoirs. One well-known nonequilibrium effect is the Ludwig-Soret effect, also called thermal diffusion, which describes the migration of molecules in a mixture caused by the presence of a thermal gradient [5, 6]. This effect plays a crucial role in many processes in nature, although the role in biological organisms has not been clarified sufficiently. The Ludwig-Soret effect has been known for a long time, and its phenomenological description has been developed in the framework of nonequilibrium thermodynamics [7]. Recently, studies to clarify the molecular mechanisms of the Ludwig-Soret effect for biological materials have been carried out extensively in dilute aqueous solutions. One of the significant features of biological materials such as saccharide, DNA, and proteins is the negative Soret coefficient, which means that the solute molecules migrate to the hot side of fluid. This observation contradicts the generally known idea that the heavier component migrates to the cold region of the fluid. In the second section, recent developments in the study of the Ludwig-Soret effect for aqueous solutions of biopolymers will be described.

5.2 Diffusion Experiments by Means of Dynamic Light Scattering

5.2.1 Basics of Scattered Light

Light is a nonperturbative probe to obtain information about the structure and dynamics of molecules. Maxwell's equations describe a transverse electromagnetic wave that oscillates in both space and time. The electric field of light $\mathbf{E}_i(\mathbf{r}, t)$ at location $\mathbf{r}$ and time t has the form

$$\mathbf{E}_i(\mathbf{r}, t) = \mathbf{n}_i E_0 \exp\{i(\mathbf{k}_i \cdot \mathbf{r})\}\exp(-i\omega_i t), \tag{5.1}$$

where $\mathbf{n}_i$ is a unit vector in the direction of the incident electric field, E_0 is the amplitude, $\mathbf{k}_i$ is the wave vector, and ω_i is the angular frequency. The distance between adjacent maxima of waves defines the wavelength λ_0 of light, and the magnitude of the wave vector is defined as $|\mathbf{k}_i| = 2\pi n/\lambda_0$, where n is the index of refraction of the medium. The angular frequency ω_i is related to the speed of light c as $\omega_i = 2\pi c/\lambda_0 = |\mathbf{k}_i|c/n$.

In the medium, a local dielectric constant $\boldsymbol{\varepsilon}(\mathbf{r}, t)$ is described by the average dielectric constant ε_0 ($\varepsilon_0 = n^2$) while assuming a nonabsorbing, nonconducting, and nonmagnetic medium,

$$\boldsymbol{\varepsilon}(\mathbf{r}, t) = \varepsilon_0 \mathbf{I} + \delta\boldsymbol{\varepsilon}(\mathbf{r}, t). \tag{5.2}$$

Here, $\delta\boldsymbol{\varepsilon}(\mathbf{r}, t)$ implies the fluctuation of the dielectric constant, and $\mathbf{I}$ is the unit tensor.

The component of the scattered electric field $E_s(R, t)$ at a detector with a large distance R from the scattering volume is expressed as

$$E_s(R, t) = -\frac{{q_s}^2 E_0}{4\pi R \varepsilon_0} \exp\{i(q_s R - \omega_i t)\} \int_V d\mathbf{r} \exp\{i(\mathbf{q} \cdot \mathbf{R})\}[\mathbf{n}_s \cdot \delta\boldsymbol{\varepsilon}(\mathbf{r}, t) \cdot \mathbf{n}_i] \tag{5.3}$$

$$= -\frac{{q_s}^2 E_0}{4\pi R \varepsilon_0} \exp\{i(q_s R - \omega_i t)\} \delta\varepsilon_{is}(\mathbf{r}, t), \tag{5.4}$$

where $\delta\varepsilon_{is}(\mathbf{q}, t) = \mathbf{n}_s \cdot \delta\boldsymbol{\varepsilon}(\mathbf{q}, t) \cdot \mathbf{n}_i = \int_V \mathrm{d}\mathbf{r} \exp\{i(\mathbf{q} \cdot \mathbf{R})\}[\mathbf{n}_s \cdot \delta\boldsymbol{\varepsilon}(\mathbf{r}, t) \cdot \mathbf{n}_i]$ is used with the definition of $\delta\boldsymbol{\varepsilon}(\mathbf{q}, t) \equiv \int_V d\mathbf{r} \exp\{i(\mathbf{q} \cdot \mathbf{R})\} \delta\boldsymbol{\varepsilon}(\mathbf{r}, t)$.The time correlation function of the scattered electric field is defined as $G^{(1)}(\tau) \equiv \langle {E_s}^*(R, 0) E_s(R, \tau) \rangle$ and is obtained as

$$G^{(1)}(\tau) = \frac{{q_s}^4 I_0}{16\pi^2 R^2 {\varepsilon_0}^2} \exp(-i\omega_i \tau) \langle {\delta\varepsilon_{is}}^*(\mathbf{q}, 0) \delta\varepsilon_{is}(\mathbf{q}, \tau) \rangle, \tag{5.5}$$

$$I_0 = |E_0|^2. \tag{5.6}$$

Here, $\langle \cdots \rangle$ means the average over them in time, which in turn is

$$\langle E^*(R, 0) E(R, t) \rangle = \lim_{T \to \infty} \frac{1}{T} \int_{-\frac{T}{2}}^{\frac{T}{2}} E^*(R, t) E(R, t + \tau) \mathrm{d}t. \tag{5.7}$$

The scattered light intensity $I_s(t)$ at the detector with time t is the square of the electric field, $I_s(t) = |E_s(R, t)|^2$; thus, it is described with a power spectrum

$$I(\mathbf{q}, \omega_s, R) = \frac{{q_s}^4 I_0}{16\pi^2 R^2 {\varepsilon_0}^2} \frac{1}{2\pi} \int_{-\infty}^{\infty} \mathrm{d}t \langle {\delta\varepsilon_{is}}^*(\mathbf{q}, 0) \delta\varepsilon_{is}(\mathbf{q}, \tau) \rangle \exp\{i(\omega_s - \omega_i)t\}. \tag{5.8}$$

For studies of biological materials in solvents, the polarization $\boldsymbol{\alpha}$ is useful instead of the dielectric constant. When interactions among the molecules are assumed to be negligible, the polarization tensor α_j of the j-th component has the form

$$\delta\varepsilon_{\mathrm{is}}(\mathbf{r},t)=\sum_j \alpha_{j,\mathrm{is}}(t)\delta\big(\mathbf{r}-\mathbf{r}_j(t)\big). \tag{5.9}$$

The number density of molecules $\rho(\mathbf{r},t)$ is equivalent to $\sum_j \delta\big(\mathbf{r}-\mathbf{r}_j(t)\big)$; thus, the fluctuation of density $\delta\rho=\rho-\langle\rho\rangle$ holds:

$$\delta\rho(\mathbf{r},t)=\int d\mathbf{r}\delta\rho(\mathbf{r},t)\exp(i\mathbf{q}\cdot\mathbf{r}). \tag{5.10}$$

The dynamic structural factor $S(\mathbf{q},\tau)=\langle\delta\rho^*(\mathbf{q},0)\delta\rho(\mathbf{q},\tau)\rangle$ is then related to the correlation function of the scattered electric field by combining Eqs. (5.5), (5.9), and (5.10) as

$$G^{(1)}(\tau)=\frac{{q_{\mathrm{s}}}^4 I_0}{16\pi^2R^2{\varepsilon_0}^2}\exp(-i\omega_i\tau)\langle\alpha_{j,\mathrm{is}}(0)\alpha_{k,\mathrm{is}}(\tau)\rangle\langle\delta\rho^*(\mathbf{q},0)\delta\rho(\mathbf{q},\tau)\rangle, \tag{5.11}$$

$$G^{(1)}(\tau)=A\exp(-i\omega_i\tau)\langle\alpha_{j,\mathrm{is}}(0)\alpha_{k,\mathrm{is}}(\tau)\rangle S(\mathbf{q},\tau). \tag{5.12}$$

Here, the proportionality constant is substituted as

$$A=\frac{{q_{\mathrm{s}}}^4 I_0}{16\pi^2R^2{\varepsilon_0}^2}. \tag{5.13}$$

The Fourier transformation of $S(\mathbf{q},\tau)$ is

$$S(\mathbf{q},\omega)=\frac{1}{2\pi}\int_{-\infty}^{\infty}S(q,\tau)\exp(-i\omega t)d\tau, \tag{5.14}$$

which is also called the dynamic structural factor. It is revealed that the correlation function is the inverse Fourier transform of the dynamic structural factor.

5.2.2 Photon Correlation Spectroscopy

The photon correlation spectroscopy measures the fluctuation of photocounts, which involves information of molecular motion. The scattered light intensity is the square of electric field ($I_s(t)=|E_{\mathrm{s}}(R,t)|^2$). When the scattering due to solvent molecules is assumed to be negligible and a single point detection at the detector surface with perfect coherence of scattered light is also assumed, the autocorrelation function of the intensity-intensity correlation function of the scattered light is written as $\langle I(t)I(t+\tau)\rangle = \langle I(0)I(\tau)\rangle \equiv G^{(2)}(\tau)$, where the starting time of data acquisition is arbitrary. Thus, the intensity-intensity correlation function $G^{(2)}(\tau)$ is

$$G^{(2)}(\tau) = \langle \mathbf{E}^*(0)\mathbf{E}(0)\mathbf{E}^*(\tau)\mathbf{E}(\tau)\rangle \tag{5.15}$$

$$= \langle \mathbf{E}^*(0)\mathbf{E}(0)\rangle\langle \mathbf{E}^*(\tau)\mathbf{E}(\tau)\rangle + \langle \mathbf{E}^*(0)\mathbf{E}(\tau)\rangle\langle \mathbf{E}^*(\tau)\mathbf{E}(0)\rangle \tag{5.16}$$

$$= \left|G^{(1)}(0)\right|^2 + \left|G^{(1)}(\tau)\right|^2 \tag{5.17}$$

$$= \left|G^{(1)}(0)\right|^2\left(1 + \left|g^{(1)}(\tau)\right|^2\right). \tag{5.18}$$

Here, $g^{(1)}(\tau) = G^{(1)}(\tau)/G^{(1)}(0)$ is the normalized correlation function of the scattered electric field. The normalized correlation function of the scattered light $g^{(2)}(\tau)$ is expressed as

$$g^{(2)}(\tau) = \frac{\langle I(0)I(\tau)\rangle}{\langle I(0)\rangle^2} = \frac{G^{(2)}(\tau)}{\left|G^{(1)}(0)\right|^2} \tag{5.19}$$

$$= 1 + \left|g^{(1)}(\tau)\right|^2. \tag{5.20}$$

Compared with Eq. (5.12), $g^{(1)}(\tau)$ is equivalent to the dynamic structural factor $S(\mathbf{q},\tau)$. Equation (5.20) is known as the Siegert relation.

5.2.3 Diffusion in Dilute Solution

To understand the decay behavior of diffusive particles (Brownian particles), a dilute solution in which the particle obeys Fick's second law of diffusion is assumed. Particles move more slowly than solvent molecules and contribute to a fluctuating field at the detector. The scattering function is related to the probability distribution $G_s(\mathbf{R},t)$ of particles with displacement $\mathbf{R}$ of particles in time t,

$$\frac{\partial}{\partial t}G_s(\mathbf{R},t) = D\nabla^2 G_s(\mathbf{R},t). \tag{5.21}$$

Here, D is the coefficient of self-diffusion of particles. The diffusing particles are expressed with the form $F_s(\mathbf{q},t) = \langle \exp[i\mathbf{q}\cdot\{\mathbf{r}_i(t) - \mathbf{r}_i(0)\}]\rangle$, which is related to the probability distribution

$$G_s(\mathbf{R},t) = \langle \delta[\mathbf{R} - \{\mathbf{r}_i(t) - \mathbf{r}_i(0)\}]\rangle, \tag{5.22}$$

and the spatial Fourier transform of $G_s(\mathbf{R},t)$ has the form $\int d^3R \exp(i\mathbf{q}\cdot\mathbf{R})$ $\langle \delta[\mathbf{R} - \{\mathbf{r}_i(t) - \mathbf{r}_i(0)\}]\rangle$.

Thus, the inverse Fourier transform gives

$$G_s(\mathbf{R}, t) = \frac{1}{(2\pi)^3}\int d^3R\exp(-i\mathbf{q}\cdot\mathbf{R})F_s(\mathbf{q}, t). \tag{5.23}$$

By comparison with Eq. (5.21), the following equation is obtained:

$$\frac{\partial}{\partial t}F_s(\mathbf{q}, t) = -q^2DF_s(\mathbf{q}, t). \tag{5.24}$$

The solution of the equation is

$$g^{(1)}(q, t) \propto F_s(\mathbf{q}, t) = \exp(-q^2Dt) \tag{5.25}$$

$$= \exp(-t/\tau_D). \tag{5.26}$$

Here, τ_D is the decay time $\tau_D = (q^2D)^{-1}$. This implies that the decay behavior of the correlation function depends on the scattering vector. That is, the plot of $1/\tau_D$ vs q^2 gives a straight line for Brownian particles under a noninteractive condition. The inverse of decay time $\Gamma(= 1/\tau_D)$ is called the decay rate. In practice, in the recent setup of the DLS experiment, homodyne detection of photocounting is often adopted, and the real system is not perfectly coherent owing to the scattering volume in the sample and a finite detector area. In this case, Eq. (5.20) is derived,

$$g^{(2)}(\tau) = B\left\{1 + \beta\left|g^{(1)}(\tau)\right|^2\right\}, \tag{5.27}$$

where B is a baseline, and β refers to a coherency factor depending on the setup.

For a spherical particle with radius a and made of optically isotropic material, the Stokes-Einstein relation is useful for determining radius a from diffusion coefficient D:

$$D = \frac{k_BT}{6\pi\eta a}, \tag{5.28}$$

where k_B is the Boltzmann's constant, and η is the viscosity of the solvent. In the case of a randomly coiled polymer chain in solvent, the hydrodynamically equivalent radius R_H is obtained instead of particle radius a.

The experimental setups for carrying out dynamic light scattering are commercially available, and the recent suppliers are, for instance, ALV, Brookhaven, Horiba, Malvern, Otsuka, and Wyatt. The conventional setup for static light scattering can be utilized for photon counting experiments, where a correlator is necessary to obtain the correlation function.

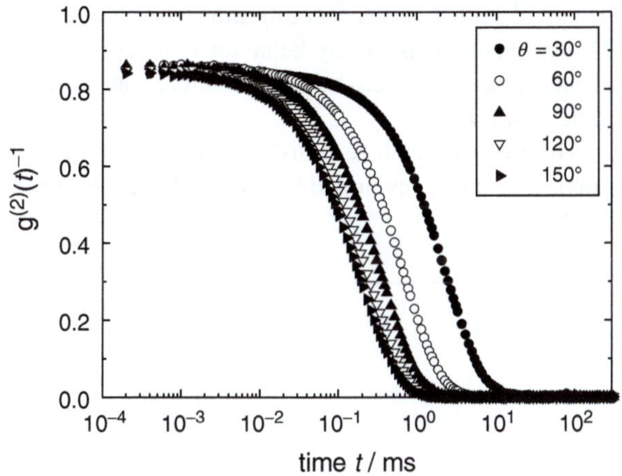

Fig. 5.1 Correlation function $g^{(2)}(t)$ of PNIPAM in ethanol (1.0 gL^{-1}) at 20.0 °C

5.2.4 Diffusion of Polymer at Uniform Temperature

Figure 5.1 shows the correlation function of scattered light $g^{(2)}(t)$ for poly(*N*-isopropylacrylamide) [PNIPAM] in ethanol (1.0 g/L). In an ideal case, the decay behavior of the correlation function is obtained by the fitting using the exponential function:

$$g^{(1)}(t) = \exp(-t/\tau_D). \tag{5.29}$$

For samples with polydispersion of particle size, the single exponential form gives no good fitting results. Thus, the cumulant expansion is often used to obtain the average decay time and the distribution factors of decay time.

$$\ln\left|g^{(1)}(t)\right| = -K_1 t + (K_2/2!)t^2 - (K_3/3!)t^3 \cdots \tag{5.30}$$

Here, K_1 corresponds to the average decay rate (Γ), and the second term $K_2/2$ and the third term $K_3/6$ correlate respectively with the variance and the skewness of the distribution.

For a high-concentration sample of polymer solutions, the sum of two exponential functions is often used, or the stretched exponential form is also useful to obtain decay time τ_D.

$$g^{(1)}(t) = \exp\left[-(t/\tau_D)^\beta\right] \tag{5.31}$$

where β ($0 < \beta \leq 1$) is the indicator of the broadness of distribution. In the case of gels, the power law relation gives good fitting results in the long delay time region.

$$g^{(1)}(t) = (1 + t/\tau')^{-\phi} \tag{5.32}$$

where τ' and ϕ are the cutoff time and the power law exponent, respectively. This means that the decay behavior is independent of the time scale, i.e., the gel cluster obtains a self-similar structure on a large scale compared with the molecular size.

The inverse Laplace transformation program (constrained regularization program, CONTIN) developed by Provencher is also used to fit the correlation function [8, 9]

$$g^{(1)}(t) = \int G(\tau)\exp(-t/\tau)\mathrm{d}\tau \tag{5.33}$$

$$= \int \tau G(\tau)\exp(-t/\tau)\mathrm{d}(\ln\tau) \tag{5.34}$$

with $\int G(\tau)\mathrm{d}\tau = 1$. This indicates that $g^{(1)}(t)$ is the Laplace transformation of the distribution function of decay time $G(\tau)$. It is necessary, in experiments and analyses, to consider the appropriate function to fit the correlation function, and preliminary information on the samples such as shape and concentration will be helpful in selecting the functions. It is also necessary to consider the usage of the CONTIN program whether or not the reliable fitting results are obtained.

Figure 5.2 shows the plot of Γ vs q^2 of PNIPAM in ethanol. It is a straight line intersecting the origin. The decay rate Γ is obtained by a least-squares fitting using Eq. (5.30) for the correlation function shown in Fig. 5.1. According to Eqs. (5.25) and (5.26), the slope in Fig. 5.2 corresponds to the diffusion coefficient of PNIPAM in ethanol. The example of the distribution function of decay time $G(\tau)$ in Eq. (5.33), obtained by the CONTIN program, is shown in Fig. 5.3.

5.2.5 Gel Formation Process

Studies of the dynamic process of network formation (gelation) are important for revealing the structure and properties of gels, since the structure of a three-dimensional network depends on the kinetics of junction formation. Fibrin gel is composed of associated fibrins and is the primary structural component of blood clots. Fibrinogen is a rod-shaped protein with a molecular weight of 3.4×10^5 and plays essential roles in various pathological processes, including hemostasis, thrombosis, and adhesion and aggregation of platelets.

The characteristics of the sol-to-gel transition obtained so far by the light scattering method can be summarized as follows: (1) power law behavior in $g^{(2)}(t)$ at the limit of decrease in β and increase in τ of the stretched exponential formula in Eq. (5.31), (2) appearance of a long-time tail in $G(\tau)$, (3) beginning of the decrease in the coherence factor in $g^{(2)}(t)$ (nonergodicity), and (4) rapid increase in scattered light intensity and its oscillating behavior. (1) to (3) are related

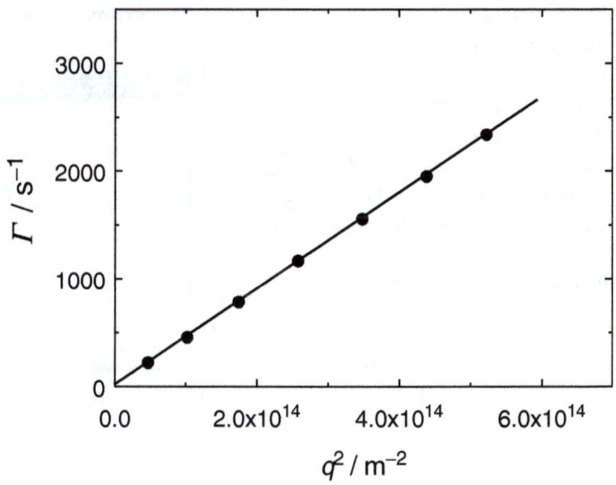

Fig. 5.2 The decay rate Γ vs q^2 of PNIPAM in ethanol ($1.0\ gL^{-1}$) at 20.0 °C

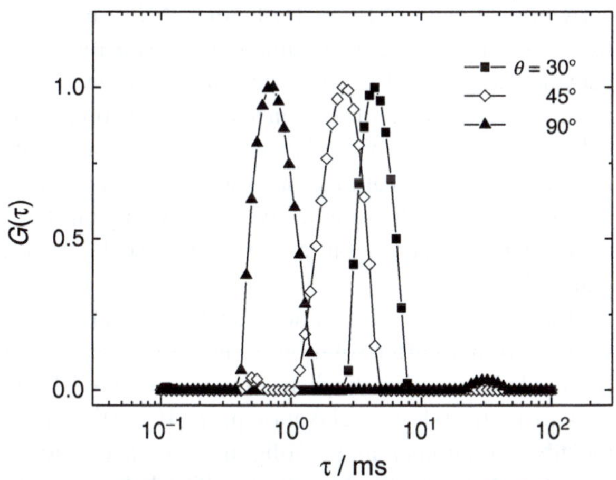

Fig. 5.3 Decay time distribution function of PNIPAM in ethanol ($1.0\ gL^{-1}$) at 20.0 °C

to dynamic light scattering and indicate the utility of the dynamic light scattering method in the study of sol-to-gel transition phenomenon [10].

Figure 5.4 shows typical double logarithmic plots of the correlation function $g^{(2)}(t) - 1$ as a function of delay time t for 2.2 g/L fibrinogen in a buffer solution [11]. The gelation process is initiated by the addition of thrombin that cleaves peptide fragments on fibrinogen. The numbers in the figure indicate the elapsed time t_e (min) of the gel formation process after the addition of thrombin. The existence of a characteristic decay time less than 220 min means that the relaxation time of a growing fibrin cluster is finite, i.e., the system is in the sol state. However, $g^{(2)}(t)$ begins to show a long-time tail in the range of long delay time with a gradual decrease in β in Eq. (5.31) at elapsed time $t_e > 260$ min. For $t_e < 260$ min, the stretched exponential formula shows good agreement with the long-time tail of the

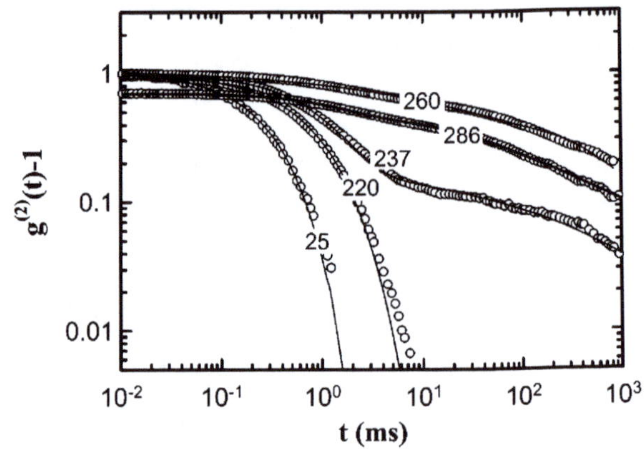

Fig. 5.4 Double logarithmic plots of the correlation function $g^{(2)}(t) - 1$ as a function of the delay time t for fibrinogen solution measured at $\theta = 30°$ with the wavelength of incident laser operated at $\lambda = 488$ nm. The numbers in the figure refer to the elapsed time of gelation process in minutes (Reprinted with permission from Kita et al. [11]. Copyright 2002 American Chemical Society)

correlation function. In contrast, for $t_e \geq 286$ min, the power law relation can be fitted well in the long delay time region. This means that the decay at $t_e \geq 286$ min became independent of the time scale, i.e., the clusters composed of fibrin fibers attain a self-similar structure on a large scale compared with the molecular size. The behavior at $t_e \geq 286$ min indicates that the formation of the fibrin gel is completed at this time. The initial amplitude of $g^{(2)}(t)$, the coherence factor, is an indicator of the ergodic nature of the sample. When some frozen inhomogeneity exists, as a result of cluster formation, on a very large scale, this amplitude should decrease markedly.

The decay time distribution function $\tau G(\tau)$ for the gelling process of 2.2 g/L fibrinogen in a buffer solution is shown in Fig. 5.5 [11]. The peak of $\tau G(\tau)$ of the fast mode varies from the fastest relaxation (3×10^{-1} to 10^0 ms) with a sharp single peak to another slower relaxation peak (10^0-10^1 ms). The protofibril formation from the fibrin monomer and/or oligomer is clearly identified by the appearance of crossover between the two modes. The behaviors of $\tau G(\tau)$ for those initial time regions of $t_e < 220$ min correspond to the molecular weight of the polymer becoming greater with time; that is, fibrin oligomers become larger, and the formation of protofibrils proceeds. Therefore, substantial entanglements among protofibrils are generated at an elapsed time of $t_e = 220$ min, and the cooperative diffusional motion due to entanglements appears. At $t_e = 237$ min, another peak appears at $\tau = 3 \times 10^2$-10^3. These results indicate that the lateral aggregation of protofibrils occurs after sufficient protofibril formation. The peak at the longest relaxation time shifts toward a slower relaxation time for $237 \leq t_e \leq 286$ min. This suggests that lateral aggregation to form fibers occurs and that the network formation proceeds by fiber growth due to the lateral aggregation of protofibrils. At $t_e = 286$ min, $\tau G(\tau)$ showed the slowest relaxation mode, and the time corresponded to the gelation time determined by the crossover behavior of $g^{(2)}(t)$. The slowest peak of $\tau G(\tau)$ should originate from the gelling clusters.

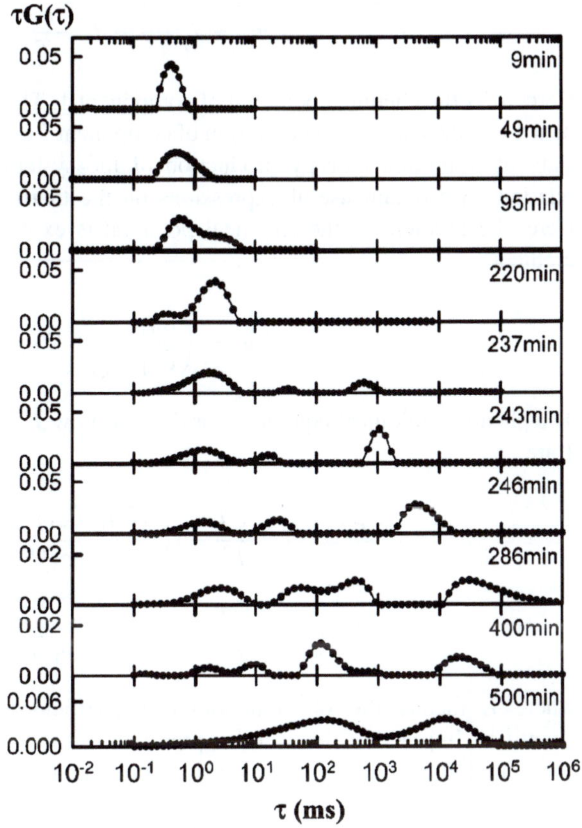

Fig. 5.5 Decay tune distribution function of fibrinogen in a buffer as a function of elapsed time of gelation process (Reprinted with permission from Kita et al. [11]. Copyright 2002 American Chemical Society)

5.3 Diffusion Under Temperature Gradient

5.3.1 Thermal Diffusion

In this section, we consider the diffusion process under a nonuniform temperature condition over the system where the mixture is assumed to be confined to closed reservoirs. For simplicity, a binary mixture with a stable temperature gradient is considered. Here, the pressure of the system is assumed to be uniform and convection is neglected. When a stable temperature gradient is applied into the fluid mixture, a concentration gradient arises. This is called thermal diffusion, also referred as the Ludwig-Soret effect, and thermodiffusion [7]. According to the phenomenological treatment of irreversible thermodynamics, the local entropy production σ is related to the heat flow $\boldsymbol{J}_q'$ and the flux $\boldsymbol{J}_1$ of one of the components in the binary mixture:

$$\sigma = -J_q^{'}\frac{\nabla T}{T^2} - J_1\frac{\nabla \mu_1}{c_2 T}. \tag{5.35}$$

Here, μ_1 is the chemical potential of component 1. The concentration c_1 of component 1 is related to the concentration of component 2 c_2 as $c_1 + c_2 = 1$. Owing to the state of minimum entropy production of the system, using Onsager reciprocal relations, we obtain useful expressions on the thermal diffusion for the system. Here, the gradient of the chemical potential is expressed with the concentration gradient

$$\nabla \mu_1 = \left(\frac{\partial \mu_1}{\partial c_1}\right)_{p,T,c_2} \nabla c_1. \tag{5.36}$$

The phenomenological equations for the heat flow J_q' and the flux J_1 of component 1 are

$$J_q^{'} = -L_{qq}\frac{\nabla T}{T^2} - L_{q1}\frac{(\partial \mu_1/\partial c_1)_{p,T}}{(1-c_1)T}\nabla c_1, \tag{5.37}$$

$$J_1 = -L_{1q}\frac{\nabla T}{T^2} - L_{11}\frac{(\partial \mu_1/\partial c_1)_{p,T}}{(1-c_1)T}\nabla c_1, \tag{5.38}$$

where L means the phenomenological coefficients for respective forces or quantities. In practice, the useful expressions of these coefficients are the diffusion coefficient D and the thermal diffusion coefficient D_{T}, which are expressed as

$$D = \frac{L_{11}}{\rho c_2 T}\left(\frac{\partial \mu_1}{\partial c_1}\right)_{p,T}, \tag{5.39}$$

$$D_{\mathrm{T}} = \frac{L_{1q}}{\rho c_1 c_2 T^2}. \tag{5.40}$$

The equation for the transport of component 1 is recalled:

$$J_1 = -\rho D \nabla c_1 - \rho c_1 c_2 D_T \nabla T. \tag{5.41}$$

At the steady state, the concentration gradient is developed in the binary mixture of interest, where the flux of components apparently vanishes ($J_1 = 0$). This leads to the expression of the magnitude of the concentration gradient generated by the temperature gradient as

$$\nabla c_1 = -\frac{c_1 c_2 D_T}{D}\nabla T. \tag{5.42}$$

Here, the Soret coefficient $S_{\mathrm{T}} \equiv D_{\mathrm{T}}/D$ is often used to study the thermal diffusion:

$$S_T \equiv \frac{D_T}{D} = \frac{1}{c_2 T (\partial \mu_1 / \partial c_1)_{p,T}} \frac{L_{1q}}{L_{11}}. \tag{5.43}$$

In the above explanations, the Dufour effect, i.e., the Dufour coefficient $L_{q1}/(\rho c_1 c_2 T^2)$, is neglected because the effect conventionally is small.

There are several experimental methods for measuring the Soret and thermal diffusion coefficients, for example, laser beam deflection, thermogravitational column, thermal field flow fractionation, and thermal diffusion forced Rayleigh scattering. The development of experimental methods has been promoted by many researchers, and recent reviews on the apparatus and principles of measuring thermal diffusion coefficient can be found elsewhere [12–14]. In the following section, the experimental results of thermal diffusion forced Rayleigh scattering (TDFRS) are primarily introduced. The details of the TDFRS method can be found in references [15, 16].

5.3.2 Thermal Diffusion of Aqueous Systems

Thermal diffusion of aqueous solutions of biopolymers is a new research target in Nano/Micro Biorheology. Many studies of molecular dynamics have been carried out from the aspects of thermodynamics and statistical physics where understanding of the behavior of biological molecules has been thoroughly examined. However, most processes occurring in everyday life are nonequilibrium phenomena, which have attracted researchers for a long time. In this section, we focus on nonisothermal systems to study the transport behavior of biological molecules and related materials. Here, we focus on the effect of a stable temperature gradient in liquids. One well-known nonequilibrium effect is the Ludwig-Soret effect, also called thermal diffusion or thermophoresis. Although the Ludwig-Soret effect has been known for a long time [5, 6], for example, the Soret coefficient of PVA in an aqueous system has been reported [17], the understanding at molecular levels of the Ludwig-Soret effect in liquid mixtures and polymer solutions is still lacking.

5.3.2.1 Water/Ethanol Mixture

A binary mixture composed of low molecular weight substances is one of the basic systems for studying the Ludwig-Soret effect. The Soret coefficient of water in ethanol has been obtained by some groups independently [18–29]. The Soret coefficient of water in ethanol (or ethanol in water) depends on the concentration of the mixture (Fig. 5.6). The Soret coefficients experimentally obtained by several techniques agree with each other; a sign change of the Soret coefficient is observed at the weight fraction of water of 0.7 in the investigated temperature range of 20–25 °C. For a water-rich composition, the Soret coefficient of water is negative, which means that the water molecules migrate to the hot side of the mixture, and ethanol molecules migrate to the cold side. For water-poor compositions, the Soret coefficient of water is positive, implying that water molecules migrate to the cold side of the mixture, while ethanol molecules simultaneously migrate to the hot side.

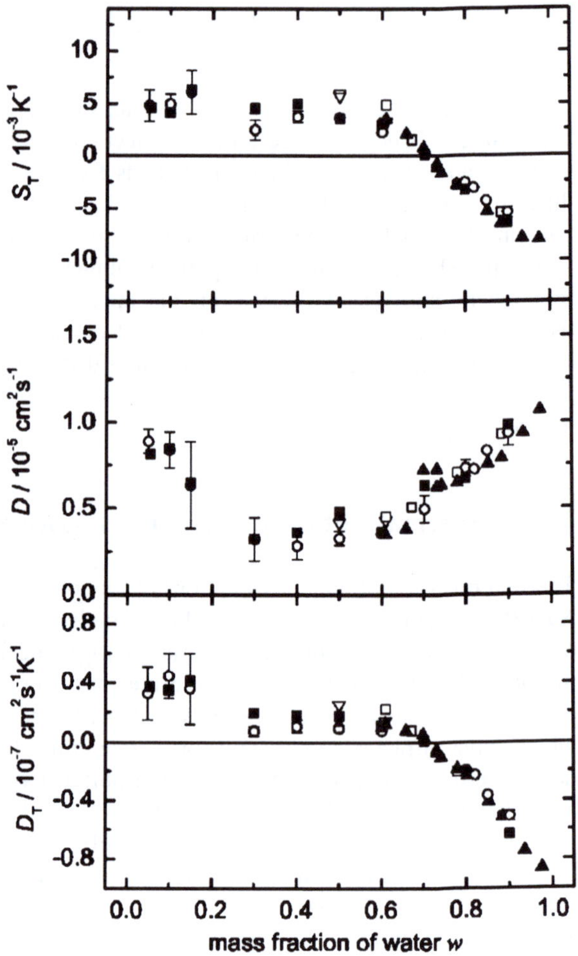

Fig. 5.6 The Soret coefficient, diffusion coefficient, and thermal diffusion coefficient of water in ethanol as a function of mass fraction of water w (Reprinted with permission from Kita et al. [20]. Copyright 2004, AIP Publishing LLC)

Theoretical models in which the molecular associations are taken into account predict this behavior successfully. The thermodynamic quantities, such as the thermal expansion or the activation energy, relating to viscosity show a correlation with the Soret coefficient. It is also mentioned that hydrogen bonds play a key role for such an associated mixture of water and polar molecules.

5.3.2.2 PEG in Water

Poly(ethylene glycol) [PEG] in water is a standard system for studying the solution properties of water-soluble polymers. Characterizations of PEG as well as the phase behavior of PEG solutions have been carried out extensively [30, 31]. The Ludwig-Soret effect of PEG in water has been studied by Chan et al. as a function of molecular weight [32]. Figure 5.7 shows the Soret coefficients s (K^{-1}) (in weight) and σ (K^{-1}) (in mole), the thermal diffusion coefficient D_T (m^2 s^{-1} K^{-1}), and the

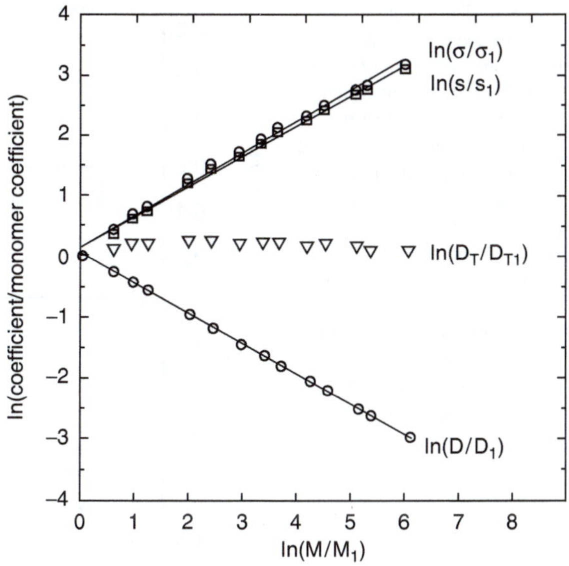

Fig. 5.7 The Soret coefficients s (K^{-1}) (in weight) and σ (K^{-1}) (in molar), the thermal diffusion coefficient D_T ($m^2\ s^{-1}\ K^{-1}$), and the mutual diffusion coefficient D ($m^2\ s^{-1}$) of PEG in water with the concentration being 0.200 mol ethylene groups per kilogram of water. M_1 refers to the mass of ethylene glycol (Reprinted from Chan et al. [32]. The figure is originally published in *Journal of Solution Chemistry* and reuses with kind permission from Springer Science and Business Media)

mutual diffusion coefficient D ($m^2\ s^{-1}$) of PEG in water with the concentration of 0.200 mol ethylene groups per kilogram of water obtained by the thermogravitational cell. All of them are normalized by the respective values of ethylene glycol (monomer). The thermal diffusion coefficient is independent of the molecular weight of PEG. In contrast, the Soret coefficient and the mutual diffusion coefficient increase and decrease with increasing molecular weight, respectively. The obtained results of s and D show a relationship to the molecular weights $M^{0.53}$ and $M^{0.52}$, respectively. The signs of the Soret and the thermal diffusion coefficient are positive, implying that the solute PEG migrates to the cold side of the mixture. Giglio and Vendramini reported a negative Soret coefficient for poly(vinyl alcohol) in water [17]. These results mean that water-soluble polymers show both positive and negative thermal diffusion behaviors, although the systems of poly(styrene) [PS], poly(methylmethacrylate) [PMMA], and other organic polymers are known to show only a positive Soret coefficient [33–38]. In the scaling law, the end-to-end distance R and the number of repeating units of polymer n have the relation $\langle R^2 \rangle^{1/2} \sim n^{\nu}$. Here, the exponent ν is known to be from 1/2 for a Gaussian chain to Flory's exponent of 3/5. Owing to the scaling law and the Stokes-Einstein equation, the molecular weight dependence of the Soret coefficient is expected to have the relation $S_T \sim n^{\nu}$. This is ascertained by experiments using PS solutions in regard to the hydrodynamic radius R_h and the molecular weight of polymer [33, 37, 39].

de Gans et al. showed the Soret coefficient of poly(ethylene oxide) [PEO] in the mixed solvent, water/ethanol [20, 40–42]. The Ludwig-Soret effect is investigated at low concentrations of PEO (1.0 and 5.0 gL^{-1} PEO in the solvent) as a function of

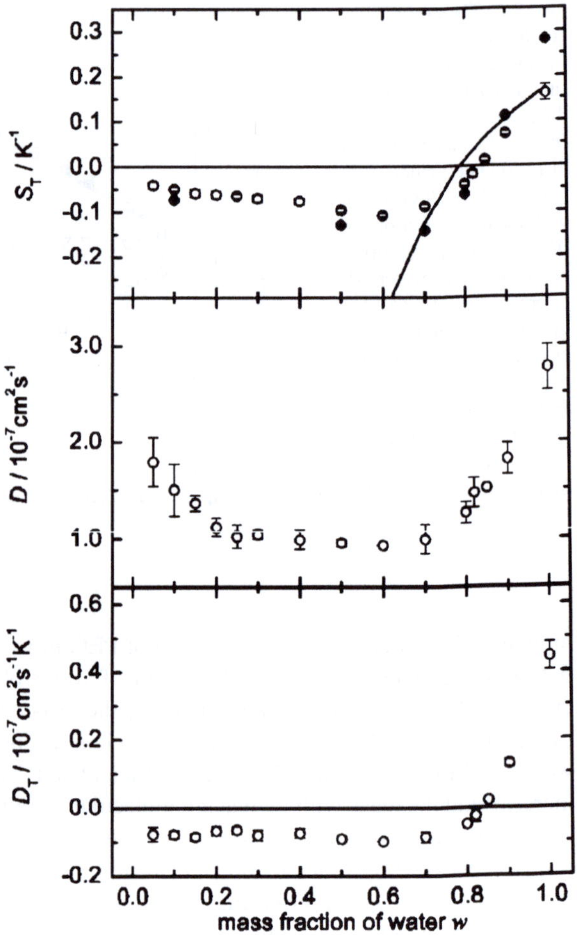

Fig. 5.8 The Soret, diffusion, and thermal diffusion coefficient of PEO in the mixed solvent, water/ethanol plotted against the weight fraction of water of the mixed solvent. Filled and open symbols refer to the polymer concentration of 1.0 gL^{-1} and 5.0 gL^{-1}, respectively. The *curve* refers to the results of a model calculation (Reprinted with permission from Kita et al. [20]. Copyright 2004, AIP Publishing LLC)

the composition of the water/ethanol mixture. As shown in Fig. 5.8, the Soret coefficient of PEO is positive in the water-rich range of the mixed solvent, i.e., PEO migrates to the cold side, similar to the results of Chan [32]. With increasing ethanol concentration, the Soret coefficient decreases and becomes zero at the weight fraction of 0.83 for water in the water/ethanol mixed solvent and shows negative values with a further increase in ethanol concentration. The negative Soret coefficients mean that the solute PEO migrates to the hot side of the fluid. A theoretical calculation with a model of the PEO in the mixed solvent ethanol and water explains the experimental results qualitatively, using the compressibility and hydrogen bonding between PEO and water molecules in the systems [41–43].

5.3.2.3 PNIPAM Solutions

The poly(*N*-isopropylacrylamide) [PNIPAM] chain in water has the θ-temperature of 30.6 °C, where the second virial coefficient is zero, and shows the coil-globule

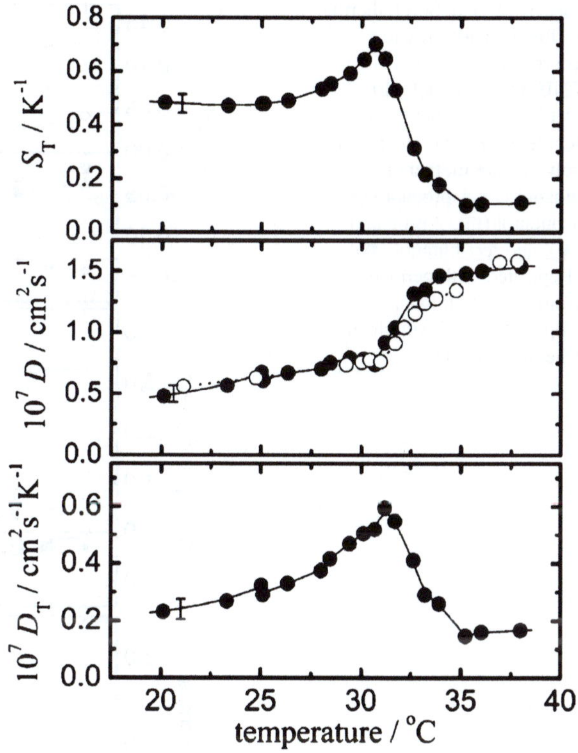

Fig. 5.9 The Soret, diffusion, and thermal diffusion coefficient of 1.0 gL^{-1} poly(*N*-isopropylacrylamide) [PNIPAM] in water in the vicinity of the oil-globule transition temperature, in a dilute concentration range of PNIPAM in water which has the θ-temperature at 30.6 °C, where the second virial coefficient is zero, and shows the coil-globule transition by heating (Reprinted with permission from Kita and Wiegand [45]. Copyright 2005 American Chemical Society)

transition upon heating [44]. The Soret coefficient of PNIPAM in water in the vicinity of the coil-globule transition temperature, in a dilute concentration range, is shown in Fig. 5.9 [45]. The Soret coefficient remains almost constant below 25 °C, although it shows a maximum at 30.7 °C with a steep decrease and a plateau at higher temperatures. The thermal diffusion coefficient D_T shows similar behavior to S_T, although D_T shows an apparent increase with increasing temperature below the θ-temperature. The positive S_T of PNIPAM indicates that polymers migrate to the cold side (typical for organic polymer solutions). At the θ-temperature, the concentration gradient is enhanced. In the consequence of the coil-globule transition of PNIPAM, segment-solvent interactions are favored at low temperatures (a good solvent condition), whereas at high temperatures (a poor solvent), the segment-segment and water-water contacts are favored owing to the liberation of water molecules from segments; thus, the chain shrinks to a compact structure. This picture for a dilute solution of PNIPAM indicates that the thermal diffusion behavior depends significantly on interactions among polymer segments and solvent molecules.

In contrast to the aqueous system, the Ludwig-Soret effect of PNIPAM in alcohols in a dilute concentration range shows a sign inversion behavior upon heating. Figure 5.10 shows the Soret, thermal diffusion, and mutual diffusion

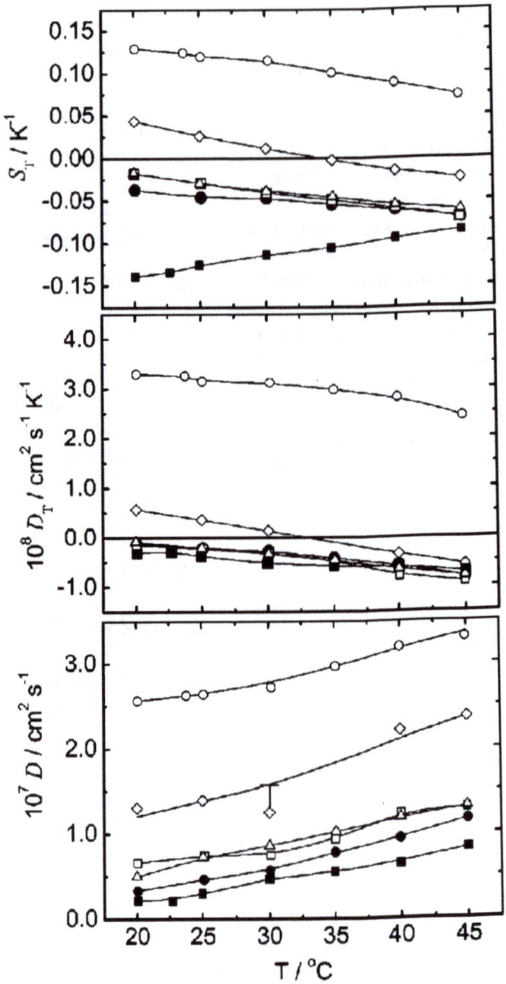

Fig. 5.10 The Soret, thermal diffusion, and diffusion coefficient of 10.0 gL^{-1} PNIPAM in monohydric alcohols as a function of temperature. The alcohols as solvents are methanol (○), ethanol (◇), 1-propanol (△), 1-butanol (□), 2-propanol (●), and *tert*-butanol (■) (Reprinted with permission from Kita et al. [47]. Copyright 2007 American Chemical Society)

coefficients of PNIPAM in monoalcohols as functions of temperature [46, 47]. In methanol, S_T and D_T of PNIPAM decrease with increasing temperature, and their signs are always positive at the studied temperatures. In ethanol, the signs of S_T and D_T become negative at 35 °C upon heating. The slopes of S_T and D_T against temperature are negative, except in the case of *tert*-butanol. This result indicates that the mechanisms of the Ludwig-Soret effect are different for *tert*-butanol compared with the other alcohol solutions. *tert*-Butanol might have a different association mechanism with PNIPAM segments owing to its structural hindrance by the OH group. Also, the mechanisms of the Ludwig-Soret effect of PNIPAM in alcohols are different from those of PNIPAM in water (Fig. 5.9).

In contrast to water and alcohol systems, aprotic organic solvents, such as dimethyl sulfoxide (DMSO), *N,N*-dimethylformamide (DMF), methyl ethyl ketone

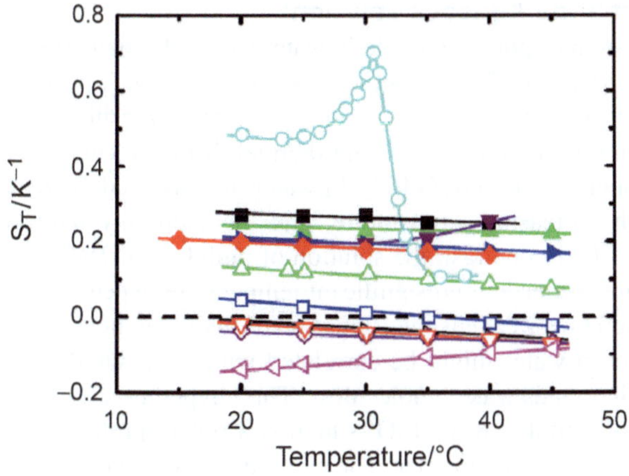

Fig. 5.11 The Soret coefficient of PNIPAM solutions as a function of temperature. Filled symbols refer to aprotic solvents as DMSO (▲), DMF (▶), MEK (■), THF (▼), and acetone (◆). Open symbols are associating solvents as water (○) and alcohols methanol (△), ethanol (□), 1-poropanol (▷) , 2-propanol (◇), 1-butanol (▽), and *tert*-butanol (◁)

(MEK), tetrahydrofuran (THF), and acetone, are used for studying the Ludwig-Soret effect of PNIPAM. Figure 5.11 shows the Soret coefficient of PNIPAM in aprotic organic solvents as well as in water and alcohols. The Soret coefficient of PNIPAM in aprotic solvents is positive with almost no temperature dependence. The result reveals that the temperature dependence of the Ludwig-Soret effect of PNIPAM is system dependent. The solubilities of aprotic organic solvents and alcohols are good for PNIPAM, although the aprotic solvent systems have positive Soret coefficients and alcohol systems show sign inversion. Water is the good solvent in a lower temperature range, whereas it becomes a poor solvent in the higher temperature range above the θ-temperature. These results suggest that complex associations of PNIPAM segments with solvent molecules determine the magnitude and the slope against temperature of the thermal diffusion behavior.

The Soret and thermal diffusion coefficients of PNIPAM microgels in water show the transition behavior from the swollen state to the globular state in the vicinity of the θ-temperature, wherein the thermal response is similar to that of the PNIPAM linear chain [48, 49]. The system is relevant for the study of the Ludwig-Soret effect with respect to the polymer nature and colloid nature, since the system changes its behavior from polymer-like associations (dominant role of segment-solvent interactions) to colloid-like associations (particle-particle interactions) upon heating. The Ludwig-Soret effect of the PNIPAM microgels in water has been reported [50, 51]. The Soret coefficient of PNIPAM microgels in water shows a peak at 31 °C. This behavior is similar to that of linear PNIPAM in water. The interfacial area exposed to solvent molecules changes markedly upon heating; accordingly, the structure of water surrounding the segments changes as well.

5.3.2.4 Saccharide Aqueous Solutions

The binary system of polysaccharide in water shows the sign inversion behavior of the Soret and thermal diffusion coefficients from negative to positive upon heating. Here, the concentration of polysaccharide in water is in the dilute range. Figure 5.12 shows the Soret, mutual diffusion, and thermal diffusion coefficients of 5.0 gL^{-1} pullulan in water and in DMSO [52]. In water, the Soret coefficient S_T of pullulan increases with increasing temperature, and the thermally induced sign change occurs at 41.7 °C. In contrast, the solution of pullulan in DMSO shows a positive S_T for all temperatures and no significant temperature dependence. An increase in solution temperature weakens the formation of hydrogen bonds; thus, the negative S_T for pullulan in water might be correlated with the strength of the interactions among pullulan and water molecules. This hypothesis is supported by the observations of pullulan in DMSO, which is a polar aprotic solvent not forming hydrogen bonds with the polymer. The sign inversion behavior is also observed in proteins, DNA, SDS micelles, and colloidal suspensions [14, 52–59], although

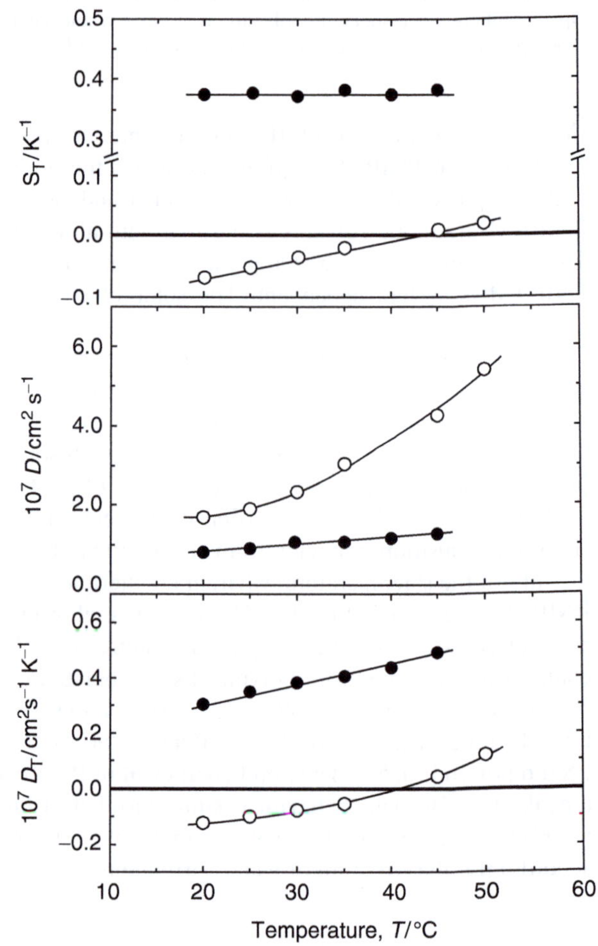

Fig. 5.12 The Soret, mutual diffusion, and thermal diffusion coefficient of pullulan in water (○) and pullulan in DMSO (●) as a function of temperature (Reprinted with permission from Kishikawa et al. [52]. Copyright 2010 American Chemical Society)

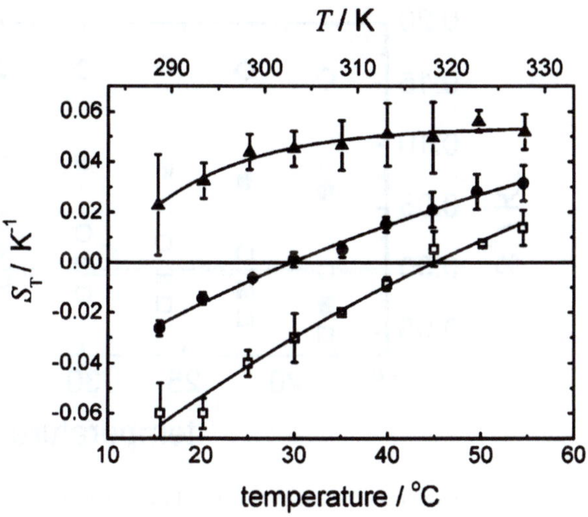

Fig. 5.13 The Soret coefficient of dextran in water (□), dextran in 2 M urea/water (●), and 5 M urea/water (▲) as a function of temperature (Reprinted with permission from Sugaya et al. [55]. Copyright 2006 American Chemical Society)

polysaccharide in water is the only binary system to show the sign inversion from negative to positive upon heating. PNIPAM in ethanol is another binary system that exhibits the sign inversion, but the slopes of the Soret and thermal diffusion coefficients against temperature are opposite, i.e., the sign inversion takes place from positive to negative upon heating of PNIPAM in ethanol. The opposite slope of the Soret coefficient against temperature implies that the mechanisms of thermal diffusion are different and might be related to the different molecular associations between polymers and solvent molecules.

Similar results with pullulan in water were observed in the system of 5.0 gL^{-1} dextran in water and the ternary system of 5.0 gL^{-1} dextran in urea/water [55]. Both pullulan and dextran are composed of glucose as the basic constituent, but the glycosidic bonds differ from each other. Figure 5.13 shows the Soret coefficient of dextran in water and has a similar temperature dependence of the Soret coefficient as that of pullulan in water. Interestingly, the Soret coefficient of dextran increases markedly with the addition of urea. The changes in the sign of S_T are observed at 0 and 2 M urea/water solvents. The negative Soret coefficient tends to become positive upon the addition of urea; this effect is analogous to the results of an increase in temperature. The strength of hydrogen bonds is weakened by the addition of a polar molecule, urea, which implies that urea molecules destroy the local structures of water. That is, the addition of urea has a similar contribution to the elevation of temperature on the sign change behavior.

5.3.2.5 Temperature Effect on Sign Inversion on Soret Coefficient

In addition to the polysaccharide aqueous solutions mentioned in Sect. 3.2.4, aqueous polymer systems often show a negative Soret coefficient on the low-temperature side. For example, the ternary system of PEO in the mixed solvent of ethanol/water of certain compositions shows the sign inversion behavior of S_T from negative to positive upon heating (Fig. 5.14) [20]. The Soret coefficient of

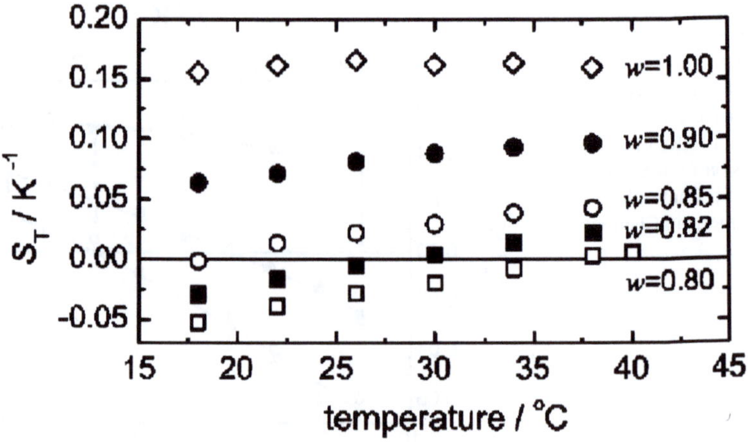

Fig. 5.14 The Soret coefficient of 5.0 gL^{-1} PEO in water/ethanol mixed solvents as a function of temperature. *w* means the weight fraction of water in the mixed solvent (Reprinted with permission from Kita et al. [20]. Copyright 2004, AIP Publishing LLC)

lysozyme in buffer solutions is known to show the sign inversion from negative to positive [53]. The temperature dependence of the Soret coefficient $S_T(T)$ is described by the following equation proposed by Piazza and coauthors [53]:

$$S_{\mathrm{T}}(T) = S_{\mathrm{T}}{}^{\infty}\left[1 - \exp\left(\frac{T_{\mathrm{inv}} - T}{T_0}\right)\right], \tag{5.44}$$

where $S_{\mathrm{T}}{}^{\infty}$ is the Soret coefficient at the high-temperature limit, T_{inv} is the temperature where the Soret coefficient changes its sign, and T_0 indicates the strength of temperature effects. The equation describes the experimental results very well for biopolymer solutions such as proteins [53, 54], polysaccharides [52, 55], and DNA [56], as well as the colloidal system [14, 57], micelle solutions [58], and fd-virus suspension [59]. The Debye screening length and thermal expansion factor are related to the negative sign of the Soret coefficient in the case of charged systems.

As for further considerations regarding the temperature dependence of the Soret coefficient, a scaled temperature dependence of S_{T} for colloidal systems is proposed as follows [60]:

$$\frac{S_{\mathrm{T}}\left(\widetilde{T}\right)}{S_{\mathrm{T}}{}^{\infty}} = 1 - \exp\left[A\left(1 - \widetilde{T}\right)\right]. \tag{5.45}$$

Here, $\widetilde{T}$ is defined as $\widetilde{T} = T/T_{\mathrm{inv}}$ and A is a dimensionless parameter $A = T_{\mathrm{inv}}/T_0$. Although the physical meaning of parameter A is still an open question, it is considered that the equation describes a universal behavior of the temperature dependence of S_{T} for colloidal systems. The effect of temperature on the thermal diffusion coefficient D_{T} is discussed in the next section.

5.3.2.6 Molecular Weight and Temperature Effects for Saccharides in Water

Aqueous solutions of saccharides are one of the interesting groups of systems for studying the Ludwig-Soret effect, since saccharides have crucial roles in biological systems. Systematic studies of changing the isomeric structure and the molecular weight have been carried out [52, 55, 61–64]. The Ludwig-Soret effect with respect to the molecular weight from monomers to high molecular weight polymers is studied extensively in the systems of poly(styrene) [PS] and poly(methyl methacrylate) [PMMA] [33–38]. Contrary to aqueous systems showing the sign inversion behavior upon heating, in saccharide aqueous solutions, the Soret coefficient is studied as a function of molecular weight [61–63].

Figure 5.15 shows the Soret, thermal diffusion, and mutual diffusion coefficients for saccharide aqueous solutions [63]. The monosaccharide is glucose [Glc]; glucose is the basic constituent of all other saccharide samples. As oligosaccharides, maltotriose and maltohexaose are studied. Maltotriose is composed of α-*D*-(1–4)-linked glucose [Glc(α1–4)Glc(α1–4)Glc]. Maltohexaose is also composed of α-*D*-(1–4)-linked glucose [Glc(α1–4)Glc(α1–4)Glc(α1–4)Glc(α1–4)Glc (α1–4)Glc]. Pullulan is composed of α-*D*-(1–6)-linked maltotriose. Dextran is composed of α-*D*-(1–6)-linked glucose with some short α-*D*-(1–3)-linked glucose branching units.

All samples show an increase in the Soret, thermal diffusion, and mutual diffusion coefficients with increasing temperature. Glucose and maltotriose in water have positive Soret and thermal diffusion coefficients in the investigated temperature range, although they are expected to have a negative Soret coefficient at lower temperatures. For all other systems, the Soret and thermal diffusion coefficients change their sign from negative to positive with increasing temperature. The temperature at which the sign changes increases with the increasing molecular weight of saccharides. The magnitude and the slope of the Soret coefficient for polymers are apparently larger than those of glucose and oligosaccharides. In contrast, the magnitude and the slope of the thermal diffusion coefficient for polymers become smaller with increasing molecular weight. The Soret coefficients at several temperatures are plotted as a function of molecular weight (Fig. 5.16). The curves are guides for the eye, because there is no theoretical expression available that describes the molecular weight dependence of the Soret coefficient for aqueous systems. The behavior of the Ludwig-Soret effect differs significantly from that in PS solutions [37, 38, 65], which may be due to the complex association between hydroxyl groups in saccharides and water molecules.

Usually, the thermal diffusion coefficient for aqueous systems is treated as having a linear relation with temperature. Further discussions on the thermal diffusion coefficient as a function of temperature have not been carried out, although the temperature dependence of the Soret coefficient for aqueous polymer systems and colloidal systems is well described in Eqs. (5.44) and (5.45), respectively. The thermal diffusion coefficient of saccharides in water is plotted against the temperature difference $\Delta T = T_{\mathrm{inv}} - T$ in Fig. 5.17. It is found that all values fall on a single curve, not a straight line. According to the results for saccharide aqueous

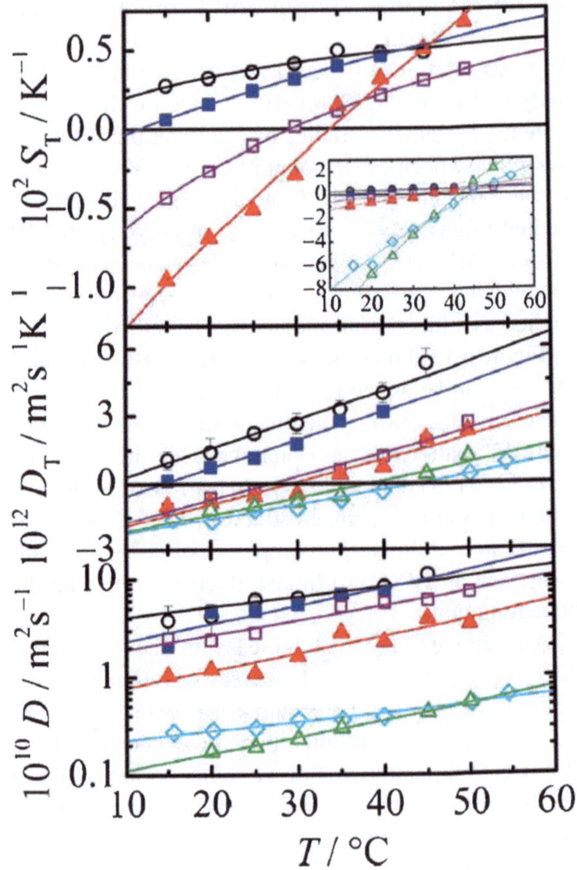

Fig. 5.15 The Soret, thermal diffusion, and mutual diffusion coefficients for aqueous solutions of glucose (○), maltotriose (■), maltohexaose(□), pullulan-4 k (▲), dextran (◇), and pullulan-440 k (△) (Reproduced from Ref. [63] with permission from the PCCP Owner Societies)

solutions, the temperature dependence of the thermal diffusion coefficient is proposed to follow the equation:

$$D_T(T) = D_{T0}\left[1 - \exp\left(\frac{\Delta T}{T_1}\right)\right]. \tag{5.46}$$

Here, D_{T0} represents the minimum D_T, T_1 is an indicator of the strength of the temperature effect, and $\Delta T = T_{inv} - T$ is the distance of the sign inversion temperature T_{inv}. This equation for $D_T(T)$ has the same form as Eq. (5.44). The result indicates that the curvature of $D_T(T)$ is not affected by the mass or the size of solute molecules in the case of saccharides in water. It is of great interest to study $D_T(T)$ to elucidate whether the context of Eq. (5.46) can be used for other systems.

5.3.2.7 Other Biological Systems

The Ludwig-Soret effect for biological molecules has been studied extensively in oligosaccharide [61, 62], nucleotide [66], DNA [56, 67–69], and proteins [53, 54,

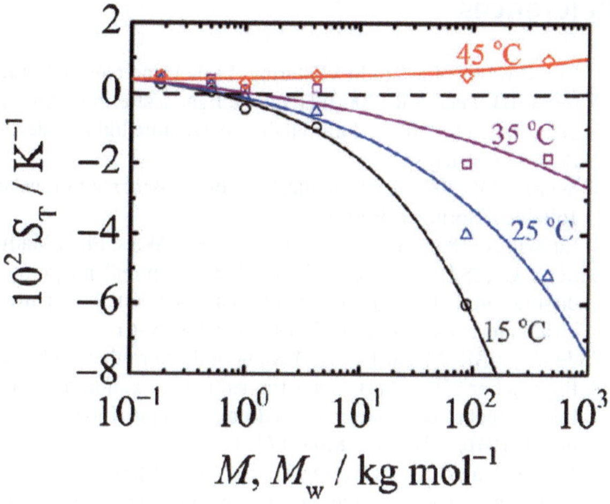

Fig. 5.16 Soret coefficient for aqueous solutions of glucose, maltotriose, maltohexaose, pullulan-4 k, dextran, and pullulan-440 k as a function of molecular weight (Reproduced from Ref. [63] with permission from the PCCP Owner Societies)

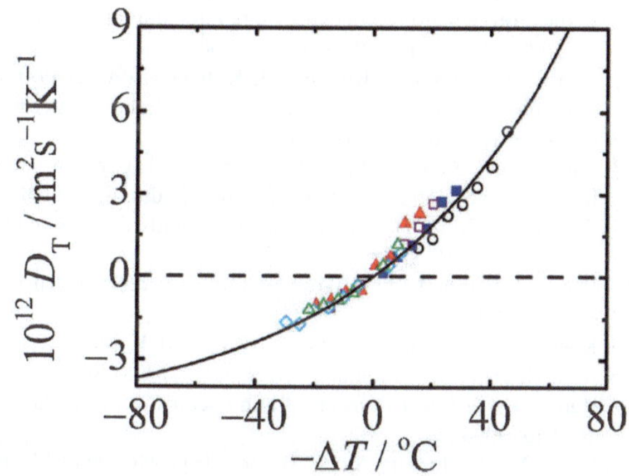

Fig. 5.17 Thermal diffusion coefficient for aqueous solutions of glucose, maltotriose, maltohexaose, pullulan-4 k, dextran, and pullulan-440 k as a function of the temperature difference ΔT ($=T_{inv} - T$) (Reproduced from Ref. [63] with permission from the PCCP Owner Societies)

70, 71]. Virus suspensions also show a sign inversion of the Soret coefficient [59]. In addition to the fundamental research on the Ludwig-Soret effect of biopolymers, studies on the application aspects of biochemical research have been carried out concerning the antibody affinity [70, 71] and development of PCR methodology [68]. These recent results on the Ludwig-Soret effect for biological systems are driven by the developments of experimental methods, as mentioned above, although information is still inadequate for understanding the mechanisms of the Ludwig-Soret effect. As for other fundamental research systems, there are numerous reports on colloidal systems [14, 57, 72–77] and aggregated systems such as micelle solutions [16, 60, 78–81]. Simulation and theoretical studies have also been carried out extensively [21, 27, 35, 36, 82–84]. For readers interested in these hot topics, refer to the recently published books or book series [85, 86], reviews, and publications cited above.

References

1. Chu B (2007) Lase light scattering: basic principles and practice, 2nd edn. Dover, New York
2. Berne BJ, Pecora R (2000) Dynamic light scattering. Dover, New York
3. Schmitz KS (1990) An introduction to dynamic light scattering by macromolecules. Academic Press, San Diego
4. Schaertl W (2007) Light scattering from polymer solutions and nanoparticle dispersions: with 16 tables. Springer, Berlin
5. Ludwig S (1856) Sitzungsber Preuss Akad Wiss. Phys Math Kl 20:539
6. Soret C (1879) Sur l'état d'équilibre que prend au point de vue de sa concentration une dissolution saaline primitivement homogène dont deux parties sont portées a des températures différentes. Arch Sci Phys Nat Lyenève 3:48–61
7. de Groot SR, Mazur P (1984) Non-equilibrium thermodynamics. Dover, New York
8. Provencher SW (1982) A constrained regularization method for inverting data represented by linear algebraic or integral equations. Comput Phys Commun 27(3):213–227. http://dx.doi.org/10.1016/0010-4655(82)90173-4
9. Provencher SW (1982) CONTIN: a general purpose constrained regularization program for inverting noisy linear algebraic and integral equations. Comput Phys Commun 27(3):229–242. http://dx.doi.org/10.1016/0010-4655(82)90174-6
10. Norisuye T, Shibayama M, Tamaki R, Chujo Y (1999) Time-resolved dynamic light scattering studies on gelation process of organic–inorganic polymer hybrids. Macromolecules 32 (5):1528–1533. doi:10.1021/ma981306h
11. Kita R, Takahashi A, Kaibara M, Kubota K (2002) Formation of fibrin gel in fibrinogen-thrombin system: static and dynamic light scattering study. Biomacromolecules 3 (5):1013–1020. doi:10.1021/bm025545v
12. Srinivasan S, Saghir MZ (2011) Experimental approaches to study thermodiffusion – a review. Int J Therm Sci 50(7):1125–1137. http://dx.doi.org/10.1016/j.ijthermalsci.2011.02.022
13. Wiegand S (2004) Thermal diffusion in liquid mixtures and polymer solutions. J Phys Condens Matter 16(10):R357
14. Piazza R, Parola A (2008) Thermophoresis in colloidal suspensions. J Phys Condens Matter 20 (15):153102
15. Koehler W, Schaefer R (2000) <APS_2000_Kohler.pdf> Adv Polym Sci 151:1–59
16. Ning H, Datta S, Sottmann T, Wiegand S (2008) Soret effect of nonionic surfactants in water studied by different transient grating setups. J Phys Chem B 112(35):10927–10934. doi:10.1021/jp800942w
17. Giglio M, Vendramini A (1977) Soret-Type Motion of Macromolecules in Solution. Phys Rev Lett 38(1):26–30
18. Kolodner P, Williams H, Moe C (1988) Optical measurement of the Soret coefficient of ethanol/water solutions. J Chem Phys 88(10):6512–6524
19. Dutrieux JF, Platten JK, Chavepeyer G, Bou-Ali MM (2002) On the measurement of positive Soret coefficients. J Phys Chem B 106(23):6104–6114. doi:10.1021/jp013945r
20. Kita R, Wiegand S, Luettmer-Strathmann J (2004) Sign change of the Soret coefficient of poly (ethylene oxide) in water/ethanol mixtures observed by thermal diffusion forced Rayleigh scattering. J Chem Phys 121(8):3874–3885. doi:10.1063/1.1771631
21. Nieto-Draghi C, Avalos JB, Rousseau B (2005) Computing the Soret coefficient in aqueous mixtures using boundary driven nonequilibrium molecular dynamics. J Chem Phys 122 (11):114503. doi:10.1063/1.1863872
22. Platten JK, Bou-Ali MM, Blanco P, Madariaga JA, Santamaria C (2007) Soret coefficients in some water–methanol, water-ethanol, and water-isopropanol systems. J Phys Chem B 111 (39):11524–11530. doi:10.1021/jp074206z
23. Wittko G, Köhler W (2007) On the temperature dependence of thermal diffusion of liquid mixtures. Europhys Lett 78(4):46007. doi:10.1209/0295-5075/78/46007

24. Eslamian M, Saghir MZ (2009) Microscopic study and modeling of thermodiffusion in binary associating mixtures. Phys Rev E 80(6):061201. doi:10.1103/PhysRevE.80
25. Abbasi A, Saghir MZ, Kawaji M (2009) A new approach to evaluate the thermodiffusion factor for associating mixtures. J Chem Phys 130(6):064506. doi:10.1063/1.3076926
26. Koniger A, Meier B, Kohler W (2009) Measurement of the Soret, diffusion, and thermal diffusion coefficients of three binary organic benchmark mixtures and of ethanol-water mixtures using a beam deflection technique. Philos Mag 89(10):907–923. doi:10.1080/14786430902814029
27. Artola PA, Rousseau B, Galliero G (2008) A new model for thermal diffusion: kinetic approach. J Am Chem Soc 130(33):10963–10969. doi:10.1021/ja800817f
28. Zhang KJ, Briggs ME, Gammon RW, Sengers JV (1996) Optical measurement of the Soret coefficient and the diffusion coefficient of liquid mixtures. J Chem Phys 104(17):6881–6892. doi:10.1063/1.471355
29. Torres JF, Komiya A, Henry D, Maruyama S (2013) Measurement of Soret and Fickian diffusion coefficients by orthogonal phase-shifting interferometry and its application to protein aqueous solutions. J Chem Phys 139(7):074203. doi:10.1063/1.4817682
30. Saeki S, Kuwahara N, Nakata M, Kaneko M (1976) Upper and lower critical solution temperatures in poly (ethylene glycol) solutions. Polymer 17(8):685–689
31. Paduano L, Sartorio R, D'Errico G, Vitagliano V (1998) Mutual diffusion in aqueous solution of ethylene glycol oligomers at 25 [degree]C. J Chem Soc Faraday Trans 94(17):2571–2576. doi:10.1039/A803567I
32. Chan J, Popov J, Kolisnek-Kehl S, Leaist D (2003) Soret coefficients for aqueous polyethylene glycol solutions and some tests of the segmental model of polymer thermal diffusion. J Solut Chem 32(3):197–214. doi:10.1023/a:1022925216642
33. Zhang KJ, Briggs ME, Gammon RW, Sengers JV, Douglas JF (1999) Thermal and mass diffusion in a semidilute good solvent-polymer solution. J Chem Phys 111(5):2270–2282. doi:10.1063/1.479498
34. Schimpf ME, Giddings JC (1987) Characterization of thermal diffusion in polymer solutions by thermal field-flow fractionation: effects of molecular weight and branching. Macromolecules 20(7):1561–1563. doi:10.1021/ma00173a022
35. Semenov SN, Schimpf ME (2005) Molecular thermodiffusion (thermophoresis) in liquid mixtures. Phys Rev E 72(4):041202. doi:10.1103/PhysRevE.72.041202
36. Zhang M, Muller-Plathe F (2006) The Soret effect in dilute polymer solutions: influence of chain length, chain stiffness, and solvent quality. J Chem Phys 125(12):124903. doi:10.1063/1.2356469
37. Stadelmaier D, Kohler W (2008) From small molecules to high polymers: investigation of the crossover of thermal diffusion in dilute polystyrene solutions. Macromolecules 41(16):6205–6209. doi:10.1021/ma800891p
38. Würger A (2009) Molecular-weight dependent thermal diffusion in dilute polymer solutions. Phys Rev Lett 102(7):078302
39. Rauch J, Kohler W (2005) On the molar mass dependence of the thermal diffusion coefficient of polymer solutions. Macromolecules 38(9):3571–3573. doi:10.1021/ma050231w
40. de Gans BJ, Kita R, Muller B, Wiegand S (2003) Negative thermodiffusion of polymers and colloids in solvent mixtures. J Chem Phys 118(17):8073–8081. doi:10.1063/1.1563601
41. de Gans BJ, Kita R, Wiegand S, Luettmer-Strathmann J (2003) Unusual thermal diffusion in polymer solutions. Phys Rev Lett 91(24):245501. doi:10.1103/PhysRevLett.91.245501
42. Luettmer-Strathmann J (2003) Two-chamber lattice model for thermodiffusion in polymer solutions. J Chem Phys 119(5):2892–2902
43. Luettmer-Strathmann J (2005) Lattice model for thermodiffusion in polymer solutions. Int J Thermophys 26(6):1693–1707. doi:10.1007/s10765-005-8589-0
44. Kubota K, Fujishige S, Ando I (1990) Single-chain transition of poly(N-isopropylacrylamide) in water. J Phys Chem 94(12):5154–5158. doi:10.1021/j100375a070

45. Kita R, Wiegand S (2005) Soret coefficient of poly(N-isopropylacrylamide)/water in the vicinity of coil-globule transition temperature. Macromolecules 38(11):4554–4556. doi:10.1021/ma050526+
46. Kita R, Kircher G, Wiegand S (2004) Thermally induced sign change of Soret coefficient for dilute and semidilute solutions of poly(N-isopropylacrylamide) in ethanol. J Chem Phys 121 (18):9140–9146. doi:10.1063/1.1803535
47. Kita R, Polyakov P, Wiegand S (2007) Ludwig-Soret effect of poly(N-isopropylacrylamide): temperature dependence study in monohydric alcohols. Macromolecules 40(5):1638–1642. doi:10.1021/ma0621831
48. Kratz K, Hellweg T, Eimer W (2001) Structural changes in PNIPAM microgel particles as seen by SANS, DLS, and EM techniques. Polymer 42(15):6631–6639. doi:10.1016/s0032-3861(01)00099-4
49. Kogure H, Nanami S, Masuda Y, Toyama Y, Kubota K (2005) Hydration and dehydration behavior of N-isopropylacrylamide gel particles. Colloid Polym Sci 283(11):1163–1171. doi:10.1007/s00396-005-1303-8
50. Kita R, Sugaya R, Kogure H, Kubota K (2006) Thermal diffusion of PNiPAM nano gel-particle at the theta-temperature. Trans Mater Res Soc Jpn 31(3)
51. Wongsuwarn S, Vigolo D, Cerbino R, Howe AM, Vailati A, Piazza R, Cicuta P (2012) Giant thermophoresis of poly(N-isopropylacrylamide) microgel particles. Soft Matter 8 (21):5857–5863. doi:10.1039/c2sm25061f
52. Kishikawa Y, Wiegand S, Kita R (2010) Temperature dependence of Soret coefficient in aqueous and nonaqueous solutions of pullulan. Biomacromolecules 11(3):740–747. doi:10.1021/bm9013149
53. Iacopini S, Piazza R (2003) Thermophoresis in protein solutions. Europhys Lett 63 (2):247–253. doi:10.1209/epl/i2003-00520-y
54. Piazza R, Iacopini S, Triulzia B (2004) Thermophoresis as a probe of particle-solvent interactions: the case of protein solutions. Phys Chem Chem Phys 6(7):1616–1622. doi:10.1039/b312856c
55. Sugaya R, Wolf BA, Kita R (2006) Thermal diffusion of dextran in aqueous solutions in the absence and the presence of urea. Biomacromolecules 7(2):435–440. doi:10.1021/bm050545r
56. Duhr S, Braun D (2006) Why molecules move along a temperature gradient. Proc Natl Acad Sci U S A 103(52):19678–19682. doi:10.1073/pnas.0603873103
57. Dhont JKG, Wiegand S, Duhr S, Braun D (2007) Thermodiffusion of charged colloids: single-particle diffusion. Langmuir 23(4):1674–1683. doi:10.1021/la062184m
58. Piazza R, Guarino A (2002) Soret effect in interacting micellar solutions. Phys Rev Lett 88 (20):208302. doi:10.1103/PhysRevLett.88.208302
59. Blanco P, Kriegs H, Lettinga MP, Holmqvist P, Wiegand S (2011) Thermal diffusion of a stiff rod-like mutant Y21M fd-virus. Biomacromolecules 12(5):1602–1609. doi:10.1021/bm2000023
60. Vigolo D, Buzzaccaro S, Piazza R (2010) Thermophoresis and thermoelectricity in surfactant solutions. Langmuir 26(11):7792–7801. doi:10.1021/la904588s
61. Blanco P, Kriegs H, Arlt B, Wiegand S (2010) Thermal diffusion of oligosaccharide solutions: the role of chain length and structure. J Phys Chem B 114(33):10740–10747. doi:10.1021/jp104534m
62. Klein M, Wiegand S (2011) The Soret effect of mono-, di- and tri-glycols in ethanol. Phys Chem Chem Phys 13(15):7059–7063. doi:10.1039/c1cp00022e
63. Kishikawa Y, Shinohara H, Maeda K, Nakamura Y, Wiegand S, Kita R (2012) Temperature dependence of thermal diffusion for aqueous solutions of monosaccharides, oligosaccharides, and polysaccharides. Phys Chem Chem Phys 14(29):10147–10153
64. Shinohara H, Kita R, Shinyashiki N, Yagihara S, Kabayama K, Inazu T (2013) Temperature dependent study of thermal diffusion for aqueous solutions of α-, β-, and γ- cyclodextrin. AIP Conf Proc 1518:710–713. doi:10.1063/1.4794663

65. Stadelmaier D, Kohler W (2009) Thermal diffusion of dilute polymer solutions: the role of chain flexibility and the effective segment size. Macromolecules 42(22):9147–9152. doi:10.1021/ma901794k
66. Wang Z, Kriegs H, Wiegand S (2012) Thermal diffusion of nucleotides. J Phys Chem B 116 (25):7463–7469. doi:10.1021/jp3032644
67. Braun D, Libchaber A (2002) Trapping of DNA by thermophoretic depletion and convection. Phys Rev Lett 89(18):188103
68. Mast CB, Braun D (2010) Thermal trap for DNA replication. Phys Rev Lett 104(18):188102. doi:10.1103/PhysRevLett.104.188102
69. Reineck P, Wienken CJ, Braun D (2010) Thermophoresis of single stranded DNA. Electrophoresis 31(2):279–286. doi:10.1002/elps.200900505
70. Wienken CJ, Baaske P, Rothbauer U, Braun D, Duhr S (2010) Protein-binding assays in biological liquids using microscale thermophoresis. Nat Commun 1:100. doi:10.1038/ncomms1093
71. Lippok S, Seidel SAI, Duhr S, Uhland K, Holthoff H-P, Jenne D, Braun D (2012) Direct detection of antibody concentration and affinity in human serum using microscale thermophoresis. Anal Chem 84(8):3523–3530. doi:10.1021/ac202923j
72. Iacopini S, Rusconi R, Piazza R (2006) The "macromolecular tourist": Universal temperature dependence of thermal diffusion in aqueous colloidal suspensions. Eur Phys J E 19(1):59–67. doi:10.1140/epje/e2006-00012-9
73. Braibanti M, Vigolo D, Piazza R (2008) Does thermophoretic mobility depend on particle size? Phys Rev Lett 100(10):108303. doi:10.1103/PhysRevLett.100.108303
74. Wurger A (2006) Heat capacity-driven inverse Soret effect of colloidal nanoparticles. Europhys Lett 74(4):658–664. doi:10.1209/epl/i2005-10579-x
75. Wurger A (2010) Thermal non-equilibrium transport in colloids. Rep Prog Phys 73 (12):126601. doi:10.1088/0034-4885/73/12/126601
76. Jiang H-R, Yoshinaga N, Sano M (2010) Active motion of a Janus particle by self-thermophoresis in a defocused laser beam. Phys Rev Lett 105(26):268302
77. Wuerger A (2009) Temperature dependence of the soret motion in colloids. Langmuir 25 (12):6696–6701. doi:10.1021/la9001913
78. Arlt B, Datta S, Sottmann T, Wiegand S (2010) Soret effect of n-octyl beta-d-glucopyranoside (C(8)G(1)) in water around the critical micelle concentration. J Phys Chem B 114 (6):2118–2123. doi:10.1021/jp907988r
79. Ning H, Kita R, Kriegs H, Luettmer-Strathmann J, Wiegand S (2006) Thermal diffusion behavior of nonionic surfactants in water. J Phys Chem B 110(22):10746–10756. doi:10.1021/jp0572986
80. Ning H, Wiegand S, Kita R (2006) Soret effect in a nonionic surfactant system. Progr Colloid Polym Sci 133:111–115. doi:10.1007/3-540-32702-9_18
81. Maeda K, Kita R, Shinyashiki N, Yagihara S (2013) Ludwig-Soret effect of non-ionic surfactant aqueous solution studied by beam deflection method. AIP Conf Proc 1518 (1):428–431
82. Bielenberg JR, Brenner H (2005) A hydrodynamic/Brownian motion model of thermal diffusion in liquids. Phys A Stat Mech Appl 356(2–4):279–293. http://dx.doi.org/10.1016/j.physa.2005.03.033
83. Brenner H, Bielenberg JR (2005) A continuum approach to phoretic motions: thermophoresis. Phys A Stat Mech Appl 355(2–4):251–273. http://dx.doi.org/10.1016/j.physa.2005.03.020
84. Rousseau B, Nieto-Draghi C, Avalos JB (2004) The role of molecular interactions in the change of sign of the Soret coefficient. Europhys Lett 67(6):976–982. doi:10.1209/epl/i2004-10136-3
85. Kohler W, Wiegand S (2002) Thermal nonequilibrium phenomena in fluid mixtures. Springer, Berlin
86. Srinivasan S, Saghir MZ (2013) Thermodiffusion in multicomponent mixtures. Springer, New York

[illegible]

[illegible] (1993) [illegible]

[illegible]

Diffusion Measurements of Water and Polymers in Hydrogels by Pulsed Field Gradient NMR

6

Shingo Matsukawa and Tom Brenner

Abstract

The basic theory and techniques of diffusion measurements by pulsed field gradient NMR are described, and experimental results for solutions and gels of poly(*N*,*N*-dimethylacrylamide), carrageenans, agar, and agarose are introduced and analyzed to give physical pictures for the gels. Discussion of experimental results for water and probe diffusion in synthetic polymer and polysaccharide gels and solution was offered. Relaxation times for the macromolecules and water give information on tumbling motion. The diffusion coefficient of probe molecules in hydrocolloid systems provided the information on the translational mobility of molecules, which can be used to infer the structure of the gel network. By comparing the results with other experimental techniques, a clear picture emerges, with clear correspondence of the microscopic events, namely, aggregation, polymer immobilization, and subsequent effects on molecular flexibility and probe diffusion, with the macroscopic (bulk) events, namely, gelation. The hydrodynamic shielding length, ξ, which represents within the mean field hydrodynamic approach the mesh size of the network, as a parameter that determines D/D_0 of probe molecules, has been discussed in detail. This parameter ξ was used to describe quantitatively the evolving structure of the gel network.

Keywords

Hydrogel • Polysaccharide • Gradient NMR • Diffusion coefficient • Hydrodynamic screening length

---

S. Matsukawa (✉) • T. Brenner
The Graduate school of Marine Science and Technology,
Tokyo University of Marine Science and Technology, Tokyo, Japan
e-mail: matsukaw@kaiyodai.ac.jp

R. Kita, T. Dobashi (eds.), *Nano/Micro Science and Technology in Biorheology*,
DOI 10.1007/978-4-431-54886-7_6

6.1 Introduction

The diffusion of molecules in hydrogels depends on intermolecular interactions with the network chains. The diffusion coefficient D of the molecules decreases through hydrodynamic interaction [1–5] and via hydrogen bonding [5] or hydrophobic interaction [6]. In synthetic polymer gels, the network size, which affects hydrodynamic interaction, depends on the polymer concentration [4, 7]. Conversely, many polysaccharides form network structures via chain aggregation. If a probe molecule is involved in the aggregation or otherwise stick to the gel network structure, its mobility is restricted, resulting in a marked decrease in D; otherwise, the molecule diffuses in the network interspace. In this latter case, D is affected by hydrodynamic interaction with the network and the change in number of entanglements due to aggregation. As a consequence, D varies even if the nominal host polymer concentration is constant throughout the aggregation and gelation process.

The tumbling and translational motions of polysaccharides change markedly during aggregation and gelation. NMR measurements provide information on the mobility of molecules in gel. The spin-lattice relaxation time (T_1) and the spin-spin relaxation time (T_2) reflect the local mobility and flexibility of the molecules, respectively, and D measured by field gradient NMR reflects the displacement of a molecule by self-diffusion.

In this chapter, we introduce the basic theory and techniques of diffusion measurements by pulsed field gradient NMR in Sect. 6.2 and experimental results for solutions and gels of poly(*N*,*N*-dimethylacrylamide) [4], carrageenans [8, 9], agar [10], and agarose [11] in Sect. 6.3.

6.2 Diffusion Coefficient Measurements

An application of field gradient attaches a spatial information to the NMR signal. Therefore, the gradient can produce a spatial distribution of nuclei, that is, NMR imaging. When two field gradients for the diphase and rephase are applied, the NMR signal decays due to the displacement of nuclei during the interval between the two field gradients. This allows calculation of the diffusion coefficient for Fickian diffusion in free space. In this section, the use of field gradient for diffusion measurements is described.

The Larmor precession frequency depends on the magnetic field experienced by the nucleus and is therefore dependent on nuclei position when the field gradient is present. The spatially dependent Larmor frequency ω ($\mathbf{r}$) at the position $\mathbf{r}$ under a spatially linear field gradient $\mathbf{g}$ is expressed as follows:

$$\omega(\mathbf{r}) = \gamma(H_0\mathbf{g}\mathbf{r}) = \omega_0 + \gamma\,\mathbf{g}\mathbf{r} \tag{6.1}$$

where H_0 is the externally applied magnetic field and $\mathbf{g} = 0$ at $\mathbf{r} = 0$. It follows that the phase angle ϕ is a function of the spatial position of the nucleus. If the field gradient duration is δ, then ϕ is given with

$$\phi(r) = \gamma \mathbf{g} \mathbf{r} \delta \tag{6.2}$$

The distance in the direction of **g** where $\phi(\mathrm{r}) = 2\,\pi$ is

$$q^{-1} = 2\pi/\gamma\, g\, \delta \tag{6.3}$$

q^{-1} is the characteristic length scale of the field gradient. For example, $q^{-1} = 235$ μm for $g = 10$ G/cm with $\delta = 1$ ms. When the sample size, or the size of detection area, is several times larger than q^{-1}, the total signal intensity vanishes because of dephasing.

For diffusion coefficient measurements, a second field gradient is applied in order to rephase the dephased magnetization. Figure 6.1 shows a typical pulse sequence with two rectangular-shaped pulsed field gradients (PFG) along the z axis [12] and the dephasing and rephasing behavior of the magnetization when individual spins do not change positions in the interval Δ between the two PFGs. (a) The magnetizations are aligned along the y axis by an r.f. π/2 pulse. (b) Under the first PFG, the magnetizations precess at an angular velocity of $\gamma\, \mathbf{g}\, \mathbf{r}$ corresponding to the individual positions in the rotating frame. (c) At the end of the first PFG, the

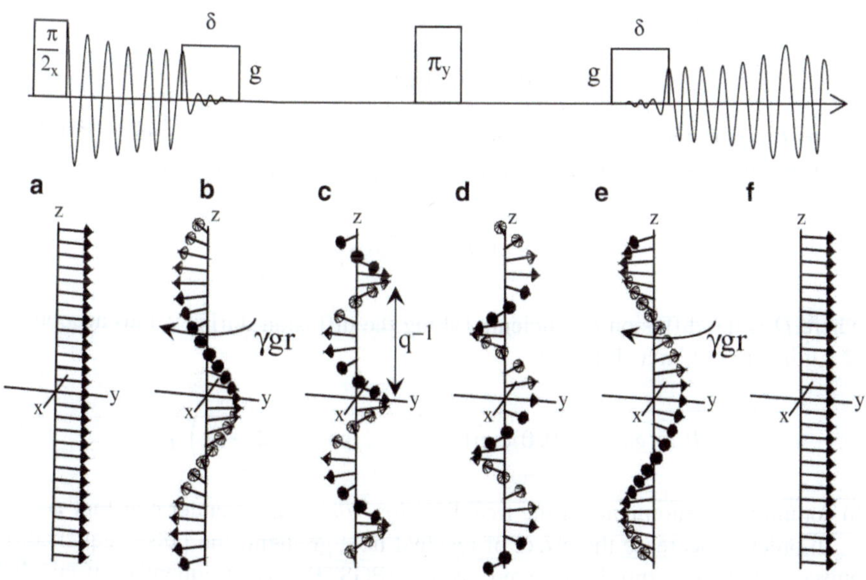

Fig. 6.1 A typical pulse sequence with two pulsed field gradients of rectangular shape and the dephasing and rephasing behavior of the magnetization. (**a**) The magnetizations are aligned along the y axis by an r.f. π/2 pulse. (**b**) Under the first PFG, the magnetizations precess at the angular velocity of $\gamma \mathbf{gr}$ corresponding to the z coordinate. (**c**) At the end of the first PFG, the magnetizations are spirally twisted at a pitch of q^{-1}. (**d**) An r.f. π pulse along the y axis rotates the individual magnetizations along the direction of y axis through 180°. (**e**) Under the second PFG, the individual magnetizations precess at the same angular velocity with that under the first PFG. (**f**) At the end of the second PFG, the magnetizations are aligned along the y axis

magnetizations are spirally twisted at a pitch of q^{-1}. (d) The application of an r.f. π pulse along the y axis rotates the individual magnetizations 180° about the direction of y axis, which yields a mirror-symmetrical arrangement of the magnetizations with respect to the y-z plane. (e) Under the second PFG, the individual magnetizations precess at the same angular velocity with that under the first PFG. (f) At the end of the second PFG, the magnetizations are aligned along the y axis. When the nucleus is displaced Δz in the z direction during Δ, its phase angle is

$$\varphi(\Delta z) = 2\pi \frac{\Delta z}{q^{-1}} = \gamma g \delta \Delta z \tag{6.4}$$

The echo signal intensity $I(2\tau, g\ \delta)$ at 2τ is proportional to the vector sum of magnetizations in the sample, therefore expressed as follows:

$$\mathrm{I}(2\tau, g\delta) = \mathrm{I}(2\tau, 0) \int \cos\left(\phi(\Delta z)\right) \rho(\mathbf{r}) p(\mathbf{r}, \Delta z) d\mathbf{r} d\Delta z \tag{6.5}$$

where $\rho(\mathbf{r})$ is the density of the nucleus and is constant for homogeneous sample, $p(\mathbf{r}, \Delta z)$ is the probability of the displacement during Δ for the nucleus at $\mathbf{r}$, and $I(2\tau,0)$ is the total signal intensity without PFG and expressed as follows:

$$I(2\tau, 0) = I(0, 0)\exp(-2\tau/T_2) \tag{6.6}$$

where $I(0,0)$ is the initial signal intensity just after the r.f. π/2 pulse. For free diffusion in an isotropic medium, $p(\mathbf{r}, \Delta z)$ becomes a Gaussian distribution

$$p(r, \Delta z) = (4\pi D\Delta)^{-1/2}\exp\left(-\frac{\Delta z^2}{4D\Delta}\right) \tag{6.7}$$

where D is the diffusion coefficient. Taking the diffusion during δ into account, $I(2\tau, g\delta)$ is rewritten as follows:

$$I(2\tau, g\delta) = I(0, 0)\exp\left(-\frac{2\tau}{\mathrm{T}_2} - (\gamma g\delta)^2 D\left(\Delta - \frac{\delta}{3}\right)\right) \tag{6.8}$$

In common measurements of the free diffusion, $g\delta$ is varied under constant Δ.

In order to decrease the effect of residual field gradient, the pulse sequence of pulsed gradient stimulated spin echo (PGSTE) is frequently used for D measurements [13–17]. The diffusion coefficient values are determined from the decay of echo signal intensities, expressed as follows:

$$I(2\tau_2 + \tau_1 g\delta) = I(2\tau_2 + \tau_1, 0)\exp\left[-\gamma^2\delta^2 g^2(\Delta - \delta/3)D\right] \tag{6.9}$$

where $I(2\tau_2 + \tau_1, g\delta)$ and $I(2\tau_2 + \tau_1, 0)$ are echo signal intensities at $t = 2\tau_2 + \tau_1$ with and without the field gradient pulse, respectively, γ is the gyromagnetic ratio of the measured nucleus, g is the field gradient strength, δ is the duration of the gradient pulse and Δ is the diffusion time, equal to the interval between two field gradient pulses. Note that $I(2\tau_2+\tau_1, 0)$ has decayed from the initial intensity, $I(0, 0)$, (i.e., the signal intensity immediately after the first $\pi/2$ rf pulse) by T_1 and T_2 relaxations:

$$I(2\tau_2 + \tau_1, 0) = I(0, 0)\exp[-2\tau_2/T_2 - \tau_1/T_1] \tag{6.10}$$

From Eq. 6.10 it follows that a decrease in the relaxation times leads to a decrease in $I(2\tau_2+\tau_1,0)$.

6.3 Diffusion of Water and Polymers in Hydrogels

6.3.1 Molecular Diffusion in Poly(Dimethylacrylamide) Gel

Most synthetic polymer gels have chemically cross-linked network structures. As a result the polymer concentration is usually homogeneous above a length scale of several nm. Conversely, physically cross-linked gels such as polysaccharide gels, where aggregations are responsible for gelation, have concentration fluctuations over larger length scales.

In this section we look at the diffusion of water and a probe polymer, polyethylene glycol (PEG), in a gel of poly(dimethylacrylamide) (PDMAA), a synthetic polymer.

6.3.1.1 Diffusion of HDO in PDMAA

The diffusion coefficient of HDO (D_{HDO}) in a PDMAA gel was determined by the PGSE ^{1}H NMR method at 303 K varying the swelling ratio Q, which is inversely proportional to the concentration of the network polymer, and the Mw of PEG added as a probe. As seen from Fig. 6.2, D_{HDO} increases as Q is increased but is almost independent of the Mw of PEG. The Q-dependence of D_{HDO} is strong at low Q values and is much more moderate at high Q values. The D_{HDO} value in the large-Q region asymptotically approaches that for HDO in neat D_2O (2.22×10^{-9} m^{2}s^{-1}). We can conclude that intermolecular interactions between water and the polymer network restrict the translational motion of water. We also see that the diffusion of HDO is independent of the Mw of PEG. We can understand more clearly the dynamical behavior of water in the gel system by appealing to the following modified free volume theory equation:

$$D_{HDO} = D_{HDO,0}\ \exp\left[Q^{-1}/\left(Q^{-1}f_{solv} - f_{solv}^2/\beta\right)\right] \tag{6.11}$$

where f_{solv}, β, and $D_{HDO,0}$ are the fractional free volume of deuterated water, the proportionality constant in the free volume theory, and D_{HDO} in neat solvent, respectively. The solid curve obtained from a least-squares fitting of the experimental data to Eq. 6.11 is shown in Fig. 6.2. The theoretical fit agrees well with the experimental data. From this result, the diffusion coefficient of HDO in the neat

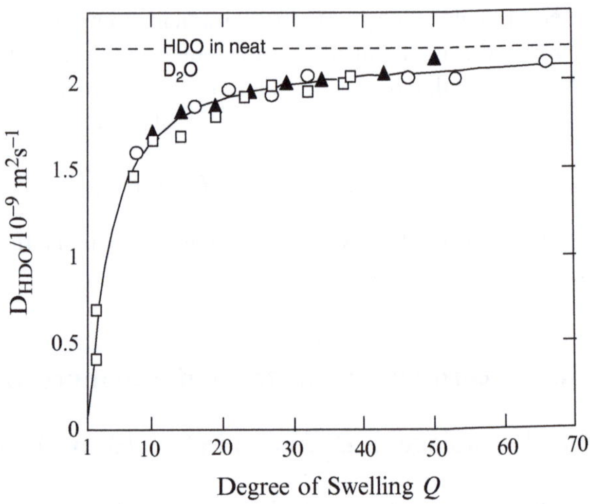

Fig. 6.2 Dependence of the diffusion coefficient of water molecule (D_{HDO}) on the degree of swelling (Q) in a PDMAA gel containing PEG with $Mw = 4{,}250$ (□), 10,890 (▲), and 20,000 (O) at 303 K

solvent ($Q \to$ infinity) is $D_{HDO,0} = 2.16 \times 10^{-9}\ m^2s^{-1}$, which is lower than HDO in neat D_2O ($2.22 \times 10^{-9}\ m^2s^{-1}$). This means that the diffusion coefficient for HDO in neat D_2O is different from the Q-extrapolated value to the infinite Q given by the modified free volume theory for the Q-dependence of the diffusion coefficient. The diffusion coefficient for HDO in bulk PDMAA ($Q \to 1$) obtained from Fig. 6.2 was $2.5 \times 10^{-11}\ m^2s^{-1}$. Clearly, the translational motion of HDO in the limit $Q \to 1$ (bulk polymer) is restricted compared to that in gels. Further, the rapid decrease of D_{HDO} in the small-Q region can be explained by a decrease in free volume.

6.3.1.2 Diffusion of PEG in PDMAA

Assuming that diffusion of a hard-sphere probe molecule is restricted mainly by hydrodynamic interactions with the host polymer, its mobility can be related to its radius and the screening length of the polymer network. A mean field treatment of the hydrodynamic model yields

$$D/D_0 = \exp(-\kappa R) \tag{6.12}$$

where D is the diffusion coefficient of the probe in the host solution or gel, D_0 is the diffusion coefficient of the probe polymer in the neat solvent, R the radius of the spherical probe, and κ^{-1} the hydrodynamic screening length of polymer chain. When κ^{-1} is equivalent to R, $D/D_0 = e^{-1} \approx 0.37$. This approach has been used to analyze the diffusion of probe molecules in semidilute polymer solutions [2] and polymer gels [3]. It has also been employed to analyze the self-diffusion of random coil polymers in solutions [18] and gels [19], in which κ^{-1} is large enough that the diffusion of random coil polymers is controlled by their hydrodynamic radius R_H, as opposed to highly restricted reptation, where besides R_H internal dynamics play a part. When κ^{-1} is large enough, the hydrodynamic interactions are more important than topological constraints, and R in Eq. 6.12 can be replaced with R_H. When the topological constraints cannot be neglected, compared with the interchain hydrodynamic interactions, D becomes larger than the value expected from Eq. 6.12. When the diffusion coefficient of an isolated PEG ($D_{PEG,0}$) is determined from the diffusion coefficient of PEG in dilute aqueous solution($D_{PEGsoln}$), it is necessary to take into account the change of local friction of polymer, which is estimated from the change of the diffusion coefficient of the solvent [20]. Then we have

$$D_{PEG,0} = D_{PEG,soln}/(D_{HDO,neat}/D_{HDO}) \tag{6.13}$$

where $D_{HDO,neat}$ is the diffusion coefficient of HDO in D_2O.

The diffusion coefficient of PEG (D_{PEG}) of PEG in PDMAA gels was determined by the PGSE 1H NMR method at 303 K for different Q values and Mw values of PEG. The D_{PEG} values are plotted against Q in Fig. 6.5a. It was found that D_{PEG} increases as Q is increased and also depends on the Mw of PEG. The D_{PEG} value in the gel is smaller than in a 1 wt% aqueous PEG solution ($D_{PEG,soln}$), as shown in Fig. 6.5b. We can conclude that the translational motion of PEG in the gel, compared to that in dilute solution, is restricted through intermolecular interaction with the polymer network. This restriction becomes stronger with increasing Mw of the probe. The diffusion of PEG in the gel can be analyzed using Eq. 6.12. The relationship between the hydrodynamic screening length (k^{-1}) and the concentration of network polymer (c) or the degree of swelling (Q) is expressed by

$$k^{-1} \sim c^{u} = Q^{-u} \tag{6.14}$$

where $-u$ is in the range 0.5–1.0 for different polymers [1]. Substitution of Eqs 6.13 and 6.14 into Eq. 6.12 gives

$$\ln(-\ln(D_{PEG}/D_{PEG,0})) = u\ln Q + \ln R_H \tag{6.15}$$

where

$$R_H = kT/6\pi\eta s D_{PEG} \tag{6.16}$$

Using Eq. 6.16, Eq. 6.15 is rewritten as follows:

$$\ln(-\ln(D_{PEG}/D_{PEG,0})) = u\ln Q + \ln(kT/6\pi\eta s) \tag{6.17}$$

In Fig. 6.3, the $\ln[(-\ln(D_{PEG}/D_{PEG,0}))\, D_{PEG,soln}]$ is plotted against $\ln Q$. A straight line with a slope of -0.71 is obtained, which yields $u = -0.71$. Therefore, we have

$$k^{-1} \sim Q^{0.71} = c^{-0.71} \tag{6.18}$$

Several theoretical treatments have been proposed to predict the scaling exponent u, most of them taking into account the network structure and interaction with the solvent [1]. de Gennes [21] proposed the relationship $k^{-1} \sim c^{-0.75}$ for flexible polymer chains diffusing in a host polymer network in a good solvent. The experimental results for the random coil PEG in a PDMAA are close to de Gennes' prediction.

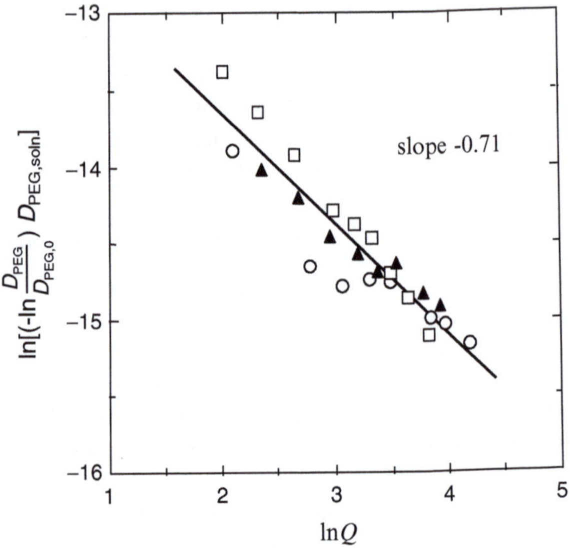

Fig. 6.3 Plot of ln[(−ln ($D_{PEG}/D_{PEG,0}$) $D_{PEG,soln}$] against ln Q at 303 K. M_ws of PEGs used are 4,250 (□), 10,890 (▲), and 20,000 (O)

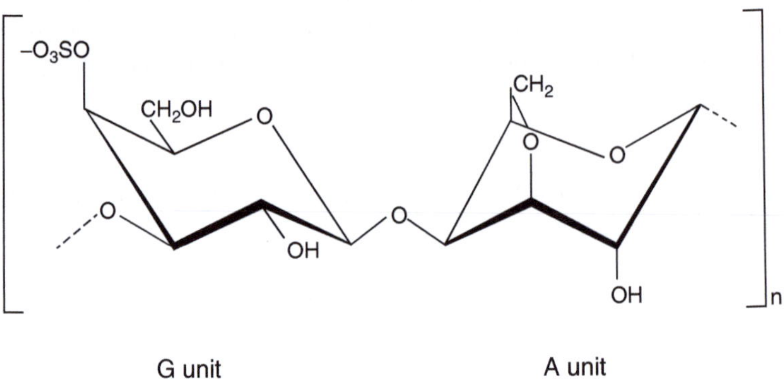

Fig. 6.4 Chemical structure of κ-carrageenan

6.3.2 Gelation Mechanism and Molecular Diffusion in Carrageenan Gels

κ-Carrageenan comes from a family of linear water-soluble polysaccharides extracted from different species of marine red algae. The chemical structure of κ-carrageenan is characterized by an alternating disaccharide composed of α-1, 3-linked galactose-4-sulfate and β-1, 4-linked 3, 6-anhydrogalactose (Fig. 6.4). It is largely used in the food [22], pharmaceutical [23, 24], and cosmetic [25] industries as a gelling and thickening agent, texture enhancer, or stabilizer. The gelling properties of κ-carrageenan have been extensively studied by many researchers [25–31]. The gelling process is generally accepted as a two-step model involving a coil-to-helix conformational transition followed by aggregation of the ordered molecules to form an infinite network [32, 33]. This aggregation leads to a decrease in the solute (random coil) concentration of κ-carrageenan, and therefore, molecular diffusion of probe molecules should be enhanced. This diffusion behavior can provide microscopic insights into the gelation mechanism of κ-carrageenan.

Recently, the diffusion of probe molecules in κ-carrageenan gels has attracted much attention since it is closely related to a variety of applications, such as release of drugs from gels and release of flavor compounds encapsulated in foods [34–37].

In this section we describe a method to estimate hydrodynamic mesh size of k-carrageenan solutions and gels from diffusion measurements of pullulan, added as a probe. The flexibility of the carrageenan chain during aggregation and gelation is also discussed on the basis of ^{1}H T_2 measurements.

6.3.2.1 Pulsed Field Gradient ^{1}H Spectra of Pullulan

Figure 6.5a shows the stacked ^{1}H spectra of a 0.1 % pullulan aqueous solution. The signal intensity of pullulan decayed with increasing g, reflecting the diffusion of pullulan.

Figure 6.5b shows the pulsed field gradient ^{1}H spectra of a 1 % κ-carrageenan solution containing 0.1 % pullulan at various temperatures. The peaks at 5.36 and 5.17 ppm, which are assigned to the anomeric proton of pullulan and the C1 proton on the β-1, 4-linked 3, 6-anhydrogalactose unit of κ-carrageenan, respectively, were chosen for the analysis. The corresponding signals are indicated by p and κ,

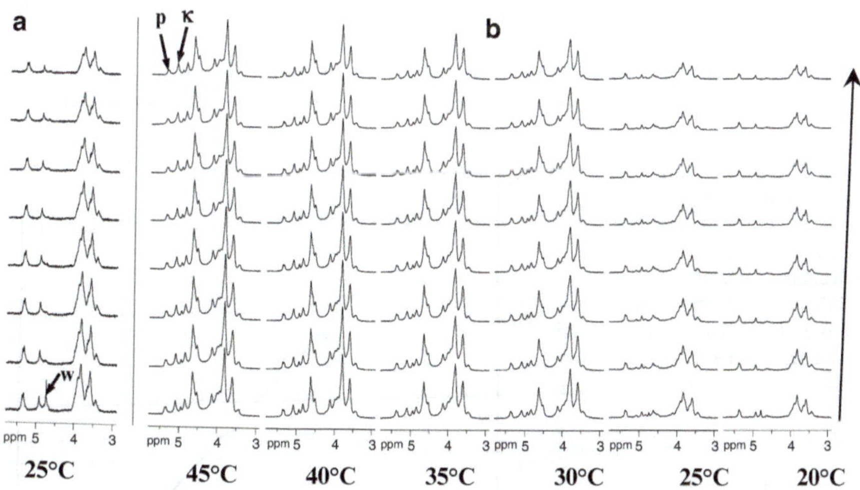

Fig. 6.5 (**a**) Stacked PGSTE ^{1}H spectra of a 0.1 % solution of pullulan with $M_w = 10.7 \times 10^4$ g/mol at 25 °C as a function of the field gradient strength g (g was varied from 1.8 to 6.0 T/m in the vertical direction). The residual water peak is indicated by a "w"; (**b**) Stacked PGSTE ^{1}H spectra of a 1 % κ-carrageenan solution containing 0.1 % pullulan with $M_w = 10.7 \times 10^4$ g/mol as a probe polymer during the cooling process. The peaks selected for the diffusion analysis of the pullulan and κ-carrageenan are indicated by "p" and "κ," respectively. The field gradient strength, g, was varied from 2.0 to 7.0 T/m in the vertical direction

respectively. The water peak at 4.7 ppm was completely eliminated due to the fast diffusion of water.

The intensity of the peak from the anomeric proton of pullulan at 5.36 ppm, which is separated from the peaks of κ-carrageenan, decayed with increasing g and indicates the diffusion of pullulan in the gel structure formed by κ-carrageenan.

Based on Eq. 6.9, a semilogarithmic plot of $I_{\text{pull}}(\text{g})$ as a function of $\gamma^2\delta^2 g^2(\Delta\text{-}\delta/3)$ gives a straight line with a slope of $-D$ if the diffusant has a single diffusion process. I_{pull} (g) values at 25 °C in D_2O and in κ-carrageenan gels, obtained from the data Fig. 6.5, are represented in semilogarithmic plots in Fig. 6.3 and lie on a straight line, indicating that pullulan undergoes single-mode diffusion in both cases. The diffusion coefficient of pullulan, D_{pull}, can be similarly obtained. The diffusion coefficient of pullulan in the κ-carrageenan solution is smaller than that in neat D_2O. This means that the diffusion of pullulan is restricted by the κ-carrageenan network. This effect [38, 39] will be discussed in detail later.

6.3.2.2 Diffusion of Pullulan in Neat Solvent

D_{pull} values in 0.1 % solution of pullulan ($D_{\text{pull},0}$) were measured for pullulan with various M_{w} at 25 °C and plotted against M_{w} in Fig. 6.6. If the diffusion follows Zimm dynamics, that is, there is no draining of solvent molecules from the interior of the polymer, then, the relationship between M_{w} and the diffusion coefficient in a dilute solution, D_0, can be expressed as [40]

$$D_0 = aM_{\text{w}}^{-\nu}. \tag{6.19}$$

where a is a constant related to the segment size of the polymer chain and ν is a scaling exponent that depends on the polymer-solvent system. As seen in Fig. 6.6,

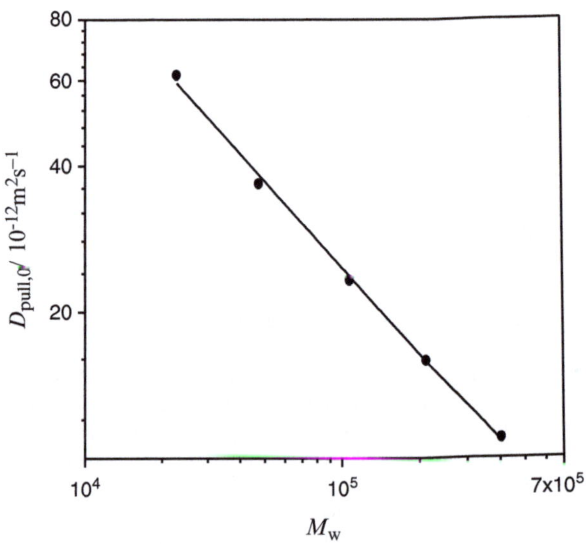

Fig. 6.6 Double-logarithmic plot of the diffusion coefficient of pullulan in D_2O, $D_{\text{pull},0}$ as a function of the molecular weight M_{w} of pullulan at 25 °C (●). The line is a fit of equation of $D_0 = a\,M_{\text{w}}^{\nu}$ to the data

the double-logarithmic plot of $D_{\text{pull,0}}$ as a function of M_w is essentially linear, and the experimental result gives the following equation:

$$D_{\text{pull,0}} = 2.25 \times 10^{-8} M_w^{-0.59} \left(m^2 s^{-1}\right). \tag{6.20}$$

The value of the exponent, $\nu = 0.59$, is close to that for flexible polymers in a good solvent. The experimental results obtained by Nisihinari et al. yielded $\nu = 0.51$ [41]. The difference is most likely attributed to the method used to determine $D_{\text{pull,0}}$,which was determined in a 0.1 % solution, whereas Nisihinari et al. used the intercept of the concentration dependence of the pullulan diffusion coefficient at $c \to 0$ using photon correlation spectrometry.

In dilute solution, the diffusivity of pullulan can be described using the Stokes-Einstein equation. Then, the hydrodynamic radius, R_H at 25 °C, can be determined from $D_{\text{pull,0}}$:

$$R_H = k_B T^\circ / 6\pi\eta_{T^\circ} D_{\text{pull,0}} = 8.82 \times 10^{-3} M_w^{0.59} (nm). \tag{6.21}$$

where k_B is Boltzmann's constant, T° =298 K (=25 °C), η_{T° is the viscosity of heavy water at T°. $R_{H,T}$ was assumed to be constant in the temperature range of the measurements. $D_{\text{pull,0}}$ at temperatures above 30 °C showed diffusion time dependence, indicating the effect of convection [8]. Therefore, $D_{\text{pull,0}}$ was determined as a function of T assuming a constant R_H:

$$D_{\text{pull,0}} = D_{\text{pull,0,}T^\circ} (T/T^\circ)/(\eta_T/\eta_{T^\circ}) \tag{6.22}$$

where $D_{\text{pull,0,}T^\circ}$ is $D_{\text{pull,0}}$ at 25 °C and given by Eq. 6.20, T° =298 K (=25 °C), η_T is the viscosity of heavy water at T.

6.3.2.3 Structural Changes of κ-Carrageenan During Gelation

The temperature dependence of the echo signal intensities of κ-carrageenan without a gradient $I_{\text{kappa}}(0)$ was determined from fits of experimental data to Eq. 6.1. The temperature dependence is shown in Fig. 6.7, where $I_{\text{kappa}}(0)$ values were normalized with $I_{\text{kappa}}(0)$ at 50 °C.

In Fig. 6.5b At high temperature, $I_{\text{kappa}}(0)$ decreased slightly with decreasing temperature, which is most likely a result of an increase in the viscosity upon cooling, which leads to a decrease in the relaxation times T_1 and T_2 and therefore an increase in $I_{\text{kappa}}(0)$ according to Eq. 6.10. At the sol-to-gel transition temperature T_{sg} (28 °C, determined by the falling ball method [10, 42]), $I_{\text{kappa}}(0)$ decreased sharply. This decrease in $I_{\text{kappa}}(0)$ reflects a decrease of the number of κ-carrageenan chains in the random coil conformation due to formation of double helices and helix aggregates. This helix formation strongly restricts the molecular motion of κ-carrageenan and leads to a very short relaxation time [10, 43]. The diffusion coefficient of κ-carrageenan, D_{kappa}, increased markedly as the temperature decreased below T_{sg} (data not shown). The aggregation of κ-carrageenan chains leads to decreased hydrodynamic restriction, as will be discussed later for

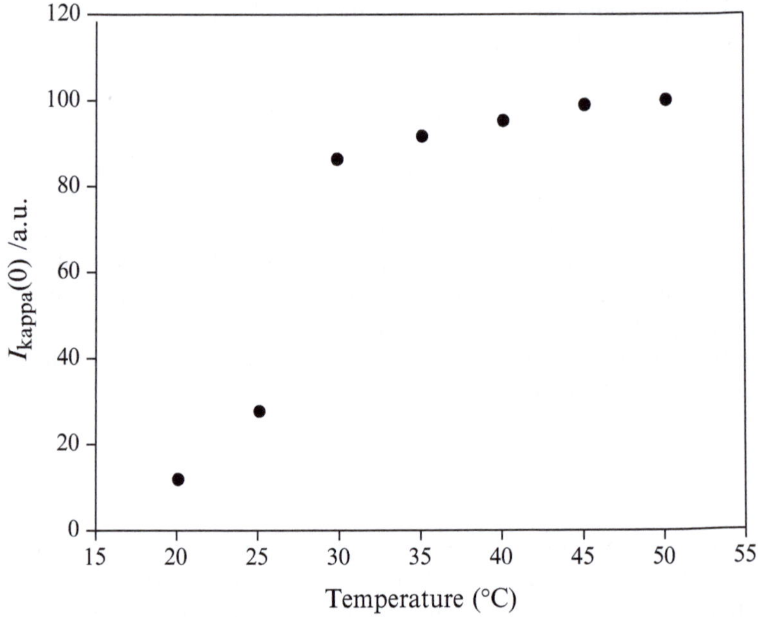

Fig. 6.7 The temperature dependence of the echo signal intensity of κ-carrageenan without a gradient $I_{kappa}(0)$ during the cooling process

diffusion of the probe. However, the increase in D_{kappa} was stronger than expected from this effect. Gel permeation chromatography (GPC) investigations revealed that the κ-carrageenan used in this study has a wide molecular weight distribution which can be roughly divided into two populations. The results suggest that the smaller κ-carrageenan chains did not aggregate and remained the only chains for which D_{kappa} was measurable below T_{sg}. Preferential aggregation of larger chains has also been observed for gellan and agarose [10, 43]. Only about 10 % of the κ-carrageenan chains remained in random coil conformation at 20 °C.

6.3.2.4 Diffusion of Pullulan in κ-Carrageenan

Aggregation of κ-carrageenan, which leads to a decrease in the number of entanglements, was expected to affect the mobility of the probe molecules. This effect was studied using pullulan samples of various Mw. The diffusion coefficient of these pullulan samples in κ-carrageenan is plotted as a function of temperature in Fig. 6.8(a). At high temperature, D_{pull} decreased slightly with decreasing temperature because the molecular mobility decreased. Below T_{sg}, D_{pull} increased with decreasing temperature, indicating an increase in the molecular mobility. As discussed above, the κ-carrageenan chains undergo extensive aggregation that begins around T_{sg}. The formation of thick aggregated bundles and network structures by solute κ-carrageenan chains is thought to result in an increase in the network mesh size [10, 39, 43]. The diffusion of pullulan should be restricted by

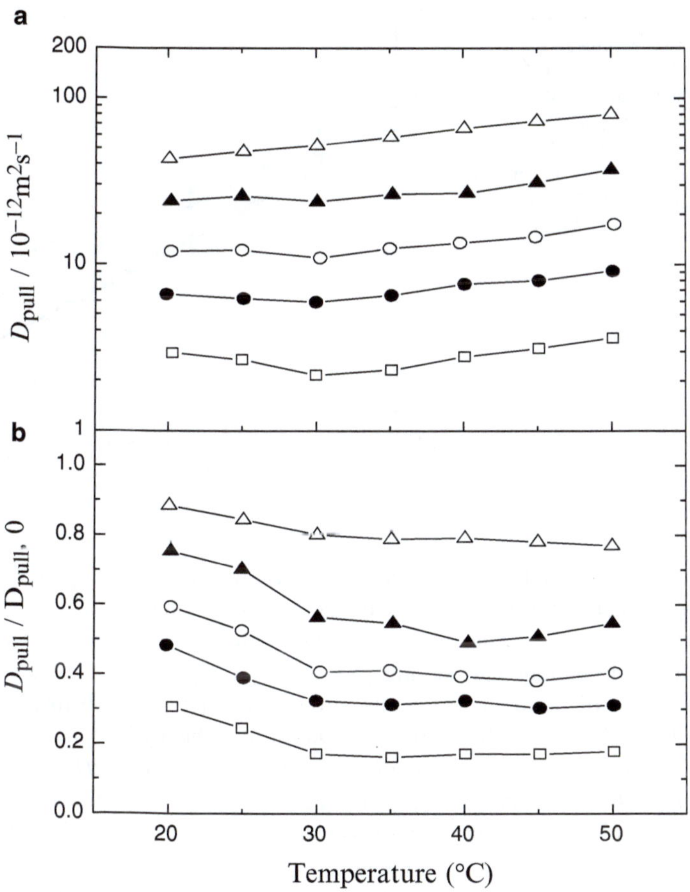

Fig. 6.8 The temperature dependence of (**a**) the diffusion coefficients of pullulan, D_{pull}, and (**b**) the ratio $D_{\text{pull}}/D_{\text{pull,0}}$ in the 1 % κ-carrageenan solutions for pullulan with different molecular weights: $M_w = 40.4 \times 10^4$ g/mol (□), $M_w = 21.2 \times 10^4$ g/mol (●), $M_w = 10.7 \times 10^4$ g/mol (○), $M_w = 4.73 \times 10^4$ g/mol (▲), and $M_w = 2.28 \times 10^4$ g/mol (△). *Lines* are drawn only as a visual aid

hydrodynamic interactions with both random coil κ-carrageenan chains and with the network of aggregates. Therefore, the restriction became smaller as number of κ-carrageenan in the random coil conformation decreased and the network mesh size increased. This is a corollary of the increase in size of aggregates with decreasing temperature.

6.3.2.5 Hydrodynamic Screening Length in a κ-Carrageenan Solution

As shown in Fig. 6.8a, D of pullulan diffusing in κ-carrageenan decreases strongly with increasing Mw. To quantify the effect of Mw on the diffusion behavior, we look at the diffusion coefficient (D_{Pull}) normalized with its value in the neat solvent ($D_{\text{Pull,0}}$). Values of $D_{\text{Pull}}/D_{\text{Pull,0}}$ are plotted as a function of temperature in Fig. 6.8b.

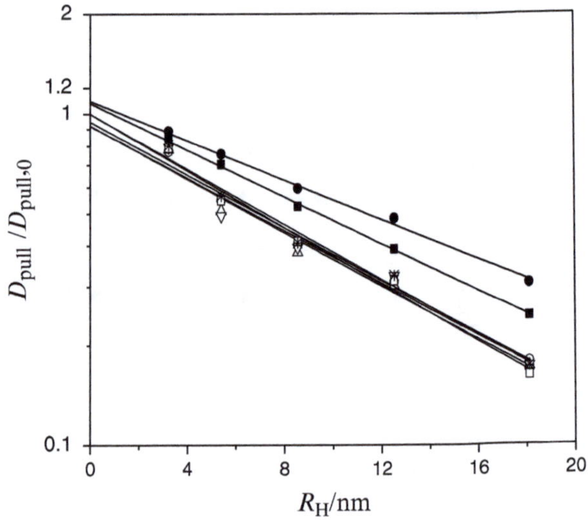

Fig. 6.9 Semilogarithmic plots of $D_{pull}/D_{pull,0}$ as a function of the hydrodynamic radius of pullulan, R_H, at 20 °C (●), 25 °C (■), 30 °C (✳), 35 °C (□), 40 °C (▽), 45 °C (△), and 50 °C (○). The lines are fits of equation of $D_{Pull}/D_{Pull,0} = \exp(-R_H/\xi)$ to the data

The ratio is less than 1 over the entire temperature range. During cooling, $D_{Pull}/D_{Pull,0}$ was nearly constant at high temperatures but increased with decreasing temperature below T_{sg}. This means that the degree of restriction deceased. It is evident that throughout the cooling process, $D_{Pull}/D_{Pull,0}$ values decreased with increasing Mw of pullulan, indicating stronger restriction on diffusion of larger probes. The relationship between $D_{Pull}/D_{Pull,0}$ and the hydrodynamic screening length ξ is given by Eq. 6.12. Semilogarithmic plots of $D_{Pull}/D_{Pull,0}$ vs. R_H are shown in Fig. 6.9. We can obtain ξ from the inverse of the slope linear fits of data in this representation. Note that an increase in ξ is reflected in smaller slopes. The slope became smaller below T_{sg}, indicating an increase in ξ. The estimated ξ values are summarized in Fig. 6.10. A virtually constant value, $\xi = 9.5$ nm, was found above T_{sg}. ξ increased with decreasing temperature below T_{sg} with values of 11.9 nm at 25 °C and 14.4 nm at 20 °C, respectively. These findings were interpreted using the hydrodynamic approach, which assumes that the decrease in entanglement density will lead to an increase in ξ due to aggregation of the κ-carrageenan chains. An equivalent statement is that the pore size between aggregates increases as the aggregation proceeds and the number of random coil chains decreases. This interpretation agrees with the I_{kappa} (0) results shown in Fig. 6.7.

Note that the intercepts of the plot in Fig. 6.9 were not equal to 1 and were also shifted to higher values below T_{sg}. This behavior is attributed to the differences between the microviscosity of the network interspace and the pure solvent. The local viscosity is increased by the presence of non-aggregated κ-carrageenan chains that are smaller than the pullulan chains. These chains cannot contribute to the hydrodynamic restriction on the probe mobility as a network but can increase the microviscosity felt by pullulan. The increase of the intercept most likely indicates that the local viscosity has decreased due to aggregation of these small

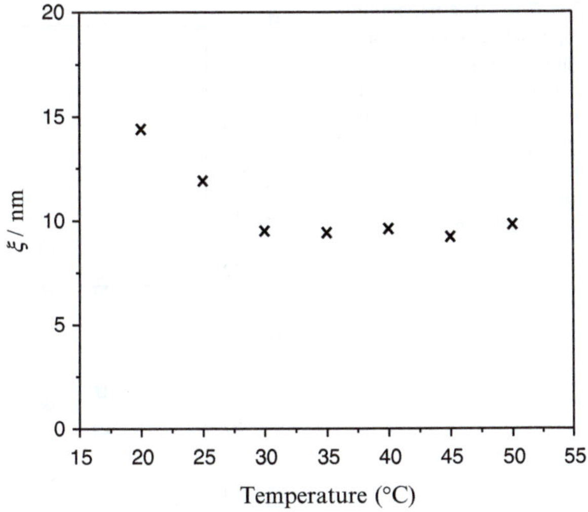

Fig. 6.10 The temperature dependence of the hydrodynamic screening length, ξ, in a 1 % κ-carrageenan solution, estimated from the results shown in Fig. 6.9

κ-carrageenan chains below T_{sg}. However, the reason for a value greater than 1 is not well understood.

6.3.2.6 Translational and Rotational Mobility of κ-Carrageenan (CarPol13)

The spin-spin relaxation time T_2 of κ-carrageenan was determined using the CPMG method. The data at all temperatures lie on a straight line in the semilogarithmic plots of the echo intensity against the echo time and were therefore analyzed assuming a monomodal relaxation, and the fits to Eq. 6.4 yield the ^{1}H T_2 values as well as the signal intensity at echo time $\tau = 0$ in CPMG measurements, $I_{kap,CPMG}(0)$. The corresponding results are shown in Fig. 6.3. Values of $I_{kap,CPMG}(0)$ decreased slightly in the temperature range from 55 to 35 °C (stage 1). In this stage, the T_2 values are in the range 35–50 ms (Fig. 6.11b), indicating that the κ-carrageenan chains have high segmental mobility. The values of $I_{kap,CPMG}(0)$ decreased steeply with decreasing temperature from 35 to 25 °C (stage 2), but the semilogarithmic plots still showed monomodal relaxatio; see Fig. 6.5b. The gelation temperature T_{sg} determined by the falling ball method for a 2 % κ-carrageenan solution is 35 °C. We therefore concluded that the steep decrease in the signal intensity in stage 2 is due to the formation of κ-carrageenan helix aggregates with highly restricted segmental mobility and extremely short T_2. Consequently, the corresponding signal of aggregates was not detected before the first echo time of 4 ms. Namely, only the non-aggregated κ-carrageenan chains (solute κ-carrageenan in random coil conformation) contributed to the CPMG echo signal intensity. It follows that $I_{kap,CPMG}(0)$ is proportional to the solute fraction of κ-carrageenan chains, so that the fraction of soluble κ-carrageenan chains at each temperature can be estimated by comparing $I_{kap,CPMG}(0)$ to $I_{kap,CPMG}(0)$ at the highest temperature. The aggregation of κ-carrageenan chains continued with decreasing temperature and at the lowest

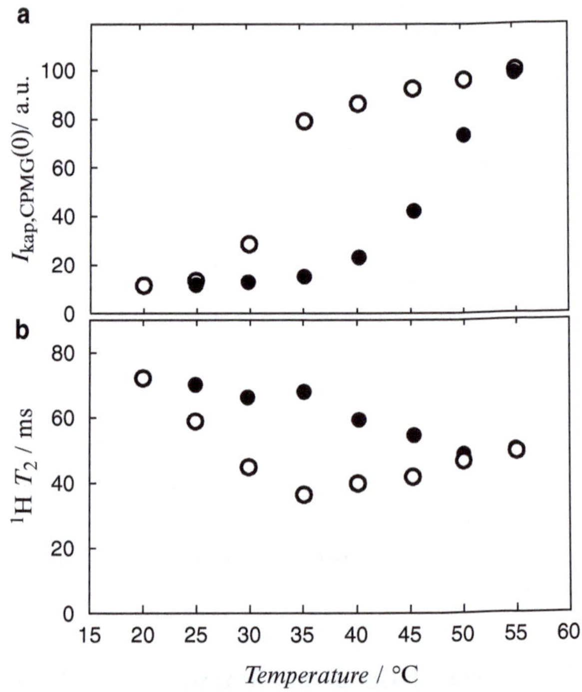

Fig. 6.11 The temperature dependence of (**a**) signal intensity of κ-carrageenan at echo time $\tau = 0$ of the CPMG measurements and (b) ^{1}H spin-spin relaxation times T_2 of a 2 % κ-carrageenan solution during cooling (○) and heating (●)

temperature measured, about 10 % of soluble κ-carrageenan chains were left in the solution. In stage 2, the T_2 values of the solute increased with decreasing temperature, indicating an increase in molecular mobility. The most plausible explanation for this increase in T_2 is a decrease in the microscopic viscosity following the incorporation of κ-carrageenan chains aggregates into the gel network with a concomitant decrease in the number of random coil κ-carrageenan chains. On heating, $I_{\text{kap,CPMG}}(0)$ showed an increase with increasing temperature above the melting temperature T_{gs} (ca. 43 °C, determined by the falling ball method), indicating melting of the network and aggregates. This was accompanied with a decrease in T_2, attributed to the increase in microviscosity.

6.3.3 Gelation Mechanism in Agar Gels

Agar is a sulfated galactan extracted from marine red algae and is widely used as a texturing agent in food and nonfood applications [44]. The wide use of agar is based on its unique ability to form strong gels in aqueous solutions. On cooling, the gelation of agar solutions is initiated by a coil-to-helix transition, followed by the aggregation of the helices to form a network structure, i.e., a gel. Agar is composed of agarose and agaropectin, the former consisting of neutral polysaccharides with a high gelling ability and the latter consisting of ionic polysaccharides with a low gelling ability [44]. The ratio of agarose to agaropectin varies depending on the source. This ratio strongly affects the physicochemical, mechanical, and rheological properties of agar. In this section, the molecular mobility of agar chains is discussed, and the hydrodynamic mesh size in agar solution was estimated from the diffusion of a dendrimer added as a probe polymer.

6.3.3.1 Molecular Mobility

PFG-STE ^{1}H NMR spectra were measured for a 2.3 wt% agar solution at various temperatures. The area of the agar peaks at chemical shifts from 3.0 to 5.8 ppm was analyzed. The logarithm of the integrated peak area was essentially linear when plotted against $\gamma^2\delta^2g^2(\Delta\text{-}\delta/3)$ throughout the investigated temperature range. The agar sample used is polydisperse. Despite the polydispersity, deviations of the experimental data from straight line fits were small at all temperatures. Therefore, the D of agar (D_{agar}) was determined by fitting the data points using Eq. (6.9), which yields a number averaged D and also the $I(0)$ of $I_{agar}(0)$.

The temperature dependence of $I_{agar}(0)$ and D_{agar} in a 2.3 wt% agar solution is shown in Fig. 6.12, where $I_{agar}(0)$ values are normalized with $I_{agar}(0)$ at 60 °C. During cooling, $I_{agar}(0)$ remained constant from 60 to 50 °C (stage 1) and decreased sharply from 50 to 35 °C (stage 2).

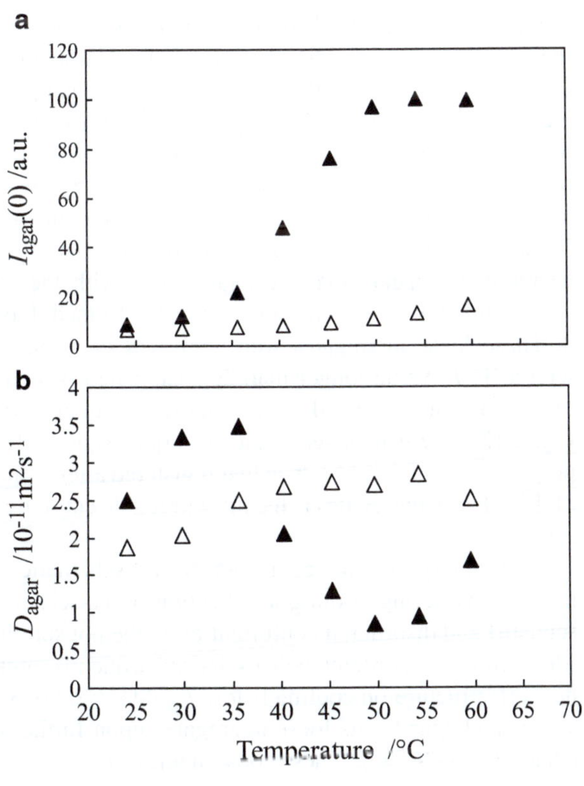

Fig. 6.12 Temperature dependences of (**a**) the ^{1}H intensities for agar at $g = 0$, $I_{agar}(0)$, and (**b**) the diffusion coefficient for agar, D_{agar}, in the 2.3 wt% agar solution during cooling (*closed triangles*) and heating (*open triangles*)

The polysaccharide chains of agar have a rather large persistence length when dissolved in water at high temperature. This relative stiffness leads to a random conformation that we will refer to as a semistiff coil. The semistiff coils have some degree of segmental mobility with much longer ^{1}H T_2 values (ca. 81–103 ms) relative to the echo time $2\tau_2$ (ca. 4 ms) and also in longer ^{1}H T_1 values (ca. 1 s) relative to the echo time τ_1 (ca. 8 ms), giving a practically undecayed intensity $I_{agar}(0)$. In contrast, agar helix aggregates have strongly restricted segmental mobility because of their rigid structure [45], resulting in short ^{1}H T_2 values and leading to disappearance of their signal during the echo times. As a consequence, only the semistiff coils contributed to $I_{agar}(0)$. The decrease in $I_{agar}(0)$ during stage 2 therefore indicates a decrease in the number of random and semistiff coils or, in other words, an increase in the number of helix aggregates [46] that form a three-dimensional gel network. This interpretation is consistent with T_{sg} from dynamic rheological measurements, 45 °C. In stage 1, however, D_{agar} decreased slightly due to a decrease in molecular mobility with decreasing temperature. We should note that D_{agar} at 60 °C could have been overestimated due to convection, which can become comparable to the small displacements under diffusion for large molecules [47–50]. In stage 2, D_{agar} increased significantly with decreasing temperature. As evidenced by the change in $I_{agar}(0)$, a large fraction of the agar chains formed helix aggregates in this stage. Agar chains have varying gelling ability, which is markedly higher for agarose and lower for agaropectin because of the presence of ionic groups on the latter [44]. If chains with higher Mw are preferentially aggregated, the molecular weight distribution of the non-aggregated agar should decrease, causing an increase in the measured D. This conjecture was supported by GPC measurements carried out on liquid squeezed from the 2.3 wt% agar gel. As shown in Fig. 6.5, the molecular weight distribution for of the squeezed liquid, which was heated at 95 °C for 1 h to dissolve aggregates, was markedly lower than that of the native agar sample. $I_{agar}(0)$ gradually decreased on cooling from 35 to 24 °C (stage 3), and $I_{agar}(0)$ at 24 °C was about 10 % of that at 60 °C, where D_{agar} started to decrease. It appears that the agar chains with the lowest aggregating capacity, which did not aggregate above 35 °C, formed loose aggregates during stage 3. These loose aggregates were expected to display certain molecular mobility with a ^{1}H T_2 value longer than $2\tau_2$ (ca. 4 ms), which allowed them to contribute to the echo signal after the echo times. This in agreement with the GPC results of Fig. 6.13, where it is seen that the liquid squeezed from the gel had contained polymers of a higher Mw than that registered after being heated to 95 °C. Formation of these loose aggregates is the most probable cause for the decrease in D_{agar} during stage 3.

In summary, we suggest in Fig. 6.14 a schematic view of the conformation of agar chains during cooling and heating. Polysaccharide chains in agar exist in a semistiff and disordered conformation in the hot solution state (stage 1, 60–50 °C). These coils form ordered helices that subsequently aggregate laterally, generating a network structure on cooling below T_{sg} (stage 2, 50–35 °C). A fraction of the non-aggregated agar forms loose aggregates upon further decrease in the temperature (stage 3, 35–24 °C). These loose aggregates dissociate with relative ease upon

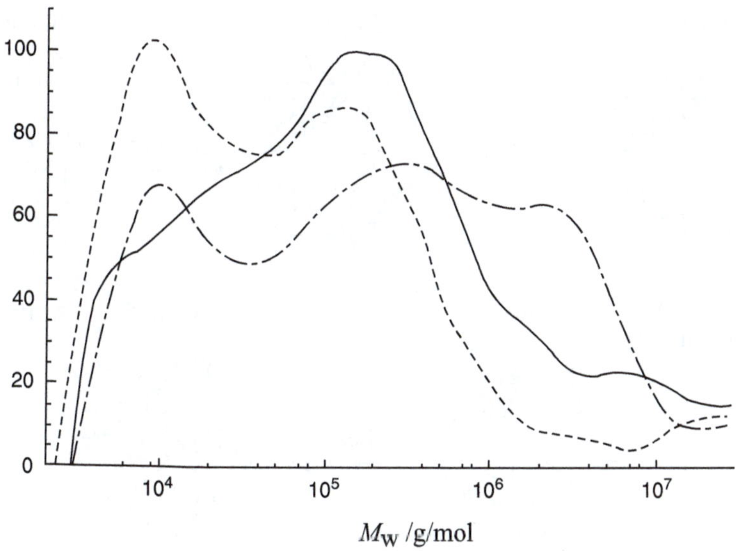

Fig. 6.13 Molecular weight distribution for the original agar (*solid line*), for the solute agar squeezed from the agar gel (*dot-dash line*), and for the solute agar after heating in solution at 95 °C for 1 h (*dashed line*). Experimental conditions are listed in Table 6.1

Table 6.1 Experimental conditions for GPC measurements on different samples

Samples	Temperature/°C			
	Squeezing	Setting	Inlet	Column
(a) Prepared by dissolving agar powder in water	-	95	60	60
(b) Squeezed from 2.3 wt% agar gel	25	r.t.	30	30
(c) Squeezed from 2.3 wt% agar gel and heated at 95 °C for 1 h	25	95	60	60

reheating (stage 4, 24–40 °C). Finally, a fraction of the helix aggregates melts gradually when the temperature is increased further (stage 5, 40–60 °C).

6.3.3.2 Estimating the Hydrodynamic Mesh Size in Agar Solutions and Gels

The effects of the macromolecular aggregation events described above on macromolecular diffusion were studied by measuring the diffusion coefficient of a dendrimer, D_{dend}. In the PFG-STE ^{1}H NMR spectra of a 2.3 wt% agar solution containing 0.1 wt% dendrimer, the peaks of the dendrimer ethylene protons (2.01–3.03 ppm) are well separated from other peaks and were thus chosen for the analysis. D_{dend} shows a decrease in stage 1, a slight increase in stage 2, and a decrease in stage 3. We again look at the normalized dendrimer diffusion coefficient, $D_{dend}/D_{dend,0}$, which is plotted as a function of temperature in Fig. 6.15. This ratio was constant during stage 1, increased in stage 2, and leveled off in stage 3. On

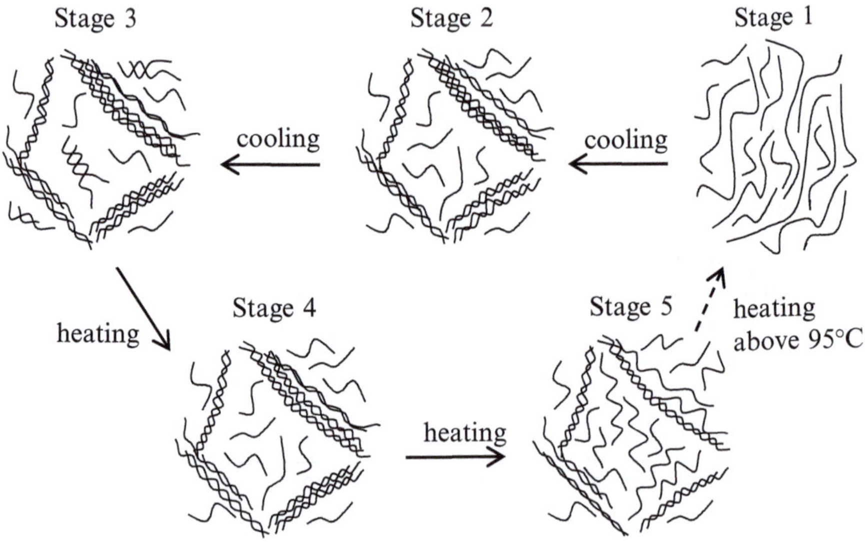

Fig. 6.14 Schematic depiction of the conformation of agar molecules at different stages during cooling and heating

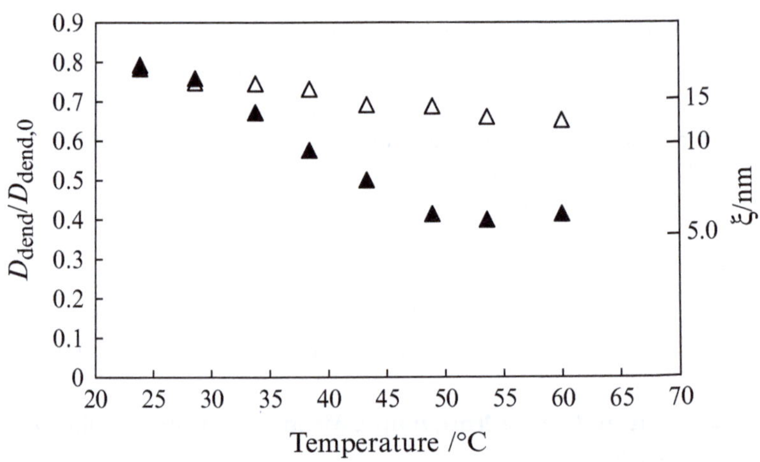

Fig. 6.15 Temperature dependence of $D_{dend}/D_{dend,0}$ in the 2.3 wt% agar solution during cooling (*closed triangles*) and heating (*open triangles*). The corresponding ξ values, calculated using $D_{inHost}/D_{inPure} = \exp(-R_H/\xi)$, are displayed on the right-hand axis

reheating, $D_{dend}/D_{dend,0}$ remained almost constant in stage 4 and decreased slightly in stage 5. These results check with the aggregation process as described above. It seems that the number of entanglements is the dominant factor determining probe diffusion in agar solutions and gels.

The ξ values calculated from $D_{dend}/D_{dend,0}$ using Eq. 6.12 are displayed on the right-hand axis in Fig. 6.15. The restriction was mostly from non-aggregated agar chains. Therefore, ξ should reflect the hydrodynamic mesh size of the non-aggregated agar.

6.3.4 Gelation Mechanism and Molecular Diffusion in Agarose Gel

Agarose is a linear and essentially sulfate-free polysaccharide with a sugar skeleton consisting of alternating 1,3-linked β-D-galactopyranose and 1,4-linked 3,6 anhydro-α-L-galactopyranose units [45, 51] and is widely used for its gelling properties. In hot solution, agarose chains assume a random, semistiff coil conformation [52–54]. Upon cooling below the gelation temperature, the coils reorder to form helices that subsequently aggregate into a three-dimensional network and form a gel. Although the helical structure is generally thought to be responsible for imparting restricted molecular motion, there is disagreement on whether the proposed high-ordered structure is a single helix [53, 55] or a double helix [45]. Due to the presence of thick helix aggregates, agarose gels are typically rigid and turbid and show considerable hysteresis between the melting and setting temperatures [56]. The formation of aggregates leads to a decrease of entanglement density, which leads to a decrease in the microviscosity. For this reason, the aggregation affects strongly molecular diffusion in gels [10, 43].

To shed further light on the aggregation and gelation of agarose, a highly water-soluble COONa dendrimer was introduced into solutions containing various concentrations of agarose to study the effect of agarose concentration and aggregation on probe diffusion.

6.3.4.1 Molecular Mobility of Agarose Chains

Figure 6.16 shows the temperature dependence of $I_{aga}(0)$ (determined from the PGSTE and CPMG measurements), ^{1}H T_2 (determined from the 2.2 wt% agarose solution), and D_{aga} (determined from the 2.1 wt% agarose solution containing 0.1 wt% dendrimer). The $I_{aga}(0)$ values obtained from the PGSTE and CPMG measurements were normalized on the basis of their respective values at 55 and 58 °C. During cooling, the $I_{aga}(0)$ values, determined from both measurements, remained relatively constant on decreasing the temperature from 55 to 50 °C (stage 1), steeply decreased from 50 to 35 °C (stage 2) and gradually decreased below 35 °C (stage 3).

In the CPMG measurements, we observed the decay of the echo signals, which provided information on the molecular mobility of the agarose chain. The semilogarithmic plot of the CPMG echo intensity against the echo time revealed linear relationships throughout stage 1 (Fig. 6.17), yielding ^{1}H T_2 values of ca. 70 ms, suggesting that the agarose chains have high segmental mobility with a narrow distribution. This high mobility corresponds to the random coil conformation of agarose at high temperature. In stage 2, the echo intensity became slightly smaller and the decay could still be described with a single component, yielding ^{1}H T_2

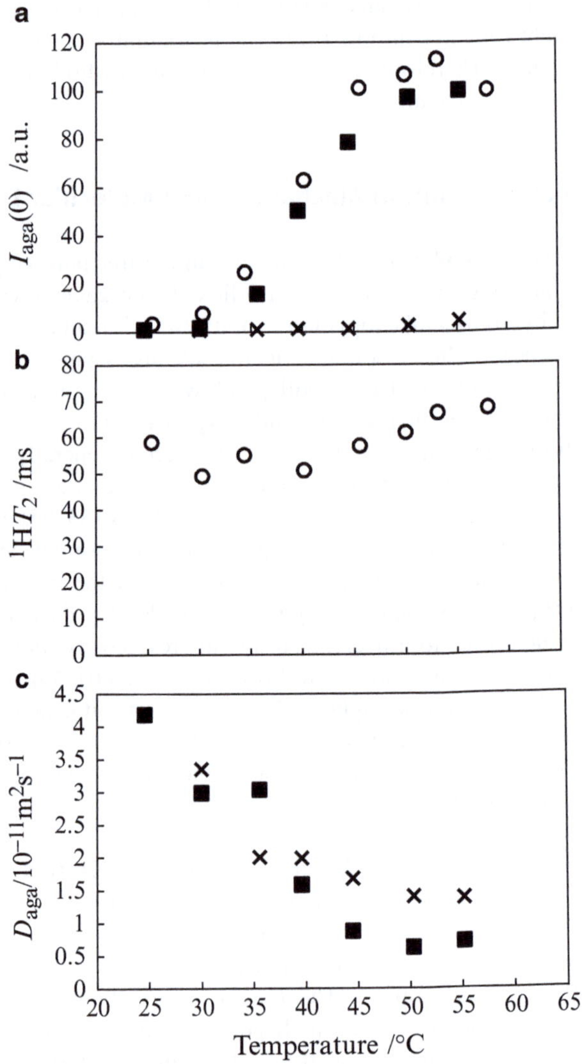

Fig. 6.16 Temperature dependence of (**a**) $I_{aga}(0)$ obtained by PGSTE on cooling (*filled squares*) and on reheating(*crosses*) and by CPMG on cooling (*open circles*), (**b**) ^1H T_2 by CPMG on cooling, and (**c**) D_{aga} values of agarose obtained by PGSTE on cooling (*filled squares*) and on reheating (*crosses*). For this experiment, a 2.1 wt% agarose solution was used

values of ca. 60 ms. We consider this decrease of the echo intensity to owe to formation of aggregates with highly restricted segmental mobility. As already described for agarose, we can approximate the fraction of non-aggregated polymer chains from $I_{aga}(0)$. It is also possible that the non-aggregated agarose chains have high segmental mobility with a narrow distribution similar to that in stage 1. The $I_{aga}(0)$ value, obtained from the PGSTE measurements, which has an echo time of ca. 10 ms, showed the same behavior as $I_{aga}(0)$ from the CPMG measurements. It should be noted that $I_{aga}(0)$ began to decrease around 50 °C and continued to decrease down to ca. 30 °C, passing through T_{sg} of 41 °C. This result suggests that helix aggregates were formed above T_{sg}, and a three-dimensional network

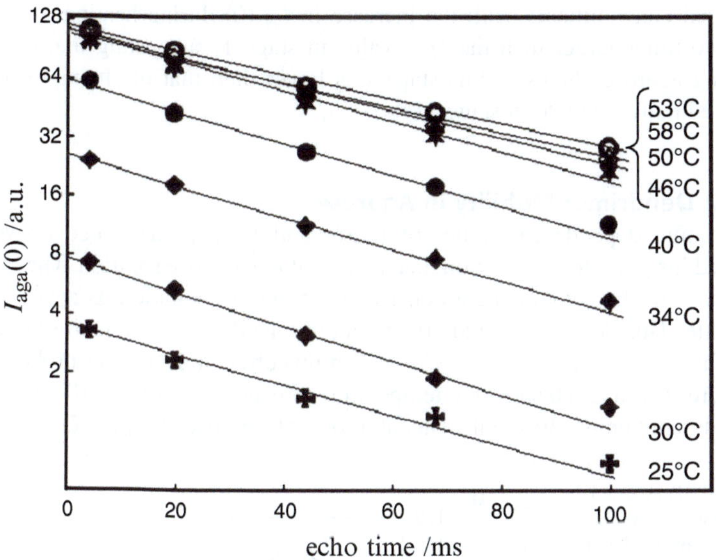

Fig. 6.17 Semilogarithmic plots of peak intensities in the CPMG echoes against the echo time for the 2.2 wt% agarose solution measured at various temperatures during cooling

structure was formed when about half of the agarose chains had aggregated. From the observed gradual decrease of $I_{aga}(0)$ in stage 3, we conclude that aggregation of agarose takes place over a very wide range of temperatures. $I_{aga}(0)$ at 30 °C was only 2 % of that at 55 °C, which reflects the high gelling ability of agarose, as almost all of the agarose chains are aggregated at 30 °C. A schematic representation of the aggregation events is given in Fig. 6.14.

As seen in Fig. 6.16c, the diffusion coefficient of agarose chains (D_{aga}) remained constant during cooling in stage 1 but markedly increased in stage 2. This is because of agarose aggregation in stage 2. Like in the case of κ-carrageenan, however, this increase is stronger than expected. The agarose used in this study has a very wide molecular weight distribution, with an Mw/Mn value of 3.48. Again, the most likely explanation for the strong increase in D_{aga} is the preferential aggregation of the larger polymer chains, leaving the smaller ones to diffuse at lower temperature [10]. The D_{aga} values in stage 3 were more than five times higher than D_{aga} values in stage 1. ^{1}H T_2 values, on the other hand, showed only a small change on cooling, suggesting opposing effects of the decrease in temperature and molecular weight, where the latter has a weaker effect on the segmental mobility than that on the mobility of the whole molecule.

During reheating, $I_{aga}(0)$ slightly increased for 2.1 wt% agarose but remained low, up to ca. 5 % of that at 55 °C (Fig. 6.16a), indicating thermal stability of the helix aggregate. These findings are consistent with the high thermal stability (melting temperature of 88 °C) of the gel as determined from the rheological measurement using the falling ball method (data not shown). Although D_{aga}

decreased concomitantly with the increase in $I_{aga}(0)$ during heating, it remained over two times larger than the D_{aga} value in stage 1, suggesting that the size of diffusing agarose chains in this stage was lower than that of chains of the fully dissolved native sample in stage 1.

6.3.4.2 Dendrimer Mobility in Agarose

Temperature dependence studies of $I_{aga}(0)$ and D_{aga}, as described above, have provided insight into the gelation mechanism of the agarose solution. They indicate a decrease in the entanglement density in the agarose solution as agarose chains aggregate. This decrease should affect not only the diffusion of agarose chains but also of macromolecular probes. We studied this effect a dendrimer probe.

Figures 6.18a and b show the temperature dependencies of $I_{dend}(0)$ and D_{dend} in solutions containing different concentrations of agarose. $D_{dend,0}$ (D_{dend} in dilute

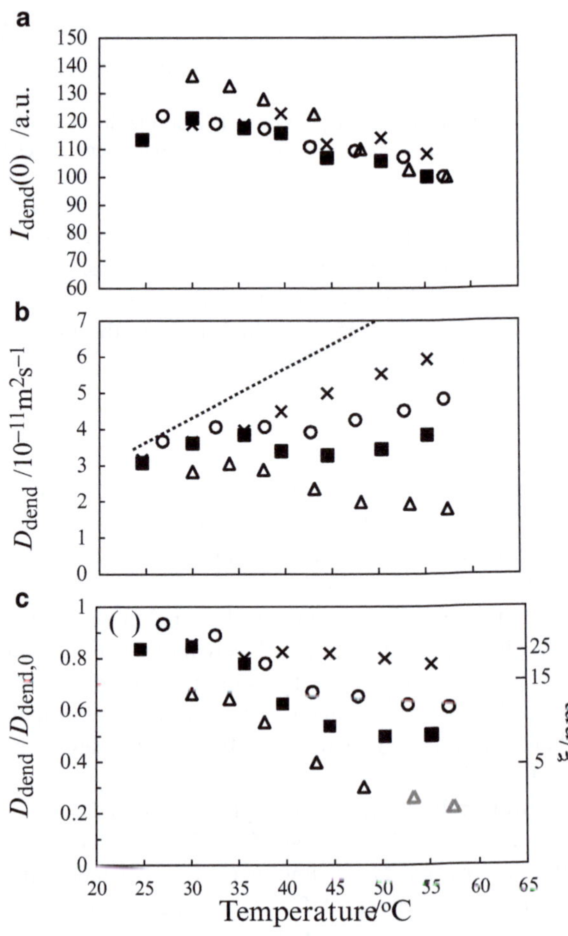

Fig. 6.18 Temperature dependence of (**a**) ^{1}H intensities of the dendrimer at $g = 0$, $I_{dend}(0)$, (**b**) diffusion coefficient of the dendrimer, D_{dend}, and (**c**) $D_{dend}/D_{dend,0}$ ratio. Data were obtained from 1.1 wt% (*open circles*), 2.1 wt% (*closed squares*), and 4.0 wt% (*open triangles*) agarose solutions during cooling. Results obtained during the heating process of 2.1 wt% agarose (*crosses*) are also displayed. The *dashed line* represents $D_{dend,0}$ values of 0.1 % dendrimer in D_2O at different temperatures

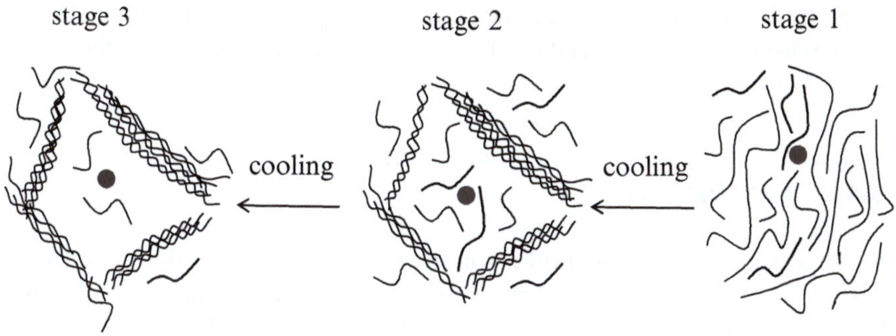

Fig. 6.19 Schematic representation of the microscopic environment of the dendrimer at different stages during cooling of the agarose gel. Agarose chains are indicated using lines and dendrimer is indicated using closed sphere

D_2O solution) values calculated with the reported viscosity of D_2O [57] are also shown (see dashed line in Fig. 6.18b). For all agarose solutions used in this study, $I_{\mathrm{dend}}(0)$ exhibited a gradual and continuous increase on decreasing the temperature. This increase in $I_{\mathrm{dend}}(0)$ was probably caused by an increase in T_1 and T_2 relaxation times of the dendrimer, probably due to the decrease in microviscosity as agarose chains progressively aggregate. This results in diminished decay during the echo times of the PGSTE sequence and therefore higher $I_{\mathrm{dend}}(0)$ values. The continuous change in $I_{\mathrm{dend}}(0)$ indicates that the dendrimer is not involved in the aggregation and gelation of agarose and that it remains dissolved without sticking to the polysaccharide network. In stage 1, D_{dend} in 1.1 % and 2.1 % agarose solutions decreased with decreasing temperature, suggesting a decrease in the molecular mobility of the dendrimer, which can be explained through the effect of temperature on the kinetic energy and the viscosity. D_{dend} in 4 wt% agarose solution in stage 1 was markedly lower in the 1.1 % and 2.1 % agarose solutions, probably because in addition to the hydrodynamic interaction, obstruction effects became more important due to the higher agarose concentration. In the 4 wt% agarose solution, D_{dend} slightly increased with decreasing temperature. The reason of this behavior is not clear at this moment.

For all the agarose concentration studied, D_{dend} slightly increased in stage 2 and then decreased in stage 3. This observed increase in D_{dend} in stage 2 should correspond, as in the case of agar and κ-carrageenan, to a decrease in entanglement density due to aggregation and gelation. Figure 6.19 provides a schematic representation of the changes in the microscopic environment of the dendrimer during cooling. The ratio $D_{\mathrm{dend}}/D_{\mathrm{dend},0}$ is plotted as a function of temperature in Fig. 6.18c. Symbols indicated in gray for D_{dend} in the 4 wt% agarose solution at high temperature symbolize values of D_{dend} where we would expect an influence by a different mechanism, namely, obstruction or entanglement with agarose chains. This plot shows that the ratio $D_{\mathrm{dend}}/D_{\mathrm{dend},0}$ increased steeply in stage 2, indicating the decrease in the degree of restriction on probe diffusion. This plot also shows that the ratio $D_{\mathrm{dend}}/D_{\mathrm{dend},0}$ for 2.1 wt% agarose remained fairly constant during reheating, which agrees with the behavior of $I_{aga}(0)$. Thus, both Fig. 6.16c and 6.18c reflect the thermal stability of aggregated agarose, especially at the higher concentrations.

6.4 Conclusions

We have demonstrated that NMR is a useful tool for the investigation of molecular mobility in hydrogel systems. Discussion of experimental results for water and probe diffusion in synthetic polymer and polysaccharide gels and solution was offered. Relaxation times for the macromolecules and water give information on tumbling motion. The diffusion coefficient of probe molecules in hydrocolloid systems provided the information on the translational mobility of molecules, which can be used to infer the structure of the gel network. By comparing the results with other experimental techniques, a clear picture emerges, with clear correspondence of the microscopic events, namely, aggregation, polymer immobilization, and subsequent effects on molecular flexibility and probe diffusion, with the macroscopic (bulk) events, namely, gelation. We discussed in detail the hydrodynamic shielding length, ξ, which represents within the mean field hydrodynamic approach the mesh size of the network, as a parameter that determines D/D_0 of probe molecules. This parameter can be used to describe quantitatively the evolving structure of the gel network.

References

1. Cukier R (1984) Diffusion of Brownian spheres in semidilute polymer solutions. Macromolecules 17:252–255
2. Phillies GDJ, Malone C, Ullmann K, Ullmann GS, Rollings J, Yu L (1987) Probe diffusion in solutions of log-chain polyelectrolytes. Macromolecules 22:2280–2289
3. Cameron RE, Jalil MA, Donald AM (1994) Diffusion of bovine serum-albumin in amylopectin gels measured using Fourier-Transform Infrared. Microspectrosc Macromol 27:2708–2713
4. Matsukawa S, Ando I (1996) A study of self-diffusion of molecules in polymer gel by pulsed-gradient spin-echo 1H NMR. Macromolecules 29:7136
5. Matsukawa S, Ando I (1997) Study of self-diffusion of molecules in polymer gel by pulsed-gradient spin-echo 1H NMR 2. Intermolecular hydrogen-bond interaction. Macromolecules 30:8310
6. Matsukawa S, Ando I (1999) Study of self-diffusion of molecules in polymer gel by pulsed-gradient spin-echo 1H NMR. 3. Stearyl itaconimide/N, N-dimethylacrylamide copolymer gels. Macromolecules 32:1865
7. Matsukawa S, Yasunaga H, Zhao C, Kuroki S, Kurosu H, Ando I (1999) Diffusion processes in polymer gels as studied by pulsed field-gradient spin-echo spectroscopy. Prog Polym Sci 24 (7):995–1044
8. Zhao Q, Matsukawa S (2012) Estimation of hydrodynamic screening length in κ-carrageenan aqueous system through probe diffusion using gradient. Polym J 44:901–906
9. Zhao Q, Brenner T, Matsukawa S (2013) Molecular mobility and microscopic structure changes in κ-carrageenan solutions studied by gradient NMR. Carbohydr Polym 95:458–464
10. Dai B, Matsukawa S (2012) NMR studies of the gelation mechanism and molecular dynamics in agar solutions. Food Hydrocoll 26:181–186

11. Dai B, Matsukawa S (2013) Elucidation of gelation mechanism and molecular interactions of agarose in solution by 1H-NMR. Carbohydr Res 365:38–45
12. Stejskal EO, Tanner JE (1965) Spin diffusion measurements: spin echoes in the presence of a time dependent field gradient. J Chem Phys 42:288
13. Karger J, Pfeifer H, Heink W (1988) Principles and applications of self-diffusion measurements by nuclear magnetic resonance. Adv Magn Reson 12:1
14. Callaghan PT (1991) Principles of nuclear magnetic resonance microscopy. Clarendon, Oxford, p 93
15. Price WS (1997) Pulsed field gradient NMR as a tool for studying translational diffusion, Part I. Basic theory. Concepts Magn Reson 9:299–336
16. Price WS (2009) Diffusion and its measurements. In: NMR study of translational motion, 1st ed. Cambridge University Press, New York. pp 1–68
17. Langevin D, Rondelez FP (1978) Sedimentation of large colloidal particles through semidilute polymer solutions. Polymer 19:875–882
18. Wheeler LM, Lodge TP (1989) Tracer diffusion of linear polystyrenes in dilute, semidilute and concentrated polyvinyl methyl ether solutions. Macromolecules 22:3399–3408
19. Rotstein NA, Lodge TP (1992) Tracer diffusion of linear polystyrene in poly(vinyl methyl ether) gels. Macromolecules 25:1316–1325
20. Nemoto N, Landry MR, Noh I, Kitano T, Wesson J, Yu H (1985) Concentration dependence of self diffusion coefficient by forced Rayleigh scattering: polystyrene in tertahydrofuran. Macromolecules 18:308–310
21. de Gennes PG (1976) Dynamics of entangled polymer solution II. Inclusion of hydrodynamic interaction. Macromolecules 9:594–598
22. Imeson AP (2000) Chapter 5: Carrageenan. In: Phillips GO, Williams PA (eds) Handbook of hydrocolloids. Woodhead Publishing, Cambridge, pp 95–100
23. Makino K, Idenuma R, Murakami T, Ohshima H (2001) Design of a rate and time-programming drug release device using a hydrogel: pulsatile drug release from κ-carrageenan hydrogel device by surface erosion of the hydrogel. Colloids Surf B: Biointerfaces 20:355–359
24. Garcia AM, Ghaly ES (1996) Preliminary spherical agglomerates of water soluble drug using natural polymer and cross-linking technique. J Control Release 40:179–186
25. Campanella L, Roversi R, Sammartino MP, Tomassetti M (1998) Hydrogen peroxide determination in pharmaceutical formulation and cosmetics using a new catalase biosensor. J Pharm Biomed Anal 18:105–116
26. Hjerde T, Smidsrød O, Christensen BE (1999) Analysis of the conformational properties of κ and ι-carrageenan by size-exclusion chromatography combined with low-angle laser light scattering. Biopolymers 49:71–80
27. MacArtain P, Jacquier JC, Dawson KA (2003) Physical characteristics of calcium induced κ-carrageenan networks. Carbohydr Polym 53(4):395–400
28. Mangione MR, Giacomazza D, Bulone D, Martorana V, Cavallaro G, San Biagio PL (2005) K^+ and Na^+ effects on the gelation properties of κ-carrageenan. Biophys Chem 113:129–135
29. Mangione MR, Giacomazza D, Bulone D, Martorana V, San Biagio PL (2003) Thermoreversible gelation of κ-carrageenan: relation between conformational transition and aggregation. Biophys Chem 104:95–105
30. Sankalia MG, Mashru RC, Sankalia JM, Sutariya VB (2006) Stability improvement of alpha-amylase entrapped in kappa-carrageenan beads: Physicochemical characterization and optimization using composite index. Int J Pharm 312:1–14
31. Nono M, Nicolai T, Durand D (2010) Gel formation of mixtures of κ-carrageenan and sodium caseinate. Food Hydrocoll 25:750–757
32. Rochas C, Rinaudo M (1984) Mechanism of gel formation in κ-carrageenan. Biopolymers 23:735–745
33. Takemasa M, Chiba A (2001) Gelation mechanism of κ- and ι-carrageenan investigated by correlation between the strain-optical coefficient and the dynamic shear modulus. Macromolecules 34:7427–7434

34. Walther B, Lorén N, Nydén M, Hermansson AM (2006) Influence of κ-carrageenan gel structures on the diffusion of probe molecules determined by transmission electron microscopy and NMR diffusometry. Langmuir 22:8221–8228
35. Lorén N, Shtykova L, Kidman S, Jarvoll P, Nydén M, Hermansson AM (2009) Dendrimer diffusion in κ-carrageenan gel structures. Biomacromolecules 10:275–284
36. Phillips RJ (2000) A hydrodynamic model for hindered diffusion of proteins and micelles in hydrogels. Biophys J 79:3350–3354
37. Amsden B (1998) Solute diffusion in hydrogels. An examination of the retardation effect. Polym Gels Networks 6:13–43
38. Amsden B (1998) Solute diffusion in hydrogels. An examination of the retardation effect. Polym Gels Networks 6(13)
39. Mstsukawa S, Sagae D, Mogi A (2009) Molecular diffusion in polysaccharide gel systems as observed by NMR. Progr Colloid Polym Sci 136:171
40. Doi M, Edward SF (1986) The theory of polymer dynamics. Oxford University Press, New York
41. Nishinari K, Kohyama K, Williams PA, Phillips GO, Burchard W, Ogino K (1991) Solution properties of Pullulan. Macromolecules 24:5590
42. Takahashi A, Sakai M, Kato T (1980) Melting temperature of thermally reversible gel: VI effect of branching on the sol–gel transition of polyethylene gels. Polym J 12:335
43. Shimizu M, Brenner T, Liao RQ, Matsukawa S (2012) Diffusion of probe polymer in gellan gum solutions during gelation process studied by gradient NMR. S. Matsukawa. Food Hydrocoll 26:28
44. Nussinovitch A (1997) Hydrocolloid applications: gum technology in the food and other industries, 1st edn. Chapman & Hall, London (Chapter 1)
45. Arnott S, Fulmer S, Scott WE (1974) The agarose double helix and its function in agarose gel structure. J Mol Biol 90:269–284
46. Aymard P, Martin D, Plucknett K, Foster T, Clark A, Norton L (2001) Influence of thermal history on the structural and mechanical properties of agarose gels. Biopolymers 59:131–144
47. Matsukawa S, Sagae D, Mogi A (2009) Molecular diffusion in polysaccharide gel systems as observed by NMR. Progr Colloid Polym Sci 136:171–176
48. Price WS (2009) NMR studies of translational motion, 1st ed. Cambridge University Press, New York (chapter 7)
49. Déléris I, Andriot I, Gobet M, Moreau C, Souchon I, Guichard E (2010) Determination of aroma compound diffusion in model food systems: comparison of macroscopic and microscopic methodologies. J Food Eng 100:557–566
50. Walderhaug H, Söderman O, Topgaard D (2010) Self-diffusion in polymer systems studied by magnetic field-gradient spin-echo NMR methods. Prog Nucl Magn Reson Spectrosc 56:406–425
51. Labropoulos KC, Niesz DE, Danforth SC, Kevrekidis PG (2002) Dynamic rheology of agar gels: theory and experiments. Part I. Development of a rheological model. Carbohydr Polym 50:393–406
52. Norton IT, Goodall DM, Austen KRJ, Morris ER (1986) Dynamics of molecular organization in agarose sulfate. Biopolymers 25:1009–1029
53. Guenet JM, Brulet A, Rochas C (1993) Agarose chain conformation in the sol state by neutron-scattering. Int J Biol Macromol 15:131–132
54. Ramzi M, Rochas C, Guenet JM (1998) Structure-properties relation for agarose thermoreversible gels in binary solvents. Macromolecules 31:6106–6111
55. Foord SA, Atkins EDT (1989) New X-ray diffraction results from agarose: extended single helix structures and implications for gelation mechanism. Biopolymers 28:1345–1365
56. Mohammed ZH, Hember MWN, Richardson RK, Morris ER (1998) Kinetic and equilibrium processes in the formation and melting of agarose gels. Carbohydr Polym 36:15–26
57. Matsunaga N, Nagashima A (1983) Transport properties of liquid and gaseous D2O over a wide range of temperature and pressure. J Phys Chem Ref Data 12:933–966

Rheological Basis of Magnetic Resonance Elastography

7

Takao Yamamoto

Abstract

Magnetic resonance elastography (MRE) has been invented to measure the elastic properties of the living body on the basis of magnetic resonance imaging. From the viewpoint of medical practice, MRE is regarded as a palpation method. MRE has the advantage over other types of elastography in that the stiffness in all areas of the living body including deep areas inaccessible to the physician's hand can be measured *quantitatively* in principle. Hence, by MRE, a complete viscoelastic map of the living body can be obtained. This advantage indicates that MRE is a powerful tool for determining the viscoelastic structures of nonuniform gels beyond the field of medicine. Unfortunately, the present MRE devices cannot perfectly realize the potential ability of the idea of MRE. Further improvements of the instruments for the MRE system and of the analysis algorithm for measured MRE data are necessary. Toward the construction of an ideal MRE system, we are trying to introduce the concept of rheology in MRE and to develop a purely theoretical approach to MRE. We consider what kind of physical quantity is actually measured and what kind of principle connects the measured quantities and the viscoelastic properties of the living body. The amplitude of a steady transverse wave excited in the living body is measured by MRE. The viscoelastic properties of the living body are described in terms of the Voigt model. The relationship between the viscoelastic parameters and the amplitude of the transverse wave is shown. The reconstruction formula of the map of the viscoelastic parameters from the data of the measured quantity is derived. The theoretical base of transverse wave excitation by a longitudinal wave, which enables us to measure the rigidity in deep areas

T. Yamamoto (✉)
Faculty of Science and Technology, Gunma University,
Tenjin-cho 1-5-1, Kiryu, Gunma 376-8515, Japan
e-mail: tyam@gunma-u.ac.jp

R. Kita, T. Dobashi (eds.), *Nano/Micro Science and Technology in Biorheology*,
DOI 10.1007/978-4-431-54886-7_7

practically, is also developed. A method of measuring the stiffness of a uniaxial system is introduced.

Keywords
MRE • Living body • Nonuniformity • Viscoelasticity • Voigt model

7.1 Introduction

Magnetic resonance elastography (MRE) [1] has been invented by Muthupillai et al. [2] to measure elastic properties of the living body on the basis of magnetic resonance imaging (MRI). Its core ideas are as follows (See Fig.7.1):

1. Transverse waves are created in the living body by an external periodic force acting on surface of the living body.
2. By applying a sinusoidally oscillating magnetic field with a gradient called the motion-sensitizing gradient (MSG) on the living body, the waveform of transverse waves rippling in the living body is measured by the phase shift of the magnetic resonance (MR) signal, where transverse waves excited in the living body and the MSG should oscillate in the same phase.
3. From the waveform derived from the phase shift, the stiffness of the living body is obtained *position-dependently*.

From the viewpoint of its hard ware, the MRE system is simply an extended version of the MRI system; it is fabricated by retrofitting an MRI system with a vibration exciter for exciting transverse waves and an MSG generator. However, the physical quantities measured by MRI and MRE differ in quality. By MRI, the material distribution (more precisely, the proton density distribution) in the living body is measured; thus, MRI is grouped with computed tomography (CT) and

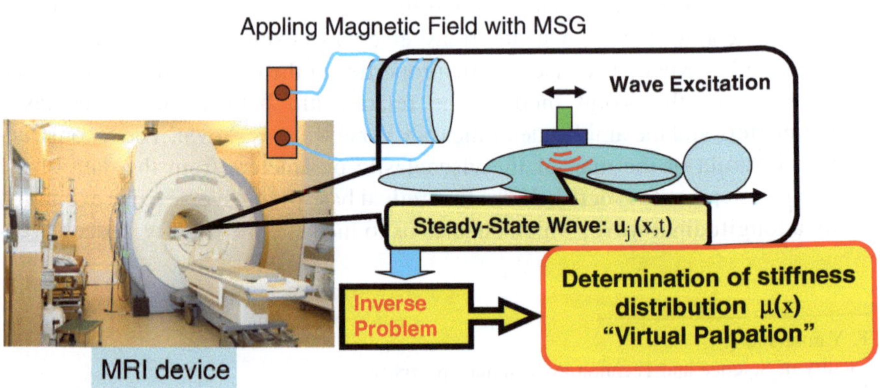

Fig. 7.1 Schematic view of MRE system. The MRE system is fabricated by retrofitting an MRI system with a vibration exciter for exciting transverse waves and an MSG generator

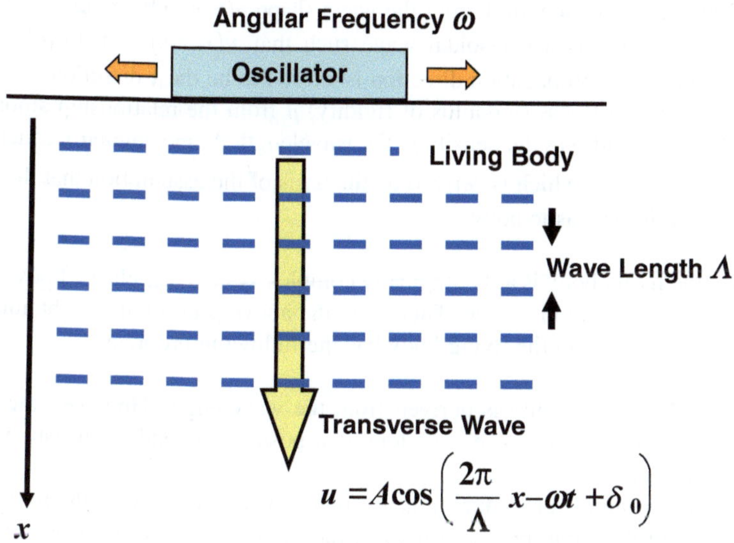

Fig. 7.2 Excitation of a transverse sinusoidal wave in the living body. Assume that the displacement of a transverse wave excited in the living body at position x and time t is expressed by the sinusoidal form with amplitude A: $u(x,t) = A\cos\left(\frac{2\pi}{\Lambda}x - \omega t + \delta_0\right)$. From the wavelength Λ and the angular frequency ω of the sinusoidal wave, the stiffness (modulus of the rigidity) μ of the living body is expressed by $\mu = \rho\left(\Lambda\frac{\omega}{2\pi}\right)^2$ on the basis of a linear elastic body picture, where ρ is the density of the living body

X-ray photography. From the viewpoint of medical practice, MRI is regarded as a visual diagnostic tool. On the other hand, by MRE, the stiffness distribution in the living body is measured. MRE is regarded as a palpation method. Palpation is an effective method of detecting tumors. However, it is restricted to body areas accessible to the physician's hand. Obtaining quantitative diagnostic outcomes by palpation is difficult. To develop strengths and eliminate weaknesses of palpation, various types of elastography for disease diagnosis have been developed [3]. MRE has the advantage over other types of elastography in that the stiffness in all areas of the living body can be measured quantitatively in principle; namely, a complete viscoelastic parameters map of the living body can be obtained.

The original procedure of MRE for measuring the stiffness is as follows (See Fig.7.2):

(P1) Generate a transverse wave with an angular frequency ω in the living body using a vibration exciter and a magnetic field with MSG whose phase is matched to that of the vibration exciter.

(P2) Obtain the phase shift θ of the MR signal induced by the transverse wave position-dependently.

(P3) Derive the wave shape $u(x, y, z)$ at the position (x, y, z) using the proportional relation between the wave shape and the phase shift $u \propto \theta$.

(P4) Obtain the wavelength Λ from the wave shape $u(x, y, z)$ by assuming that the transverse wave is a sinusoidal wave such that $u(x, y, z) = A\cos\left(\frac{2\pi}{\Lambda}x + \delta_0\right)$, where the wave propagation direction is chosen to be the x-direction.

(P5) Obtain the stiffness (modulus of rigidity) μ from the relationship among the modulus of rigidityμ, the density ρ, the wavelength Λ, and angular frequency ω, i.e., $\mu = \rho\left(\Lambda\frac{\omega}{2\pi}\right)^2$, which is derived on the basis of the assumption that the living body is a linear elastic body.

Since the living body is a rheologically complex system, the elastic body picture of the living body is too simple. Therefore, the above procedure for obtaining the stiffness distribution in the living body has the following problems.

1. The modulus of rigidity is derived from the wavelength. However, the wavelength of a transverse wave is so long that position-dependent measurement is quite difficult.
2. In the above simple procedure, it is assumed that all the points of the living body oscillate in the same phase. However, the viscosity shifts the phase position-dependently.
3. A transverse wave cannot reach deep areas of the living body since the viscosity damps the amplitude of the transverse wave rapidly.
4. Since the living body is soft, it is difficult for the vibration exciter to excite accurately a sinusoidal vibration on the living body surface.

Further improvements of the instruments for the MRE system and of the analysis algorithm of the MR signal could resolve some of the above programs. Actually, various efforts to improve the instruments [4] and algorithm [5] have been made and some MRE devices are actually used in clinical practice [1, 6]. The aim of these efforts is to develop excellent medical diagnostic devices. This aim is, of course, the starting point of the development of MRE, but it might limit our options of improving the MRE device. To avoid patient discomfort, measurement time should be as short as possible. To utilize existing MRI devices effectively, a small-scale modification of MRI devices is acceptable. Besides its clinical usage, the MRE device is useful for the nondestructive measurement of the viscoelastic properties of soft materials in general [7, 8]. MRE can therefore become a powerful tool for determining the viscoelastic structure of nonuniform gels and for observing the characteristic slow dynamics in gels. MRE is expected to contribute to the progress of gel science. To expand the fields in which MRE is utilized beyond the field of medicine, it is necessary to consider what kind of physical quantity is actually measured and what kind of principle relates the measured quantity to the viscoelastic properties, from the viewpoint of basic science. The approach from basic science would also promote the development of MRE as a medical diagnostic device. Here, we try to introduce the concept of rheology in MRE and to develop a purely theoretical approach to MRE.

This chapter is organized as follows. First, the quantities measured by MRE are explained. In MRE, the measurable quantity is the phase shift of the MR signal. The relationship between the phase shift and the motion of the concerned point in the living body is determined independently of the details of the MRE device. The relationship among the motion, MSG, and the phase shift is explained. The shape of standing waves in the living body is visualized from the phase shift data. The visualization mechanism is explained.

To connect the shape of a wave excited in the living body to the viscoelastic structure, the modeling of the living body based on Newton's equation of motion is required. Next, the modeling of the living body is discussed. The living body is regarded as a viscoelastic body and its viscoelastic structure is described in terms of the Voigt model. Of course, a more complex model can be adopted. In complex models, however, many viscoelastic parameters must be introduced. The multitude of such parameters disturbs the unique determination of viscoelastic properties. Therefore, we choose the Voigt model, which is the simplest of all viscoelastic models. Here, we must note that all viscoelastic parameters are "defined" through the modeling based on the Voigt model. The Voigt picture of the living body is described by Lagrangian formalism for general versatility and simplicity in the mathematical treatment of such formalism.

By the above modeling, the motion at each point in the living body can be described in terms of viscoelastic parameters. In principle, the values of the viscoelastic parameters can be derived from measured data of the motion. All the quantities expressing the mechanical properties of materials such as rigidity and viscosity are determined through this "inverse" problem. Unfortunately, in MRE, the motion cannot be measured, but the shape of an excited steady-state wave can. The solution of the inverse problem based on the sinusoidal-wave assumption cannot be adopted because of the long wavelength [Problem (I)] and the viscosity of the living body [Problem (II)]. A novel method of solving the inverse problem and of reconstructing the map of the modulus of rigidity in the living body from the measured data is required. The reconstruction formula necessary to comply with this requirement is explained.

As a way of resolving Problem (III), the method of exciting a transverse wave using a longitudinal wave has been invented. A theoretical basis of transverse wave excitation by a longitudinal wave is also developed. For hard body organs and tissues such as bone and tendons, the viscoelastic properties are anisotropic and often uniaxial. Finally, a method of measuring the rigidity of a uniaxial system is introduced.

7.2 Visualization of Wave Shape by MSG

7.2.1 Original Idea for MRE

The original idea of Muthupillai et al. [2] is briefly reviewed.

Let us denote the displacement of an infinitesimal element of the living body at a position $\mathbf{x} = (x_1, x_2, x_3)$ at a time t by

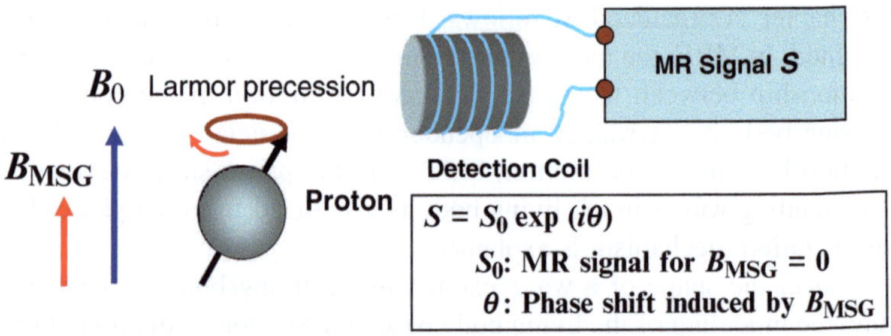

Fig. 7.3 Phase shift of the MR signal induced by the magnetic field with MSG and excited transverse wave. The magnetic field with MSG $\mathbf{B}_{MSG}$ superimposed on the static magnetic field $\mathbf{B}_0$ that induces the Larmor precession of protons and the transverse wave excited in the living body change the MR signal from S_0 to $S = S_0\exp(i\theta)$. From the phase shift θ, the shape of the transverse wave is obtained

$$\mathbf{u}(\mathbf{x},t) = (u_1(\mathbf{x},t), u_2(\mathbf{x},t), u_3(\mathbf{x},t)). \tag{7.1}$$

Let us choose the direction of the static magnetic field $\mathbf{B}_0$ that induces the Larmor precession of protons as the x_3-direction: $\mathbf{B}_0 = (0, 0, B_0)$. The magnetic field superimposed on the static magnetic field $\mathbf{B}_0$ for inducing MSG is given by

$$\mathbf{B}_{MSG}(\mathbf{x},t) = (0, 0, \mathbf{G}(t)\cdot\mathbf{x}). \tag{7.2}$$

Superimposing the magnetic field $\mathbf{B}_{MSG}$ on the magnetic field $\mathbf{B}_0$, we obtain the phase shift of MR signal at $\mathbf{x}$,

$$\theta(\mathbf{x}) = \Gamma\int_0^\tau \mathbf{G}(t)\cdot\mathbf{u}(\mathbf{x},t)dt, \tag{7.3}$$

where τ is the "measurement time" and Γ is a constant proportional to proton density (See Fig. 7.3).

Let us assume that we generate a transverse sinusoidal wave with an angular frequency ω, a wave vector $\mathbf{k}$, and a phase constant δ_0 in the living body by a vibration exciter, whose displacement is expressed by

$$\mathbf{u}(\mathbf{x},t) = \mathbf{A}\cos(\mathbf{k}\cdot\mathbf{x} - \omega t + \delta_0). \tag{7.4}$$

We oscillate the gradient vector $\mathbf{G}(t)$ at the angular frequency ω harmonically and synchronize it with the vibration motion of the vibration exciter as

$$\mathbf{G}(t) = \mathbf{G}_0\cos\omega t. \tag{7.5}$$

Choosing the measurement time of an integral multiple of the period of the harmonic oscillation, $\tau = 2\pi N/\omega$ (N is a positive integer), we obtain the phase shift as

$$\theta(\mathbf{x}) = \mathbf{G}_0 \cdot \mathbf{A}\frac{\pi N \Gamma}{\omega} \cos\left(\mathbf{k}\cdot\mathbf{x} + \delta_0\right). \tag{7.6}$$

The above expression shows that the position dependence of the phase shift indicates a snapshot of the sinusoidal wave, i.e., the wave shape. Then, the wave motion is visualized by means of MSG.

In terms of the wave vector **k** obtained from the wave shape, the wavelength is expressed as $\Lambda = 2\pi/|\mathbf{k}|$. Assuming a linear isotropic elasticity in the living body, the displacement satisfies the wave equation

$$\Delta\mathbf{u} - \frac{1}{c^2}\frac{\partial^2\mathbf{u}}{\partial t^2} = 0, \tag{7.7}$$

where c is the propagation velocity of the wave and is given by

$$c = \sqrt{\frac{\mu}{\rho}}. \tag{7.8}$$

In the above, ρ is the density of the living body and is estimated as $\rho \approx 1$ g/cm^3. Substituting Eq. (7.4) into Eq. (7.7), we obtain the dispersion relation

$$\omega = 2\pi c/\Lambda \tag{7.9}$$

and obtain the expression of the rigidity

$$\mu = \rho\left(\Lambda\frac{\omega}{2\pi}\right)^2 = \rho\left(\frac{\omega}{|\mathbf{k}|}\right)^2. \tag{7.10}$$

This expression allows us evaluate rigidity.

7.2.2 Utilization of Visualization Ability of MSG

The abovementioned method of deriving rigidity is very simple, but the validity of the result strongly depends on the wave shape of Eq. (7.4). However, the sinusoidal-wave assumption is not always valid. The nonuniformity of rigidity and/or viscosity makes the amplitude **A**, the phase constant δ_0, and the wave vector **k** depend on the position **x**. Hence, the displacement is generally described as

$$\mathbf{u}(\mathbf{x}, t) = \mathbf{A}(\mathbf{x})\cos\left(\mathbf{k}(\mathbf{x})\cdot\mathbf{x} - \omega t + \delta_0(\mathbf{x})\right). \tag{7.11}$$

Substituting Eqs. (7.5) and (7.11) into Eq. (7.3), we obtain the phase shift

$$\theta(\mathbf{x}) = \mathbf{G}_0 \cdot \mathbf{A}(\mathbf{x}) \frac{\pi N \Gamma}{\omega} \cos\left(\mathbf{k} \cdot \mathbf{x} + \delta_0(\mathbf{x})\right). \tag{7.12}$$

This expression shows that the phase shift indicates a snapshot of the wave even if the wave is not sinusoidal. MSG can visualize the wave shape generally. Note, however, that the relationship between the wave shape expressed by Eq. (7.12) and the rigidity $\mu(\mathbf{x})$ is not clear since the velocity expression Eq. (7.8) and the dispersion relation Eq. (7.9) no longer hold. To derive the precise relationship between wave shape and rigidity, the modeling of the rheological structure of the living body is necessary. The transverse wave expressed by Eq. (7.11) is rewritten as

$$\mathbf{u}(\mathbf{x}, t) = \boldsymbol{\varphi}(\mathbf{x}) \cos \omega t + \boldsymbol{\psi}(\mathbf{x}) \sin \omega t, \tag{7.13}$$

where

$$\begin{aligned} \boldsymbol{\varphi}(\mathbf{x}) &= \mathbf{A}(\mathbf{x}) \cos\left(\mathbf{k}(\mathbf{x}) \cdot \mathbf{x} + \delta_0(\mathbf{x})\right) \\ \boldsymbol{\psi}(\mathbf{x}) &= \mathbf{A}(\mathbf{x}) \sin\left(\mathbf{k}(\mathbf{x}) \cdot \mathbf{x} + \delta_0(\mathbf{x})\right), \end{aligned} \tag{7.14}$$

Using the two types of MSG, the phases of whose oscillation differ by only $\pi/2$, i.e.,

$$\begin{cases} \mathbf{G}(t) = \mathbf{G}_0 \cos \omega t \\ \mathbf{G}'(t) = \mathbf{G}_0 \sin \omega t, \end{cases} \tag{7.15}$$

we can obtain two types of phase shift:

$$\begin{aligned} \theta_0(\mathbf{x}) &= \frac{\pi N \Gamma}{\omega} \mathbf{G}_0 \cdot \boldsymbol{\varphi}(\mathbf{x}) \\ \theta_{\pi/2}(\mathbf{x}) &= \frac{\pi N \Gamma}{\omega} \mathbf{G}_0 \cdot \boldsymbol{\psi}(\mathbf{x}). \end{aligned} \tag{7.16}$$

The above expression shows that we can obtain the rigidity $\mu(\mathbf{x})$ position-dependently by determining the relationship between the two amplitudes of the vibration, i.e., $\boldsymbol{\varphi}(\mathbf{x})$ and $\boldsymbol{\psi}(\mathbf{x})$, and the rheological structure of the living body. Hence, the visualization ability of MSG can be effectively utilized by modeling the rheological structure of the living body.

7.3 Modeling of Living Body

7.3.1 Voigt Model Picture

The amplitude of the excited transverse wave is about 1–100 μm and the wavelength is about a few centimeters. Since the displacement and the gradient of the displacement are very small, the nonlinear effect on the rheology of the living body is negligible. Therefore, the rheological properties of the living body are described in terms of a linear viscoelastic structure. Let us describe the viscoelastic structure of the living body on the basis of the Voigt model. Introducing the strain tensor defined as

$$u_{ij}(\mathbf{x}) = \frac{1}{2}\left(\frac{\partial u_i}{\partial x_j} + \frac{\partial u_j}{\partial x_i}\right) \tag{7.17}$$

and expressing the density by $\rho(\approx 1\ \text{g/cm}^3)$ and the elasticity tensor at $\mathbf{x}$ by $\lambda_{ijkl}(\mathbf{x})$, we express the elastic property of the living body by the Lagrangian [9, 10]

$$L = \int_\Omega d^3\mathbf{x}\left[\frac{1}{2}\rho\sum_{j=1}^{3}\left(\dot{u}_j(\mathbf{x})\right)^2 - \frac{1}{2}\sum_{i,j,k,l=1}^{3}\lambda_{ijkl}(\mathbf{x})u_{ij}(\mathbf{x})u_{kl}(\mathbf{x})\right], \tag{7.18}$$

where Ω is the space region occupied by the living body and $\dot{u}_j(\mathbf{x})$ is the j-th component of the velocity (the time derivative of the j-th component of the displacement $\dot{u}_j = \partial u_j/\partial t$). Note the elasticity tensor has the following symmetry property:

$$\lambda_{ijkl} = \lambda_{jikl} = \lambda_{ijlk} = \lambda_{klij}. \tag{7.19}$$

In terms of the viscosity tensor $\eta_{ijkl}(\mathbf{x})$, the viscous property of the living body is expressed by the dissipation function given by [11]

$$D = \int_\Omega d^3\mathbf{x}\left[\frac{1}{2}\sum_{i,j,k,l=1}^{3}\eta_{ijkl}(\mathbf{x})\dot{u}_{ij}(\mathbf{x})\dot{u}_{kl}(\mathbf{x})\right], \tag{7.20}$$

where $\dot{u}_{ij}$ is the time derivative of the strain tensor:

$$\dot{u}_{ij}(\mathbf{x}) = \frac{1}{2}\left(\frac{\partial \dot{u}_i}{\partial x_j} + \frac{\partial \dot{u}_j}{\partial x_i}\right). \tag{7.21}$$

Note that the viscosity tensor has the following symmetry property:

$$\eta_{ijkl} = \eta_{jikl} = \eta_{ijlk} = \eta_{klij}. \tag{7.22}$$

The rheological structure of the living body is expressed by the elasticity tensor "field" $\lambda_{ijkl}(\mathbf{x})$ and the viscosity tensor "field" $\eta_{ijkl}(\mathbf{x})$. From the viewpoint of rheology, palpation is performed to obtain the map of the viscoelastic parameters: $\lambda_{ijkl}(\mathbf{x})$ and $\eta_{ijkl}(\mathbf{x})$.

7.3.2 Equation of Motion

The equation of motion of the living body is obtained from the Lagrange equation [12]

$$\frac{d}{dt}\frac{\delta L}{\delta \dot{u}_j(\mathbf{x},t)} - \frac{\delta L}{\delta u_j(\mathbf{x},t)} + \frac{\delta D}{\delta \dot{u}_j(\mathbf{x},t)} = 0, \tag{7.23}$$

where $\delta L/\delta u_j$ and $\delta L/\delta \dot{u}_j$ stand for the functional derivatives. We can easily calculate all the terms on the left-hand side of the above equation and obtain

$$\frac{d}{dt}\frac{\delta L}{\delta \dot{u}_j(\mathbf{x},t)} = \frac{d}{dt}\ \rho \dot{u}_j(\mathbf{x},t) = \rho \ddot{u}_j(\mathbf{x},t), \tag{7.24}$$

where $\ddot{u}_j = \partial \dot{u}_j/\partial t$ and

$$\begin{aligned}
\frac{\delta L}{\delta u_j(\mathbf{x},t)} &= -\int_\Omega d^3\mathbf{x}'\frac{1}{2}\sum_{i,k,l,m}\frac{\delta}{\delta u_j(\mathbf{x},t)}(\lambda_{iklm}u_{ik}u_{lm}) \\
&= -\int_\Omega d^3\mathbf{x}'\frac{1}{2}\sum_{i,k,l,m}\sum_{\alpha,\beta}\frac{\delta u_{\alpha\beta}(\mathbf{x}',t)}{\delta u_j(\mathbf{x},t)}\frac{\partial}{\partial u_{\alpha\beta}}(\lambda_{iklm}u_{ik}u_{lm}) \\
&= -\int_\Omega d^3\mathbf{x}'\sum_{l,m}\sum_{\alpha,\beta}\frac{\delta u_{\alpha\beta}(\mathbf{x}',t)}{\delta u_i(\mathbf{x},t)}\lambda_{\alpha\beta lm}\left(\mathbf{x}'\right)u_{lm}\left(\mathbf{x}',t\right)
\end{aligned} \tag{7.25}$$

$$\begin{aligned}
\frac{\delta D}{\delta \dot{u}_i(\mathbf{x},t)} &= \int_\Omega d^3\mathbf{x}'\frac{1}{2}\sum_{i,k,l,m}\frac{\delta}{\delta \dot{u}_i(\mathbf{x},t)}(\eta_{iklm}\dot{u}_{ik}\dot{u}_{lm}) \\
&= \int_\Omega d^3\mathbf{x}'\frac{1}{2}\sum_{i,k,l,m}\sum_{\alpha,\beta}\frac{\delta \dot{u}_{\alpha\beta}(\mathbf{x}',t)}{\delta \dot{u}_i(\mathbf{x},t)}\frac{\partial}{\partial \dot{u}_{\alpha\beta}}(\eta_{iklm}\dot{u}_{ik}\dot{u}_{lm}) \\
&= \int_\Omega d^3\mathbf{x}'\sum_{l,m}\sum_{\alpha,\beta}\frac{\delta \dot{u}_{\alpha\beta}(\mathbf{x}',t)}{\delta \dot{u}_i(\mathbf{x},t)}\eta_{\alpha\beta lm}\left(\mathbf{x}'\right)\dot{u}_{lm}\left(\mathbf{x}',t\right)
\end{aligned} \tag{7.26}$$

where

$$\frac{\partial}{\partial u_{\alpha\beta}}\sum_{i,k,l,m}\frac{1}{2}\lambda_{iklm}u_{ik}u_{lm} = \sum_{l,m}\lambda_{\alpha\beta lm}\left(\mathbf{x}'\right)u_{lm}\left(\mathbf{x}',t\right) \equiv \sigma_{\alpha\beta}\left(\mathbf{x}',t\right) \tag{7.27}$$

is the stress tensor and

$$\frac{\partial}{\partial \dot{u}_{\alpha\beta}}\sum_{i,k,l,m}\frac{1}{2}\eta_{iklm}\dot{u}_{ik}\dot{u}_{lm} = \sum_{l,m}\eta_{\alpha\beta lm}\left(\mathbf{x}'\right)\dot{u}_{lm}\left(\mathbf{x}',t\right) \equiv \sigma'_{\alpha\beta}\left(\mathbf{x}',t\right) \tag{7.28}$$

is the dissipative stress tensor (the share stress tensor). From

$$\begin{aligned}\frac{\delta u_{\alpha\beta}\left(\mathbf{x}',t\right)}{\delta u_i(\mathbf{x},t)} &= \frac{\delta}{\delta u_i(\mathbf{x},t)}\frac{1}{2}\left(\frac{\partial u_\alpha}{\partial x'_\beta}+\frac{\partial u_\beta}{\partial x'_\alpha}\right)\\ &= \frac{1}{2}\left(\delta_{i\alpha}\frac{\partial}{\partial x'_\beta}\delta^3\left(\mathbf{x}'-\mathbf{x}\right)+\delta_{i\beta}\frac{\partial}{\partial x'_\alpha}\delta^3\left(\mathbf{x}'-\mathbf{x}\right)\right)\end{aligned} \tag{7.29}$$

and

$$\begin{aligned}\frac{\delta \dot{u}_{\alpha\beta}\left(\mathbf{x}',t\right)}{\delta \dot{u}_i(\mathbf{x},t)} &= \frac{\delta}{\delta \dot{u}_i(\mathbf{x},t)}\frac{1}{2}\left(\frac{\partial \dot{u}_\alpha}{\partial x'_\beta}+\frac{\partial \dot{u}_\beta}{\partial x'_\alpha}\right)\\ &= \frac{1}{2}\left(\delta_{i\alpha}\frac{\partial}{\partial x'_\beta}\delta^3\left(\mathbf{x}'-\mathbf{x}\right)+\delta_{i\beta}\frac{\partial}{\partial x'_\alpha}\delta^3\left(\mathbf{x}'-\mathbf{x}\right)\right),\end{aligned} \tag{7.30}$$

we have

$$\begin{aligned}&\int_\Omega d^3\mathbf{x}'\sum_{l,m}\sum_{\alpha,\beta}\frac{\delta u_{\alpha\beta}\left(\mathbf{x}',t\right)}{\delta u_i(\mathbf{x},t)}\lambda_{\alpha\beta lm}\left(\mathbf{x}'\right)u_{lm}\left(\mathbf{x}',t\right)\\ &= \int_\Omega d^3\mathbf{x}'\sum_k\frac{\partial\delta^3\left(\mathbf{x}'-\mathbf{x}\right)}{\partial x'_k}\sum_{l,m}\lambda_{iklm}\left(\mathbf{x}'\right)u_{lm}\left(\mathbf{x}',t\right)\\ &= -\int_\Omega d^3\mathbf{x}'\delta^3\left(\mathbf{x}'-\mathbf{x}\right)\sum_k\frac{\partial}{\partial x'_k}\sum_{l,m}\lambda_{iklm}\left(\mathbf{x}'\right)u_{lm}\left(\mathbf{x}',t\right)\\ &= -\sum_k\frac{\partial}{\partial x_k}\sum_{l,m}\lambda_{iklm}(\mathbf{x})u_{lm}(\mathbf{x},t)\end{aligned} \tag{7.31}$$

and

$$\begin{aligned}&\int_\Omega d^3\mathbf{x}'\sum_{l,m}\sum_{\alpha,\beta}\frac{\delta \dot{u}_{\alpha\beta}\left(\mathbf{x}',t\right)}{\delta \dot{u}_i(\mathbf{x},t)}\eta_{\alpha\beta lm}\left(\mathbf{x}'\right)\dot{u}_{lm}\left(\mathbf{x}',t\right)\\ &= \int_\Omega d^3\mathbf{x}'\sum_k\frac{\partial\delta^3\left(\mathbf{x}'-\mathbf{x}\right)}{\partial x'_k}\sum_{l,m}\eta_{iklm}\left(\mathbf{x}'\right)\dot{u}_{lm}\left(\mathbf{x}',t\right)\\ &= -\int_\Omega d^3\mathbf{x}'\delta^3\left(\mathbf{x}'-\mathbf{x}\right)\sum_k\frac{\partial}{\partial x'_k}\sum_{l,m}\eta_{iklm}\left(\mathbf{x}'\right)\dot{u}_{lm}\left(\mathbf{x}',t\right)\\ &= -\sum_k\frac{\partial}{\partial x_k}\sum_{l,m}\eta_{iklm}(\mathbf{x})\dot{u}_{lm}(\mathbf{x},t),\end{aligned} \tag{7.32}$$

where $\delta^3(\mathbf{x})$ is the three-dimensional delta function and δ_{ij} is the Kronecker delta. Then, the equation of motion is given by

$$\rho \ddot{u}_i(\mathbf{x},t) - \sum_k \frac{\partial}{\partial x_k} \sum_{l,m} \lambda_{iklm}(\mathbf{x}) u_{lm}(\mathbf{x},t) - \sum_k \frac{\partial}{\partial x_k} \sum_{l,m} \eta_{iklm}(\mathbf{x}) \dot{u}_{lm}(\mathbf{x},t) = 0. \quad (7.33)$$

In soft body organs and tissues such as the liver, the viscoelastic properties are isotropic. In hard body organs and tissues such as bone and tendons, the viscoelastic properties are anisotropic and often uniaxial. Here, we pay attention to soft body organs. For the isotropic system, the numbers of individual components of the stress tensor and share stress tensor are both two, and the tensors are respectively expressed as

$$\lambda_{ijkl}(\mathbf{x}) = \lambda(\mathbf{x})\delta_{ij}\delta_{kl} + 2\mu(\mathbf{x})\delta_{ik}\delta_{jl} \quad (7.34)$$

$$\eta_{ijkl}(\mathbf{x}) = \tilde{\varsigma}(\mathbf{x})\delta_{ij}\delta_{kl} + 2\eta(\mathbf{x})\delta_{ik}\delta_{jl}. \quad (7.35)$$

The parameters λ and μ are called Lame's constants. The parameter μ is the modulus of rigidity and λ is related to the bulk modulus K as $K = \lambda + \frac{2}{3}\mu$. The parameter η is the share viscosity and the parameter $\tilde{\zeta}$ is related to the bulk viscosity ζ as $\zeta = \tilde{\zeta} + \frac{2}{3}\eta$. The equation of motion is rewritten as

$$\begin{aligned} \rho \ddot{u}_i(\mathbf{x},t) &- \frac{\partial}{\partial x_i}\lambda(\mathbf{x}) \sum_k u_{kk}(\mathbf{x},t) - 2\sum_k \frac{\partial}{\partial x_k}[\mu(\mathbf{x})u_{ik}(\mathbf{x},t)] \\ &- \frac{\partial}{\partial x_i}\tilde{\zeta}(\mathbf{x}) \sum_k \dot{u}_{kk}(\mathbf{x},t) - 2\sum_k \frac{\partial}{\partial x_k}[\eta(\mathbf{x})\dot{u}_{ik}(\mathbf{x},t)] = 0. \end{aligned} \quad (7.36)$$

Since the living body is "soft" (thus, $K >> \mu$), share deformation is much more easily caused than compression and expansion deformation. Therefore, the incompressibility,

$$\sum_i u_{ii}(\mathbf{x},t) = 0, \quad (7.37)$$

is naturally assumed. We also have the time derivative of this expression:

$$\sum_i \dot{u}_{ii}(\mathbf{x},t) = 0. \quad (7.38)$$

Assuming the above incompressibility, we obtain the equation of motion:

$$\rho \ddot{u}_i(\mathbf{x},t) - 2\sum_k \frac{\partial}{\partial x_k}[\mu(\mathbf{x})u_{ik}(\mathbf{x},t)] - 2\sum_k \frac{\partial}{\partial x_k}[\eta(\mathbf{x})\dot{u}_{ik}(\mathbf{x},t)] = 0. \quad (7.39)$$

In MRE, motion in a steady state is analyzed. The motion is obtained using the equation under the boundary condition on the surface area $\partial\Omega$ of the living body determined by the external vibration force $F_i(\mathbf{r}',t)$ $(i = 1,2,3)$ acting on the surface area A ($\mathbf{r}' \in A \subset \partial\Omega$) of the living body, that is,

$$u_i(\mathbf{r},t) = U\big(\mathbf{r}; \{F_i(\mathbf{r}',t)\}_A\big), \tag{7.40}$$

where $\mathbf{r} \in \partial\Omega$.

7.4 Reconstruction of Viscoelastic Parameters

Let us discuss the reconstruction formula for the viscoelastic parameters of the living body from the phase shift of the MR signal on the basis of the Voigt model picture.

When the surface area A of the living body is excited by a perfect sinusoidal external force with an angular frequency ω

$$F_i(\mathbf{r}',t) = F_i^0(\mathbf{r}')\cos\omega t, \tag{7.41}$$

the excited surface area A of the living body vibrates sinusoidally at the angular frequency ω;

$$u_i(\mathbf{r},t) = \phi_i(\mathbf{r})\cos\omega t(\mathbf{r} \in A). \tag{7.42}$$

In this case, all the points of the living body vibrate at the angular frequency ω. Therefore, the solution of Eq. (7.39) can be assumed to be

$$u_i(\mathbf{x},t) = \phi_i(\mathbf{x})\cos\omega t + \psi_i(\mathbf{x})\sin\omega t. \tag{7.43}$$

Inserting the above expression into Eq. (7.39), we obtain the following simultaneous equations with respect to the two types of "amplitude," namely, $\phi_i(\mathbf{x})$ and $\psi_i(\mathbf{x})$:

$$-\rho\omega^2\phi_i(\mathbf{x}) - 2\sum_k \frac{\partial}{\partial x_k}[\mu(\mathbf{x})\phi_{ik}(\mathbf{x})] - 2\omega\sum_k \frac{\partial}{\partial x_k}[\eta(\mathbf{x})\psi_{ik}(\mathbf{x})] = 0, \tag{7.44}$$

$$-\rho\omega^2\psi_i(\mathbf{x}) - 2\sum_k \frac{\partial}{\partial x_k}[\mu(\mathbf{x})\psi_{ik}(\mathbf{x})] + 2\omega\sum_k \frac{\partial}{\partial x_k}[\eta(\mathbf{x})\phi_{ik}(\mathbf{x})] = 0, \tag{7.45}$$

where

$$\begin{aligned} \phi_{ij}(\mathbf{x}) &= \frac{1}{2}\left(\frac{\partial\phi_i}{\partial x_j} + \frac{\partial\phi_j}{\partial x_i}\right) \\ \psi_{ij}(\mathbf{x}) &= \frac{1}{2}\left(\frac{\partial\psi_i}{\partial x_j} + \frac{\partial\psi_j}{\partial x_i}\right). \end{aligned} \tag{7.46}$$

In medical practice, a perfect map of $\mu(\mathbf{x})$ is not required; only the values of $\mu(\mathbf{x})$ in the region of interest (ROI) are required. Thus, we can neglect the regions showing changes in the tissue structures in which $\partial\mu(\mathbf{x})/\partial x_j \neq 0$ and $\partial\eta(\mathbf{x})/\partial x_j \neq 0$, and we pay attention to the regions in which

$$\begin{aligned} \partial\mu(\mathbf{x})/\partial x_j &\approx 0 \\ \partial\eta(\mathbf{x})/\partial x_j &\approx 0. \end{aligned} \tag{7.47}$$

The simultaneous equations, Eqs. (7.44) and (7.45), are simplified as

$$-\rho\omega^2\phi_i(\mathbf{x}) - 2\mu(\mathbf{x})\sum_k \frac{\partial}{\partial x_k}\phi_{ik}(\mathbf{x}) - 2\omega\eta(\mathbf{x})\sum_k \frac{\partial}{\partial x_k}\psi_{ik}(\mathbf{x}) = 0 \tag{7.48}$$

$$-\rho\omega^2\psi_i(\mathbf{x}) - 2\mu(\mathbf{x})\sum_k \frac{\partial}{\partial x_k}\psi_{ik}(\mathbf{x}) + 2\omega\eta(\mathbf{x})\sum_k \frac{\partial}{\partial x_k}\phi_{ik}(\mathbf{x}) = 0. \tag{7.49}$$

In terms of ϕ_i and ψ_i, the incompressibility shown by Eq. (7.37) is expressed as

$$\begin{cases} \sum_i \phi_{ii}(\mathbf{x}) = 0 \\ \sum_i \psi_{ii}(\mathbf{x}) = 0. \end{cases} \tag{7.50}$$

Then,

$$\begin{aligned} \sum_k \frac{\partial}{\partial x_k}\phi_{ik}(\mathbf{x}) &= \frac{1}{2}\left(\sum_k \frac{\partial^2\phi_i}{\partial x_k^2} + \sum_k \frac{\partial}{\partial x_k}\frac{\partial\phi_k}{\partial x_i}\right) \\ &= \frac{1}{2}\left(\sum_k \frac{\partial^2\phi_i}{\partial x_k^2} + \frac{\partial}{\partial x_i}\sum_k \frac{\partial\phi_k}{\partial x_k}\right) \\ &= \frac{1}{2}\left(\sum_k \frac{\partial^2\phi_i}{\partial x_k^2} + \frac{\partial}{\partial x_i}\sum_k \phi_{kk}\right) = \frac{1}{2}\Delta\phi_i \end{aligned} \tag{7.51}$$

$$\begin{aligned} \sum_k \frac{\partial}{\partial x_k}\psi_{ik}(\mathbf{x}) &= \frac{1}{2}\left(\sum_k \frac{\partial^2\psi_i}{\partial x_k^2} + \sum_k \frac{\partial}{\partial x_k}\frac{\partial\psi_k}{\partial x_i}\right) \\ &= \frac{1}{2}\left(\sum_k \frac{\partial^2\psi_i}{\partial x_k^2} + \frac{\partial}{\partial x_i}\sum_k \frac{\partial\psi_k}{\partial x_k}\right) \\ &= \frac{1}{2}\left(\sum_k \frac{\partial^2\psi_i}{\partial x_k^2} + \frac{\partial}{\partial x_i}\sum_k \psi_{kk}\right) = \frac{1}{2}\Delta\psi_i, \end{aligned} \tag{7.52}$$

where $\Delta = \sum_{k=1}^{3} \partial^2/\partial x_k^2$. Therefore, the simultaneous equations for the amplitudes $\phi_i(\mathbf{x})$ and $\psi_i(\mathbf{x})$ finally lead to

$$-\rho\omega^2\phi_i(\mathbf{x}) - \mu(\mathbf{x})\Delta\phi_i(\mathbf{x}) - \omega\eta(\mathbf{x})\Delta\psi_i(\mathbf{x}) = 0 \tag{7.53}$$

$$-\rho\omega^2\psi_i(\mathbf{x}) - \mu(\mathbf{x})\Delta\psi_i(\mathbf{x}) + \omega\eta(\mathbf{x})\Delta\phi_i(\mathbf{x}) = 0. \tag{7.54}$$

When the data of the amplitudes $\phi_i(\mathbf{x})$ and $\psi_i(\mathbf{x})$ are obtained, we have the rigidity and viscosity using the following expressions:

$$\mu(\mathbf{x}) = -\frac{\rho\omega^2(\phi_i(\mathbf{x})\Delta\phi_i(\mathbf{x}) + \psi_i(\mathbf{x})\Delta\psi_i(\mathbf{x}))}{(\Delta\phi_i(\mathbf{x}))^2 + (\Delta\psi_i(\mathbf{x}))^2}. \tag{7.55}$$

$$\eta(\mathbf{x}) = -\frac{\rho\omega(\phi_i(\mathbf{x})\Delta\psi_i(\mathbf{x}) - \psi_i(\mathbf{x})\Delta\phi_i(\mathbf{x}))}{(\Delta\phi_i(\mathbf{x}))^2 + (\Delta\psi_i(\mathbf{x}))^2}. \tag{7.56}$$

Therefore, from the vibration amplitudes in ROI, the rigidity and viscosity in ROI are derived.

Choosing MSG along the x_3-direction and applying two types of the magnetic field inducing MSG with the angular frequency ω and the phases different by only $\pi/2$, i.e.,

$$\begin{cases} B_{\mathrm{MSG}}(\mathbf{x}, t) = G_0 x_3 \cos \omega t \\ B'_{\mathrm{MSG}}(\mathbf{x}, t) = G_0 x_3 \sin \omega t \end{cases}, \tag{7.57}$$

we can obtain the two types of phase shift of the MR signal: the phase shift induced by B_{MSG},

$$\theta_0(\mathbf{x}) = \frac{\pi N \Gamma}{\omega} G_0 \phi_3(\mathbf{x}), \tag{7.58}$$

and that induced by B'_{MSG},

$$\theta_{\pi/2}(\mathbf{x}) = \frac{\pi N \Gamma}{\omega} G_0 \psi_3(\mathbf{x}). \tag{7.59}$$

In terms of the phase shifts, the viscoelastic parameters are given by

$$\mu(\mathbf{x}) = -\frac{\rho\omega^2\left(\theta_0(\mathbf{x})\Delta\theta_0(\mathbf{x}) + \theta_{\pi/2}(\mathbf{x})\Delta\theta_{\pi/2}(\mathbf{x})\right)}{(\Delta\theta_0(\mathbf{x}))^2 + \left(\Delta\theta_{\pi/2}(\mathbf{x})\right)^2} \tag{7.60}$$

$$\eta(\mathbf{x}) = -\frac{\rho\omega\left(\theta_0(\mathbf{x})\Delta\theta_{\pi/2}(\mathbf{x}) - \theta_{\pi/2}(\mathbf{x})\Delta\theta_0(\mathbf{x})\right)}{(\Delta\theta_0(\mathbf{x}))^2 + \left(\Delta\theta_{\pi/2}(\mathbf{x})\right)^2}. \tag{7.61}$$

The above expressions show that the map of the viscoelastic parameters in ROI can be reconstructed using the phase shift data. Note that the two magnetic fields cannot be applied at the same time. This means that the measurement time of MRE is longer than that of MRI.

Phase shifts are obtained as numerical data. Therefore, the derivative in Eqs. (7.60) and (7.61) should be performed numerically. The numerical derivative is very sensitive to errors (or noises) in the phase shift data. To avoid the influence of errors, integral-type expressions of Eqs. (7.60) and (7.61) have been proposed [13]. Let us assume that μ and η are homogeneous in the region $D \in \Omega$ and consider the function $v(\mathbf{r})$ satisfying the relations

$$\begin{cases} v(\mathbf{r}) = 0 \\ \dfrac{\partial v(\mathbf{r})}{\partial n} = 0 \end{cases} \tag{7.62}$$

at the boundary of D, where $\partial/\partial n$ denotes the normal derivative. An example of the function $v(\mathbf{r})$ is given by

$$v(\mathbf{r}) = \exp\left(\frac{R^2}{|\mathbf{r}-\mathbf{x}|^2 - R^2}\right), \tag{7.63}$$

where we choose a spherical region with a center $\mathbf{x}$ and a radius R as the region D:

$$D = \{\mathbf{r}\,|\,|\mathbf{r}-\mathbf{x}_0| \leq R\}. \tag{7.64}$$

Multiplying both sides of Eqs. (7.53) and (7.54) by v, integrating both sides over the region D and using Green's theorem, we obtain

$$-\rho\omega^2 \int_D \phi_i(\mathbf{r})v(\mathbf{r})d^3\mathbf{r} - \mu(\mathbf{x})\int_D \phi_i(\mathbf{r})\Delta v(\mathbf{r})d^3\mathbf{r} - \omega\eta(\mathbf{x})\int_D \psi_i(\mathbf{r})\Delta v(\mathbf{r})d^3\mathbf{r} = 0 \tag{7.65}$$

$$-\rho\omega^2 \int_D \psi_i(\mathbf{r})v(\mathbf{r})d^3\mathbf{r} - \mu(\mathbf{x})\int_D \psi_i(\mathbf{r})\Delta v(\mathbf{r})d^3\mathbf{r} + \omega\eta(\mathbf{x})\int_D \phi_i(\mathbf{r})\Delta v(\mathbf{r})d^3\mathbf{r} = 0. \tag{7.66}$$

Using the above equations and the relationships expressed by Eqs. (7.58) and (7.59), we translate the expressions (7.60) and (7.61) in differential form to the following expressions in integral form:

$$\mu(\mathbf{x}) = -\rho\omega^2 \frac{\int_D \theta_0(\mathbf{r})v(\mathbf{r})d^3\mathbf{r}\int_D \theta_0(\mathbf{r})\Delta v(\mathbf{r})d^3\mathbf{r} + \int_D \theta_{\pi/2}(\mathbf{r})v(\mathbf{r})d^3\mathbf{r}\int_D \theta_{\pi/2}(\mathbf{r})\Delta v(\mathbf{r})d^3\mathbf{r}}{\left[\int_D \theta_0(\mathbf{r})\Delta v(\mathbf{r})d^3\mathbf{r}\right]^2 + \left[\int_D \theta_{\pi/2}(\mathbf{r})\Delta v(\mathbf{r})d^3\mathbf{r}\right]^2} \tag{7.67}$$

$$\eta(\mathbf{x}) = -\rho\omega \frac{\int_D \theta_0(\mathbf{r})v(\mathbf{r})d^3\mathbf{r} \int_D \theta_{\pi/2}(\mathbf{r})\Delta v(\mathbf{r})d^3\mathbf{r} - \int_D \theta_{\pi/2}(\mathbf{r})v(\mathbf{r})d^3\mathbf{r} \int_D \theta_0(\mathbf{r})\Delta v(\mathbf{r})d^3\mathbf{r}}{\left[\int_D \theta_0(\mathbf{r})\Delta v(\mathbf{r})d^3\mathbf{r}\right]^2 + \left[\int_D \theta_{\pi/2}(\mathbf{r})\Delta v(\mathbf{r})d^3\mathbf{r}\right]^2}. \tag{7.68}$$

The above integral-type reconstruction formula is insensitive to noises since noises are smeared out by the integral.

7.5 Transverse Wave Excitation by Longitudinal Wave

In soft materials such as the living body, a transverse wave damps rapidly. Thus, a transverse wave excited at the surface of the living body hardly reaches deep areas of the living body. A longitudinal wave, however, gradually damps even in soft materials. If a longitudinal wave excited at the surface can induce a transverse wave at the deep area in the living body, we can easily excite a steady-state transverse wave in the deep area of the living body. It has been experimentally shown that transverse wave excitation by a longitudinal wave is possible [14]. Here, a theoretical basis of the excitation method is developed.

For soft materials, the values of the two Lame's constants are quite different, that is, $\lambda >> \mu$. Let us introduce the parameter characterizing the magnitude of the difference:

$$g = \frac{\max_{\mathbf{x}\in\Omega} \{\lambda(\mathbf{x})\}}{\max_{\mathbf{x}\in\Omega} \{\mu(\mathbf{x})\}}. \tag{7.69}$$

Let us also express $\lambda(\mathbf{x})$ as

$$\lambda(\mathbf{x}) = g\hat{\lambda}(\mathbf{x}). \tag{7.70}$$

$\hat{\lambda}(\mathbf{x})$ is comparable in magnitude to the rigidity $\mu(\mathbf{x})$. The parameter g stands for the ratio of the bulk modulus K to the share modulus μ since $\lambda \approx K$ for the living body. The system is incompressible when $g = \infty$. The magnitude g^{-1} stands for the degree of the effect of compressibility. We rewrite Eq. (7.36) in terms of the amplitudes ϕ_i and ψ_i and the parameters μ, $\hat{\lambda}$, and g:

$$\begin{aligned} &-\rho\omega^2\phi_i(\mathbf{x}) - g\frac{\partial}{\partial x_i}\left(\hat{\lambda}(\mathbf{x})\sum_l \phi_{ll}(\mathbf{x})\right) - 2\sum_k \frac{\partial}{\partial x_k}[\mu(\mathbf{x})\phi_{ik}(\mathbf{x})] \\ &- \omega\frac{\partial}{\partial x_i}\left(\tilde{\varsigma}(\mathbf{x})\sum_l \psi_{ll}(\mathbf{x})\right) - 2\omega\sum_k \frac{\partial}{\partial x_k}[\eta(\mathbf{x})\psi_{ik}(\mathbf{x})] = 0 \end{aligned} \tag{7.71}$$

$$
\begin{aligned}
&-\rho\omega^2\psi_i(\mathbf{x}) - g\frac{\partial}{\partial x_i}\left(\hat{\lambda}(\mathbf{x})\sum_l \psi_{ll}(\mathbf{x})\right) - 2\sum_k \frac{\partial}{\partial x_k}[\mu(\mathbf{x})\psi_{ik}(\mathbf{x})] \\
&+\omega\frac{\partial}{\partial x_i}\left(\varsigma(\mathbf{x})\sum_l \phi_{ll}(\mathbf{x})\right) + 2\omega\sum_k \frac{\partial}{\partial x_k}[\eta(\mathbf{x})\phi_{ik}(\mathbf{x})] = 0.
\end{aligned}
\tag{7.72}
$$

Let us use the perturbation method and expand the amplitudes $\phi_i(\mathbf{x})$ and $\psi_i(\mathbf{x})$ in a power series of g^{-1} as

$$
\begin{cases}
\phi_i(\mathbf{x}) = \sum_{j=0}^{\infty} g^{-j}\phi_i^{(j)}(\mathbf{x}) \\
\psi_i(\mathbf{x}) = \sum_{j=0}^{\infty} g^{-j}\psi_i^{(j)}(\mathbf{x}).
\end{cases}
\tag{7.73}
$$

Inserting the expansion form Eq. (7.73) into Eqs. (7.71) and (7.72) and assuming that the equations are satisfied by $\phi_{\mathrm{i}}^{(\mathrm{j})}(\mathbf{x})$ and $\psi_{\mathrm{i}}^{(\mathrm{j})}(\mathbf{x})$ for any g according to the conventional perturbation procedure, we obtain a series of equations satisfied by $\phi_i^{(\mathrm{j})}(\mathbf{x})$ and $\psi_i^{(\mathrm{j})}(\mathbf{x})$ with respect to each power of g. The first power of g is the highest power of g in the series of equations, and the equations with respect to the highest power are given by

$$
-g\frac{\partial}{\partial x_i}\left(\hat{\lambda}(\mathbf{x})\sum_l \phi_{ll}^{(0)}(\mathbf{x})\right) = 0
\tag{7.74}
$$

$$
-g\frac{\partial}{\partial x_i}\left(\hat{\lambda}(\mathbf{x})\sum_l \psi_{ll}^{(0)}(\mathbf{x})\right) = 0.
\tag{7.75}
$$

Since the displacement is measured from the equilibrium point, we have

$$
\sum_l \phi_{ll}^{(0)}(\mathbf{x}) = 0
\tag{7.76}
$$

$$
\sum_l \psi_{ll}^{(0)}(\mathbf{x}) = 0
\tag{7.77}
$$

Hence, the zeroth-order components (the lowest-order components) in the expansion form Eq. (7.73) should satisfy the incompressible condition; the zeroth-order components in Eq. (7.73) are the amplitudes of the transverse wave excited in the living body. The higher-order components $\phi_i^{(\mathrm{j})}(\mathbf{r})$ and $\psi_i^{(\mathrm{j})}(\mathbf{r})(j \geq 1)$ express the compressible properties and vanish in the zero compressibility limit ($g \to \infty$).

Using Eqs. (7.71) and (7.72), we obtain the equations with respect to the zeroth power of g:

$$-\rho\omega^2\phi_i^{(0)}(\mathbf{x}) - 2\sum_k \frac{\partial}{\partial x_k}\left[\mu(\mathbf{x})\phi_{ik}^{(0)}(\mathbf{x})\right] - 2\omega\sum_k \frac{\partial}{\partial x_k}\left[\eta(\mathbf{x})\psi_{ik}^{(0)}(\mathbf{x})\right] = P_\phi(\mathbf{x}) \tag{7.78}$$

$$-\rho\omega^2\psi_i^{(0)}(\mathbf{x}) - 2\sum_k \frac{\partial}{\partial x_k}\left[\mu(\mathbf{x})\psi_{ik}^{(0)}(\mathbf{x})\right] + 2\omega\sum_k \frac{\partial}{\partial x_k}\left[\eta(\mathbf{x})\phi_{ik}^{(0)}(\mathbf{x})\right] = P_\psi(\mathbf{x}), \tag{7.79}$$

where

$$\begin{cases} P_\phi(\mathbf{x}) = \dfrac{\partial}{\partial x_i}\left[\hat{\lambda}(\mathbf{x})Q_\phi(\mathbf{x})\right] \\ P_\psi(\mathbf{x}) = \dfrac{\partial}{\partial x_i}\left[\hat{\lambda}(\mathbf{x})Q_\psi(\mathbf{x})\right] \end{cases} \tag{7.80}$$

with

$$\begin{cases} Q_\phi(\mathbf{x}) = \sum_l \phi_{ll}^{(1)}(\mathbf{x}) \\ Q_\psi(\mathbf{x}) = \sum_l \psi_{ll}^{(1)}(\mathbf{x}). \end{cases} \tag{7.81}$$

Equations (7.78) and (7.79) are regarded as the equation of motion for the incompressible components $\phi_{\mathrm{i}}^{(0)}(\mathbf{x})$ and $\psi_{\mathrm{i}}^{(0)}(\mathbf{x})$ excited by the "volume force" per unit volume:

$$F_i(\mathbf{x}, t) = \hat{\lambda}(\mathbf{x})\left[Q_\phi(\mathbf{x})\cos\omega t + Q_\psi(\mathbf{x})\sin\omega t\right]. \tag{7.82}$$

The volume force is expressed by the compressible component corresponding to the longitudinal wave component. Hence, Eqs. (7.78) and (7.79) show that the longitudinal wave with amplitudes $\phi_i^{(1)}(\mathbf{x})$ and $\psi_i^{(1)}(\mathbf{x})$ induces the transverse wave with amplitudes of $\phi_i^{(0)}(\mathbf{x})$ and $\psi_i^{(0)}(\mathbf{x})$.

The quantities Q_ϕ and Q_ψ denote the magnitude of compression or dilatation. Since those spatial variations are gradual, we can approximate them as

$$\begin{cases} P_\phi = \dfrac{\partial\hat{\lambda}(\mathbf{x})}{\partial x_i}Q_\phi(\mathbf{x}) + \hat{\lambda}(\mathbf{x})\dfrac{\partial Q_\phi(\mathbf{x})}{\partial x_i} \approx \dfrac{\partial\hat{\lambda}(\mathbf{x})}{\partial x_i}Q_\phi(\mathbf{x}) \\ P_\psi = \dfrac{\partial\hat{\lambda}(\mathbf{x})}{\partial x_i}Q_\psi(\mathbf{x}) + \hat{\lambda}(\mathbf{x})\dfrac{\partial Q_\psi(\mathbf{x})}{\partial x_i} \approx \dfrac{\partial\hat{\lambda}(\mathbf{x})}{\partial x_i}Q_\psi(\mathbf{x}). \end{cases} \tag{7.83}$$

This expression shows that the inhomogeneity of the bulk modulus converts the longitudinal wave to the transversal wave. Finally we obtain the following equations satisfied by the incompressible component:

$$\begin{aligned} -\rho\omega^2\phi_i^{(0)}(\mathbf{x}) - \frac{\partial\hat{\lambda}(\mathbf{x})}{\partial x_i}Q_\phi(\mathbf{x}) - 2\sum_k \frac{\partial\mu(\mathbf{x})}{\partial x_k}\phi_{ik}^{(0)}(\mathbf{x}) - \mu(\mathbf{x})\sum_k \frac{\partial^2\phi_i^{(0)}(\mathbf{x})}{\partial x_k^2} \\ - 2\omega\sum_k \frac{\partial\eta(\mathbf{x})}{\partial x_k}\psi_{ik}^{(0)}(\mathbf{x}) - \omega\eta(\mathbf{x})\sum_k \frac{\partial^2\psi_i^{(0)}(\mathbf{x})}{\partial x_k^2} = 0 \end{aligned} \tag{7.84}$$

$$\begin{aligned} \rho\omega^2\psi_i^{(0)}(\mathbf{x}) - \frac{\partial\hat{\lambda}(\mathbf{x})}{\partial x_i}Q_\psi(\mathbf{x}) - 2\sum_k \frac{\partial\mu(\mathbf{x})}{\partial x_k}\psi_{ik}^{(0)}(\mathbf{x}) - \mu(\mathbf{x})\sum_k \frac{\partial^2\psi_i^{(0)}(\mathbf{x})}{\partial x_k^2} \\ + 2\omega\sum_k \frac{\partial\eta(\mathbf{x})}{\partial x_k}\phi_{ik}^{(0)}(\mathbf{x}) + \omega\eta(\mathbf{x})\sum_k \frac{\partial^2\phi_i^{(0)}(\mathbf{x})}{\partial x_k^2} = 0 \end{aligned} \tag{7.85}$$

Neglecting the regions showing changes in the tissue structures in which $\partial\mu(\mathbf{x})/\partial x_j \neq 0$ and $\partial\eta(\mathbf{x})/\partial x_j \neq 0$ and paying attention to only the regions in which $\partial\mu(\mathbf{x})/\partial x_j \approx 0$ and $\partial\eta(\mathbf{x})/\partial x_j \approx 0$, we obtain the following equations satisfied by the zeroth-order amplitude:

$$-\rho\omega^2\phi_i^{(0)}(\mathbf{x}) - \mu(\mathbf{x})\Delta\phi_i^{(0)} - \omega\eta(\mathbf{x})\Delta\psi_i^{(0)} = 0 \tag{7.86}$$

$$-\rho\omega^2\psi_i^{(0)}(\mathbf{x}) - \mu(\mathbf{x})\Delta\psi_i^{(0)} + \omega\eta(\mathbf{x})\Delta\phi_i^{(0)} = 0. \tag{7.87}$$

Since the bulk modulus is much larger than the rigidity in soft materials, the approximation $g^{-1} \approx 0$ (or $g \approx \infty$) is suitable; with great accuracy, the amplitude is expressed in terms of only the incompressible component as

$$\begin{cases} \phi_i(\mathbf{x}) \cong \phi_i^{(0)}(\mathbf{x}) \\ \psi_i(\mathbf{x}) \cong \psi_i^{(0)}(\mathbf{x}). \end{cases} \tag{7.88}$$

The amplitude of the longitudinal wave is negligible and the wave detected by the phase shift of the MR signal is the transversal wave:

$$\begin{aligned} u_i(\mathbf{x},t) = \phi_i(\mathbf{x})\cos\omega t + \psi_i(\mathbf{x})\sin\omega t = \sum_{j=0}^{\infty} g^{\ j}\phi_i^{(j)}(\mathbf{x})\cos\omega t + \sum_{j=0}^{\infty} g^{-j}\psi_i^{(j)}(\mathbf{x})\sin\omega t \\ \cong \phi_i^{(0)}(\mathbf{x})\cos\omega t + \psi_i^{(0)}(\mathbf{x})\sin\omega t. \end{aligned} \tag{7.89}$$

Hence, even if the wave is excited longitudinally, the reconstruction formula Eqs. (7.60) and (7.61) (or Eqs. (7.67) and (7.68)) are valid. This shows that the rigidity can be measured by longitudinal wave excitation.

7.6 Uniaxial Organs and Tissues

In hard body organs and tissues such as bone and tendons, the viscoelastic properties are anisotropic and often uniaxial. The relationship between the phase shift of the MRI signal and the viscoelastic parameters in a uniaxial system is discussed. Let us choose the x_3 -axis along the symmetric axis. The Lagrangian and dissipation functions of the uniaxial system are respectively given by

$$L = \int_\Omega d^3\mathbf{x}\Big[\frac{1}{2}\rho\sum_{j=1}^{3}(\dot{u}_j(\mathbf{x}))^2 - \frac{1}{2}\sum_{a,b,c,d=1,2}\lambda^{(1-2)}_{abcd}(\mathbf{x})u_{ab}(\mathbf{x})u_{cd}(\mathbf{x}) - \frac{1}{2}\kappa_1(\mathbf{x})u_{33}^2(\mathbf{x}) - 2\kappa_2(\mathbf{x})u_{33}(\mathbf{x})\sum_{a=1,2}u_{aa}(\mathbf{x}) - 2\kappa_3(\mathbf{x})\sum_{a=1,2}\left(u_{a3}^2(\mathbf{x}) + u_{3a}^2(\mathbf{x})\right)\Big] \tag{7.90}$$

and

$$D = \int_\Omega d^3\mathbf{x}\Big[\frac{1}{2}\sum_{a,b,c,d=1,2}\eta^{(1-2)}_{abcd}(\mathbf{x})\dot{u}_{ab}(\mathbf{x})\dot{u}_{cd}(\mathbf{x}) + \frac{1}{2}\xi_1(\mathbf{x})\dot{u}_{33}^2(\mathbf{x}) + 2\xi_2(\mathbf{x})\dot{u}_{33}\sum_{a=1,2}\dot{u}_{aa}(\mathbf{x}) + 2\xi_3(\mathbf{x})\sum_{a=1,2}\left(\dot{u}_{a3}^2(\mathbf{x}) + \dot{u}_{3a}^2(\mathbf{x})\right)\Big], \tag{7.91}$$

where for $a,b,c,d = 1,2,$

$$\lambda^{(1-2)}_{abcd}(\mathbf{x}) = \lambda(\mathbf{x})\delta_{ab}\delta_{cd} + 2\mu(\mathbf{x})\delta_{ac}\delta_{bd} \tag{7.92}$$

and

$$\eta^{(1-2)}_{abcd}(\mathbf{x}) = \varsigma(\mathbf{x})\delta_{ab}\delta_{cd} + 2\eta(\mathbf{x})\delta_{ac}\delta_{bd}. \tag{7.93}$$

The uniaxial symmetry requires new components of the elasticity tensor, i.e., κ_1, κ_2, and κ_3 and of the viscosity tensor, i.e., ξ_1, ξ_2, and ξ_3. The equation of motion of the uniaxial system is also obtained from the Lagrange equation Eq. (7.23). Using the expression

$$u_i(\mathbf{x}, t) = \phi_i(\mathbf{x})\cos\omega t + \psi_i(\mathbf{x})\sin\omega t, \tag{7.94}$$

the equations satisfied by the amplitudes ϕ_i and ψ_i are derived. These equations are very complex as compared with those of the isotropic system, namely, Eqs. (7.44) and (7.45). Noting that the quantities that should be measured by MRE are the moduli of rigidity in the axial direction (in the x_3-direction) κ_3 and in the x_1-x_2 plane μ, we try to simplify the complex equations.

Introducing the expansion parameters

$$\begin{cases} g_1 = \dfrac{\max\limits_{\mathbf{x}\in\Omega}\{\lambda(\mathbf{x})\}}{\max\limits_{\mathbf{x}\in\Omega}\{\mu(\mathbf{x})\}} \\ g_2 = \dfrac{\max\limits_{\mathbf{x}\in\Omega}\{\kappa_1(\mathbf{x})\}}{\max\limits_{\mathbf{x}\in\Omega}\{\kappa_3(\mathbf{x})\},} \end{cases} \tag{7.95}$$

we expand the amplitudes ϕ_i and ψ_i in the power series of g_1^{-1} and g_2^{-1} to

$$\begin{cases} \phi_i(\mathbf{x}) = \sum\limits_{j=0}^{\infty}\left[g_1^{-j}\phi_i^{(j,1)}(\mathbf{x}) + g_2^{-j}\phi_i^{(j,2)}(\mathbf{x})\right] \\ \psi_i(\mathbf{x}) = \sum\limits_{j=0}^{\infty}\left[g_1^{-j}\psi_i^{(j,1)}(\mathbf{x}) + g_2^{-j}\psi_i^{(j,2)}(\mathbf{x})\right]. \end{cases} \tag{7.96}$$

Approximating ϕ_i and ψ_i using the lowest-order terms,

$$\begin{cases} \phi_i(\mathbf{x}) \cong \phi_i^{(0,1)}(\mathbf{x}) + \phi_i^{(0,2)}(\mathbf{x}) \\ \psi_i(\mathbf{x}) \cong \psi_i^{(0,1)}(\mathbf{x}) + \psi_i^{(0,2)}(\mathbf{x}), \end{cases} \tag{7.97}$$

and neglecting the regions showing changes in the tissue structures, we obtain the anisotropic incompressible conditions

$$\begin{cases} \dfrac{\partial \phi_1}{\partial x_1} + \dfrac{\partial \phi_2}{\partial x_2} = 0 \\ \dfrac{\partial \phi_3}{\partial x_3} = 0 \end{cases} \tag{7.98}$$

and

$$\begin{cases} \dfrac{\partial \psi_1}{\partial x_1} + \dfrac{\partial \psi_2}{\partial x_2} = 0 \\ \dfrac{\partial \psi_3}{\partial x_3} = 0 \end{cases} \tag{7.99}$$

and the equations of motion for the amplitudes in the x_1-x_2 plane ($a = 1, 2$),

$$\begin{aligned} &-\rho\omega^2\phi_a(\mathbf{x}) - \mu(\mathbf{x})\Delta_2\phi_a(\mathbf{x}) - 2\kappa_3(\mathbf{x})\frac{\partial^2\phi_a(\mathbf{x})}{\partial x_3^2} \\ &-\omega\eta(\mathbf{x})\Delta_2\psi_a(\mathbf{x})\text{-}2\omega\xi_3(\mathbf{x})\frac{\partial^2\psi_a(\mathbf{x})}{\partial x_3^2} - 0 \end{aligned} \tag{7.100}$$

$$-\rho\omega^2\psi_a(\mathbf{x}) - \mu(\mathbf{x})\Delta_2\psi_a(\mathbf{x}) - 2\kappa_3(\mathbf{x})\frac{\partial^2\psi_a(\mathbf{x})}{\partial x_3^2} + \omega\eta(\mathbf{x})\Delta_2\phi_a(\mathbf{x}) + 2\omega\xi_3(\mathbf{x})\frac{\partial^2\phi_a(\mathbf{x})}{\partial x_3^2} = 0 \tag{7.101}$$

and those in the x_3-direction

$$-\rho(\mathbf{x})\omega^2\phi_3(\mathbf{x}) - 2\kappa_3(\mathbf{x})\Delta_2\phi_3(\mathbf{x}) - 2\omega\xi_3(\mathbf{x})\Delta_2\psi_3(\mathbf{x}) = 0 \tag{7.102}$$

$$-\rho(\mathbf{x})\omega^2\psi_3(\mathbf{x}) - 2\kappa_3(\mathbf{x})\Delta_2\psi_3(\mathbf{x}) + 2\omega\xi_3(\mathbf{x})\Delta_2\phi_3(\mathbf{x}) = 0, \tag{7.103}$$

where

$$\Delta_2 = \frac{\partial^2}{\partial x_1^2} + \frac{\partial^2}{\partial x_2^2}. \tag{7.104}$$

To measure the rigidities of the uniaxial system, in addition to MSG in the x_3-direction, MSG in the x_1-x_2 plane is required. Therefore, the following magnetic fields for inducing MSG in the x_1-direction are introduced:

$$\begin{cases} B''_{MSG}(\mathbf{x}, t) = G_0 x_1 \cos\omega t \\ B'_{MSG}(\mathbf{x}, t) = G_0 x_1 \sin\omega t \end{cases}. \tag{7.105}$$

The phase shifts induced by B''_{MSG} and B'''_{MSG} are respectively given by

$$\begin{cases} \vartheta_0(\mathbf{x}) = \dfrac{\pi N\Gamma}{\omega} G_0\phi_1(\mathbf{x}) \\ \vartheta_{\pi/2}(\mathbf{x}) = \dfrac{\pi N\Gamma}{\omega} G_0\psi_1(\mathbf{x}) \end{cases} \tag{7.106}$$

In terms of the phase shifts expressed by Eqs. (7.58), (7.59), and (7.106), the rigidities are expressed as

$$\kappa_3(\mathbf{x}) = -\frac{1}{2}\frac{\rho\omega^2(\theta_0(\mathbf{x})\Delta_2\theta_0(\mathbf{x}) + \theta_{\pi/2}(\mathbf{x})\Delta_2\theta_{\pi/2}(\mathbf{x}))}{(\Delta_2\theta_0(\mathbf{x}))^2 + (\Delta_2\theta_{\pi/2}(\mathbf{x}))^2} \tag{7.107}$$

and

$$\begin{aligned} \mu = Z_0(\mathbf{x})\big[&- \rho\omega^2(\vartheta_0(\mathbf{x})\Delta_2\vartheta_0(\mathbf{x}) + \vartheta_{\pi/2}(\mathbf{x})\Delta_2\vartheta_{\pi/2}(\mathbf{x})) \\ &- 2\kappa_3(\mathbf{x})\left(\Delta_2\vartheta_0(\mathbf{x})\frac{\partial^2\vartheta_0(\mathbf{x})}{\partial x_3^2} + \Delta_2\vartheta_{\pi/2}(\mathbf{x})\frac{\partial^2\vartheta_{\pi/2}(\mathbf{x})}{\partial x_3^2}\right) \\ &- 2\omega\xi_3(\mathbf{x})\left(\Delta_2\vartheta_0(\mathbf{x})\frac{\partial^2\vartheta_{\pi/2}(\mathbf{x})}{\partial x_3^2} - \Delta_2\vartheta_{\pi/2}(\mathbf{x})\frac{\partial^2\vartheta_0(\mathbf{x})}{\partial x_3^2}\right)\big] \end{aligned} \tag{7.108}$$

with

$$\xi_3(\mathbf{x}) = -\frac{1}{2}\frac{\rho\omega\left(\theta_0(\mathbf{x})\Delta_2\theta_{\pi/2}(\mathbf{x}) - \theta_{\pi/2}(\mathbf{x})\Delta_2\theta_0(\mathbf{x})\right)}{\left(\Delta_2\theta_0(\mathbf{x})\right)^2 + \left(\Delta_2\theta_{\pi/2}(\mathbf{x})\right)^2} \tag{7.109}$$

and

$$Z_0(\mathbf{x}) = \left[\left(\Delta_2\vartheta_0(\mathbf{x})\right)^2 + \left(\Delta_2\vartheta_{\pi/2}(\mathbf{x})\right)^2\right]^{-1}. \tag{7.110}$$

The above reconstruction formula is exactly correct when $g_1 >> 1$ and $g_2 >> 1$. In uniaxial tissues, however, g_2 is not much greater than one. Therefore, the formula is an approximate one. The differentiation-type reconstruction formula Eqs. (7.107), (7.108), (7.109), (7.100), and (7.110) can be translated to an integral-type reconstruction formula even for the uniaxial system.

References

1. For a recent review, see Mariappan YK, Glaser K.J, Ehman RL (2010) Magnetic resonance elastography: a review. Clin Anat 23:497–511
2. Muthupillai R, Lomas DJ, Rossman PJ, Greenleaf JF, Manduca A, Ehman RL (1995) Magnetic resonance elastography by direct visualization of propagating acoustic strain waves. Science 269:1854–1857
3. For a review, see Parker KJ, Doyley MM, Rubens DJ (2011) Imaging the elastic properties of tissue: the 20 year perspective. Phys Med Biol 56:R1–R29
4. Uffmann K, Ladd ME (2008) Actuation systems for MR elastography. IEEE Eng Med Biol 27:28–34
5. Manduca A, Oliphant TE, Dresner MA, Mahowald JL, Kruse SA, Amromin E, Felmlee JP, Greenleaf JF, Ehman RL (2001) Magnetic resonance elastography: non-invasive mapping of tissue elasticity. Med Image Anal 5:237–254
6. As an example, it is noted that diagnostic accuracy of MRE for assessing hepatic fibrosis has been verified. See Yin M, Talwalker JA, Glaser KJ, Manduca A, Grimm RC, Rossman PJ, Fidler JL, Ehman RL (2007) Assessment of hepatic fibrosis with magnetic resonance elastography. Clin Gastroenterol Hepatol 5:1207–1213
7. Suga M, Aga T, Minato K (2004) Development of a magnetic resonance elastic microscope system. Proc 26th Annu Int Conf IEEE EMBS: 1025–1027
8. Tadano S, Fujisaki K, Suzuki H, Takao S, Suga M, Kajiwara I, Yamamoto T, Jiang Y, Nakamura G (2012) Excitation system for magnetic resonance elastography using micro MRI. J Biomech Sci Eng 7:463–474
9. Landau LD, Lifshitz EM (1976) Mechanics, 3rd edn (trans: Sykes JB, Bell JS). Elsevier, Amsterdam, §8
10. Landau LD, Lifshitz EM (1970) Theory of elasticity, 2nd edn (trans: Sykes JB, Reid WH). Pergamon, Oxford, §10
11. Landau LD, Lifshitz EM (1970) Theory of elasticity, 2nd edn (trans: Sykes JB, Reid WH). Pergamon, Oxford, §34
12. Goldstein H (1980) Classical mechanics, 2nd edn. Addison-Wesley, Cambridge, MA, pp 21–24

13. Suga M, Miura H, Fujiwara T, Yamamoto T, Tanaka T, Yu Q, Arai K, Gong JP, Nakamura G, Ikehira H, Obata T (2009) Inversion algorithm by integral type reconstruction formula for magnetic resonance elastography. Proc Int Soc Magn Reson Med 17:2506
14. Suga M, Obata T, Hirano M, Tanaka T, Ikehira H (2007) Magnetic resonance elastography to observe deep areas. Proc 29th Annu Int Conf IEEE EMBS: 2599–2602

Dynamics of Water, Biomaterials, and Skin Investigated by Means of Dielectric Relaxation Spectroscopy

8

Shin Yagihara

Abstract

Recent developments in experimental techniques of dielectric spectroscopy are based on the new concept of broadband dynamics observed by using a broadband dielectric spectroscopy (BDS) measuring system. BDS techniques are particularly effective for studying water structures in biological systems. Water molecules interacting with biomolecules show various behaviors of collective modes, and these behaviors are closely related to the macroscopic and microscopic properties and functions of biomaterials. This chapter discusses the basic concept of the dynamic behaviors of water structures obtained from experimental and analytical methods based on BDS. Observation techniques based on BDS solve some of the problems of observing dynamics. The physical picture of water structures and the interpretation of relaxation phenomena are suggested from dynamic hydrogen bonding (HB) networks, and the dynamic properties are also explained on the basis of the fractal concept. Although this concept is not perfect, it is a suitable approach for characterizing the dynamic behaviors of HB networks. Although the fractal approach has been applied to complex materials using various techniques, this approach also suggests the existence of a problem related to length scales in the observation of dynamic behaviors. Complementary approaches with various observation techniques will be very effective for solving such problems and for suggesting future applications.

Keywords

Microwave dielectric spectroscopy • Broadband dynamics • Observation problems of dynamics • Complementary techniques of observation • Fractal concept of water structures

S. Yagihara (✉)
Department of Physics, Tokai University, 4-1-1 Kitakaname, Hiratsuka, Kanagawa 259-1292, Japan
e-mail: yagihara@keyaki.cc.u-tokai.ac.jp

R. Kita, T. Dobashi (eds.), *Nano/Micro Science and Technology in Biorheology*,
DOI 10.1007/978-4-431-54886-7_8

8.1 Introduction

The dynamic structures and collective behaviors of molecules in materials are reflected in their dielectric relaxation phenomena. Dielectric spectroscopy techniques have greatly developed in the last few decades in spite of their long history since the establishment of dielectric spectroscopy as a measuring technique based on Debye's research on the molecular concept of the dipole moment in the experimental work of Mizushima [1]. The International Dielectric Society was established in 2001 [2], after the closure of the Division of Dielectrics in the Gordon Conference owing to the death of Prof. Cole of Brown University [3]. The most important concept in recent dielectric studies is broadband dynamics obtained by broadband dielectric spectroscopy (BDS) measurements [4] making it possible to clarify the structures and properties of molecular complex materials in which molecules are hierarchically aggregated and structured in various length and time scales.

8.2 Various Water Structures and Slow Dynamics

8.2.1 Broadband Dynamics of Water Structures and Observation Problems

Figure 8.1 shows typical water structures in various aqueous systems observed using BDS techniques. The large relaxation process observed at approximately 20 GHz is due to the dynamic behavior of collective water molecules. A precise

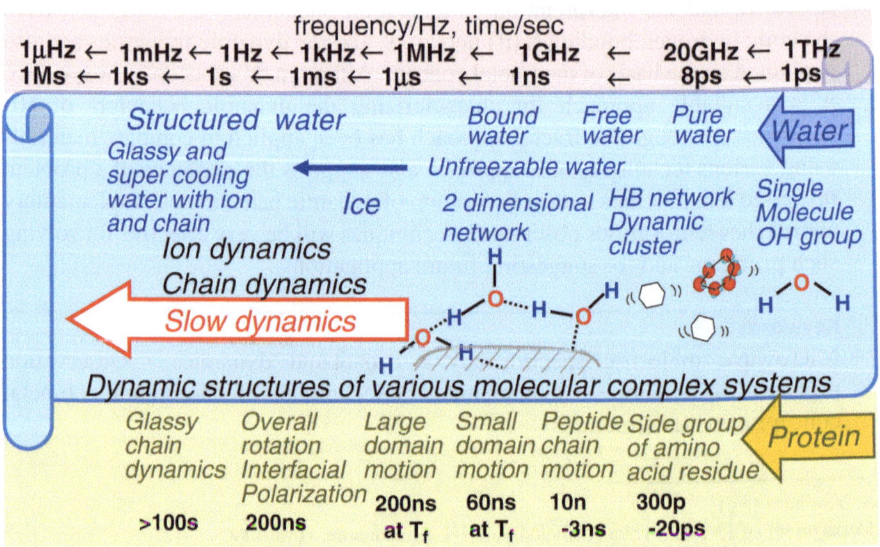

Fig. 8.1 Various water structures and slow dynamics in aqueous materials. Examples of dynamics for protein molecules are also shown

measurement carried out for pure water at 25 °C suggests a relaxation time of 8.3 ps and a the relaxation strength of 78.3 for a single relaxation process called the Debye relaxation process [5]. It is also suggested that water undergoes a small relaxation process at a higher frequency of approximately 160 GHz corresponding to a relaxation time of 1.0 ps [6].

The dynamics of water molecules interacting with biopolymers in biological systems are generally restricted, and the restriction makes the relaxation frequency shift to a lower frequency. The cooperative dynamics may include not only the water molecules but also the chain dynamics of biopolymers, the overall rotation and local motion of atomic groups at various length scales, and those of counter ions. The relaxation frequencies of these dynamics are thus dependent on the time scale of the collective interactions. In the case of aqueous solutions of immunoglobulin, relaxation processes due to the overall rotation and motion of some domains, such as hinge-bending motion or elbow motion, have been recognized at low frequencies as explained in Sect. 8.4.1.1. Even in these protein dynamics, the ion and water molecules surrounding the protein surface cooperatively interact.

Figure 8.1 shows the water structures observed for various aqueous mixtures such as DNA, liposomes, cells, tissues, and cement with processes due to the interfacial polarization and chain dynamics of solute molecules. These various water structures and physical properties are characterized in extremely wide time and space scales. BDS is one of the most effective experimental techniques for studying the dynamic behaviors of water through the correlation of dipole moments of water and interacting molecules. A general picture of various water structures can be deduced from BDS measurements.

The observation techniques used for dynamic molecular complex systems often include serious problems [7]. Figure 8.2 explains one of the problems from a physical viewpoint of observation. The dynamic structures of complex systems have a characteristic time, i.e., the lifetime. All observation techniques have a characteristic time scale. If the lifetime of the structure is sufficiently longer than the characteristic time scale of the observation technique, the instantaneous structure can be correctly recorded by the measuring system. On the other hand, if the lifetime of the dynamic structure is shorter than the characteristic time scale of the observation, the result obtained is simply the average during the observation period, and dynamic structures with a short lifetime apparently disappear, as shown in Fig. 8.2. For example, the technique of differential scanning calorimetry (DSC) has a characteristic time scale of 100–1,000 s, which is not the scanning rate of the temperature. If the time scale is not suitably taken into account, the dynamic structures cannot be correctly characterized. Since water has various structures with different lifetimes from ps to ms order, misinterpretation of the observation easily leads to incorrect water structures.

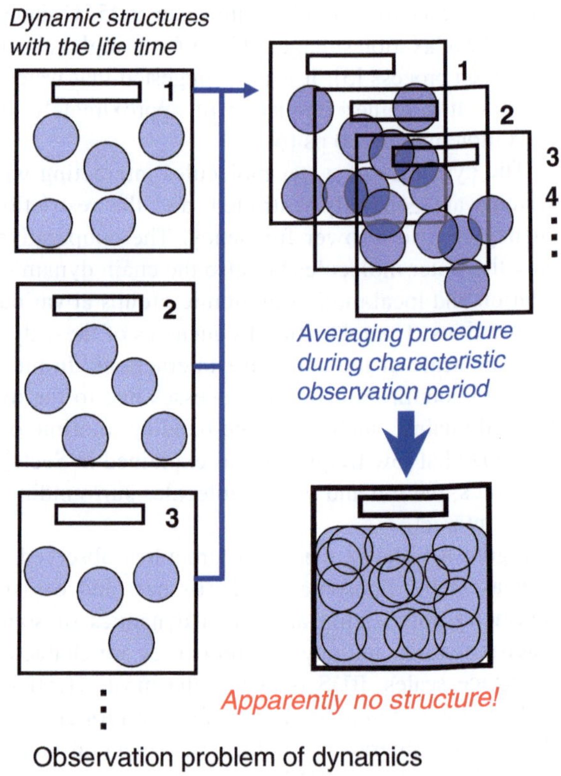

Fig. 8.2 Observation problem of dynamics of molecular complex systems (Adapted from Yagihara [7] with permission from the Nanofiber Society. Copyright 2013)

8.2.2 Hydrogen Bonding Network of Aqueous Systems

The liquid structure of pure water has recently been understood in terms of the behaviors of its hydrogen bonding (HB) network. The HB network obtained by molecular dynamics (MD) simulation is shown as green dotted lines in Fig. 8.3. An MD simulation of 500 water molecules (TIP3P) in NVT ensembles was performed using Discovery Studio 3.0 (Accelrys) at the computing center of Tokai University with the particle mesh Ewald and SHAKE methods. The calculation was performed at intervals of 1 fs over 1 ns in the equilibrium state at 300 K. The length and angle were analyzed for 500 hydrogen bonds (steps, 10,000; time step, 0.001 ps; target temperature, 300, 310, 320 K; type, NVT; frequency of saved results, 100) forming a flexible HB network. The snapshot of the HB network expanding to all of molecules in Fig. 8.3 suggests the existence of intermolecular interactions with long-range correlation throughout the HB network.

The dielectric relaxation approach suggests the suitability of the concept of the HB network for explaining some of the experimental results for water in various aqueous systems [8]. The relaxation time for a dipole relaxation process due to the reorientation of motional units with a high molecular weight shows a discontinuous temperature dependence at the freezing temperature in the Arrhenius plot since larger motional units have a larger excluded volume effect. For example, in the case

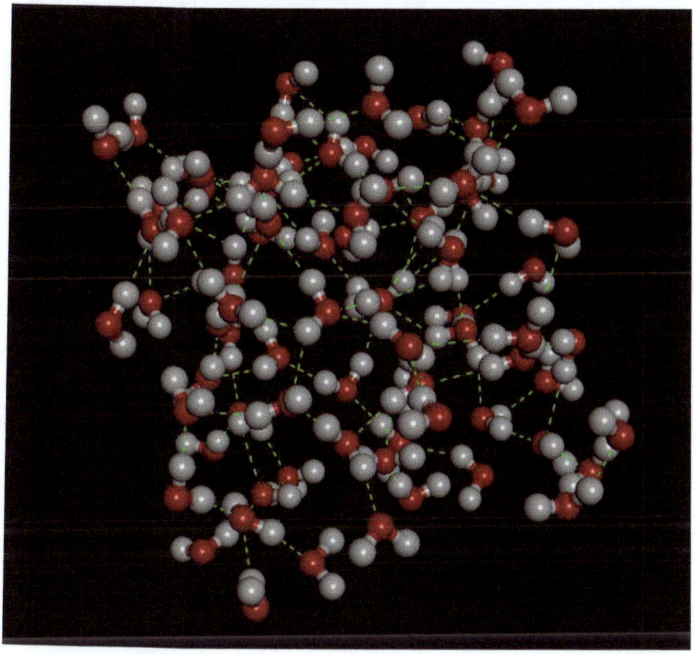

Fig. 8.3 HB network obtained by MD simulation. MD simulation was performed by Discovery Studio 3.0 (Accelrys) at the computing center of Tokai University

of a polymer blend of poly(vinyl pyrrolidone) (PVP) and poly(ethylene glycol) (PEG), the local chain dynamics of PVP show discontinuous behavior, but those of PEG show continuous behavior, since the local chain motion of the oxide polymer (PEG) is much more flexible because of the oxygen atoms. In the case of liquid crystal molecules with an anisotropic and rigid shape, benzene solutions of 4-cyano-4′-pentylbiphenyl (5CB) exhibit continuous behavior for molecular rotations around the long axis but discontinuous rotations around the short axis because of the large excluded volume effect. The Arrhenius plot for the relaxation process due to the chain dynamics of poly(vinyl acetate) (PVAc) in benzene solution shows both a continuous change and a discontinuous change in the slope, which corresponds to the apparent activation energy. On the other hand, the Arrhenius plot for PEG chain dynamics shows not only a discontinuous change in the activation energy but also a discontinuous change in the relaxation time. This means that the HB interaction has a considerably different molecular mechanism from the van der Waals interaction for its dynamics because of the network structure of both solute and solvent molecules, and small water molecules behave as a large motional unit.

This clearer model of HB network behavior is supported by the results of dielectric spectroscopy as shown in Fig. 8.4 [7, 9]. In this figure, the relaxation

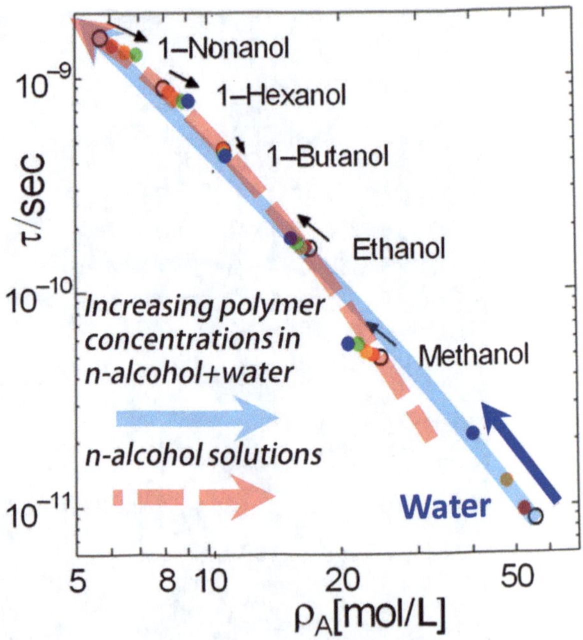

Fig. 8.4 The relaxation time for solvent dynamics vs. HB acceptor density. *Open circles* indicate values for the pure solvents and *closed circles* were obtained for 10, 20, 30, and 40 wt% polymer solutions (Adapted from Yagihara [7] with permission from the Nanofiber Society. Copyright 2013)

time of the relaxation process due to liquid solvents, such as water and various n-alcohols in solutions of PVP at various concentrations, is plotted against the HB acceptor density, which was simply estimated from the chemical structures. The relative behaviors of the trajectories in the figure are not affected by the approximate estimation. Figure 8.4 suggests that a high HB density reduces the relaxation time since the rate of exchange of HB in the network is higher. Small arrows indicate the direction of the trajectory with increasing polymer concentration for each solvent. The HB densities of the n-alcohols and the repeat unit VP appear to be the same for 1-butanol and 1-pentanol. The plots obtained for all solutions lie on the same line. This result means that the mechanism of solvent molecule dynamics in n-alcohol solutions of PVP can be explained from the HB network. However, the plots for different alcohols such as dihydric and trihydric alcohols are not located on this line. In that sense, it is also possible to remove plots for aqueous solutions from the line in Fig. 8.4, as shown by the dotted line. This ambiguity does not contradict the molecular mechanism of the HB network mentioned above.

8.2.3 Fractal Concept of Water Structures

It is not easy to devise an effective approach to describing water structures from the concept of an HB network. Although the relaxation time characterizes the average behavior of the dynamics, the dynamic behaviors of an HB network cannot be explained by only the average behavior. Thus, the new concept of the fractal behavior of water structures is useful. The expression of the relaxation process for fractal water structures considers not only the average behavior but also the fluctuations evaluated using the relaxation time distribution [10–12]. To evaluate this distribution in aqueous mixtures, if an exponential hyperbolic-type dipolar correlation function corresponding to the Cole–Cole equation is compared with the memory function, the exponent and the spatial–temporal fractal dimension are found to be in good agreement. The Cole–Cole relaxation function is here expressed as

$$\varepsilon^* - \varepsilon_\infty = \frac{\Delta\varepsilon}{1 + (j\omega\tau)^\beta} \tag{8.1}$$

$$0 < \beta \le 1, \tag{8.2}$$

where ε^* is the complex dielectric permittivity, ε_∞ is the high-frequency limit of the dielectric constant, $\Delta\varepsilon$ is the relaxation strength, j is the imaginary unit, ω is the angular frequency, τ is the relaxation time, and β is the relaxation time distribution parameter. Debye's single relaxation process is expressed by setting $\beta = 1$. The Cole–Cole relaxation function has been applied to the relaxation process of water in polymer aqueous solutions and also to the process induced by confining molecular liquids in porous systems. The common feature of the molecules contributing to the symmetric loss spectrum is that they are spatially confined and exist under a geometrical constraint [13]. Monte Carlo simulations of the random walk of a particle in a space with a geometrical constraint of the fractal structure also lead to a symmetric Cole–Cole type relaxation function [14].

Both the relaxation time, τ, and the distribution parameter, β, had been treated as independent quantities until our recently developed fractal approach, although the parameters τ and β are closely related to each other in the case of water structures. The conventional investigation carried out without considering the relationship between τ and β might be a result of the fact that the dynamics of water molecules have not been directly observed so far because of the lack of a suitable observation technique. The space fractal dimension used to scale the volume division size of the relaxation unit, reflected in Eq. (8.1), can be used to obtain the relationship between the relaxation time and the exponential in the Cole–Cole relaxation function. This new concept suggests that the fluctuation of the characteristic time of the dynamics, β, can be explained by the geometrical self-similarity of the polymer network [12] as follows:

$$\beta = \frac{d_G}{2} \frac{\ln(\tau\omega_s)}{\ln(\tau/\tau_0)}, \tag{8.3}$$

where d_G is the spatial fractal dimension for relaxed units interacting with the surrounding environment, τ_o is the cutoff of the scaling time, and ω_s is given by

$$\omega_s = 2d_{\mathrm{E}} G^{2/d_{\mathrm{G}}} D_s / R_0^2, \tag{8.4}$$

with Euclidean dimension d_E geometrical coefficient G (usually taking a value of about unity), self-diffusion coefficient D_s, and cutoff size of the scaling in the space R_o.

Fractal analysis with Eq. (8.3) is performed by considering the τ-β diagram. The values of τ and β obtained from experimental results for the water content dependence of the relaxation process due to water are plotted and the hyperbolic behaviors of the trajectories of the plots are evaluated. Fractal water structures are found to be characterized by the trajectory reflecting the characteristic slow dynamics of water for each material. By examining various aqueous systems, some pattern of the dominant area was shown for each kind of molecular behavior. Dispersion and gel systems appear to have water structures with a lower fractal dimension. The solution and dispersion systems clearly appear in different areas of the diagram. In other words, this fractal analysis is essentially different from conventional analysis, in which the relaxation time and its fluctuations are compared at the same water content. It is natural to consider that the same concentration will not have the same effect on the slow dynamics in different aqueous systems. Thus, conventional analytical methods are not suitable for the comparison of solution and dispersion aqueous systems. Some examples of the application of fractal analysis are given in the following sections.

8.3 Experimental Techniques of Dielectric Spectroscopy for Biomaterials

8.3.1 Broadband Dielectric Spectroscopy for Various Water Structures

BDS involves the use of different subsystems to carry out measurements in different frequency regions. For example, in the BDS equipment used at Tokai University, the frequency range of all subsystems is from 1 μHz to 50 GHz. All subsystems are classified into two kinds of measuring systems for the frequency domain and time domain measurements of response functions. The time and frequency response functions are related to each other in accordance with linear response theory [15]. Cole expressed a general relationship with the dielectric relation function as the frequency response function and the dipole correlation function as the time response function using linear response theory [16]. Generally, for biomaterials, frequency domain measurements of the time response function result in less damage to electrodes, but simple and quick procedures can be used in time domain measurements. The more suitable of the two experimental techniques is selected to obtain the time and frequency response functions.

8.3.2 Time Domain Reflectometry Method in Microwave Frequency Region

The time domain reflectometry (TDR) method is a useful and effective technique for observing water structures in the GHz frequency region. It has also been developed as a nondestructive measuring tool for biological systems including the living body through easy and simple treatments. TDR is a time domain spectroscopy (TDS) method, which was first developed in electrical engineering, for example, as a technique to detect damage to undersea cables by comparing the transmitted and reflected signals from damaged cables with the incident pulse [17, 18]. This technique was subsequently applied to dielectric measurements in the physicochemical analysis of materials [19–22]. Cole and coworkers rearranged the equations used in the TDS method to obtain more general mathematical expressions without approximations restricting measurements by numerical analysis using computers [23, 24]. A different method of TDR measuring techniques was also developed [25–27], in which the difference between reflected pulses from unknown and reference samples was analyzed with a precision two orders of magnitude higher than that in the conventional direct method. The TDR method has also been refined through the use of newer devices, greater computer power, high-quality microwave cables, better-designed electrodes, new analytical methods, and so forth, since the basic establishment of its architecture [28–31].

A typical example of a TDR setup for relaxation measurements in the frequency range from 0.1 to 30 GHz is shown in Fig. 8.5a. The main frame of a digitizing oscilloscope (HP54120B, Hewlett-Packard) and a four channel test set (HP54124A, Hewlett-Packard) are employed here. A step pulse with a rapidly increasing voltage

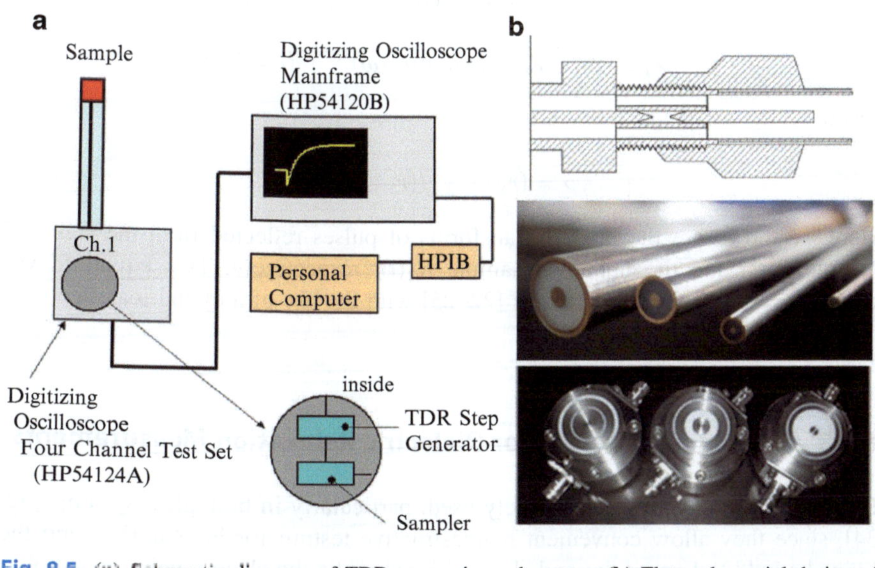

Fig. 8.5 (**a**) Schematic diagram of TDR measuring subsystem. (**b**) Flat-end coaxial design of electrodes (*upper*) and semirigid coaxial electrodes (*middle*), triaxial electrodes (*below*)

step having a 35 ps rise time and a 200 mV pulse height is applied via a flexible coaxial cable to a sample probe, in which the flat-end transmission line is terminated by the sample, and the pulse reflected from the sample is observed and analyzed. The highest frequency of such measurements has recently been extended to 50 GHz.

The reflected signal is observed by the electrodes in the setup and averaged. One of the difficulties in the use of flat-end electrodes is obtaining good contact between the surfaces of the electrodes and the materials. The good contact should always be confirmed during measurements by a careful procedure. The complex permittivity of the sample can be expressed by the basic equation in the TDR method [23, 24] as

$$\varepsilon^* = \frac{c'}{j\omega\gamma d}\frac{v_0 - r}{v_0 + r}X\cot X, \tag{8.5}$$

$$X = \left(\frac{\omega d}{c}\right)\sqrt{\varepsilon^*}, \tag{8.6}$$

where c' is the speed of propagation in the coaxial line; v_0 and r are the Laplace transforms of the incident and reflected pulse waveforms, respectively, X cot X expresses the multiple reflections from the interfaces between the sample and air and the coaxial line geometry; d is the length of the inner electrode; and γd is the effective electric length of electrodes. Using a dielectric material with a known complex permittivity ε^*_S as a standard sample, the complex permittivity of an unknown sample ε^*_X can be expressed as

$$\varepsilon_X^*(\omega) = \varepsilon_S^*(\omega)\frac{1 + \{(cf_S)/[j\omega(\gamma d)\varepsilon_S^*(\omega)]\}\rho f_X}{1 + \{[j\omega(\gamma d)\varepsilon_S^*(\omega)]/(cf_S)\}\rho f_S}, \tag{8.7}$$

$$f_i = Z_i \cot Z_i, \quad Z_i = (\omega d/c)\varepsilon_i^*(\omega)^{1/2} \quad (i = X\,\mathrm{or}\,S), \tag{8.8}$$

where

$$\rho = (r_S - r_X)/(r_S + r_X). \tag{8.9}$$

Here, r_S and r_X are the Fourier transforms of pulses reflected from the standard sample $R_S(t)$ and the unknown sample $R_X(t)$, respectively. This equation was originally derived for TDR studies [22–25] with the use of a cylindrical electrode [32, 33].

8.3.3 Flat-End Electrodes for Dielectric Relaxation Measurements

Flat-end electrodes have been widely used, particularly in biological systems [32, 34], since they allow convenient nondestructive testing for biomaterials and the living body [32]. Even if flat-end electrodes cannot fix the electric field for samples, any ambiguity can be corrected by a suitable procedure using a reference sample.

A typical schematic drawing of a flat-end coaxial electrode used at higher frequencies is shown in Fig. 8.5b (upper). Examples of electrodes modified from semirigid coaxial cables that are directly connected to the TDR-sampling head via an SMA connector are shown in Fig. 8.5b (middle). An example of triaxial electrodes used at low frequencies is also shown in Fig. 8.5b (lower). The actual size of each electrode is listed in Table 8.1.

The effective measuring depth depends on the size of the flat-end probe since the field contour lines depend on this size [35]. Naito et al. suggested that the effective measuring depth in the case of human skin is approximately $1/2\gamma d$ [36–38]. The characteristic depth of the electric field inside a material has been examined using various definitions. The dielectric constant, $\varepsilon_X(l)$, obtained from measurements with a flat-end electrode perpendicularly applied to the surface of a Teflon block sunk in a liquid at a frequency of approximately 1 GHz, which is the lower side of the relaxation frequency of the liquid was described as

$$\varepsilon_L - \varepsilon_X(l) = (\varepsilon_L - \varepsilon_T)\exp\left(-\frac{l}{L_0}\right), \tag{8.10}$$

where ε_L and ε_T are the static dielectric constants for the liquid and for Teflon, respectively, l is the gap length between the surface of the Teflon block and the electrode, and L_0 is the characteristic measuring depth of the electric field in the material. The L_0 values thus obtained with acetone as the liquid are shown in Table 8.1. The L_0 value obtained from Eq. (8.10) is dependent on the combination of materials used for the liquid and block. The definition of the characteristic measuring depth is applicable to triaxial electrodes used at lower frequencies as listed in Table 8.1. The examination of Eq. (8.10) suggests that the two exponential terms reflecting short-scale dipolar and long-scale ionic behaviors are characterized by two different L_0 values.

8.4 Water Structures and Various Dynamics

8.4.1 Dynamic Structures Observed in Suspensions of Biological Materials

8.4.1.1 Proteins: Hierarchical Structures

Figure 8.6 shows the molecular weight, M_n, dependence of the relaxation time for a process due to the overall rotation of proteins and other molecules in 5 wt% aqueous solutions. The Debye–Stokes–Einstein equation for the rotational motion of spheres with radius a in a liquid with viscosity η is

$$\begin{aligned}\tau &= \frac{4\pi a^3\eta}{kT}\\ &\propto M_n,\end{aligned} \tag{8.11}$$

where k is Boltzmann's constant and T is the absolute temperature. Thus, the slope of the straight line in Fig. 8.6 should be unity, in good agreement with the value of

Table 8.1 Examples of characteristic diameters of flat-end coaxial and triaxial electrodes with the electric length of penetration depth of the electric field determined for a two-layer model with acetone–Teflon system

Type	Code	Diameter of outer conductor/mm	Diameter of dielectric/mm	Diameter of inner conductor/mm	Electric length L_0/mm	Inner Diameter of G electrode/mm	Outer Diameter of G electrode/mm	Large scale electric length L_{0L}/mm
Semi-rigid electrodes	S09	0.86	0.66	0.20	0.0806	–	–	–
	S12	1.19	0.94	0.29	0.110	–	–	–
	S22	2.20	1.68	0.51	0.215	–	–	–
	S36	3.58	2.98	0.92	0.319	–	–	–
	S64	6.35	5.31	1.63	0.577	–	–	–
Triaxial electrodes	K3	30.00	20.00	18.00	1.351	34.00	40.00	–
	K2	30.00	19.00	10.00	1.586	32.00	40.00	4.642
	K4	30.00	25.00	5.00	1.851	31.00	40.00	6.101

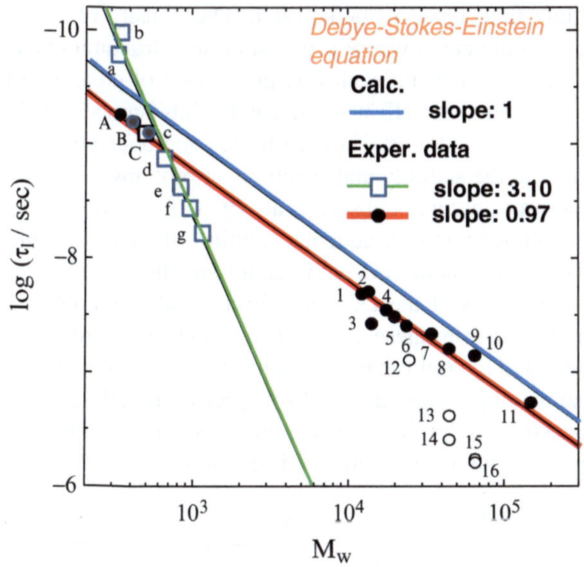

Fig. 8.6 The relaxation time for the process due to the overall rotation of globular protein and other molecules vs. molecular weight. Aqueous solutions of (*a*) trehalose (342), (*b*) maltose (360), (*c*) maltotriose (504), (*d*) maltoteraose (667), (*e*) maltopentaose (829), (*f*) maltohexaose (991), (*g*) maltoheptaose (1153); (*A*) adenosine monophosphate (347), (*B*) adenosine diphosphate (427), (*C*) adenosine triphosphate (507); (*1*) cytochrome C (12384), (*2*) ribonuclease A (13,700), (*3*) lysozyme (14,300), (*4*) myoglobin (17,800), (*5*) trypsin inhibitor (20,100), (*6*) trypsin (24,000), (*7*) pepsin (34,700), (*8*) ovalbumin (45,000), (*9*) hemoglobin (65,750), (*10*) albumin (66,000), (*11*) γ-globulin (150,000), (*12*) α-chymotrypsin (25,000) with 8 M-urea (U), (*13*) ovalbumin with 8 M-U, (*14*) ovalbumin with 1 wt% mercaptoethanol and 8 M-U, (*15*) albumin with 8 M-U, (16) albumin 1 wt% mercaptoethanol and 8 M-U. Numbers in parentheses mean the molecular weight

0.97 obtained from the experimental results for proteins and nucleotides. On the other hand, the slope of 3.10 obtained for oligosaccharides is considered to be due to the linear shape of the molecules. Plots obtained for proteins denatured by urea lie on the larger-relaxation-time side of the straight line, indicating their larger radii than the native values. When the density of proteins is assumed to be 1.3 g/cm^3 and the viscosity of water is 1 cP or larger, the relaxation times calculated for native proteins indicate that their radii are 10 % smaller than the experimental values. Since the same result has also been obtained by X-ray analysis [39], the existence of bound water surrounding the protein molecules is suggested. The relaxation strength associated with the electric dipole moment for the overall rotation does not only depend on the molecular weight. Takashima calculated the dipole moment from structures obtained by X-ray and nuclear magnetic resonance (NMR) studies and confirmed good agreement with the values obtained from dielectric measurements [40].

Nakasako reported that water molecules located at suitable positions on the surface of proteins can be clearly observed by low-temperature X-ray analysis at 100 K [39]. Corresponding results were also obtained by Higo from MD simulations [41], in which water molecules surrounding proteins were oriented for a longer time than those in bulk water. The density of water molecules is particularly high in the clefts where active sites and structural changes in proteins are often found [42]. Water molecules on the surface of protein molecules cannot withstand the high density of HB networks and exhibit cooperatively slow dynamics with peptide chains and ions. These restricted and bound water molecules may be closely related to the stability and functions of proteins.

Dielectric absorption curves observed at −9 °C for 5 wt% immunoglobulin G (IgG) aqueous solutions and an aqueous solution of two fragments, F_{ab} and F_c, obtained by the hydrolysis of IgG with papain are shown in Fig. 8.7a, b, respectively. The contributions of the DC conductivity and electrode polarization have already been removed from these curves. Large relaxation processes such as those due to bulk water and overall rotations emphasized by ion behaviors are prohibited below the freezing temperature. By employing conventional fitting procedures, four relaxation processes for IgG and three processes for F_{ab} and F_c solutions were indicated. The thus obtained permittivity is given by

$$\begin{aligned}\varepsilon^* - \varepsilon_\infty = {} & \frac{\Delta\varepsilon_{l1}}{1 + j\omega\tau_{l1}} + \frac{\Delta\varepsilon_{l2}}{1 + j\omega\tau_{l2}} + \frac{\Delta\varepsilon_{\mathrm{m}}}{\left[1 + (j\omega\tau_{\mathrm{m}})^{\beta_{\mathrm{m}}}\right]^{\alpha_{\mathrm{m}}}} \\ & + \frac{\Delta\varepsilon_{\mathrm{h}}}{1 + (j\omega\tau_{\mathrm{h}})^{\beta_{\mathrm{h}}}}, \quad 0 < \alpha_{\mathrm{m}}, \beta_{\mathrm{m}}, \beta_{\mathrm{h}} \le 1\end{aligned} \tag{8.12}$$

where the subscripts l1, l2, m, and h indicate the relaxation processes l1, l2, m, and h, respectively, as shown in Fig. 8.7 [43, 44]. These relaxation processes were

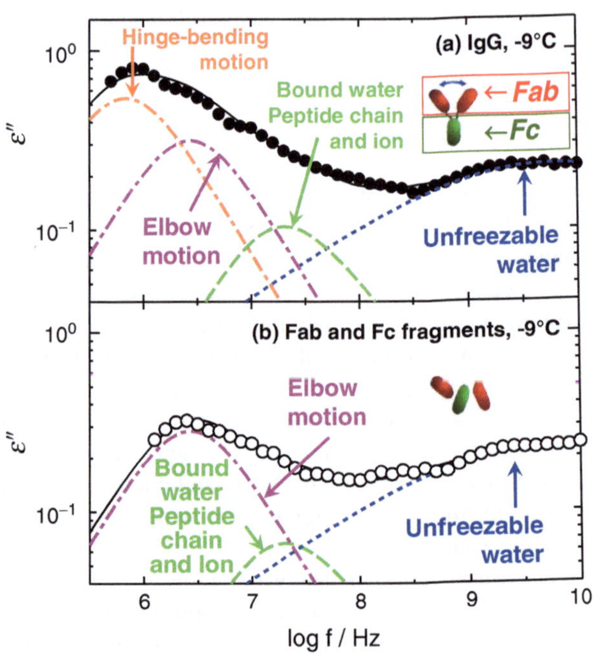

Fig. 8.7 Dielectric absorption curves for (**a**) 5 wt % aqueous solution of IgG at −9 °C and (**b**) 5 wt % aqueous solution of Fab and Fc fragment mixture at −9 °C

identified as the dynamic behaviors of the unfreezable water around 5 GHz (h process), the peptide chain with bound water and ions at approximately 20 MHz (m process), the elbow motion of two F_{ab} fragments at approximately 3 MHz (l2 process), and the hinge-bending motion between F_{ab} fragments at approximately 800 kHz (l1 process). Domain motion such as elbow and hinge-bending motion [45, 46] is also correlated with the ion behavior. When the hinge-bending motion vanished due to hydrolysis by papain, the low-frequency relaxation process l2 for the mixture of F_{ab} and F_c fragments was only caused by the elbow motion.

Peptide chain motion is usually restricted in native proteins and is invisible in dielectric relaxation curves owing to the large contributions of interacting water molecules and ion behavior to the permittivity. To confirm the existence of peptide chain motion, dielectric relaxation measurements were performed for a glass egg [47]. The decreasing HB network density with decreasing amount of water reduces the hydrophobic and hydrophilic interactions of globular protein and induces a globule–coil transition. Simultaneously, the gel–glass transition is promoted. Actually the glass transition phenomenon has also been observed by performing DSC for the glass egg [47]. The change in the water structure in the gel of the heat-denatured globular protein is significant, and the diffusion of water molecules is gradually restricted in the shrinking networks of the gel [48, 49]. Similar results have been reported for keratin–water mixtures [37].

As explained in Sect. 8.2.3, the τ-β diagram obtained for a process due to the dynamic behaviors of the rotational diffusion of water molecules is a possible tool for discussing a universal treatment for water structures. Figure 8.8 shows the τ-β diagram for protein solutions and various kinds of cheeses [50]. The plots for the present samples are shown in the diagram in an extended lengthwise form, and they lie in the area for dispersion systems. The values of the fractal dimension obtained were 0.03 ± 2.45 for BSA, 0.03 ± 1.70 for CEWL, and 0.22 ± 0.89 for OVA. These values are much smaller than those obtained for ordinary aqueous solution systems.

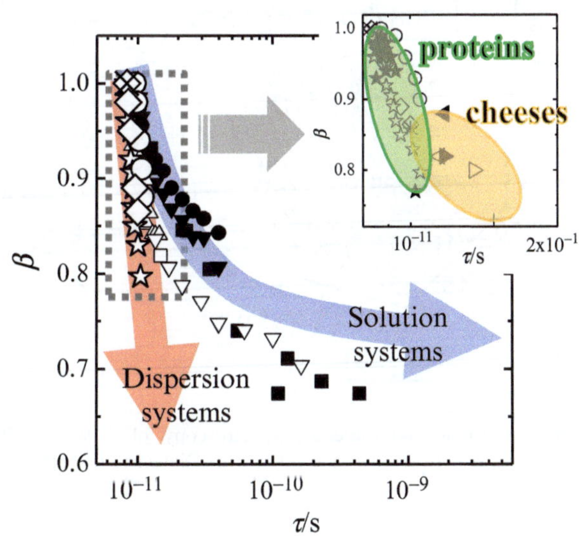

Fig. 8.8 The τ-β diagram of the relaxation time versus the relaxation time distribution parameter of Cole–Cole equation for water(×), globular proteins; BSA (○), CEWL (♦), OVA (☆), synthesized polymers; PEG (●), PVP (■), PVME (▼), PVA (△), PAIA (□), and PEI (▽). The inset is the τ-β diagram for proteins (★) and cheeses; mozzarella [◀], camembert [◁], cream [*], processed [▶], gouda [▷], and red cheddar [+] (Adapted from Maruyama et al. [50] with permission from Elsevier. Copyright 2014)

The fractal dimensions obtained for water structures in aqueous polymer solutions are usually about 1.3–1.5 with an error of 0.02–0.71. Water structures with a small fractal dimension exhibit only a small increase in the relaxation time with increasing concentration. It turns out that the physical meaning of a fractal dimension of less than 1 is the HB network structure fractionized in dispersed system from the knowledge of our estimation of the fractal dimension for geometric models [51, 52]. Such a small fractal dimension should be explained from the spatiotemporal scale used in the technique for the observation and analysis of the slow dynamics. The process of the slow dynamics reflects the conformation of the polymer chains and the agglomeration structures and also the hydrogen bonding ability. Therefore, the characterization of the slow dynamics is determined not only by the absolute value of the fractal dimension but also by the geometric features of the domain in which the locus of a plot appears in the τ-β diagram.

8.4.1.2 Liposomes: Broadband Dynamics

Clarification of the dynamic behaviors of liposome particles in water is critical for the successful development of pharmaceutical liposome drug formulations. Water is essential for maintaining the liposome bilayer structure, and hydrophilic regions of lipid membranes on the surface of liposomes are exposed to the surrounding water. Thus, the temperature and frequency dependences of the dielectric constant of liposome–water mixtures have been investigated to obtain information about the dynamic structures. Figure 8.9 shows dielectric dispersion (Fig. 8.9a) and dielectric absorption (Fig. 8.9b) curves obtained by TDR in the frequency range of 100 MHz–30 GHz and by using an impedance analyzer (IA) (Hewlett-Packard 4294A) in the frequency range of 40 Hz–110 MHz for a liposome dispersion of 4 wt%

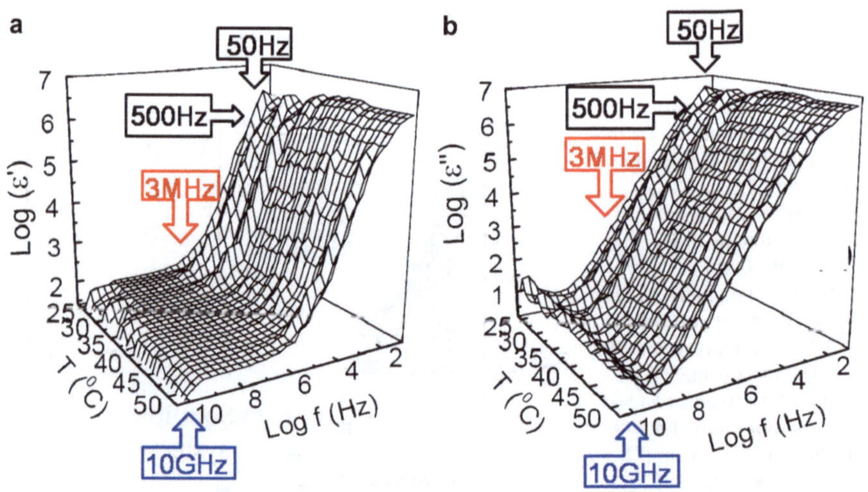

Fig. 8.9 Broadband dielectric spectroscopy of 1 wt% DPPC liposome–water at various temperatures between 25° and 50 °C. (**a**) Dielectric dispersion curves, (**b**) dielectric absorption curves

1, 2-dipalmitoyl-sn-glycero-3-phosphacholine (DPPC)–water mixture in the temperature range of 25–50 °C. The DPPC liposome shows three phases, a non-rippled gel, a rippled gel, and a liquid crystal phase, with increasing temperature below the pretransition temperature [53, 54]. In the non-rippled gel phase, the mobility of the phospholipid molecules is low. The phospholipid molecules are aligned perpendicular to the bilayer surface at about 30 °C. In the rippled gel phase, the surface of membrane takes a rippled structure with mobile head groups of phospholipid and ordered acyl chains tilting to the membrane normal. Above the main transition temperature between the rippled gel and liquid crystalline phases, the acyl chains become mobile and in a disordered state then become perpendicular to the membrane surface again. The pretransition and main transition temperatures were determined by DSC measurements to be approximately 34 and 41 °C, respectively [54].

To characterize all the relaxation processes observed in the temperature range, curve fitting was carried out for the four relaxation processes and the DC conductivity using

$$\varepsilon^* = \varepsilon_\infty + \sum_{i=1}^{4} \frac{\Delta\varepsilon_i}{1 + (j\omega\tau_i)^{\beta_i}} + \frac{\sigma_{DC}}{j\omega\varepsilon_0}, \tag{8.13}$$

where σ_{DC} is the DC conductivity and the subscript i indicates the ith relaxation process from the lower-frequency side.

Relaxation process 4, which occurs near 20 GHz and is due to so-called free water surrounding the liposomes was observed at all temperatures [55–57]. The relaxation parameters obtained for process 4 for 3 wt% DPPC–water mixtures are plotted against the temperature along with the transition temperatures in Fig. 8.10a. The relaxation strength and relaxation time show conventional behavior with increasing temperature. However, the relaxation time distribution parameter decreased above 41 °C [58–60]. This process was described by the Debye function in the gel phase and the Cole–Cole function in the liquid crystal phase above the main transition temperature. Although the water molecules surrounding the liposome surface directly interact with the head groups of the phospholipids, no abrupt change was observed at the pretransition temperature. It is considered that the interactions between the water and the lipid bilayer cooperatively bring about the increased fluctuation of water structures surrounding the liposomes in the liquid crystalline phase with the increased mobility of lipid molecules.

The dielectric behaviors of the phospholipids, liposomes, and ions were observed at lower frequencies region. Figure 8.9 shows process 3, which occurs at approximately 3 MHz; process 2, which occurs at approximately 500 Hz; and process 1, which occurs at approximately 50 Hz. Process 3 often reflects the local chain and domain dynamics of phospholipids, and process 2 is related to larger-scale structures of the membrane and liposomes with ionic behavior. The large-scale mobility of ions upon electrode polarization is usually observed in the lowest-frequency region with the largest dielectric permittivity such as in process 1. Figure 8.10b shows the temperature dependences of the relaxation time for

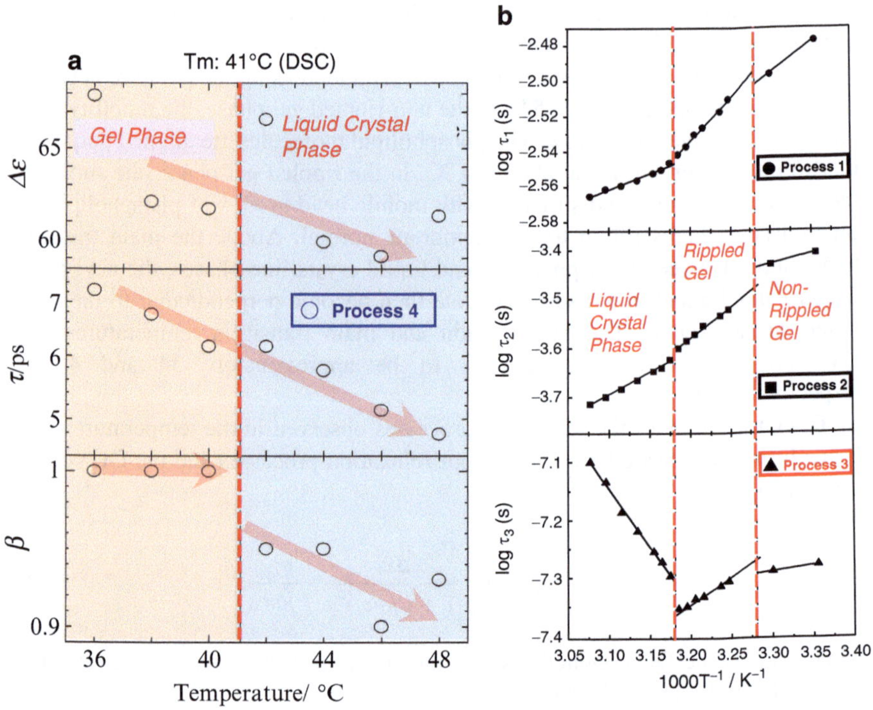

Fig. 8.10 Temperature dependences of relaxation parameters for liposome dispersion systems. The main transition temperature, 41 °C, and the pretransition temperature, 33 °C, determined by DSC measurements are shown by *dashed* lines. (**a**) The relaxation strength, $\Delta\varepsilon$; the relaxation time, τ; and the relaxation time distribution parameter, β for the process 4 of 3 wt% DPPC–water mixture. (**b**) Plot of the relaxation time against the reciprocal of absolute temperature for the processes 1–3 of 4 wt% DPPC–water mixtures (Adapted from Hosoi et al. [59] with permission from the Material Research Society of Japan. Copyright 2010)

processes 1, 2, and 3 [61]. A straight line was drawn in each phase, and discontinuous changes in the relaxation time and/or the slope are shown at the main and pretransition temperatures, respectively. The distinctive V-shaped temperature dependence shown by process 3 at the main transition temperature is also a typical characteristic feature of ferroelectric liquid crystals [62, 63]. The phospholipid molecules aligned in the liposome membrane thus exhibit ferroelectric behavior [57, 61, 64]. Process 3 is therefore attributed to the dynamics of phospholipid molecules. The behavior of ions surrounding the liposomes and electrodes in processes 2 and 1, respectively, are also changed at the transition temperatures, reflecting changes in the shape and fluidity of the liposome membrane.

8.4.1.3 Cells: Interfacial Polarization

The dielectric behaviors of various suspension systems including biological cells have been investigated by Asami and coworkers, who attempted to reduce the contribution of electrode polarization through the use of suitably designed cells, especially at lower frequencies [65–68]. Figure 8.11a shows the frequency dependences of the dielectric constant and conductivity obtained at 21 °C for suspensions of Madin–Darby canine kidney (MDCK) cells in PBS buffer. Parallel plate electrodes consisting of a small chamber and a dielectric spacer were used for dielectric measurements with an HP4294A impedance analyzer and an HP16092A spring clip fixture (Yokogawa Hewlett-Packard). The electrodes were plated with platinum black to reduce the parasitic capacitance resulting from the effect of electrode polarization. An increase in the electrode capacitance shifts the electrode polarization to lower frequencies and decreases the interference at frequencies of interest. MDCK cells were packed in the small chamber by centrifugation. The dielectric cell constant and the stray capacitance of the chamber were 0.014 pF and 3.5 pF, respectively. Dielectric measurements were performed on the day of preparation and also on the following day for each sample. Another measurement was also performed after heat treatment at 60 °C for 30 min.

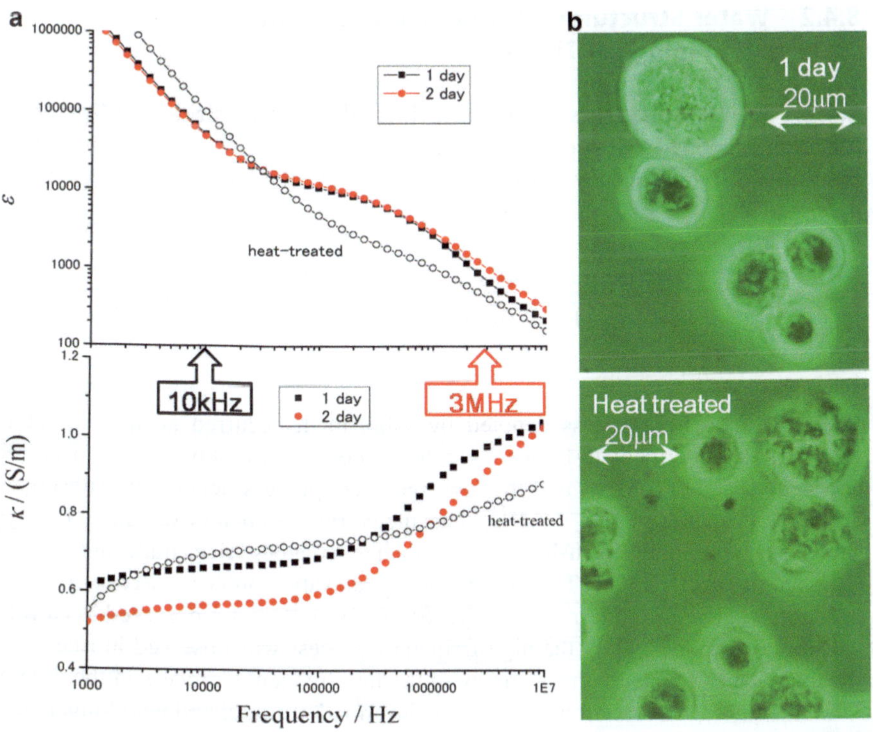

Fig. 8.11 Frequency dependence of dielectric constant and conductivity for suspensions of MDCK cells at 21 °C. Dielectric measurements were performed twice for the same sample immediately and 1 day after the preparation. First day (micrograph) (■), second day (●), heat treatment (micrograph) (○). Experimental data were obtained at laboratory of Prof. Asami with support by the Collaborative Research Program of Institute for Chemical Research, Kyoto University (grant # 2013–58)

Figure 8.11a shows two relaxation processes which occur at approximately 3 MHz and at a lower frequency of less than 10 kHz. The lower-frequency relaxation process reflects the electrode polarization. The higher-frequency process is due to interfacial polarization of the cell. The contribution of interfacial polarization is smaller after the heat treatment. Asami reported that the changes in the dielectric constant and conductivity are accompanied by the leakage of ions from the cell interior [65]. Phase contrast micrographs (Fig. 8.11b) of CMDK cells before and after the heat treatment show changes in the membrane structures because of the damage from the treatment, supporting the idea of ion leakage. The small difference in the relaxation curves observed on the first and second days suggests that the membrane structures were not significantly damaged during the 2 days. Dielectric measurement can be an effective experimental technique for evaluating not only the viability of life processes but also the morphological changes during cell differentiation [69].

8.4.2 Water Structures Observed in Foodstuffs and Biological Tissues

8.4.2.1 Water Behaviors in Biomaterials Observed Using TDR System

It has been reported that two or three relaxation processes are usually observed for foodstuffs in the frequency range from 100 kHz to 30 GHz. These processes are described as

$$\varepsilon^* - \varepsilon_\infty = \frac{\Delta\varepsilon_l}{1 + j\omega\tau_l} + \frac{\Delta\varepsilon_m}{\left[1 + (j\omega\tau_m)^{\beta_m}\right]^{\alpha_m}} + \frac{\Delta\varepsilon_h}{1 + (j\omega\tau_h)^{\beta_h}},\ 0 < \alpha_m, \beta_m, \beta_h \le 1 \tag{8.14}$$

The high-frequency process denoted by subscript h occurred at approximately 10 GHz, the intermediate-frequency process denoted by subscript m occurred between 10 and 100 MHz, and the low-frequency process denoted by subscript l occurred at approximately 1 MHz. The dielectric parameters obtained for egg white, egg yolk, chicken, fish, carrots, radishes, potatoes, rice, milk, and cheese are listed in Table 8.2 [70]. The low-frequency relaxation process was not observed in agar or cheese. The intermediate-frequency process was not observed in carrots, radishes, or potatoes. Only the high-frequency process was observed in rice. Care should be taken to ensure the accuracy of the low-frequency process in the case of TDR measurements since it is easy to miss information associated with longer-time behaviors in time domain measurements. However, it is reasonable to consider that the l- and m-processes suggest the existence of some heterogeneous structures and a mobile backbone of peptide and other chains around which interacting water molecules and ions are restricted [70, 71].

Various food-stuffs contain the overall and local mobile structures of biomolecules, cells, and tissues along with structured water and ions. BDS measurements provide information of these characteristic structures and free

Table 8.2 Dielectric relaxation parameters observed for various foodstuffs at 25 °C. Adapted from [70] with permission from John Wiley and Sons. Copyright 2006

Relaxation process	Low-frequency process				Intermediate-frequency process				High-frequency process			–	
Sample	$\Delta\varepsilon_l$	log τ_l [s]	α_l	βl	$\Delta\varepsilon_m$	log τ_m [s]	α_m	β_m	$\Delta\varepsilon_h$	log τ_h [s]	β_h	ε_∞	σ_{dc} [Sm^{-1}]
Egg white	5.0	−7.65	1.00	1.00	9.1	−8.57	0.59	0.88	62.3	−11.04	0.97	5.10	0.90
Egg yolk	329	−6.45	1.00	1.00	40.5	−7.95	0.93	0.76	37.5	−11.08	0.76	4.00	0.34
Fish	1910	−6.29	0.78	1.00	18.3	−8.59	0.55	1.00	51.2	−11.09	0.84	4.20	0.59
Fish (frozen)	670	−6.54	0.75	1.00	28.0	−8.77	0.66	0.79	45.0	−11.06	0.87	4.00	0.73
Fish (boiled)	186	−6.73	1.00	1.00	120.0	−7.79	0.82	0.78	50.4	−11.09	0.81	4.00	0.58
Chicken	1040	−6.54	0.73	1.00	12.6	−9.19	0.66	1.00	49.6	−11.07	0.87	4.10	0.53
Chicken (frozen)	670	−6.54	0.75	1.00	28.0	−8.77	0.66	0.79	45.0	−11.06	0.87	4.00	0.53
Chicken (boiled)	146	−6.68	1.00	1.00	153.0	−7.71	0.72	0.81	31.9	−10.98	0.73	4.00	0.39
Carrot	7500	−6.70	1.00	1.00	–	–	–	–	64.9	−11.02	0.96	4.00	0.02
Carrot (boiled)	336	−6.86	0.79	1.00	–	–	–	–	66.0	−11.05	0.93	4.00	0.17
Radish	30900	−6.13	1.00	0.93	–	–	–	–	65.4	−11.07	1.00	4.00	0.03
Radish (boiled)	613	−6.61	0.81	1.00	–	–	–	–	68.9	−11.05	0.97	4.00	0.55
Potato	29600	−6.29	1.00	0.89	–	–	–	–	53.3	−11.07	1.00	4.00	0.05
Potato (boiled)	247	−6.86	0.64	1.00	–	–	–	–	61.7	−10.98	0.61	4.00	0.39
Rice	–	–	–	–	–	–	–	–	9.8	−9.58	0.40	2.10	–
Milk	7.1	−7.58	1.00	1.00	8.0–8.52	–	0.69	0.80	4.0	−11.04	0.94	5.60	0.48
Cheese	–	–	–	–	7.8–8.86	–	1.00	1.00	5.3	−9.86	0.76	5.80	0.07

Adapted from Miura et al. [70] with permission from John Wiley and Sons. Copyright 2006 water, and the molecular mechanisms of physical properties and functions can be understood from the dynamic structures. Thus, it is also easily expected that BDS measurements can be used to analyze changes in dynamic structures during biological growth and aging as well as freshness, decomposition, clinical conditions, food processes, drying, freezing, cooking, and so forth.

8.4.2.2 Cheese: Behaviors of Water Observed by Complementary Techniques

Many kinds of cheese exist, which include various amounts of water. This makes cheese one of the most interesting foodstuffs for exhibiting the slow dynamics of water molecules for a wide range of water content. The slow dynamics of water structures in cheese shown in Fig. 8.8 suggest similar water structures to those in protein–water mixtures and dispersion systems. The variation of the plots obtained for various kinds of cheese are also shown in the diagram. Despite the different kinds of cheese and the range of water contents from 15 to 30 wt%, all the plots for the cheeses indicate the water structures of typical dispersion systems. However, it is interesting to note that the area of the trajectories for the cheese samples is slightly shifted toward the area of solution systems compared with protein–water mixtures. This is reasonable because proteins make up only one-third of cheese, and the other components, such as fat, have more homogeneous water structures.

It has been reported that the plot of the relaxation time distribution parameter, β, against the inverse of the self-diffusion coefficient obtained from the pulse field gradient spin-echo (PFGSE) method of NMR measurements for water molecules in various kinds of cheese resembles the figure in Fig. 8.8 [50]. The Stokes–Einstein equation describes the diffusion coefficient of a spherical motional unit in terms of its hydrodynamic radius R_{h} as follows:

$$D = \frac{kT}{6\pi\eta R_h}. \tag{8.15}$$

The comparison of Eq. (8.15) with the Debye–Stokes–Einstein equation described by Eq. (8.11) suggests that the relaxation time corresponds to the inverse of the diffusion coefficient. The $1/D$-β diagram can thus be considered as an effective tool for characterizing water structures with a time scale in the ms region, which is much larger than that of the present fractal analysis with the τ-β diagram.

The small difference in the area of the trajectory in the τ-β diagram for cheeses from that for proteins suggests a smaller fractal dimension and more homogeneous water structures. As explained in Sects. 8.2.1 and 8.2.3, these methods of analyzing structures are dependent on the time and length scales. Then $1/D$-β diagram may be different from the τ-β diagram because of the different time scales of BDS and PFGSE measurements. Although insufficiently detailed analysis of the problems of observation has been reported in recent studies some implications of these problems on experimental data for characteristic heterogeneity, particularly in water structures, have started to be discussed [50]. More investigations and greater understanding of the problems of observing the dynamic fractal structures of water are necessary.

8.4.2.3 Rice: Fractal Water Structures

It is particularly important to investigate the water structures in kernels to ensure the effective storage of grains. Using a flat-end coaxial electrode with a suitable sample holder to adjust the polished surface of a rice kernel, a TDR measuring system makes it possible to analyze the dielectric relaxation of even a single kernel [72]. There are some difficulties in achieving good contact between the surfaces of the electrode and the kernel, but dielectric measurements of small, solid, and conductive materials gave reliable results. In our previous measurements [72], the water in a rice kernel held in storage had relaxation times in the range between 100 and 600 ps and a relaxation time distribution parameter, β, of approximately 0.41. Increasing the moisture content decreased the relaxation time, τ, and increased the β value, indicating less restricted dynamics of the water molecules, and these changes are expressed as a trajectory in the τ-β diagram in Fig. 8.12. The trajectory suggests how water takes a fractal structure and the scaling behavior in rice kernels and in flour. These plots for rice kernels and flour overlap and are dispersed around the same trajectory. However, the plots for boiled rice exhibit a trajectory with a higher value of β, since the denatured structures of biological molecules in the kernels may change the water structure into a more homogeneous structure.

The trajectory for the water structures in rice kernels held in storage showed plots around the point ($\tau = 100$–600 ps, $\beta = 0.41$), and increasing the moisture content shifted the plot toward that for pure water ($\tau = 8.3$ ps, $\beta = 1.00$) via the point ($\tau = 30$ ps, $\beta = 0.6$). Compared with other aqueous systems, the dynamic

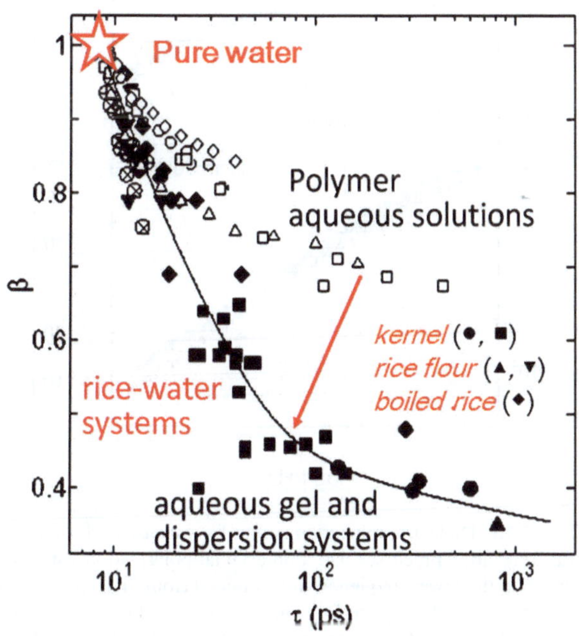

Fig. 8.12 The τ-β diagram for water, synthesized polymers, and rice with various water contents. Water (☆), rice kernel (●), rice kernels with water (■), rice flour (▲), rice flour with water (▼), boiled rice (◆). Polymer aqueous solutions; PVP (□), PEG (◇),PVME (○), PVA (⊕),PAA (⊗), PEI (△) (Adapted from Yagihara et al. [72] with permission from IOP. Copyright 2007, IOP Publishing)

structure of water in rice kernels is different from that in ordinary polymer solutions except for an aqueous solution of PAA. However, a PAA aqueous solution with high polymer concentration could not be prepared because of a low solubility of PAA. Furthermore, the present moist rice systems exhibit some trajectories in the area for dispersion systems. Thus, the water structure in moist rice characterized by the τ-β diagram shows similar fractal structures to the water included in dispersion systems. The present study suggests that more systematic and precise measurements of rice kernels will make it possible to obtain a detailed understanding of the dynamic structure of water in kernels, which will be applicable to other related research fields, such as the denaturalization of biological polymers and organs, the glass transition of complex fluids, and food science. More comparisons with other aqueous systems are expected to provide important information for these research fields.

8.4.2.4 Organs: Water Structures in Tissues

Organs purchased in city markets are not only foodstuffs but can also be models used to living organs or the body. Figure 8.13 shows the results of dielectric

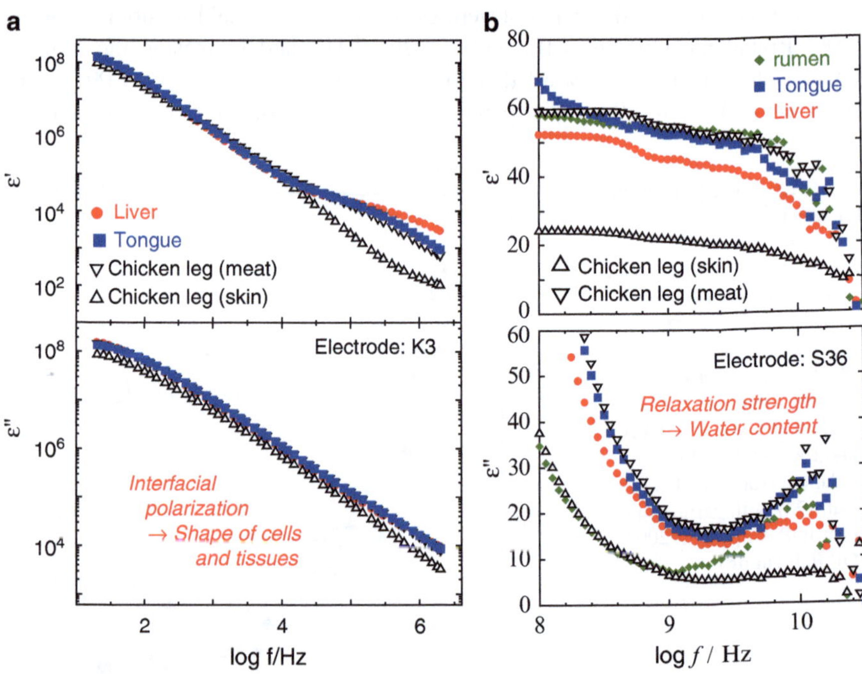

Fig. 8.13 Dielectric relaxation curves for various kinds of meats with various electrodes at 25 °C. (**a**) Relaxation processes due to interfacial polarization with contributions of the electrode polarization in the lower frequency side obtained from dielectric measurements with LCR meter and the flat-end electrode, K3. (**b**) Relaxation processes due to water with tails of low-frequency processes obtained from TDR measurements with the flat-end electrode S36

relaxation measurements performed on organs at 25 °C using LCR meter in the range of 20 Hz–1 MHz and using a K3 triaxial electrode and a TDR system with a S22 coaxial electrode in the range of 100 k–30 GHz. In addition to the very large contribution of electrode polarization at the lowest frequencies, a noteworthy process that occurred at approximately 1 MHz and other processes between the two frequency regions are shown in Fig. 8.13a. The mechanism of some of the relaxation processes shown is considered to be the interfacial polarizations of cells, tissues, and organs. Figure 8.13b shows the 10 GHz frequency process induced by water and another process at approximately 100 MHz reflecting the cooperative dynamics of molecular chains, water, and ions.

The dielectric behavior of water observed as the 10 GHz frequency process provides the water content. The smallest value was obtained from TDR measurements with the flat-end electrode placed against the skin side of a chicken leg because of the presence of subcutaneous fat. The water contents of the flesh side of the chicken leg and the rumen appeared to be larger than those for the tongue and liver. Note that the organ meat samples were more easily dried during the dielectric measurements than samples such as the chicken leg because of the large surface area of their porous structures. Thus, the water content continuously decreased, and the results were dependent on the timing of the measurements.

On the other hand, the process that occurred at MHz frequency was considered to reflect the structures of cells and tissues, and the relaxation strength was not directly affected by the water content. The process observed for the skin side of a chicken leg was smaller than that observed for the muscle side. As shown in the dielectric relaxation process induced by the interfacial polarization of cells, cells with a clear shape and boundary exhibit a larger relaxation process. Therefore, food processes such as drying, freezing, and cooking affect the relaxation process, and the effects of these food processes can be evaluated through the results of dielectrics relaxation measurements [70].

8.4.2.5 Skin: Water in Layer Structures

We have investigated the dielectric properties of human skin in vivo in the microwave frequency range using the TDR method with flat-end coaxial probes. Since the process that occurs at 10 GHz results from the reorientation of free water molecules, the free water content of skin can be quantitatively analyzed by dielectric measurements.

The free water content of finger skin increased by about 10 % after soaking in water at 37 °C for 30 min, then it monotonically decreased as the skin dried. Thus, this analytical method has been applied to the study of skin burns. The free water content of human cheek skin burned by hydrofluoric acid was significantly lower than that of normal skin, and the burned skin recovered its free water content through the healing process. In the case of human hand skin burned by heat, although the free water content was almost the same as that of normal skin immediately after being burnt, it decreased during the first 10 days of the healing process and then began to increase. Although there was only one subject tested in each experiment, it was shown that the free water content is a good indicator for

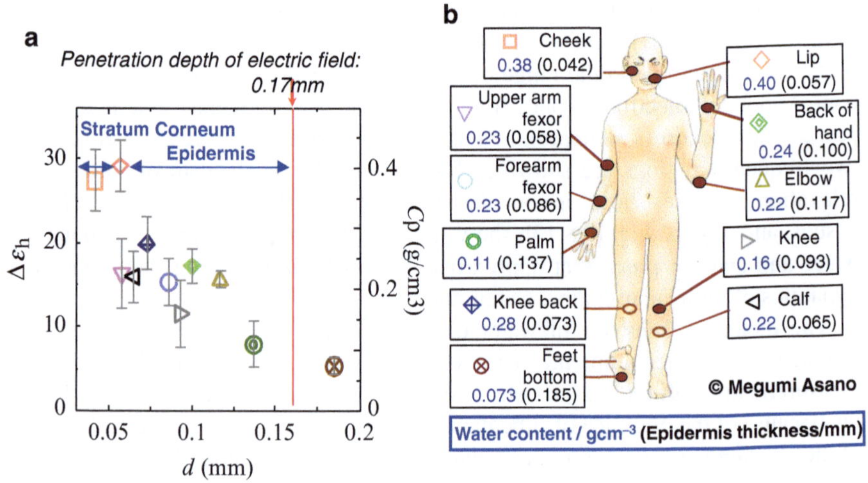

Fig. 8.14 (**a**) Plot of the relaxation strength for the high-frequency process due to water obtained from TDR measurements with flat-end probe with the electric length of an electrode (*solid line*) against thickness of stratum corneum and epidermis determined from anatomic data. (**b**) Points of TDR measurements for body parts and the thickness of epidermis and dermis

evaluating skin health and that it can be accurately monitored by dielectric spectroscopy [73].

The apparent water content obtained from the relaxation strength is dependent on the thickness of the stratum corneum and epidermis, in which the water content is lower than in deeper areas such as the dermis. A thicker epidermis has a lower water content as shown in Fig. 8.14a [74, 75]. Here, the dependence of the thickness of the epidermis on the body part was obtained from anatomic data [76]. The body parts used for the measurements are shown with their water contents and epidermis thicknesses in Fig. 8.14b. The analytical curve obtained for each electrode can determine the epidermis thickness of any part of the body. Electrodes with various electrical lengths should provide a method of analyzing the inner structures of skin. The penetration depth of the electric field was determined from Eq. (8.10) to be 0.17 mm for the electrode S22 from the experimental data for the stratum corneum and epidermis as a water-poor layers and the dermis as a water-rich layer. The difference from the value of 0.22 mm in Table 8.1 indicates the dependence of the electrical length on the material.

8.5 Future Works on Broadband Dynamics Using Complementary Techniques

The characteristic time and frequency scales of the observations of molecular dynamics are extended by BDS measurements. Compared with the BDS technique, other complementary experimental techniques of observation provide different but narrower scales as shown in Fig. 8.15 [7]. The results obtained from two different

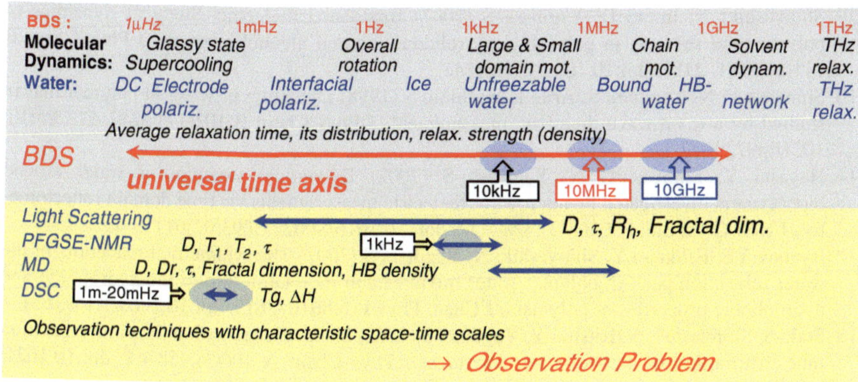

Fig. 8.15 Characteristic time and frequency scales of BDS and complimentary experimental techniques to observe dynamics of molecular complex systems (Adapted from Yagihara [7] with permission from the Nanofiber Society. Copyright 2013)

techniques without any overlap of scales cannot be compared, but both results can be easily compared, if they are indirectly compared by comparison with those of BDS. Therefore BDS provides not only a technique for observing various dynamics but also an opportunity to perform universal analysis with common and wide time and frequency scales. As described in Sect. 8.1, the thus obtained broadband dynamics provide considerable information about the detailed structures of complex systems. The difference in the molecular dynamics observed at different time scales reflects the heterogeneous structures of materials at the molecular level. Methods of analyzing these broadband dynamics should be developed to provide researchers with more effective tools.

References

1. Debye PJW (1929) Polar molecules. Dover, New York
2. The International Dielectric Society. http://permittivity.org/index.html
3. Memorial volume of for late Prof. R. H. Cole (1986). J Mol Liq 36(1). doi:10.1016/0167-7322(87)
4. Kremer F, Schoenhals A (eds) (2002) Broadband dielectric spectroscopy. Springer, Berlin
5. Barthel J, Buchner R (1991) High frequency permittivity and its use in the investigation of solution properties. Pure Appl Chem 63(10):1473–1482. doi:10.1351/pac199163101473
6. Fukasawa T, Sato T, Watanabe J, Hama Y, Kunz W, Buchner R (2005) Relation between dielectric and low-frequency Raman spectra of hydrogen-bond liquids. Phys Rev Lett 95(19):197802. doi:10.1103/PhysRevLett.95.197802
7. Yagihara S (2013) Concepts of observation problem and fractal on water structure by dielectric spectroscopy. Nanofiber (Japanese) 4(1):26–31
8. Yagihara S, Asano M, Kosuge M, Tsubotani S, Imoto D, Shinyashiki N (2005) Dynamical behavior of unfreezable molecules restricted in a frozen matrix. J Non-Cryst Solid 351(33–36):2629–2634. doi:10.1016/j.jnoncrysol.2005.03.059

9. Shinyashiki N, Imoto D, Yagihara S (2007) Broadband dielectric. Study of dynamics of polymer and solvents in poly(vinyl pyrrolidone)/normal alcohol mixtures. J Phys Chem B 111(9):2181–2187. doi:10.1021/jp065414e
10. Shinyashiki N, Yagihara S, Arita I, Mashimo S (1998) Dynamics of water in polymer matrix studied by a microwave dielectric measurement. J Phys Chem B 102(17):3249–51. doi:10.1021/jp9729627
11. Hayashi Y, Shinyashiki N, Yagihara S (2002) Dynamical structure of water around biopolymers investigated by microwave dielectric measurements via time domain reflectometry. J Non-Cryst Solid 305(1–3):328–332. doi: 10.1016/S0022-3093(02)01113-4
12. Ryabov YE, Feldman Y, Shinyashiki N, Yagihara S (2002) The symmetric broadening of the water relaxation peak in polymer-water mixtures and its relationship to the hydrophilic and hydrophobic properties of polymers. J Chem Phys 116:8610–8615. doi:10.1063/1.1471551
13. Sudo S, Shinyashiki N, Kitsuki Y, Yagihara S (2002) Dielectric relaxation time and relaxation time distribution of alcohol-water mixtures. J Phys Chem A 106(3):458–64. doi:10.1021/jp013117y
14. Fujiwara S, Yonezawa F (1995) Anomalous relaxation in fractal structures. Phys Rev E 51 (3):2278–85. doi:10.1103/PhysRevE.51.2277
15. Kubo R (1957) Statistical-mechanical theory of irreversible processes. I. General theory and simple applications to magnetic and conduction problems. J Phys Soc Jpn 12:570–586. doi:10.1143/JPSJ.12.570
16. Cole RH (1965) Correlation function theory of dielectric relaxation. J Chem Phys 42 (2):637–643. doi:10.1063/1.1695984
17. Oliver BM (1964) Time domain reflectometry. Hewlett-Packard J 15:6
18. Nicolson AM (1968) Broad-band microwave transmission. Characteristics from a single measurement of the transient. Response. IEEE Trans lnstrum Meas IM 17(4):395–402. doi:10.1109/TIM.1968.4313741
19. Fellner-Feldegg H (1969) Measurement of dielectrics in the time domain. J Phys Chem 73:616–626. doi:10.1021/j100723a023
20. Suggett A, Mackness PA, Tait MJ, Loeb HW, Young GM (1970) Dielectric relaxation studies by time domain spectroscopy. Nature 228:456–457. doi:10.1038/228456a0
21. Van Gernert MJC (1973) High-frequency time-domain methods in dielectric spectroscopy. Philips Res Rep 28:530–572
22. Cole RH (1977) Time domain reflectometry. Annu Rev Phys Chem 28:283–300. doi:10.1146/annurev.pc.28.100177.001435
23. Cole RH (1975) Evaluation of dielectric behavior by time domain spectroscopy. I. Dielectric response by real time analysis. J Phys Chem 79(14):1459–1469. doi:10.1021/j100581a022
24. Cole RH (1975) Evaluation of dielectric behavior by time domain spectroscopy. II. Complex permittivity. J Phys Chem 79(14):1469–1474. doi:10.1021/j100581a023
25. Cole RH, Mashimo S, Winsor IVP (1980) Evaluation of dielectric behavior by time domain spectroscopy. 3. Precision difference methods. J Phys Chem 84(7):786–793. doi:10.1021/j100444a017
26. Nakamura H, Mashimo S, Wada A (1982) Application of time domain reflectometry covering a wide frequency range to the dielectric study of solutions. J Appl Phys 21(3):467–474. doi:10.1143/JJAP.21.467
27. Nakamura H, Mashimo S, Wada A (1982) Easy method of TDR to obtain dielectric relaxation spectra in GHz region. J Appl Phys 21(7):1022–1024. doi:10.1143/JJAP.21.1022
28. Cole RH, Berberian JG, Mashimo S, Chryssikos G, Burns A, Tombari E (1989) Time domain reflection methods for dielectric measurements to 10 GHz. J Appl Phys 66(2):793–802. doi:10.1063/1.343499
29. Mashimo S, Umehara T, Ota T, Kuwabara S, Shinyashiki N, Yagihara S (1987) Evaluation of complex permittivity of aqueous solution by time domain reflectometry. J Mol Liq 36 (1):135–151. doi:10.1016/0167-7322(87)80036-3

30. Nozaki R, Bose TK (1990) Broadband complex permittivity measurements by time-domain spectroscopy. IEEE Trans Instrum Meas 39(6):945–951. doi:10.1109/19.65803
31. Feldman Y, Kozlovich N (1995) Time domain dielectric spectroscopy studies of macromolecular solutions. Trends Polym Sci 3(2):53–60. doi:10.1016/S0966-4793(00)88890-1
32. Yagihara S, Miura N, Hayashi Y, Miyairi H, Asano M, Yamada G, Shinyashiki N, Mashimo S, Umehara T, Tokita M, Naito S, Nagahama T, Shiotsubo M (2001) Microwave dielectric study on water structure and physical properties of aqueous systems using time domain reflectometry with flat-end cells. Subsurf Sens Technol Appl 2(1):15–29
33. Berberian JD, King E (2002) An overview of time domain spectroscopy. J Non-Cryst Solids 305(1–3):10–18. doi:10.1016/S0022-3093(02)01082-7
34. Shinyashiki N, Asaka N, Mashimo S, Yagihara S, Sasaki N (1990) Microwave dielectric study on hydration of moist collagen. Biopolymers 29(8–9):1185–91. doi:10.1002/bip.360290809
35. Lahtinen T, Nuutinen J, Alanen E (1997) Dielectric properties of the skin. Phys Med Biol 42 (7):1471–2. doi:10.1088/0031-9155/42/7/020
36. Naito S, Hoshi M, Mashimo S (1996) A method of measuring surface permittivity by microwave dielectric analysis. Rev Sci Instrum 67(10):3633–41. doi:10.1063/1.1147128
37. Naito S, Hoshi M, Mashimo S (1997) In vivo dielectric analysis of free water content of biomaterials by time domain reflectometry. Anal Biochem 251(2):163–72. doi:10.1006/abio.1997.2256
38. Naito S, Hoshi M, Yagihara S (1998) Microwave dielectric analysis of human stratum corneum in vivo. Biochim Biophys Acta 1381(3):293–304. doi:10.1016/S0304-4165(98)00041-5
39. Nakasako M (1999) Large-scale networks of hydration water molecules around bovine β-trypsin revealed by cryogenic X-ray crystal structure analysis. J Mol Biol 289(3):547–564. doi:10.1006/jmbi.1999.2795
40. Takashima S, Asami K (1993) Calculation and measurement of the dipole moment of small proteins: use of protein data base. Biopolymers 33(1):59–68. doi:10.1002/bip.360330107
41. Higo J, Kono H, Nakamura H, Sarai A (2000) Solvent density and long-range dipole field around a DNA-binding protein studied by molecular dynamics. Proteins 40(2):193–206. doi:10.1002/(SICI)1097-0134(20000801)40:2<193::AID-PROT30>3.0.CO;2-0
42. Yokomizo T, Yagihara S, Higo J (2003) Rotational motions of solvent site–dipole field around a protein. Chem Phys Lett 374(5–6):453–458. doi:10.1016/S0009-2614(03)00684-5
43. Miura N, Hayashi Y, Mashimo S (1996) Hinge-bending deformation of enzyme observed by microwave dielectric measurement. Biopolymers 39(2):183–187. doi:10.1002/(SICI)1097-0282(199608)39:2<183::AID-BIP6>3.0.CO;2-L
44. Hayashi Y, Miura N, Isobe J, Shinyashiki Y, Yagihara S (2000) Molecular dynamics of Hinge bending motion of IgG vanishing with hydrolysis by Papain. Biophys J 79(2):1023–1029. doi: http://dx.doi.org/10.1016/S0006-3495(00)76356-9
45. Hayashi Y et al (2014) personal communications
46. Karplus M, McCammon JA (2002) Molecular dynamics simulations of biomolecules. Nat Struct Biol 9(9):646–652. doi:10.1038/nsb0902-646
47. Hayashi Y, Miura N, Shinyashiki N, Yagihara S, Mashimo S (2000) Globule-coil transition of denatured globular protein investigated by a microwave dielectric technique. Biopolymers 54 (6):388–397. doi:10.1002/1097-0282(200011)54:6<388::AID-BIP30>3.0.CO;2-K
48. Tokita M, Miyoshi T, Takegoshi K, Hikichi K (1996) Probe diffusion in gels. Phys Rev E 53 (2):1823–1827. doi:10.1103/PhysRevE.53.1823
49. Yamada G, Hashimoto T, Morita T, Shinyashiki N, Yagihara S, Tokita M (2001) Dielectric study on dynamics for volume phase transition of PAAm gel in acetone-water system. Trans Mater Res Soc Jpn 26(2):701–704
50. Maruyama Y, Numamoto Y, Saito H, Kita R, Shinyashiki N, Yagihara S, Fukuzaki M (2014) Complementary analyses of fractal and dynamic water structures in protein-water mixtures and cheeses. Colloid Surf A Physicochem Eng Asp 440:42–48. doi:10.1016/j.colsurfa.2012.10.051

51. Matsumoto K, Shimazaki K, Kitamura K, Kita R, Shinyashiki N, Yagihara S (2008) Geometrical feature of fractal structure expression and characterization of aqueous mixtures by dielectric spectroscopy (Japanese). Proc Sch Sci Tokai Univ 43:127–143
52. Shimazaki K, Matsumoto K, Kita R, Shinyashiki N, Yagihara S (2010) Molecular dynamics simulation for molecular characterization analysis of aqueous solution of oligomer (Japanese). Proc Sch Sci Tokai Univ 45:113–129
53. Ladbrooke BD, Chapman D (1969) Thermal analysis of lipids, proteins and biological membranes. A review and summary of some recent studies. Chem Phys Lipids 3 (4):304–356. doi:10.1016/0009-3084(69)90040-1
54. Janiak MJ, Small DM, Shipley GG (1979) Temperature and compositional dependence of the structure of hydrated dimyristoyl lecithin. J Biol Chem 254(13):6068–6078
55. Schrader W, Kaatze U (2001) Zwitterion headgroup orientation correlation and mobility and domain structure of membranes. J Phys Chem B 105:6266–6272. doi:10.1021/jp010525t
56. Schrader W, Halstenberg S, Behrends R, Kaatze U (2003) Critical slowing in lipid bilayers. J Phys Chem B 107(51):14457–14463. doi:10.1021/jp0306489
57. Ermolina I, Lewis A, Feldman Y (2003) Dielectric properties of the bR membrane. J Phys Chem B 107(51):14537–14544. doi:10.1021/jp022682d
58. Okudaira S, Choe S, Kosuge M, Kundu SK, Yamamoto W, Kita R, Shinyashiki N, Yagihara S (2007) Dielectric study on restricted molecules of liquid crystal and liposome. Trans Mater Res Soc Jpn 32(3):823–826
59. Hosoi Y, Kita R, Shinyashiki N, Yagihara S (2010) Dielectric study of liposome dispersion using statistical analysis of Jitter. Trans Mater Res Soc Jpn 35(4):861–864
60. Chernik GG (1995) Phase equilibria in phospholipid-water systems. Adv Colloid Interf Sci 61:65–129. doi:10.1016/0001-8686(95)00262-O
61. Kundu SK, Choe S, Yamamoto W, Kita R, Yagihara S (2007) Dielectric relaxation and dynamic light scattering study of liposome in the aqueous solution. MRS Proc 1019:FF04–08
62. Kundu SK, Suzuki K, Chaudhuri BK (2003) The effect of biasing field on the soft mode in the vicinity of the SmC*-SmA phase transition of a ferroelectric liquid crystal "ZLI4851". J Appl Phys 94(4):2271–76. doi:10.1063/1.1587871
63. Kundu SK, Chaudhuri BK, Catala L, Mery S (2002) Soft mode and related behaviour in the SmA and SmC* phases of a ferroelectric liquid crystalline polymer by dielectric spectroscopy. Liq Cryst 29:837–842. doi:10.1080/02678290210133123
64. Ermolina I, Strinkovski A, Lewis A, Feldman Y (2001) Observation of liquid-crystal-like ferroelectric behavior in a biological membrane. J Phys Chem B 105(14):2673–2676. doi:10.1021/jp001054y
65. Asami K (1977) Dielectric behavior of yeast cell suspensions: effects of some chemical agents and physical treatments on the Plasma membranes and the cytoplasms. Bull Inst Chem Res Kyoto Univ 55:283–309
66. Asami K (2011) Design of a measurement cell for low-frequency dielectric spectroscopy of biological cell suspensions. Meas Sci Technol 22:085801. doi:10.1088/0957-0233/22/8/085801 (7 pp)
67. Asami K (2013) Dielectric properties of dipicrylamine-doped erythrocytes, cultured cells and lipid vesicles. Bioelectrochem J Bioelechem 92:14–21. doi:10.1016/.2013.02.003
68. Asami K (2012) Dielectric spectroscopy reveals nanoholes in erythrocyte ghosts. Soft Matter 8:3250–57. doi:10.1039/C2SM06306A
69. Daoud J, Asami K, Rosenberg L, Tabrizian M (2012) Dielectric spectroscopy for non-invasive monitoring of epithelial cell differentiation within three-dimensional scaffolds. Phys Med Biol 57(21):5097–5112. doi:10.1088/0031-9155/57/16/5097
70. Miura N, Yagihara S, Mashimo S (2003) Microwave dielectric properties of solid and liquid foods investigated by time-domain reflectometry. J Food Sci 68(4):1396–1403. doi:10.1111/j.1365-2621.2003.tb09656.x

71. Mashimo S, Kuwabara S, Yagihara S, Higasi K (1987) Dielectric relaxation time and structure of bound water in biological materials. J Phys Chem 91(25):6337–6338. doi:10.1021/j100309a005
72. Yagihara S, Oyama M, Inoue A, Asano M, Sudo S, Shinyashiki N (2007) Dielectric relaxation measurement and analysis of restricted water structure in rice kernels. Meas Sci Technol 18 (4):983–990. doi:10.1088/0957-0233/18/4/004
73. Hayashi Y, Miura N, Shinyashiki N, Yagihara S (2005) Free water content and monitoring of healing processes of skin burns studied by microwave dielectric spectroscopy *in vivo*. Phys Med Biol 50(4):599–612. doi:10.1088/0031-9155/50/4/003
74. Goto T, Hashimoto M, Shinyashiki N, Yagihara S, Hayashi Y (2006) Dielectric study on distribution of water in human skin. Trans Mater Res Soc Jpn 31:771–774
75. Hashimoto M, Goto T, Shinyashiki N, Yagihara S (2006) Interpretation of hydration structure of human skin from analysis of electrodes used in dielectric spectroscopy (Japanese). Tokai J Sport Med Sci 19:53–62
76. Yazawa S (1933) Comparative study of the skin tissue in the human body (Japanese). Med Res (Japanese) 7:1805–1834

Dynamics and Glass Transition of Aqueous Solutions of Molecular Liquid, Polymer, and Protein Studied by Broadband Dielectric Spectroscopy

9

Naoki Shinyashiki

Abstract

Broadband dielectric spectroscopy (BDS) is the most powerful tool for observing the dynamics of molecules in an extremely wide frequency range between 1 μHz and 50 GHz or more, i.e., the time window between megaseconds and picoseconds that covers the dynamics of solids to liquids. The progress of the investigations using BDS on the dynamics of aqueous solutions of hydrophilic molecular liquids, polymers, and proteins from solid to liquid states in this decade is presented in terms of the glass transition. The relaxation process, ν-relaxation, originating from the local reorientational motion of water molecules and the α-relaxation originating from the cooperative motion of hydrated solute are strongly related. The universalities of the dynamical hierarchy from fast and local motion of water molecules to slow and global cooperative motion of hydrated guest molecules extending over a wide range of glass transition temperature are successfully explained by applying the theoretical interpretation of mixtures of van der Waals liquids using the coupling model. More complicated partially crystallized aqueous protein solutions are also included.

Keywords

Dielectric relaxation • Glass transition • Water • Polymer • Protein

N. Shinyashiki (✉)
Department of Physics, School of Science, Tokai University,
4-1-1 Kitakaname, Hiratsuka, Kanagawa 259-1292, Japan
e-mail: naoki-ko@keyaki.cc.u-tokai.ac.jp

R. Kita, T. Dobashi (eds.), *Nano/Micro Science and Technology in Biorheology*,
DOI 10.1007/978-4-431-54886-7_9

9.1 Introduction

Living organisms can function only in the presence of water. In simple terms, biological materials without water are mere solids that do not allow changes in the relative positions of molecules and atoms constituting the solid materials and can only undergo thermal vibration on site. Therefore, neither material transfer nor molecular motion occurs, and the state of the materials does not change with time. With water, however, biological materials become flexible. Flexible materials can move and actively perform translation and rotation in addition to atomic and molecular vibrations. All biological activities are expressed by molecular and atomic motion and the transfer and deformation of larger structures using heat as the basic energy source. However, even water-containing flexible biological materials solidify when their temperature is reduced. For instance, water is mostly crystallized below 0 °C. For materials with low water content, crystallization does not occur, but their molecular motion decreases with decreasing temperature, and they solidify at lower temperatures. Thus, favorable water content and temperature are essential factors for sustaining biological activities.

In refrigerators, meat and fish can be preserved for a long time. As the temperature inside the refrigerator decreases, the deterioration of meat and fish is suppressed, enabling their long-term preservation. Foods containing water generally deteriorate and decay with time, whereas dried foods can be preserved for a long time at room temperature and then rehydrated for consumption by placing them in cold or hot water as necessary. Dried plant seeds at flower shops are solids and remain viable for several decades or more. Such seeds can be planted and watered in autumn, but they do not immediately germinate. They sleep through winter until they germinate in spring. Many trees contain less water in winter, but when spring approaches, they absorb water from the ground to prepare for their subsequent activities. For instance, cherry trees bloom in spring and grow fresh green leaves. These natural phenomena are exhibited by biological materials that skillfully control their functions in accordance with temperature and water content, which raises the following questions. How are biological materials related to water and temperature? What natural laws exist among them? The fundamental mechanisms of life are clarified by elucidating the motions of water and biological molecules, which always exist in the lower layers of the hierarchical structure in all the changes of nature and activities of life.

Most biological activities on Earth occur between −30 and +50 °C. However, research on water-containing materials is often carried out at lower temperatures. One reason for this is the difficulty in observing the effect of weak interactions between materials in the temperature range where biological activities take place because of their active thermal fluctuation. At low temperatures, the weak interactions between materials become strong relative to the thermal fluctuation, causing the interactions to be prominent. Therefore, a slight difference in the molecular motion near room temperature becomes distinct at low temperatures. Another reason lies in the various characteristics specific to water at low temperatures. Water reaches its maximum density under 1 atmospheric pressure at approximately 4 °C and has a melting point of 0 °C. Water has a higher density than ice and cannot exist as a liquid between 150 and 231 K, the so-called no man's land. There are two types of water structure with different densities below 150 K, and their phase transition occurs depending on the temperature and pressure. The critical point of this phase transition is considered to be in the no man's land. Many scientists have considered that relevant phenomena occur in relation to the motion of water molecules. Thus, the observation of the motion of water molecules at low temperatures is very effective for understanding the characteristics of water, which can be exploited in not only biological science but also materials science and technology.

9.2 Dielectric Relaxation of Hydrophilic Polymer in Solutions

The observation of polymer solutions by dielectric spectroscopy in the early days had been carried out on solutions containing nonpolar or weakly polar organic solvents [1–3]. In such cases, the motion of polymer chains was mainly discussed. Thus, the relaxation process observed by dielectric spectroscopy was interpreted to be due to polymer chain dynamics in viscous fluids and explained by polymer-solvent interactions, which vary with intramolecular interactions [4–6]. In contrast, in polymer solutions containing hydrogen-bonding liquids as the solvent, such as water, the motion of both polar solvent molecules and polymers can be observed as dielectric relaxation. Such molecular motion strongly depends on the polymer structure and solvent-polymer interaction. For example, if the rates of molecular motion markedly differ between a solvent and a solute, the relaxations of the solvent and solute are independently observed. However, if their rates are close, or if their molecular motions occur cooperatively, a single relaxation is observed. Therefore, it can be considered that solvent molecules adjacent to polymer chains move together with the polymer chain as a result of their interaction, whereas those distant from the polymer chains maintain their original rate of motion. Such properties of liquids cannot be explained without observing molecular motion. Sometimes, molecular dynamic models have been proposed on the basis of static observation, which cannot capture molecular motion, and their validity has not always been verified on the basis of experimental data.

When the high-frequency dielectric spectroscopy of aqueous polymer solutions became possible, the dynamics of water as a solvent was first observed [7–10], in contrast to the case of polymer solutions of nonpolar or weakly polar organic solvents. Until the 1990s, the frequency window used in the studies of the dielectric relaxation of aqueous systems was too narrow to see both the relaxation processes originated from the molecular motions of water and polymer. Another reason is that water has a large relaxation strength, and also, the contributions of electrode polarization and direct current (DC) electrical conduction by impurities and ions in water significantly contribute to the permittivity and mask the polymer

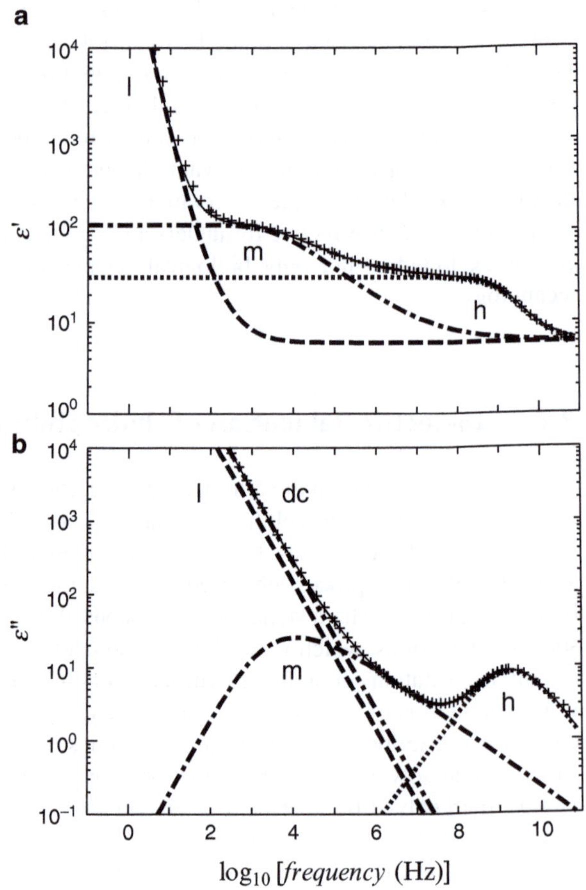

Fig. 9.1 Frequency dependences of real (**a**) and imaginary (**b**) parts of dielectric function for 65 wt% PVP-water mixture at 298 K. Pluses; data obtained by experiments. The lines were drawn using Eq. (9.1): *dotted lines* indicated h are the primary relaxation of water; *dot-dash lines* indicated m are the PVP relaxation; *dash lines* indicated l are electrode polarization; *double-dots-dashed line* is dc conductivity; solid lines, sum of all the processes. Reprinted from [47] with permission from AIP Publishing LLC

relaxation, which occurs at frequencies lower than that of water relaxation, making the direct observation of polymer relaxation difficult. As explained above, the observation of the motion of polymers in aqueous solutions is very difficult and therefore has rarely been reported.

In polymer solutions containing water or alcohol, however, the molecular motions of polymers and solvents were simultaneously observed in certain narrow ranges of concentration and temperature [11, 12]. Broadband dielectric spectroscopy (BDS) was performed at 25 °C in the frequency range of 30 GHz–40 Hz for polyvinylpyrrolidone (PVP) solutions containing monohydroxyl alcohol with 1–10 carbon atoms (n-alcohol) or water as the solvent [11] as well as for the PVP solutions of ethyleneglycol and its oligomers [12]. The PVP concentration varied between 0 and 70 wt%. Figure 9.1 shows real and imaginary parts of dielectric functions for a 65 wt% PVP-water mixture at 25 °C as an example of various PVP solutions at 25 °C. The relaxation of solvent was observed in the frequency range of

10 GHz–100 MHz for the PVP solutions with different PVP concentrations in terms of its complex permittivity at 25 °C. Another relaxation caused by the local motion of PVP molecular chains (denoted here as PVP relaxation) was observed in the frequency range of 10 MHz–10 kHz.

The dielectric constant and loss for the PVP mixtures can be described by the simple summation of three relaxation processes and a contribution of dc conductivity as

$$\varepsilon^*(\omega) = \varepsilon_\infty + \frac{\Delta\varepsilon_h}{\left\{1+\left(j\omega\tau_h^{HN}\right)^{\beta_h}\right\}^{\alpha_h}} + \Delta\varepsilon_{\mathrm{PVP}}\int_0^\infty\left(-\frac{d\Phi_{\mathrm{PVP}}}{dt}\right)\exp(-j\omega t)dt + \frac{\Delta\varepsilon_{\mathrm{ep}}}{1+\left(j\omega\tau_{\mathrm{ep}}\right)^{\beta_{\mathrm{ep}}}} - j\frac{\sigma}{\varepsilon_0\omega} \tag{9.1}$$

for

$$\Phi_{\mathrm{PVP}} = \exp\left[-\left(\frac{t}{\tau_{\mathrm{PVP}}^K}\right)^{\beta_K}\right].$$

Here, ω is the angular frequency, t is the time, j is the imaginary unit given by $j^2 = -1$, ε_0 is the dielectric constant in vacuum, ε_∞ is the limiting high-frequency dielectric constant, $\Delta\varepsilon$ is the relaxation strength, τ is the relaxation time, α and β are the asymmetric and symmetric broadening parameters ($0 < \alpha, \beta \leq 1$), respectively, β_{K} is the asymmetric broadening parameter ($0 < \beta_{\mathrm{K}} \leq 1$) of the Kohlrausch-Williams-Watts (KWW) function [13], and σ is the conductivity. The subscripts h, PVP, and ep denote the relaxation process originating from the reorientational motion of solvent molecules at higher frequency, the relaxation process of local chain motion of PVP, and electrode polarization, respectively. The superscripts HN and K denote the Havriliak-Negami equation [14] and the KWW function [13], respectively. The h-, PVP-, and ep-processes can be described well by the Havriliak-Negami equation, the KWW function, and the Cole-Cole equation [15], respectively. The relaxation times of the h- and PVP-processes were not simply determined using $\tau^{\mathrm{HN}}{}_{\mathrm{h}}$ and $\tau^{\mathrm{K}}{}_{\mathrm{PVP}}$ in Eq. (9.1). They were determined from the dielectric loss peak frequency of the h- and PVP-processes, f_{ph} and $f_{\mathrm{p\ PVP}}$, using $\tau_{\mathrm{h}} = 1/2\pi f_{\mathrm{ph}}$ and $\tau_{\mathrm{PVP}} = 1/2\pi f_{\mathrm{p\ PVP}}$, respectively. If an electric field is applied to the materials, large dielectric constants and losses due to long-range drift of ions and barrier layer formation at the electrode surfaces have been observed as electrode polarization.

PVP relaxation exhibits the following features [11, 12]:

1. The frequency at which PVP relaxation occurs is lower than that for the relaxation of the solvent and is around the frequency range in which the relaxation due to the local motion of polymer chains in nonpolar solvents occurs.

2. The relaxation intensity is almost proportional to the PVP concentration.
3. PVP relaxation exhibits an asymmetric curve that is broad on the high-frequency side of the loss peak, similarly to the relaxation due to the motion of polymer chains in nonpolar solvents.
4. The relaxation time in the infinitely diluted state is determined by the viscosity of the solvent.

These features are similar to those of the relaxation of polymer chains in nonpolar solvents. In addition, the magnitude of the effective dipole moment of the repeat unit of PVP, calculated from the strength of the PVP relaxation observed in solutions with water, mono- and polyhydroxy alcohols, and chloroform of different concentrations and the relaxation intensity of vinylpyrrolidone (VP, the monomer of PVP), is determined from the permittivity on the high-frequency side of the PVP relaxation [16]. When the permittivity on the high-frequency side is 1, which corresponds to the permittivity of the vacuum, the dipole moment of a repeat unit of PVP is in agreement with that obtained in the vacuum by quantum chemical calculation. The above experimental results led to the conclusion that the PVP relaxation observed for the solutions of various solvents has the same origin and is caused by the local fluctuation of the PVP chain. The relaxation time of the h-relaxation attributed to the reorientational motion of solvent molecules increases with increasing PVP concentration in alcohols with a small number of carbon atoms (e.g., from methanol to propanol), whereas it decreases for butanol and alcohols with a large number of carbon atoms [11]. This indicates that the relaxation time of alcohols is determined by ρ_{HB}, the density of hydrogen-bonding sites in solution, which changes with increasing PVP concentration.

Moreover, PVP solutions of chloroform, which has a lower permittivity than alcohol, are used to observe the PVP relaxation at different PVP concentrations and temperatures [17]. The results indicated that the apparent activation energy increases with decreasing temperature, which is known as the Vogel-Fulcher-Tammann-Hesse (VFTH) [18–20] temperature dependence of relaxation time. The VFTH equation is given by

$$\tau = \tau_{\infty} \exp\left(\frac{B}{T - T_0}\right) \tag{9.2}$$

where τ_{∞} (the relaxation time at an infinite high temperature), B, and T_0 are the VFTH parameters. The above characteristics are similar to those of the relaxation attributed to the local motion of polymer chains in solutions with nonpolar solvents, i.e., the α-relaxation of polymer. The above results led to the conclusion that the PVP relaxation can be attributed to the local molecular motion of PVP chains in solution.

9.3 Dynamics of Aqueous Solutions from Liquid to Glass

9.3.1 Glass Transition: General Properties of Glass Formers

Broadband dielectric spectroscopy (BDS) research on the molecular motion of various substances, such as molecular liquids and polymers in liquid to solid states, has been popular over the past two decades, for example, [21, 22]. Research has shown that the α-relaxation with a relaxation time τ_α of nanosecond order or shorter is generally observed at relatively high temperatures where substances are in the liquid state. τ_α increases with decreasing temperature, and one or more secondary relaxations on the high-frequency side of the α-relaxation appear. The multiple relaxations frequently observed in various polymers were conventionally called α-, β-, and γ-relaxations in order of increasing frequency. In addition to them, the normal mode relaxation originating from the overall rotational motion of polymer appears at a frequency lower than that of the α-relaxation if the polymer has components of dipole moment parallel to its main chain. The α-relaxation has been found to show common characteristics in various substances. The α-relaxation has an asymmetric shape of its dielectric loss plotted against the logarithm of frequency, that is, broader at high frequencies than at low frequencies with respect to the loss peak, and τ_α exhibits a VFTH temperature dependence. The VFTH temperature dependence of τ_α corresponds to the increase in the apparent activation energy of the reorientational motion of molecules with decreasing temperature. The extent of the change in the apparent activation energy, i.e., the deviation from the Arrhenius temperature dependence of τ_α, has been evaluated using the fragility [23], i.e., the steepness index [24] m. m is defined as the slope at the glass transition temperature T_g on the plane of log τ_α vs inverse temperature, $1/T$, normalized by $1/T_g$, as

$$m = \frac{d[\log_{10}(\tau)]}{d\left[{}^{T_g}/_{T}\right]} \Big|_{T=T_g} \tag{9.3}$$

A material with a large m value is fragile glass and that with a small m is strong glass. These terms of "fragile" and "strong" do not express the real mechanical strength or fragility of the materials but present the information of the degree of change in the apparent activation energy with temperature. The change in the apparent activation energy is generally interpreted to be due to the change in the cooperativity of molecular motion contributing to the relaxation process. These characteristics indicate that the α-relaxation is caused by the cooperative motion of many molecules and that the molecular motion causing the α-relaxation is the origin of the glass transition. Thermal measurements revealed that the glass transition is observed at temperatures at which $\tau_\alpha = 100$–$1{,}000$ s. Because of this, such temperatures have been generally defined as the glass transition temperature $T_{g\alpha}$ in dynamics observation. On the other hand, the relaxation time of secondary relaxations of polymers, such as β- and γ-relaxations, has weak temperature dependence than that of α-relaxation and generally exhibits an Arrhenius temperature dependence with a constant apparent activation energy. The Arrhenius equation is given by

$$\tau = \tau_{\infty Arr} \exp\left(\frac{\Delta E}{RT}\right), \tag{9.4}$$

where $\tau_{\infty Arr}$ is the pre-exponential factor corresponding to τ at infinitely high temperature, R the gas constant, and ΔE the apparent molar activation energy. Therefore, β- and γ-relaxations are considered to be caused by the intramolecular motion of, for example, freely moving side chains, which is more local than that in the case of α-relaxation. Johari and Goldstein demonstrated that β-relaxation occurs on the high-frequency side of α-relaxation, even in molecular liquids with no degree of freedom within a molecule, and that it is not always caused by the local intramolecular motion [25, 26].

In recent years, secondary relaxations have been classified into the Johari-Goldstein (JG) relaxation, as the elementary process of α-relaxation and local β-relaxation caused by other local molecular motions [27]. The primitive relaxation defined by the coupling model is the precursor of the cooperative α-relaxation [28–31], but it is a local process. The coupling model predicted the relationship between the primitive relaxation time τ_0 and the relaxation time of the α-relaxation τ_α as

$$\tau_0 = (t_c)^n (\tau_\alpha)^{1-n}. \tag{9.5}$$

Here, t_c is the crossover time from independent relaxation to cooperative relaxation and is approximately 2 ps for small molecular liquids [31] and n is the coupling constant and is the exponent of the stretched exponential function and determines the asymmetric broadening of the shape of α-relaxation in the dielectric loss spectrum in the Kohlrausch function, i.e., the time dependence of the response function of α-relaxation, as

$$\phi(t) = \exp\left[-(t/\tau_\alpha)^{1-n}\right]. \tag{9.6}$$

Equation (9.5) enables τ_0 to be calculated from τ_α using the coupling constant n. Thus, this attribute of the primitive relaxation is shared with the JG relaxation, and it can be expected that the primitive relaxation time τ_0 is approximately located near the relaxation time τ_{JG} of the JG relaxation at various temperatures and pressures [32, 33]. It was found that $\tau_0(T, P) \approx \tau_\beta(T, P)$ holds in many glass formers having the JG relaxations [32, 33]. The examination of $\tau_0(T, P) \approx \tau_\beta(T, P)$ by comparing the measured τ_β and the calculated τ_0 at ambient pressure reveals that they are nearly the same at various temperatures [32, 34]. Then the JG relaxation is a precursor or local step of the cooperative α-relaxation. Similarly to the α-relaxation, the JG relaxation is universally observed in various substances, not only in polymers but also in low-molecular-weight molecular liquids even in the case with no degree of freedom within molecules. The JG relaxation exhibits symmetric behavior with respect to the loss peak. Similarly to the α-relaxation time, the JG relaxation time shows VFTH or at least non-Arrhenius temperature dependence above T_g and Arrhenius temperature dependence with a constant apparent activation energy below $T_{g\alpha}$. Secondary relaxations other than JG relaxation are attributed to the local intramolecular motion of molecules with an internal degree of freedom, similar to the motion of part of the polymer side chains, and are not directly related to the glass transition [27].

9.3.2 Glass Transition in Water-Containing Systems

BDS measurements have been carried out in a very wide frequency range of 10 μHz–10 GHz and a temperature range of 80–298 K using various aqueous solutions containing solutes of different concentrations, in which the solution temperature can be decreased to T_g without the crystallization of water (water content ≤40 %). The solutes used were glycerol [35, 36], ethylene glycol [37, 38], ethylene glycol oligomer [39–41], poly(ethylene glycol) [36, 42, 43], propylene glycol and its oligomers [36, 44], poly(propylene glycol) [45], propanol [46], PVP [36, 43, 47–49], poly (vinyl methyl ether) (PVME) [50, 51], fructose [52], and other various materials [38, 43, 53], which have different molecular structures and a wide range of T_g values. Figure 9.2 shows dielectric loss spectra observed for a 65 wt% pentaethylene glycol-water mixture at various temperatures as an example of water mixtures. A single loss peak is observed at temperatures higher than approximately 200 K. Below 200 K, two loss peaks are clearly observed. Despite the differences in the water content and molecular structure of the solutes (including the resulting difference in T_g) in the various aqueous solutions, the relaxation attributed to the motion of water molecules observed in all the solutions showed many common characteristics [53–56]. This water relaxation is called hydration

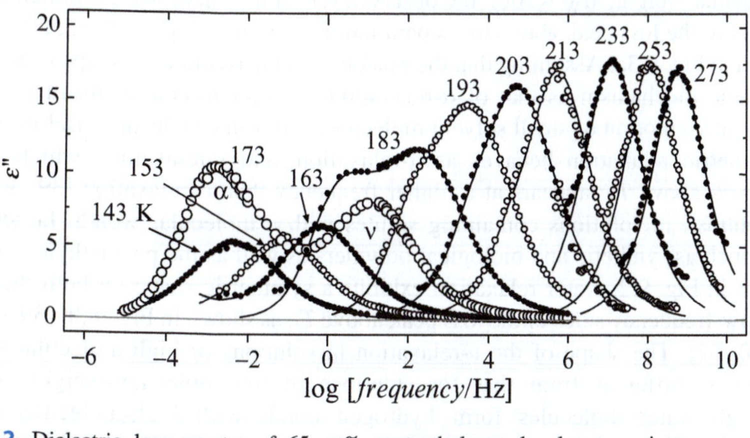

Fig. 9.2 Dielectric loss spectra of 65 wt% pentaethylene glycol-water mixture at various temperatures from 143 to 273 K. The loss peaks of the α- and ν-relaxations are indicated for the spectrum at 173 K, which is close vicinity of T_g (172 K). Data from [39] are replotted

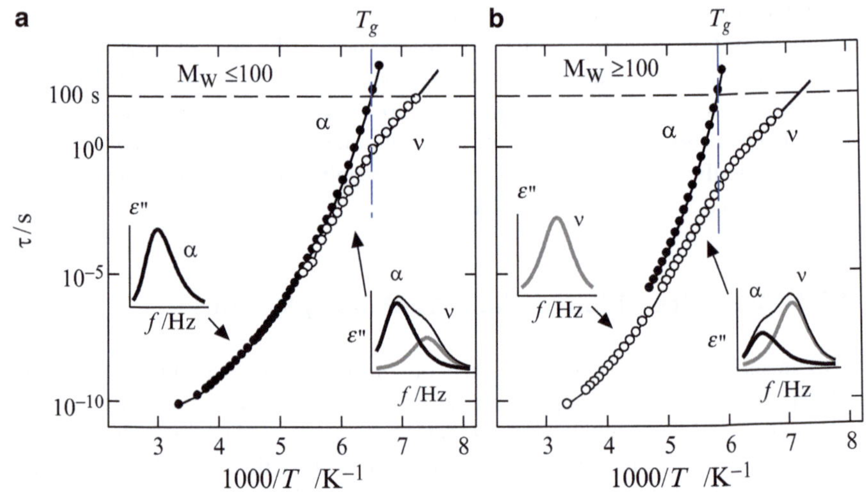

Fig. 9.3 Schematic representations of the temperature dependences of the relaxation times of the α-relaxation plotted by solid circles and the ν-relaxation plotted by open circles observed for aqueous solutions of (**a**) small solute molecule (Mw $\leq$ 100) and (**b**) large solute molecule (Mw $\geq$ 100). The plots of the relaxation time are observed for the mixtures of 35 wt% water mixtures with (**a**) diethyleneglycol and (**b**) pentaethylene glycol. *Horizontal dashed lines* at 100 s indicate the relaxation time at T_g given by the vertical dashed lines. Data from [39] are replotted

water relaxation or water-specific relaxation [57]. Here, we call it ν-relaxation [55], ν being the initial letter of νερο, which means water in Greek. It was shown that ν-relaxation is the primary dielectric relaxation of water observed in various aqueous solutions at around room temperature, and the shape of the ν-relaxation spectrum depends on the molecular weight of the solute [39, 40, 55, 58–62].

In solutions containing solutes with a low molecular weight, such as alcohols (molecular weight, $M_W \leq 100$), the observed relaxation curve was asymmetric with respect to the loss peak above the separation temperature T_S [39, 40, 55, 59–61], as shown in Fig. 9.3a. Assuming that the relaxation with asymmetric shape is based on the same mechanism as that of α-relaxation, it is interpreted to result from the cooperative motion of small solvent molecules and water molecules. Below T_c, the asymmetric relaxation behaves as α-relaxation, and ν-relaxation, which is not observed above T_S, appears at a higher frequency than α-relaxation [39, 40, 55]. In contrast, in solutions containing solutes with a molecular weight larger than 100, such as synthetic and biological polymers as well as the pentaethylene glycol shown in Fig. 9.2, water relaxation exhibits a symmetric shape on both the high- and low-frequency sides of the loss peak above T_s, as shown in Fig. 9.3b [39, 40, 55, 58–60, 62]. The shape of the ν-relaxation in solutions of high-molecular-weight solutes is different from that for solutions of low-molecular-weight solutes, although water molecules form hydrogen bonds with high-molecular-weight solutes. As shown in Fig. 9.3b, the high-molecular-weight solute molecules (M_W $\geq$100) behave as obstacles to water molecules and spatially restrict the motion of

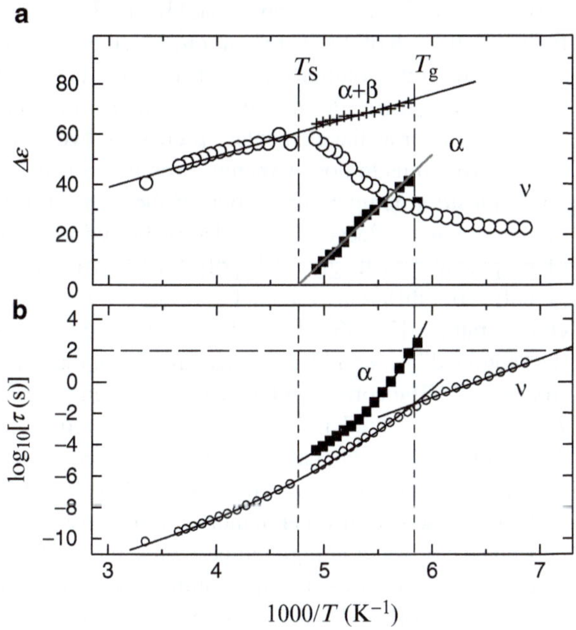

Fig. 9.4 Temperature dependences of (**a**) relaxation strengths and (**b**) relaxation times of the α (*solid squares*)- and ν (*open circles*)-relaxations observed for 65 wt % pentaethylene glycol-water mixture. *Pluses* indicate the sum of the relaxation strength of the α- and ν-relaxations. *Solid straight lines* in (**a**) were drawn by the least squares methods. *Straight* and *curved solid lines* in (**b**) were drawn by the least squares method using the Arrhenius and VFTH equations, respectively. T_S is the temperature below which two relaxation processes are observed. Data from [39] are replotted

water molecules. This spatially restricted motion of water molecules shows symmetric behavior, which is considered to be due to a mechanism similar to that which results in the symmetric behavior of the secondary relaxations attributed to the distribution of relaxation time in various local environments. Below T_S, the relaxation observed at above T_S becomes ν-relaxation; on the low-frequency side, a new α-relaxation is observed. Figure 9.4 shows temperature dependences of the relaxation times and strength for the α- and ν-relaxations, i.e., τ_α, τ_ν, $\Delta\varepsilon_\alpha$, and $\Delta\varepsilon_\nu$, respectively, for pentaethylene glycol (5EG) solutions as an example of solutions containing a solute with a high molecular weight [39, 55]. The temperature dependence of $\Delta\varepsilon_\nu$ is also weak below T_g but significantly differs above T_g, similarly to the temperature dependences of enthalpy, entropy, and volume. $\Delta\varepsilon_\alpha$ is small at high temperatures because of the less cooperative motion between water and solute molecules. When cooperative motion is enhanced with decreasing temperature below T_c, $\Delta\varepsilon_\alpha$ increases but $\Delta\varepsilon_\nu$ decreases. Below T_g, τ_ν values are similar among various aqueous solutions of different concentrations and independent of the structure and molecular weight of the solute molecules and T_g [54–56]. τ_ν has an Arrhenius temperature dependence, where the plots of relaxation time against the inverse of temperature are linear with an activation energy of approximately 50 kJ/mol. This is because the local motion of water molecules maintains behavior specific to water.

τ_ν shows different temperature dependences above and below T_g and exhibits VFTH, or at least non-Arrhenius, behavior above T_g. The same features are also clearly seen in fructose-water mixtures [52]. The significantly different temperature dependences of τ_ν and $\Delta\varepsilon_\nu$ above and below T_g are also observed in the relaxation time and strength of the JG relaxation, which is observed in various substances, not only in aqueous solutions [27]. Therefore, the motion of various substances, including polymers, in aqueous solutions is considered to result from the local motion of water as the elementary process of hydrated solute molecules [56].

The Arrhenius to non-Arrhenius crossover of the relaxation time at around the glass transition temperature is one of the most attractive features of ν-relaxation discovered in this decade. The dielectric behavior of supercooled aqueous solutions of propylene glycol, glycerol, poly-(vinylpyrrolidone) (PVP), and poly(ethylene glycol)s of different molecular weights has been examined for different concentrations [36, 43]. The sub-T_g process due to the reorientational motion of the "confined" water molecules that are part of the matrix and the matrix itself were frozen at T_g. The corresponding dielectric strength shows an appreciable change at T_g, indicating a hindered rotation of water molecules in the glassy phase [36, 43]. Furthermore, the nature of this confined water appears to be anomalous compared with those of most other supercooled confined liquids [43].

It was presented by other authors that because of the non-Arrhenius temperature dependence of water at high temperatures, the mobility of the matrix (origin of the α-relaxation) increases at temperatures above T_g, and water is able to form an extended network and relax in a cooperative manner, as in bulk water [44]. On the other hand, because of the Arrhenius temperature dependence of water at low temperatures, water behaves as if it is confined, i.e., there is local motion of water molecules in the glassy frozen matrix [44]. Then, the crossover temperature is directly related to the glass transition temperature of the mixtures [45, 48, 49, 51]. 18 different water-rich mixtures with very different hydrophilic substances, such as low-molecular-weight organic glass formers, polymers, sugars, and biopolymers (proteins and DNA), were studied by BDS, and the observed non-Arrhenius to Arrhenius crossover seems to be a general feature not only for water solutions but for dynamically asymmetric mixtures [54]. Another interpretation is that the apparent fragile-to-strong transition for supercooled confined water is due to a merged αβ-relaxation at high temperatures and a pure β-relaxation (confined water) below the crossover temperature, while the α-relaxation disappears at low temperatures owing to confinement effects [63].

The change in the temperature dependence of the relaxation time of ν-relaxation is also observed for water in an aqueous layer of 6 Å thickness (equivalent to that of two water molecules) formed by the hydration of Na vermiculite [64, 65], molecular sieves [66], silica hydrogels [67–69], and poly(2-hydroxyethyl methacrylate) [70–73]. Many such characteristics of ν-relaxation are common with those of JG relaxation observed in nonaqueous systems, such as mixtures of van der Waals liquids [56]. The above experimental results led to the conclusion that, in aqueous solutions, ν-relaxation behaves similarly to JG relaxation, although α-relaxation is attributed to the cooperative motion of solute and water molecules.

The observation of the α-relaxation originating from the segmental motion of polymer chains is generally difficult in aqueous polymer solutions since it is masked by the components of electrical conduction and electrode polarization;

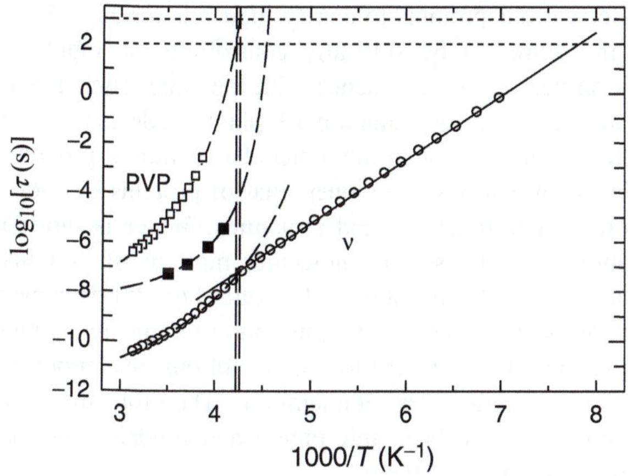

Fig. 9.5 Temperature dependence of the relaxation times observed for 65 wt% PVP-water mixture. *Open squares*; dielectric PVP relaxation (the α-relaxation), *open circles*; dielectric ν-relaxation, *solid square*; the α-relaxation of PVP observed by quasielastic neutron scattering [49]. *Dashed curves* and *straight solid line* were drawn by the least squares method using the VFTH and the Arrhenius equations, respectively. *Dashed vertical straight lines* indicate T_g at which τ_{PVP} is in the range between 100 and 1,000 s presented by two dotted *horizontal straight lines*. Data from [47, 49] are replotted

only ν-relaxation can be observed dielectrically [42, 48, 50, 54]. Temperature dependences of the relaxation process originating from the segmental motion of the polymer main chain (the α-process) were observed in PVP aqueous solution with 4 water molecules/monomeric unit of PVP [49] and in PVME aqueous solutions with 30 wt% water [51] by quasielastic neutron scattering (QENS), but not by BDS. T_g of α-relaxation was determined by DSC and the extrapolation of the relaxation time of α-relaxation to 100–1,000 s. T_g agrees well with the temperature at which the Arrhenius to non-Arrhenius crossover of ν-relaxation occurs. Recently, we have observed the α-relaxation attributed to the local main chain motion of PVP in aqueous solution by dielectric spectroscopy [47], since the PVP was deionized and the contributions of electrode polarization and dc conductivity were reduced. Figure 9.5 shows the inverse temperature dependence of the relaxation times of the α-relaxation of PVP and the ν-relaxation of water in the 65 wt% PVP-water mixture. The relaxation times of the dielectrically observed α-process of PVP [47] are somewhat larger than those determined by QENS [49], as shown in Fig. 9.5. In addition, the Arrhenius to non-Arrhenius transition of ν-relaxation is observed in close proximity to T_g of the dielectric α-relaxation.

9.3.3 Dielectric Relaxations in Protein-Water Systems

Proteins have various functions and structures in biological bodies and serve as the source of biological activities. Therefore, many scientists have studied the functions and structures of various proteins to clarify their relationships. It can be assumed that proteins in nature act by skillfully controlling the rapid change in their molecular motion near T_g in accordance with the water content and temperature, as similarly shown in the germination of plant seeds that are watered at an appropriate temperature. Although the molecular motion of protein-water systems has been observed by various techniques, that of proteins has not yet been fully elucidated, unlike their functions and structures. Water is indispensable in the molecular motion of proteins. The molecular motions of proteins have caused them to be considered as being "slaved" by water [74, 75]. However, the simultaneous observation of the dynamics of water and proteins by a single technique is difficult because the rates of their molecular motions near room temperature are expected to differ by several orders of magnitude. Therefore, the molecular motions of water and proteins have been interpreted and modeled on the basis of that observed in one or the other of them.

Generally, investigation of the dynamics in protein-water systems has been mainly carried out for water content less than 50 wt% since systems with less water can avoid the crystallization of water even below the melting temperature of water. The ν-relaxation of water is observed for various protein-water systems. It has been shown that many characteristics of ν-relaxation in protein-water systems are comparable to those of ν-relaxation observed in the aqueous solutions presented above. Although BDS can cover the wide frequency range from 1 μHz up to 100 GHz, the investigations of the protein-water system were carried out in limited frequency ranges; one is in the range up to approximately 1 MHz and another is down to approximately 1 MHz. The electric circuit for the dielectric measurements can be treated as a lumped constant circuit below 1 MHz, but must be treated as a distributed constant circuit above 1 MHz. Some equipment can cover the frequencies both above and below 1 MHz, but many experiments have been carried out for only one of them. The investigation above 1 MHz is generally focused on the dynamics of water and protein above the melting temperature of ice, and that below 1 MHz is focused on the dynamics related to the glass transition.

In the vicinity of room temperature, the relaxation process of water (ν-relaxation) has been observed at approximately 10 GHz. Several relaxation processes at frequencies in the GHz to MHz range have also been presented. The origins of the relaxations are interpreted to be bound water, trapped water, side group of protein, local chain motion of protein, overall rotation of protein molecule, and their combinations. Such discussions have been continued for several decades (see references [76, 77]), since the confirmation of the existence and the assignment of the origin of the relaxation processes in this frequency range are difficult. The water-protein systems include large contributions not only from the relaxation of water but also the DC conductivity and sometimes by electrode polarization [78]. For example, when the contribution of DC conductivity was eliminated from the dielectric loss spectra in a limited frequency range, for example, down to 1 MHz, it must be assumed that the dielectric loss spectra at the lower limit of the measured frequency range are contributed to by only DC conductivity. This estimation seems to lead to the conclusion of the presence of a dielectric loss peak at a frequency approximately one order higher than the low-frequency limit of the measurements. Therefore, we must pay the utmost attention to the data treatments.

These days, the precision of dielectric measurements in a frequency range of 1 MHz−50 GHz has been improved, and detailed discussion on the relaxation processes observed in protein (lysozyme)-water mixtures at around room temperature has been presented [76, 77]. According to the experimental results, β-, γ-, and δ-relaxations were found. The β-relaxation observed at approximately 10 MHz was attributed to the rotation of the polar protein molecules in their aqueous medium, and the γ-relaxation was attributed to the reorientational motion of the free water molecules [76, 77]. On the other hand, the nature of δ-relaxation is still controversial. One of the interpretation is that δ-relaxation is attributed to the orientation polarization of water molecules in the vicinity of the protein surface (hydration water) [77]. At a high protein concentration, δ-relaxation has a bimodal structure associated with two types of hydration water (tightly and loosely bound hydration water) [77]. Others claimed that γ-relaxation cannot be fully understood [76].

At temperatures lower than approximately 200 K, investigations of ν-relaxation in protein-water systems have been carried out at frequencies below 1 MHz, since the dielectric loss peak of ν-relaxation is in this frequency range. The relationships among glass transition, intermolecular interactions, dynamics of water and proteins, and crystallization and melting of water in protein-water mixtures at subzero temperatures for protein-water systems with an extremely wide range of water fraction between 0.1 and 0.9, which covers both uncrystallized and partially crystallized systems, have been studied using water equilibrium sorption isotherm (ESI), differential scanning calorimetry (DSC), dielectric relaxation spectroscopy (DRS), and thermally stimulated depolarization current (TSDC) [79–82]. The proteins studied were bovine serum albumin (BSA) [79, 80] and lysozyme [81] as globular proteins and elastin [82] as a fibrous protein. T_g and the melting and crystallization temperatures of water in BSA and the lysozyme-water system were clarified, as shown in Fig. 9.6, using a combination of the results obtained by DSC, DRS, and TSDC [80]. The α-relaxation associated with the glass transition of the hydrated system was observed by DRS and TSDC for water fractions equal to or higher than 7 %. A plasticization of the α-relaxation by water was observed up to a water fraction of 28 %. The plasticization stops at water fractions higher than the critical water fraction of 28 %, above which the water crystallized during cooling. As a result of the studies of the main dielectric relaxation due to uncrystallized water, common characteristic properties have been presented. The shape of the loss peak is symmetric or nearly symmetric, the temperature dependence of the relaxation time is the Arrhenius type with an apparent activation energy of 0.45–0.55 eV, and the relaxation time is longer and the strength is smaller in the system with less water content. In the protein-water systems with low hydration levels, the interrelation of the uncrystallized water and polar groups on the surface of the protein was discussed. The uncrystallized water molecules in low-hydration-level protein trigger the motion of small polar groups on the surface of the protein molecules, and these small polar groups are plasticized by water [79].

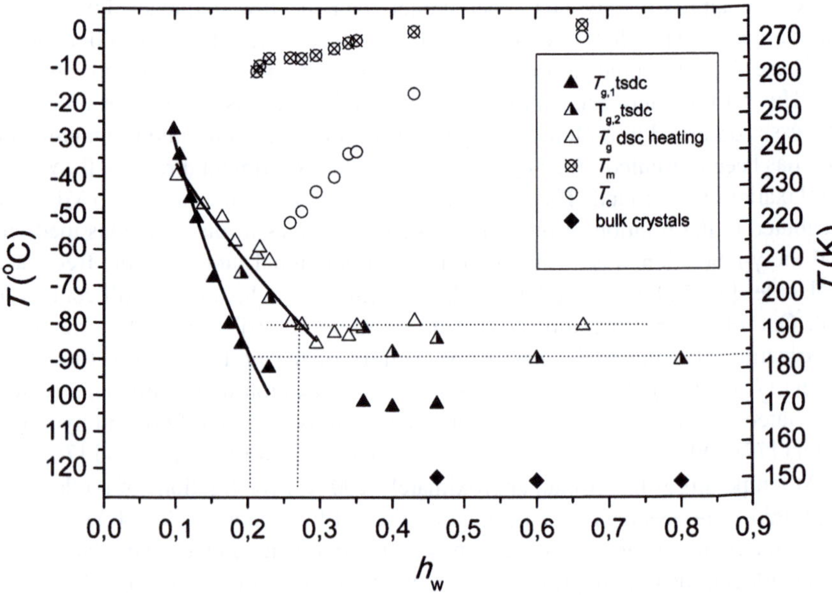

Fig. 9.6 Characteristic TSDC and DSC temperatures, indicated on the plot, against water fraction h_w (=mass of water/mass of hydrated protein). The *solid lines* are fits to the TSDC and DSC data to the Fox equation [91]. The *dotted lines* are given by guide of eyes, concerning the estimation of uncrystallized water according to the phase diagram [92]. The glass transition temperatures given by *solid triangles* $T_{g,1}$ and *half open triangle* $T_{g,2}$ and *solid diamonds* (temperature of relaxation peaks due to bulk ice crystals) were determined by TSDC. *Open triangles*; T_g (on heating process), *circles* with cross; melting temperature T_m (on heating process) of ice, and *open circles*; crystallization temperature T_C of water on cooling process were determined by DSC. Reprinted from [80] with permission from Elsevier

An adiabatic calorimetry study showed that enthalpy relaxation (glass transition) was observed at different temperatures for crystallized 20 %(w/w) bovine serum albumin (BSA) aqueous solution by measuring the heat capacity and enthalpy relaxation rate of the solution [83]. Specifically, enthalpy relaxation was observed at 100, 135, and above 180 K for BSA solutions annealed at 200–240 K after rapid cooling. The range of relaxation time that can be observed by adiabatic calorimetry is limited to a relatively long time (100–1,000 s). Therefore, BDS, which can cover wide ranges of temperature and frequency, is more effective for understanding the mechanism behind the abovementioned relaxations. We carried out BDS measurements for BSA aqueous solutions of different concentrations in a frequency range of 13 orders of magnitude, i.e., 2 mHz–1.8 GHz, and in the temperature range of 80–270 K, and observed three relaxations for frozen BSA aqueous solutions [84]. The temperatures at which relaxation time was 100 s were 100, 145, and 200 K, which were in good agreement with the T_g observed by adiabatic calorimetry. From this result, we interpreted the three relaxations as follows: on the basis of the temperature dependence of the dielectric relaxation, i.e., by representing the

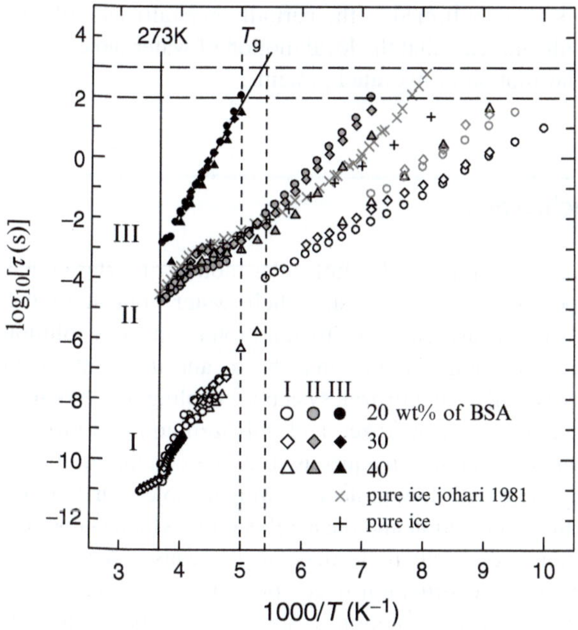

Fig. 9.7 Temperature dependences of relaxation times observed for 20, 30, and 40 wt% BSA water mixtures and pure ice. The shape of the symbols indicate the BSA concentrations as *circles*; 20 wt%, *diamonds*; 30 wt%, and *triangles*; 40 wt%. The *open*, *gray*, and *solid* symbols indicate the relaxations of I, II, and III, respectively. The relaxation of ice is represented by *cross* [85] and *plus* measured by us. The process I is very broad and cannot be described by assuming a single relaxation function, then we fit the process I with two relaxations. Then open gray plots are shown in each temperature at lower temperature region. *Vertical solid straight line* indicates the melting temperature of pure ice at 273 K. *Dotted vertical straight lines* indicate T_g of hydrated BSA at which τ of III-relaxation is in the range between 100 and 1,000 s given by *dashed horizontal straight lines*. Parts of data from [84] are replotted

molecular motion that caused the glass transition, which was observed in thermal measurements, by dielectric relaxation. The inverse temperature dependences of the relaxation times of the three relaxation process are shown in Fig. 9.7. First, the fastest relaxation with a relaxation time of 100 or 1,000 s at approximately 100 K (I-relaxation) is attributed to the reorientational motion of unfrozen water molecules in the liquid state even at low temperatures corresponding to the ν-relaxation. The second fastest relaxation with a relaxation time of 100 s at approximately 140 K, as observed in the medium-frequency range (II-relaxation), is attributed to the relaxation of ice judged by comparison with the relaxation of pure ice [85]. The slowest relaxation with a relaxation time of 100 s at approximately 200 K (III-relaxation) is attributed to the cooperative motion of protein and adjacent water molecules, which is the origin of the glass transition, i.e., the α-relaxation of hydrated protein. Note that the temperature dependence of the relaxation time of I-relaxation changes at T_g, as shown in Fig. 9.7. The paired contribution to the hierarchical structure of water-complex systems by the ν- and α-relaxations is also confirmed in the partially crystallized protein-water systems. The above results indicate that the local motion of water molecules is the precursor to the molecular motion of hydrated proteins.

9.4 Conclusions

Previously, the observation of the molecular motion of water-containing materials was mainly performed by decreasing their water content to prevent freezing. However, the BDS measurement of frozen protein aqueous solutions [79–82, 84], as mentioned above, demonstrated that the α- and ν-relaxations are observed in frozen systems as well as in unfrozen systems, enabling the consistent interpretation of these relaxations between frozen and unfrozen aqueous protein solutions. We also focused on the dynamics in aqueous solutions containing solutes with a simple structure (e.g., those with low-molecular-weight molecular liquids and polymers) and observed the ν-relaxation and the α-relaxation, which is due to the cooperative motion of water and solute molecules, which causes the glass transition. The correlation between the structural fluctuation of proteins and the water relaxation observed in aqueous protein solutions is similar to that between the behaviors of the α- and ν-relaxations observed in other aqueous solutions. We applied the theoretical interpretation of a mixture of van der Waals liquids using the coupling model to more complicated aqueous solutions [56, 86] and successfully explained the behavior of more complicated partially crystallized aqueous protein solutions [84]. Specifically, it has been demonstrated that the correlation between the α- and the JG relaxations observed in various amorphous substances mimics that between the relaxation processes of hydrated protein and water molecules in protein-water systems [84, 87–90]. This means that the local motion of water molecules serves as the elementary process of the cooperative motion of water and protein molecules that undergo local structural fluctuation and is expected to be related to the expression of the functions of proteins. From the correlation between the α- and ν-relaxations universally observed by BDS in various aqueous solutions, it is clarified that part of the spatial and temporal hierarchical structures – from the ν-relaxation attributed to the fast motion of water molecules that occurs in a small space in aqueous solutions to the α-relaxation attributed to the slow motion of hydrated polymers in a large space – is related to the dynamics of aqueous solutions. It can be easily imagined that these hierarchical structures can provide the materials for the formation of higher-order complicated hierarchical structures and functions of life.

Acknowledgments The author thanks all his collaborators for carrying out much of the work reported in this chapter.

References

1. Stockmayer WH (1967) Dielectric dispersion in solutions of flexible polymers. Pure Appl Chem 15(3–4):539–554. doi:10.1351/pac196715030539
2. Mashimo S (1976) Dielectric study of chain motion of poly(p-chlorostyrene) in dilute solution. Macromolecules 9(1):91–97. doi:10.1021/ma60049a018
3. Adachi K, Fujihara I, Ishida Y (1975) Diluent effects on molecular motions and glass transition in polymers. I. Polystyrene–toluene. J Polym Sci Polym Phys Ed 13(11):2155–2171. doi:10.1002/pol.1975.180131108
4. Ono K, Ueda K, Yamamoto M (1994) Local chain dynamics of poly(cis-1,4-isoprene) in dilute solutions studied by the fluorescence depolarization method. Polym J 26:1345–1351. doi:10.1295/polymj.26.1345
5. Mashimo S (1977) Effect of diluent on dielectric relaxation time of polymers. J Chem Phys 67 (6):2651–2658. doi:10.1063/1.435178
6. Adolf DB, Ediger MD, Kitano T, Ito K (1992) Viscosity dependence of the local segmental dynamics of anthracene-labeled polyisoprene in dilute solution. Macromolecules 25 (2):867–872. doi:10.1021/ma00028a055
7. Kaatze U (1975) Dielectric relaxation in aqueous solutions of polyvinylpyrrolidone. Adv Mol Relax Process 7:71–85
8. Kaatze U, Gottmann O, Podbielski R, Pottel R, Terveer U (1978) Dielectric relaxation in aqueous solutions of some oxygen-containing linear hydrocarbon polymers. J Phys Chem 82 (1):112–120. doi:10.1021/j100490a025
9. Jain SK, Johari GP (1988) Dielectric studies of molecular motions in the glassy states of pure and aqueous poly(vinylpyrrolidone). J Phys Chem 92(20):5851–5854. doi:10.1021/j100331a061
10. Shinyashiki N, Asaka N, Mashimo S, Yagihara S (1990) Dielectric study on dynamics of water in polymer matrix using a frequency range 10[sup 6]–10[sup 10] Hz. J Chem Phys 93 (1):760–764. doi:10.1063/1.459526
11. Shinyashiki N, Imoto D, Yagihara S (2007) Broadband dielectric study of dynamics of polymer and solvent in poly(vinyl pyrrolidone)/normal alcohol mixtures. J Phys Chem B 111(9):2181–2187. doi:10.1021/jp065414e
12. Shinyashiki N, Sengwa RJ, Tsubotani S, Nakamura H, Sudo S, Yagihara S (2006) Broadband dielectric study of dynamics of poly(vinyl pyrrolidone)-ethylene glycol oligomer blends. J Phys Chem A 110(15):4953–4957. doi:10.1021/jp0605193
13. Williams G, Watts DC (1970) Non-symmetrical dielectric relaxation behaviour arising from a simple empirical decay function. Trans Faraday Soc 66:80–85. doi:10.1039/tf9706600080
14. Havriliak S, Negami S (1967) A complex plane representation of dielectric and mechanical relaxation processes in some polymers. Polymer 8:161–210
15. Cole KS, Cole RH (1941) Dispersion and absorption in dielectrics I. Alternating current characteristics. J Chem Phys 9(4):341–351. doi:10.1063/1.1750906
16. Shinyashiki N, Miyara M, Nakano S, Yamamoto W, Ueshima M, Imoto D, Sasaki K, Kita R, Yagihara S (2013) Dielectric relaxation strength and magnitude of dipole moment of poly(vinyl pyrrolidone)in polar solutions. J Mol Liq 181:110–114. doi:10.1016/j.molliq.2013.02.020
17. Shinyashiki N, Spanoudaki A, Yamamoto W, Nambu E, Yoneda K, Kyritsis A, Pissis P, Kita R, Yagihara S (2011) Segmental relaxation of hydrophilic poly(vinylpyrrolidone) in chloroform studied by broadband dielectric spectroscopy. Macromolecules 44 (7):2140–2148. doi:10.1021/Ma102394s
18. Vogel H (1921) The law of the relation between the viscosity of liquids and the temperature. Phys Z 22:645–646
19. Fulcher GS (1925) Analysis if recent measurements of the viscosity of glasses. J Am Ceram Soc 8(339–355):789–794

20. Tammann G, Hesse W (1926) Die Abhängigkeit der Viscosität von der Temperatur bie unterkühlten Flüssigkeiten. Z Anorg Allg Chem 156(1):245–257. doi:10.1002/zaac.19261560121
21. Kremer F, Schönhals A (eds) (2003) Broadband dielectric spectroscopy. Springer, Berlin/Heidelberg. doi:10.1007/978-3-642-56120-7
22. Richert R, Blumen A (1994) Disorder effects on relaxational processes glasses, polymers, proteins. Springer, Berlin/Heidelberg. doi:10.1007/978-3-642-78576-4
23. Böhmer R, Angell C (1992) Correlations of the nonexponentiality and state dependence of mechanical relaxations with bond connectivity in Ge-As-Se supercooled liquids. Phys Rev B 45(17):10091–10094. doi:10.1103/PhysRevB.45.10091
24. Plazek DJ, Ngai KL (1991) Correlation of polymer segmental chain dynamics with temperature-dependent time-scale shifts. Macromolecules 24(5):1222–1224. doi:10.1021/ma00005a044
25. Johari GP, Goldstein M (1970) Viscous liquids and the glass transition. II Secondary relaxations in glasses of rigid molecules. J Chem Phys 53(6):2372–2388. doi:10.1063/1.1674335
26. Johari GP, Goldstein M (1971) Viscous liquids and the glass transition. III Secondary relaxations in aliphatic alcohols and other nonrigid molecules. J Chem Phys 55 (9):4245–4252. doi:10.1063/1.1676742
27. Ngai KL, Paluch M (2004) Classification of secondary relaxation in glass-formers based on dynamic properties. J Chem Phys 120(2):857–873. doi:10.1063/1.1630295
28. Tsang KY, Ngai KL (1996) Relaxation in interacting arrays of oscillators. Phys Rev E 54(4): R3067–R3070. doi:10.1103/PhysRevE.54.R3067
29. Ngai KL, Tsang KY (1999) Similarity of relaxation in supercooled liquids and interacting arrays of oscillators. Phys Rev E 60(4):4511–4517. doi:10.1103/PhysRevE.60.4511
30. Ngai KL (2001) Coupling model explanation of salient dynamic properties of glass-forming substances. IEEE Trans Dielectr Electr Insul 8(3):329–344. doi:10.1109/94.933340
31. Ngai KL, Rendell RW (1997) Basic physics of the coupling model: direct experimental evidences. In: Supercooled liquids advances and novel applications. ACS symposium series, vol 676. American Chemical Society, Washington, DC pp 45–66. doi:10.1021/bk-1997-0676.ch004
32. Ngai KL (2003) An extended coupling model description of the evolution of dynamics with time in supercooled liquids and ionic conductors. J Phys Condens Matter 15(11):S1107–S1125. doi:10.1088/0953-8984/15/11/332
33. Ngai KL, Paluch M (2003) Inference of the evolution from caged dynamics to cooperative relaxation in glass-formers from dielectric relaxation data. J Phys Chem B 107 (28):6865–6872. doi:10.1021/jp026611q
34. Hensel-Bielowka S, Ziolo J, Paluch M, Roland CM (2002) The effect of pressure on the structural and secondary relaxations in 1,1′-bis (p-methoxyphenyl) cyclohexane. J Chem Phys 117(5):2317. doi:10.1063/1.1488593
35. Sudo S, Shimomura M, Shinyashiki N, Yagihara S (2002) Broadband dielectric study of α−β separation for supercooled glycerol−water mixtures. J Non-Cryst Solids 307–310:356–363. doi:10.1016/S0022-3093(02)
36. Murthy SSN (2000) Experimental study of the dynamics of water and the phase behavior of the supercooled aqueous solutions of propylene glycol, glycerol, poly(ethylene glycol)s, and poly (vinylpyrrolidone). J Phys Chem B 104(29):6955–6962. doi:10.1021/jp9931915
37. Sudo S, Shinyashiki N, Yagihara S (2001) The dielectric relaxation of supercooled ethyleneglycol-water mixtures. J Mol Liq 90(1–3):113–120. doi:10.1016/S0167-7322(01)00113-1
38. Murthy SSN (1997) Phase behavior of the supercooled aqueous solutions of dimethyl sulfoxide, ethylene glycol, and methanol as seen by dielectric spectroscopy. J Phys Chem B 101 (31):6043–6049. doi:10.1021/jp970451e

39. Sudo S, Shimomura M, Saito T, Kashiwagi T, Shinyashiki N, Yagihara S (2002) Dielectric study on α- and β-processes in supercooled diethyleneglycol– and pentaethyleneglycol–water mixtures. J Non-Cryst Solids 305(1–3):197–203. doi:10.1016/S0022-3093(02)01094-3
40. Sudo S, Tsubotani S, Shimomura M, Shinyashiki N, Yagihara S (2004) Dielectric study of the alpha and beta processes in supercooled ethylene glycol oligomer-water mixtures. J Chem Phys 121(15):7332–7340. doi:10.1063/1.1796232
41. Sudo S, Shinyashiki N, Arima Y, Yagihara S (2008) Broadband dielectric study on the water-concentration dependence of the primary and secondary processes for triethyleneglycol-water mixtures. Phys Rev E 78(1):011501–011507. doi:10.1103/PhysRevE.78.011501
42. Sudo S, Shimomura M, Kanari K, Shinyashiki N, Yagihara S (2006) Broadband dielectric study of the glass transition in poly(ethyleneglycol)-water mixture. J Chem Phys 124 (4):044901. doi:10.1063/1.2149860
43. Tyagi M, Murthy SS (2006) Dynamics of water in supercooled aqueous solutions of glucose and poly(ethylene glycol)s as studied by dielectric spectroscopy. Carbohydr Res 341 (5):650–662. doi:10.1016/j.carres.2006.01.006
44. Cerveny S, Schwartz GA, Alegria A, Bergman R, Swenson J (2006) Water dynamics in n-propylene glycol aqueous solutions. J Chem Phys 124(19):194501. doi:10.1063/1.2198206
45. Singh LP, Cerveny S, Alegria A, Colmenero J (2011) Dynamics of water in supercooled aqueous solutions of poly(propylene glycol) as studied by broadband dielectric spectroscopy and low-temperature FTIR-ATR spectroscopy. J Phys Chem B 115(47):13817–13827. doi:10.1021/jp2073705
46. Sudo S, Tobinai S, Shinyashiki N, Yagihara S (2006) Broadband Dielectric study on glass transition of 1-propanol-water mixture. In: Tokuyama M, Maruyama S (eds) AIP Conf Proc 832:149–152. doi:10.1063/1.2204477
47. Miyara M, Sakuramatsu Y, Sasaki K, Kita R, Shinyashiki N, Yagihara S (2013) Glass transition and dynamics of poly(vinyl pyrrolidone)-water mixture. In: Tokuyama M, Oppenheim I (eds) AIP Conf Proc 1518:300–303. doi:10.1063/1.4794586
48. Cerveny S, Alegria A, Colmenero J (2008) Broadband dielectric investigation on poly(vinyl pyrrolidone) and its water mixtures. J Chem Phys 128(4):044901. doi:10.1063/1.2822332
49. Busselez R, Arbe A, Cerveny S, Capponi S, Colmenero J, Frick B (2012) Component dynamics in polyvinylpyrrolidone concentrated aqueous solutions. J Chem Phys 137 (8):084902. doi:10.1063/1.4746020
50. Cerveny S, Colmenero J, Alegría A (2005) Dielectric investigation of the low-temperature water dynamics in the poly(vinyl methyl ether)/H2O system. Macromolecules 38 (16):7056–7063. doi:10.1021/ma050811t
51. Capponi S, Arbe A, Cerveny S, Busselez R, Frick B, Embs JP, Colmenero J (2011) Quasielastic neutron scattering study of hydrogen motions in an aqueous poly(vinyl methyl ether) solution. J Chem Phys 134(20):204906. doi:10.1063/1.3592560
52. Shinyashiki N, Shinohara M, Iwata Y, Goto T, Oyama M, Suzuki S, Yamamoto W, Yagihara S, Inoue T, Oyaizu S, Yamamoto S, Ngai KL, Capaccioli S (2008) The glass transition and dielectric secondary relaxation of fructose-water mixtures. J Phys Chem B 112(48):15470–15477. doi:10.1021/jp807038r
53. Cerveny S, Schwartz G, Bergman R, Swenson J (2004) Glass transition and relaxation processes in supercooled water. Phys Rev Lett 93(24):245702. doi:10.1103/PhysRevLett.93.245702
54. Cerveny S, Alegría Á, Colmenero J (2008) Universal features of water dynamics in solutions of hydrophilic polymers, biopolymers, and small glass-forming materials. Phys Rev E 77 (3):031803. doi:10.1103/PhysRevE.77.031803
55. Shinyashiki N, Sudo S, Yagihara S, Spanoudaki A, Kyritsis A, Pissis P (2007) Relaxation processes of water in the liquid to glassy states of water mixtures studied by broadband dielectric spectroscopy. J Phys Condens Matter 19(20):205113 (205112 pp). 10.1088/0953-8984/19/20/205113

56. Capaccioli S, Ngai KL, Shinyashiki N (2007) The Johari-Goldstein beta-relaxation of water. J Phys Chem B 111(28):8197–8209. doi:10.1021/jp071857m
57. Ngai KL, Capaccioli S, Paciaroni A (2013) Nature of the water specific relaxation in hydrated proteins and aqueous mixtures. Chem Phys 424:37–44. doi:10.1016/j.chemphys.2013.05.018
58. Shinyashiki N, Sudo S, Abe W, Yagihara S (1998) Shape of dielectric relaxation curves of ethylene glycol oligomer water mixtures. J Chem Phys 109(22):9843–9847. doi:10.1063/1.477653
59. Shinyashiki N, Yagihara S, Arita I, Mashimo S (1998) Dynamics of water in a polymer matrix studied by a microwave dielectric measurement. J Phys Chem B 102(17):3249–3251. doi:10.1021/Jp9729627
60. Shinyashiki N, Yagihara S (1999) Comparison of dielectric relaxations of water mixtures of poly(vinylpyrrolidone) and 1-vinyl-2-pyrrolidinone. J Phys Chem B 103(21):4481–4484. doi:10.1021/jp983800t
61. Sudo S, Shinyashiki N, Kitsuki Y, Yagihara S (2002) Dielectric relaxation time and relaxation time distribution of alcohol–water mixtures. J Phys Chem A 106(3):458–464. doi:10.1021/jp013117y
62. Hayashi Y, Shinyashiki N, Yagihara S (2002) Dynamical structure of water around biopolymers investigated by microwave dielectric measurements using time domain reflectometry method. J Non-Cryst Solids 305(1–3):328–332. doi: 10.1016/S0022-3093(02)01113-4. Pii S0022-3093(02)01113-4
63. Swenson J, Jansson H, Bergman R (2006) Relaxation processes in supercooled confined water and implications for protein dynamics. Phys Rev Lett 96(24):247802. doi:10.1103/PhysRevLett.96.247802
64. Bergman R, Swenson J (2000) Dynamics of supercooled water in confined geometry. Nature 403(6767):283–286. doi:10.1038/35002027
65. Bergman R, Swenson J, Borjesson L, Jacobsson P (2000) Dielectric study of supercooled 2D water in a vermiculite clay. J Chem Phys 113(1):357–363. doi:10.1063/1.481800
66. Jansson H, Swenson J (2003) Dynamics of water in molecular sieves by dielectric spectroscopy. Eur Phys J E Soft Matter 12(Suppl 1):S51–S54. doi:10.1140/epjed/e2003-01-013-5
67. Cammarata M, Levantino M, Cupane A, Longo A, Martorana A, Bruni F (2003) Structure and dynamics of water confined in silica hydrogels: X-ray scattering and dielectric spectroscopy studies. Eur Phys J E Soft Matter 12(Suppl 1):S63–S66. doi:10.1140/epjed/e2003-01-016-2
68. Cupane A, Levantino M, Santangelo MG (2002) Near-infrared spectra of water confined in silica hydrogels in the temperature interval 365–5 K. J Phys Chem B 106(43):11323–11328. doi:10.1021/jp026117m
69. Schirò G, Cupane A, Pagnotta SE, Bruni F (2007) Dynamic properties of solvent confined in silica gels studied by broadband dielectric spectroscopy. J Non-Cryst Solids 353 (47–51):4546–4551. doi:10.1016/j.jnoncrysol.2007.02.083
70. Pathmanathan K, Johari GP (1990) Dielectric and conductivity relaxations in poly(hema) and of water in its hydrogel. J Polym Sci B Polym Phys 28(5):675–689. doi:10.1002/polb.1990.090280507
71. Pathmanathan K, Johari GP (1994) Relaxation and crystallization of water in a hydrogel. J Chem Soc Faraday Trans 90(8):1143–1148. doi:10.1039/FT9949001143
72. Johari GP (1996) Water's character from dielectric relaxation above its Tg. J Chem Phys 105 (16):7079. doi:10.1063/1.472509
73. Bosio L, Johari GP, Oumezzine M, Teixeira J (1992) X-ray and neutron scattering studies of the structure of water in a hydrogel. Chem Phys Lett 188(1–2):113–118. doi:10.1016/0009-2614(92)85098-U
74. Ansari A, Berendzen J, Braunstein D, Cowen BR, Frauenfelder H, Hong MK, Iben IE, Johnson JB, Ormos P, Sauke TB et al (1987) Rebinding and relaxation in the myoglobin pocket. Biophys Chem 26(2–3):337–355. doi:10.1016/0301-4622(87)80034-0

75. Fenimore PW, Frauenfelder H, McMahon BH, Parak FG (2002) Slaving: solvent fluctuations dominate protein dynamics and functions. Proc Natl Acad Sci U S A 99(25):16047–16051. doi:10.1073/pnas.212637899
76. Wolf M, Gulich R, Lunkenheimer P, Loidl A (2012) Relaxation dynamics of a protein solution investigated by dielectric spectroscopy. Biochim Biophys Acta 1824(5):723–730. doi:10.1016/j.bbapap.2012.02.008
77. Cametti C, Marchetti S, Gambi CM, Onori G (2011) Dielectric relaxation spectroscopy of lysozyme aqueous solutions: analysis of the delta-dispersion and the contribution of the hydration water. J Phys Chem B 115(21):7144–7153. doi:10.1021/jp2019389
78. Emmert S, Wolf M, Gulich R, Krohns S, Kastner S, Lunkenheimer P, Loidl A (2011) Electrode polarization effects in broadband dielectric spectroscopy. Eur Phys J B 83(2):157–165. doi:10.1140/epjb/e2011-20439-8
79. Panagopoulou A, Kyritsis A, Shinyashiki N, Pissis P (2012) Protein and water dynamics in bovine serum albumin–water mixtures over wide ranges of composition. J Phys Chem B 116(15):4593–4602. doi:10.1021/jp2105727
80. Panagopoulou A, Kyritsis A, Sabater ISR, Gomez Ribelles JL, Shinyashiki N, Pissis P (2011) Glass transition and dynamics in BSA-water mixtures over wide ranges of composition studied by thermal and dielectric techniques. Biochim Biophys Acta 1814(12):1984–1996. doi:10.1016/j.bbapap.2011.07.014
81. Panagopoulou A, Kyritsis A, Aravantinou A-M, Nanopoulos D, i Serra RS, Gómez Ribelles JL, Shinyashiki N, Pissis P (2011) Glass transition and dynamics in lysozyme–water mixtures over wide ranges of composition. Food Biophys 6(2):199–209. doi:10.1007/s11483-010-9201-0
82. Panagopoulou A, Kyritsis A, Vodina M, Pissis P (2013) Dynamics of uncrystallized water and protein in hydrated elastin studied by thermal and dielectric techniques. Biochim Biophys Acta 1834(6):977–988. doi:10.1016/j.bbapap.2013.03.015
83. Kawai K, Suzuki T, Oguni M (2006) Low-temperature glass transitions of quenched and annealed bovine serum albumin aqueous solutions. Biophys J 90(10):3732–3738. doi:10.1529/biophysj.105.075986
84. Shinyashiki N, Yamamoto W, Yokoyama A, Yoshinari T, Yagihara S, Kita R, Ngai KL, Capaccioli S (2009) Glass transitions in aqueous solutions of protein (bovine serum albumin). J Phys Chem B 113(43):14448–14456. doi:10.1021/jp905511w
85. Johari GP, Whalley E (1981) The dielectric properties of ice Ih in the range 272–133 K. J Chem Phys 75(3):1333–1340. doi:10.1063/1.442139
86. Ngai KL, Capaccioli S, Shinyashiki N (2008) The protein "glass" transition and the role of the solvent. J Phys Chem B 112(12):3826–3832. doi:10.1021/jp710462e
87. Ngai KL, Capaccioli S, Thayyil MS, Shinyashiki N (2010) Resolution of problems in soft matter dynamics by combining calorimetry and other spectroscopies. J Therm Anal Calorim 99(1):123–138. doi:10.1007/s10973-009-0500-y
88. Ngai KL, Capaccioli S, Shinyashiki N, Thayyil MS (2010) Recent progress in understanding relaxation in complex systems. J Non-Cryst Solids 356(11–17):535–541. doi:10.1016/j.jnoncrysol.2009.03.011
89. Capaccioli S, Ngai KL, Ancherbak S, Rolla PA, Shinyashiki N (2011) The role of primitive relaxation in the dynamics of aqueous mixtures, nano-confined water and hydrated proteins. J Non-Cryst Solids 357(2):641–654. doi:10.1016/j.jnoncrysol.2010.07.054
90. Ngai KL, Capaccioli S, Ancherbak S, Shinyashiki N (2011) Resolving the ambiguity of the dynamics of water and clarifying its role in hydrated proteins. Philos Mag 91(13–15):1809–1835. doi:10.1080/14786435.2010.523716
91. Fox TG (1956) Influence of diluent and of copolymer composition on the glass temperature of a polymer system. Bull Am Phys Soc 1:123–125
92. Rault J, Lucas A, Neffati R, Monleón Pradas M (1997) Thermal transitions in hydrogels of poly(ethyl acrylate)/poly(hydroxyethyl acrylate) interpenetrating networks. Macromolecules 30(25):7866–7873. doi:10.1021/ma970344i

Biorheological Aspect of Microcapsules 10

Naoko Tomita and Toshiaki Dobashi

Abstract

Microcapsules have attracted great interest in various fields including biological and medical science and technology. Since the function of microcapsules is closely related to their preparation procedure and the resulting structure and physicochemical properties of their wall membranes and cores, the relationships among these factors are important for microcapsule design. The wall membranes of microcapsules are often regarded to be in a nonequilibrium gel state. Therefore, the gel properties of wall membranes are the key to understanding such relationships. In this chapter, essential properties of microcapsules with a gel wall membrane are summarized, focusing on the rheological aspects of bio-related microcapsules. Some examples of systematic studies of microcapsules are also introduced.

Keywords

Microcapsule • Micro-rheological effect • Gel wall membrane • Biological application

N. Tomita
National Institute of Advanced Industrial Science and Technology, Energy Storage Materials Group, Energy Technology Research Institute, 16-1 Onogawa, Tsukuba, Ibaraki 305-8569, Japan

T. Dobashi (✉)
Division of Molecular Science, Faculty of Science and Technology, Gunma University, 1-5-1 Tenjin, 376-8515 Kiryu, Gunma, Japan
e-mail: dobashi@gunma-u.ac.jp

R. Kita, T. Dobashi (eds.), *Nano/Micro Science and Technology in Biorheology*,
DOI 10.1007/978-4-431-54886-7_10

10.1 Introduction

Microcapsules were first developed as functional recording and printing materials in the 1950s [1] but are now used in various fields such as biomedical, pharmaceutical, food, cosmetic, and agricultural engineering [2]. The size of these capsules is controlled by their preparation method and conditions, and capsules of nano- to millisizes are now easily available. For smaller capsules, the molecule-based bottom-up preparation method is more often used, whereas for larger capsules, the top-down preparation method is dominant for the encapsulation. Therefore, the preparation methods are very different between nanocapsules and micro-/millicapsules. In most procedures for preparing microcapsules, the first reaction occurs at the interface between different phases. For example, in encapsulation by interfacial polymerization and insolubilization, a capsule membrane is produced by a reaction of two chemical species, each in their phases coexisting at the interface [3]. If one of the chemical species is a macromolecule, some conformation change of the macromolecule can occur near the reaction point [4]. This "micro-rheological effect" on phase contact has recently been discussed for anisotropic gelation [5, 6]. The structure and physicochemical properties of the wall membrane of microcapsules are also determined by the preparation conditions. Here, we note that the wall membranes of most microcapsules for conventional use are in the gel state and that the polymer network is swollen by a large amount of aqueous medium. This is an important point when designing microcapsules, because the properties of a solid membrane are very different from those of a gel membrane. Since the gel state is typically produced when a kind of nonequilibrium state is quenched, the structure and properties of the wall membrane depend on the path of the thermodynamic state in the preparation process, i.e., often not determined only by the thermodynamic condition. Two important and sophisticated functions of microcapsules are to protect entrapped materials and to release them to the environment at the desired time. The trigger of the release is one of various stimuli such as changes in pressure, temperature, electric field, and pH, as well as abrasion and chemical potential differences [7–12]. The release mechanisms for practically used nano-/microcapsules are often explained as the disintegration of their wall membrane by stress or the flow of chemical agents against viscous resistance. Therefore, the release process can be regarded as a microscopic rheological process in nano- and micro-spaces. As a special use of microcapsules, a cell scaffold degradable by an enzyme both in vitro and in vivo is raised as an example [13]. Recently, Tanaka et al. have shown by a series of thermal measurements and in situ ATR-IR that intermediate water is indispensable for biocompatible materials such as scaffolds [14]. Therefore, even in such a case, the close relationship between surface characteristics and cell viability could be regarded as a process by interactions of the cell and scaffold through water.

Thus, rheology is closely related to various aspects of microcapsules, such as the preparation procedure for microcapsules, and their resulting characteristic structure, physicochemical properties, and functions in biological systems. The systematic research performed up to now is, however, limited to a few examples. In

Sect. 10.2, the essential properties of microcapsules are summarized from the rheological aspects, especially the preparation and functions such as the release of functional chemical agents from microcapsules. In Sect. 10.3, the relationship between different microcapsules prepared by conventional methods of interfacial polymerization and newly developed dialysis-induced cross-linking is introduced. In Sect. 10.4, recent trends in biological and medical applications of microcapsules are briefly described.

10.2 Microencapsulation and Microcapsule Characteristics

10.2.1 Rheological Effect on Microencapsulation

In microencapsulation by interfacial polycondensation in emulsion, as shown in Fig. 10.1, the size distribution of microcapsules is affected by that of water or oil droplets in emulsion. Therefore, it depends on the kind and amount of emulsifier and stirring rate. Without an emulsifier, the effect of stirring rate is more remarkable, and the size distribution curve is sharper and narrower when the stirring rate is higher [15]. At a constant stirring rate, the size distribution curve becomes narrower and sharper when the emulsifier concentration is lower. Above the threshold of the emulsifier concentration, the size distribution is constant [16]. When the polycondensation reaction rate for microencapsulation is high, the size distribution of microcapsules is narrow. In contrast, when the polycondensation rate is low, large water or oil droplets are produced by the coalescence of colliding droplets, resulting in a wide distribution of microcapsules [16].

Several methods are known to substantially produce monodisperse emulsions and therefore monodisperse microcapsules. Emulsification using Shirasu porous

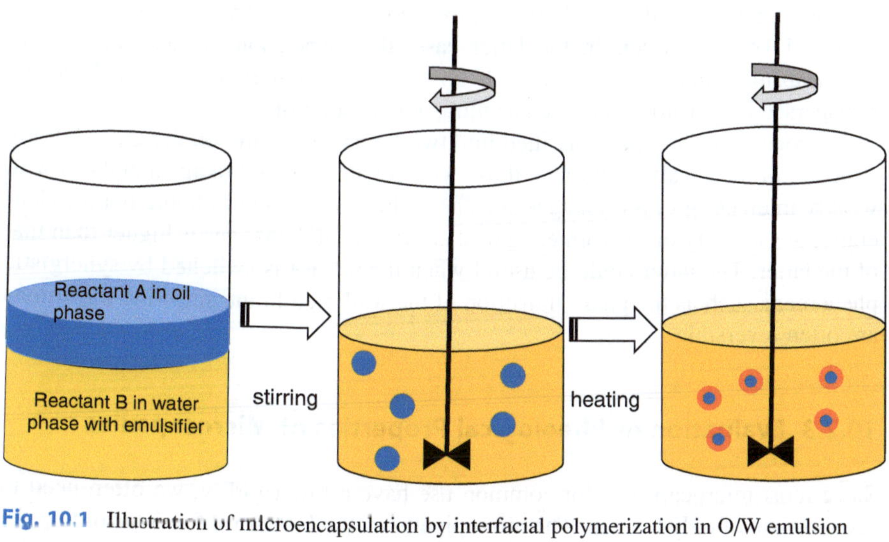

Fig. 10.1 Illustration of microencapsulation by interfacial polymerization in O/W emulsion

glass (SPG) is widely used for preparing microcapsules [17]; its use in bioscience and technology will be described in the next chapter. Monodisperse liquid droplets are also prepared by controlling the vibrational mode of the liquid column mechanically or electrically. By applying an alternating electric field on liquid flowing in air, the size of the resultant droplets is determined by the frequency of the electric field and surface tension. Monodisperse microcapsules with a standard deviation as low as 1.5 % were obtained by this method [18]. The coacervation method in microencapsulation utilizes the phase separation of polymer solutions, especially polyelectrolyte solutions. Droplets of the concentrated phase containing functional materials are fixed by different materials such as talc [$Mg_3Si_4O_{10}(OH)_2$] and extracted as a collection of microcapsules. The growth of liquid droplets in the process is induced by the chemical potential difference before and after the phase separation, and the growth rate depends on the diffusion constant of polymer molecules. Phase separation is conventionally classified into a nucleation and growth process and a spinodal decomposition process. The droplets produced by the former process have a large distribution because of the difference in the time of appearance of each nucleus; however, homogeneous and coherent droplet growth is realized in the latter process. Therefore, microencapsulation by means of spinodal decomposition can be used to design a preparation procedure for monodisperse microcapsules, and its optimum conditions are predicted by the free energy of the system [19].

10.2.2 Stability of Microcapsule Suspension

The stability of a suspension of microcapsules is mainly governed by the characteristics of their wall membrane. When the wall membrane has a net charge, the suspension is stable due to repulsive coulombic force. When it has no net charge, stability usually needs to be provided by a protective colloid due to the excluded volume effect. In the latter case, the suspension is quasi-stable, i.e., microcapsules finally aggregate with time, and the aggregates obtained can be redispersed by perturbations such as ultrasonic irradiation.

Microcapsules are also classified into two kinds depending on the cross-linking structure of their wall membrane: those with covalent cross-linkage and those with weakly interacting cross-linkages such as hydrogen and hydrophobic bonds. Generally, the stability of the former against environmental changes is higher than that of the latter. The latter could be useful when the release is switched by synergistic phenomena such as the phase transition of the wall membrane since the transitions are often reversible.

10.2.3 Evaluation of Rheological Properties of Microcapsules

Since most microcapsules for common use have a low rigidity, we often need to take account of only tension. The micropipette method is useful for determining the tension of a microcapsule wall membrane. In this method, a micromanipulator is operated with a micropipette and placed on a microcapsule to apply negative pressure, and the membrane tension is deduced from the observed length of the part withdrawn into the pipette, the radius of the capsule, and the curvature of the membrane. It was applied for estimating the membraneous tension of nylon microcapsules [20] and the deformability of erythrocytes [21]. The measurement of cell filterability using fabricated nickel mesh membranes is also useful for assessing cell deformability with high precision and has been used in studying the pathophysiology of various diseases [22]. When materials are released from microcapsules with a solid wall membrane owing to the breaking of the wall membrane by external pressure, the release features depend on microcapsule diameter and wall membrane thickness. Breaking stress increases with increasing microcapsule diameter at a constant membrane thickness, as well as with increasing membrane thickness at a constant microcapsule diameter [23].

The elastic modulus K of microcapsules in a suspension can be measured by a combination of ultrasonic velocity (V) and density (ρ) measurements. The typical frequency of ultrasonic waves used in laboratories is within 1–10 MHz, which corresponds to a wavelength range of 0.15–1.5 mm in aqueous solutions. Thus, the Rayleigh scattering approximation holds for microcapsules with a typical size of 1 μm, which is much smaller than the wavelength. Then, we can assume an additivity by volume to evaluate the adiabatic compressibility ($B_s = 1/K = 1/\rho V^2$) of the microcapsule membrane [24], which is the key parameter for the pressure sensitivity of microcapsules. As will be discussed in the next section using typical examples, the wall membrane of microcapsules is often in a swollen gel state, and the structure of the wall membrane could vary depending on the core-/wall-forming material weight ratio r [25]. Therefore, on the basis of this principle, the elastic modulus of each part of a microcapsule such as the swollen wall membrane and the inner core with a known volume fraction can be estimated [26]. The acoustic method using Brillouin scattering is used to measure microcapsule elasticity in the high frequency range. Relaxation dynamics in poly(lactide-*co*-glycolide) microcapsules with perfluorooctyl bromide was measured to be as high as 18 GHz [27].

When the inner core of microcapsules becomes nonvolatile on desiccation, the viscoelasticity of the microcapsules can be assessed by measuring the viscoelasticity of filter paper or a glass fiber blade on which microcapsule suspensions are spread [28].

Single-capsule measurement using the shell mechanics theory has been reviewed in several books [29]. The measurement of deformation of polyelectrolyte multilayer capsules has been demonstrated by this method [30]. The recent development of scanning microscopy now enables us to obtain the spatial distribution of viscoelasticity. Atomic force microscopy (AFM) developed originally for hard crystals [31] is also useful for soft hydrogels in liquid to determine the elasticity of microcapsules directly. The surface of microcapsules used as a bio-artificial pancreas was scanned, and a three-dimensional image was observed in a relatively early stage of development of AFM [32]. Recently, by combining AFM with fluorescence microscopy, the release of materials from the core of microcapsules with a polyelectrolyte multilayer membrane and plastic deformation was simultaneously measured [33].

Optical coherence microscopy (OCM) is a noninvasive noncontact in vivo imaging technique used to observe regions inside our body such as skin layers and the brain, using near-infrared light sources [34]. Using this technique, we can also obtain the elasticity distribution of a microcapsule surface. Optical coherence elastography (OCE) is one of the techniques of OCM and uses optical coherence tomography to measure the strain distribution by speckle tracking [35].

10.3 Relationship Among Preparation, Structure, Properties, and Function of Microcapsule

A large number of microencapsulation methods such as physical methods of air-suspension coating, coacervation-phase separation, centrifugal extrusion, and pan coating spray-drying and chemical methods of solvent evaporation, interfacial polymerization, and in situ polymerization have been developed. Even if limited to the use of only one of these methods, there will still be a variety of microcapsules with different structures and characteristics. Since the function of microcapsules and their mechanism depend on the structure and characteristics of microcapsules, the relationship among the preparation, structure, properties, and function of each type of microcapsule needs to be examined. Here, we consider typical conventional microcapsules prepared by interfacial polymerization in O/W or W/O emulsion: microcapsules with typical but very different characteristics, poly(urea-urethane) (PUU) microcapsules, and poly(L-lysine-*alt*-terephthalic acid) (PPL) microcapsules. A series of systematic studies from preparation to function have been performed on these microcapsules. A PUU wall membrane consists of a covalently cross-linked network with no charge. A PPL wall membrane consists of oligomers cross-linked by hydrogen bonds with a net charge depending on pH. In both phases, the reactants are low-molecular-weight compounds. When one of the reactants is composed of macromolecules, completely different phenomena occur, i.e., we can obtain microcapsule membranes with a unique anisotropic structure, since the conformation of macromolecules is modified on the contact surface, through which only low-molecular-weight compounds or ions except macromolecules diffuse, as shown in Fig. 10.2. This type of microencapsulation, i.e., dialysis-induced gelation on the surface, is also introduced.

10.3.1 Microcapsules Prepared by Interfacial Polymerization

1. Poly(urea-urethane) microcapsules

Microcapsules with poly(urea-urethane) (PUU) wall membranes are widely used as temperature-sensitive recording materials. The rate of thermal response is higher

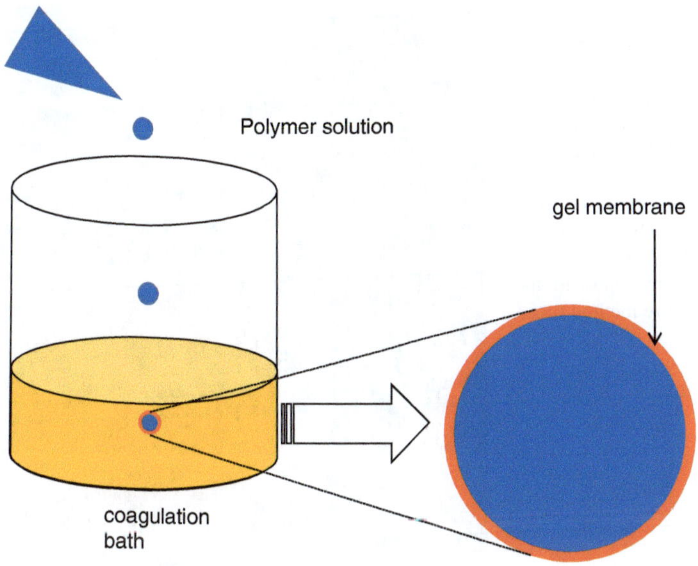

Fig. 10.2 Illustration of microencapsulation by insolubilization

when the membrane thickness is smaller and the surface area is larger. Considering typical microcapsules with a diameter of 1 μm and a membrane thickness of several tens of nanometers, the rate of temperature change is estimated to be 150 K/ms. Thus, the temperature of the membrane could be changed instantaneously above the glass transition temperature of PUU from room temperature. This allows us to control the release of functional materials such as pigments and dyes from a microcapsule core by changing temperature, resulting in a change in the wall membrane permeation coefficient. Since PUU is biocompatible, microcapsules with a PUU wall membrane are also expected to be used as biomaterials such as drug delivery carriers [36].

They can be synthesized by interfacial polymerization in O/W emulsion, e.g., triisocyanate monomers in the organic phase react with immersing water at the interface of the oil phase and water phase to produce a microcapsule membrane. On the basis of the expected release rate of chemical agents such as pigments, an appropriate set of bulky stable organic solvents has often been used in industry.

Figure 10.3 shows cross-sectional views of microcapsules for systems with different volume fractions of the wall component and core component, as observed by scanning electron microscopy by the freeze-fracture method. The surface of the microcapsule membrane is clear for (a), and the membrane seems to be melted and very soft for (b) suggesting a swollen gel state. A noninvasive method of determining the structure of microcapsules in suspensions is a scattering method and a combination of a scattering method with other techniques [37]. Microcapsules with diameters in the order of 0.1 μm were investigated by the synchrotron small-angle X-ray scattering (SAXS) method and a combination of the SAXS and light

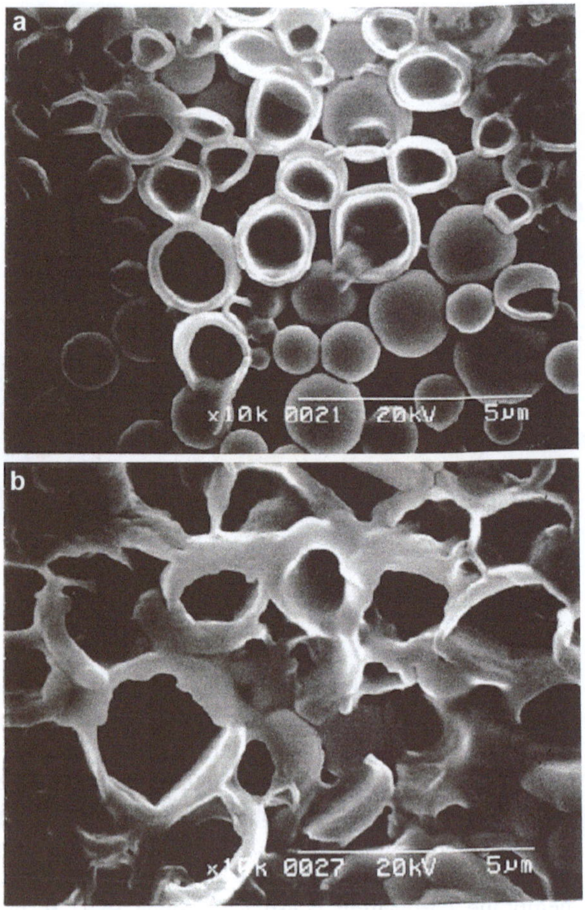

Fig. 10.3 (From Ref. [25a], Fig. 3) Cross-sectional views of PUU microcapsules with dioctyl phthalate (DOP) core prepared with different ratios of DOP and that of the wall-forming materials of triisocyanate monomer: (**a**) $\phi = 1.64$ and (**b**) 7.81, as observed by scanning electron microscopy with the freeze-fracture method for system (**a**)

scattering [38] methods, and those with diameters in the order of 1 μm and wall membrane thicknesses in the order of 0.1 μm were measured by a single-particle light scattering method [25a]. The swelling ratio of microcapsules consisting of a dioctyl phthalate core and an outer PUU membrane measured using single-particle light scattering and a freeze-fracture method in combination with electron microscopy shows that the swelling ratio of the wall membrane is as large as 1.2 [26b]. According to the theory of the swelling of a polymer network [39], the swelling ratio is related to the difference in the interaction parameter χ or the solubility parameter σ between the polymer and the solvent. According to the Fedors method [40], the σ values of a triisocyanate monomer and dioctyl phthalate are calculated to be 12.0 and 9.6 $(\mathrm{cal/cm^3})^{1/2}$, respectively. From the large difference in σ, the swelling ratio of 1.2, as determined by the light scattering method, seems to be too large. The microspheres prepared using only triisocyanate monomers did not swell so much in dioctyl phthalate. This apparent inconsistency might be related to the mechanism of interfacial polymerization and suggests that the presence of organic solvents affects the structure and physicochemical properties of microcapsule membranes. The resultant effects could be generalized to explain the mechanism of interfacial polymerization. On the other hand, such effects can be utilized to control the microcapsule structure and ultimately its function. Thus, it is of great importance to clarify the correlation of chemical cross-linking and molecular interactions between the wall-forming materials and the dispersing media in the mechanism of microencapsulation. It was also observed that the distribution of the core materials is not necessarily homogeneous. An elemental specific imaging technique in transmission electron microscopy showed that in microcapsules consisting of a PUU membrane and a triphenyl phosphate (TPP) core, TPP was found to be more concentrated in the vicinity of the outermost and innermost layers of the wall membrane [41], indicating the cross-linking density of the wall membrane also being inhomogeneous.

The glass transition temperature of a PUU microcapsule membrane depends on the core materials as a result of the plasticizing effect, when the solubility parameter of the core materials approaches that of the PUU network [25b]. A dynamic viscoelasticity study and scanning electron microscopy showed that PUU microcapsules containing tricresyl phosphate (TCP) did not form a core-shell structure but formed a swollen microsphere when the overall weight ratio of the core to the wall-forming material $r < r_{\mathrm{critical}} = 0.5$. The difference in the threshold ratio was reasonably explained by specific interactions such as the hydrogen bonding of the polymer network and the core materials through dipole moment and cohesive energy [25b]. With a core consisting of a compatible solvent, a homogeneous elastic membrane is formed by swelling, whereas with a core consisting of a less compatible solvent, the membrane becomes brittle and heterogeneous owing to phase separation. The rate of dye release from the microcapsules could be varied by more than 20-fold by varying the affinity between the core and the wall-forming materials [42].

The structural change of the wall membrane affects the mechanical properties such as the compressibility of microcapsules. An application of the ultrasonic method to suspensions of microcapsules consisting of a TCP core and an outer PUU membrane with a known volume fraction was performed, and the compressibility of the wall membrane was determined as a function of r [26].

Most of the release behaviors of microcapsules observed in experiments have been analyzed by kinetic theories based on Fick's law with the assumption that the release rate is proportional to the concentration gradient of solutes, i.e., the ideal gas approximation [2, 43]. Microcapsules are, however, considered to be more complex systems; solutes such as staining materials could be distributed in the core and the membrane of microcapsules as well as in the dispersing medium with different fractions even in the equilibrium state. Furthermore, the compositions of each domain could change in the time course of the release. Thus, it is necessary to employ more strict treatment to understand the release mechanism and to characterize the release properties of such complex systems clearly. The theoretical results indicated that the release curve is expressed by the same exponential form as that for the theoretical one under ideal gas approximation but the time constant is related

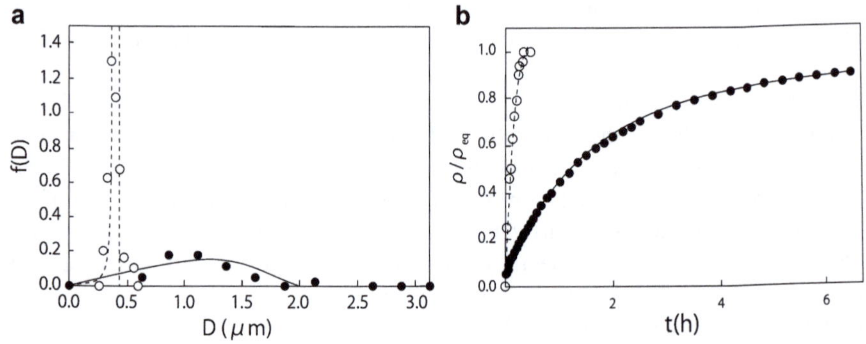

Fig. 10.4 (From Ref. [46], Figs. 3 and 4) (**a**) Diameter distribution curves for microcapsules consisting of PUU membrane and DOP core containing dye with narrow size distribution (*open circle*) and those with wide size distribution (*closed circle*) and (**b**) corresponding dye-release curves during exchange of medium from water to methanol. ρ and ρ^{eq} denote dye concentration at each time and in the equilibrium state, respectively. The *solid curve* of (**a**) is that calculated from the *dash-dotted line* of (**a**) and the release profile of (**b**)

to the concentration derivatives of the chemical potential of solutes [44]. The difference should be clear when the concentration derivatives of the chemical potential vary significantly near the critical solution point of the core solution.

All of the above theoretical analyses are valid for a collection of microcapsules with a monodisperse size distribution. In real applications, however, the release rate is not precisely controlled because of the size distribution of synthetic microcapsules. Since relaxation time depends on the size of the microcapsules, "many relaxation times" may exist for a system of polydisperse microcapsules. On the basis of an analogy to relaxation phenomena with multiple characteristic times [45], theoretical arguments showed that the release curve can be expressed with a stretched exponential function with an effective time constant. Thus, the release curve for polydisperse microcapsules can be estimated from the size distribution and release behavior of monodisperse microcapsules, as shown in Fig. 10.4 [46, 47].

Since the PUU wall membrane has no charge, the stability of PUU microcapsule suspensions can be realized owing to the entropic effect of protective colloids such as poly(vinyl alcohol-vinyl acetate). For higher stability, hydrophilic polymers are often attached covalently to the membrane. Poly(ethylene glycol)s (PEGs) are one of the suitable candidates for this purpose [48].

2. Poly(L-lysine-*alt*-terephthalic acid) microcapsule

Poly(L-lysine-*alt*-terephthalic acid) (PPL) microcapsules are prepared by copolymerizing L-lysine and terephthalic acid on the surface of aqueous droplets in W/O emulsion and replacing the oil phase with an aqueous solution, systematic studies of which were performed by Kondo, Makino, and Ohshima [49–52]. Since the component oligomers of PPL microcapsule membranes have amino groups and

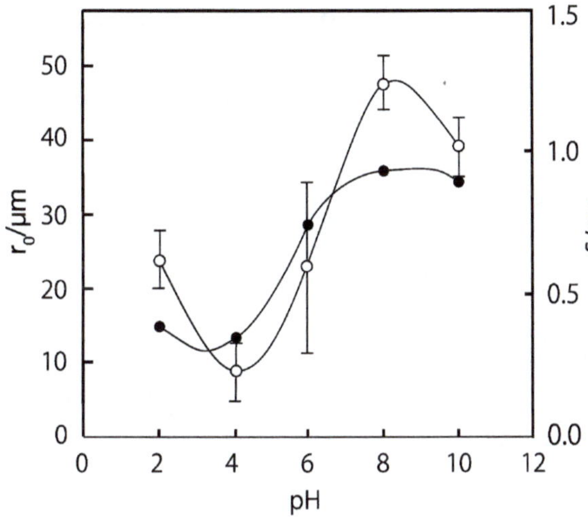

Fig. 10.5 (From Ref. [51], Fig. 7) Outer radius r_o (*closed circle*) and membrane thickness δ (*open circle*) determined by single-particle light scattering of PPL microcapsules. Error bars denote the standard deviations of δ. The standard deviations of r_o are smaller than the size of the circles

carboxylic acid groups at their terminals and many carboxylic acid groups along their skeleton [50], the polymer molecules interact with each other by hydrogen bonds, resulting in a three-dimensional polymer network to produce a microcapsule wall membrane. The electric charge of PPL microcapsule membranes changes in response to the pH of the medium, since the amino groups are protonated and carboxylic acid groups dissociate in neutral and alkaline media, while those near the isoelectric point between pH 2.3 and 2.6 charges are canceled out. The results of a single-particle light scattering method showed that PPL microcapsules with a thin membrane ($\sigma \sim 0.2$–1.2 μm) containing a buffer solution core undergo a volume phase transition as a function of medium pH [51]. Then, both the outer diameter of the microcapsules and the thickness of the wall membrane change, as shown in Fig. 10.5, resulting in the volume of PPL microcapsule membranes exhibiting marked changes in a narrow pH range between 5.0 and 6.0; thus, the size of PPL microcapsules is maximum at approximately pH 8 and minimum at approximately pH 4.5, where Makino et al. suggested the formation of a certain ion complex structure [52].

Gels that exhibit an immediate response, such as swelling or shrinking, to changes in the environment are materials of great potential and application [53]. For example, with such gels, chemo-mechanical systems, responsive artificial actuators, and carriers of drug delivery systems are expected to be developed further [54]. The characteristic time of swelling or shrinking of bulk gels is proportional to the square of the equilibrium size of the gel [55]. The behavior intrinsic to the gel could also be observed in a thin gel. Dynamic processes in the swelling and shrinking of a hydrogel membrane of PPL microcapsules are very different from those of bulk gels for the following reasons. (1) The microcapsule core serves as a proton reservoir in the swelling process (Fig. 10.6a) and as a hydroxyl anion reservoir in the shrinking process. This unique characteristic affects the transfer of the charge from the outside to the inside of the membrane. (2) The

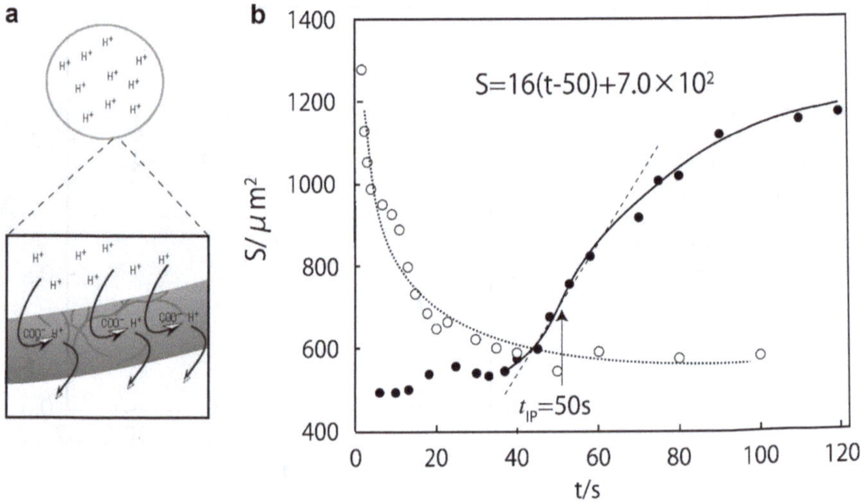

Fig. 10.6 (From Ref. [56], Figs. 1 and 2) (**a**) Illustration of initial swelling of volume phase transition of polyelectrolyte microcapsule membrane. (**b**) Time courses of the cross-sectional area *S* of PPL microcapsules during the swelling induced by a pH jump from 4.5 to 7 (*closed circle*) and during the shrinking induced by a pH jump from 7 to 4.5 (*open circle*). The *solid* and *dotted lines* are those calculated using theoretical equations, for the swelling and shrinking. t_{IP} is the time of the inflection point during the swelling

microcapsule membrane is extremely thin. Thus, collective diffusion is not time consuming, and the ion concentration gradient in the membrane is negligible. Theoretical studies showed that the dynamic processes of the swelling and shrinking of a PPL microcapsule hydrogel membrane can be characterized by two time constants related to the thin wall membrane and the large proton reservoir of the core [56]. Experimental swelling and shrinking processes are well explained by theoretical equations, as shown in Fig. 10.6b.

The viscoelastic properties of PPL microcapsules were studied as a function of medium pH, as shown in Fig. 10.7 [57]. An abrupt increase in the apparent relative viscosity of PPL microcapsule suspensions was observed in the pH range between 4.0 and 7.0 due to the increase in total particle volume concentration resulting from a sudden increase in microcapsule size. Above pH 7.0, the relative viscosity decreases, indicating deformable microcapsules. The adiabatic compressibility of PPL microcapsule membranes determined by ultrasonic measurement is lowest at pH 4.0 and increases markedly when the pH shifts away from 4.0. In contrast, the membrane density is highest at pH 4.0 and decreases as the medium pH shifts to either side of this value, implying that the microcapsules behave as compact particles at approximately pH 4.0 and as loose particles at high pH. These properties are a result of the structural change induced by the pH change.

For the pharmaceutical use of the above microcapsules, it is important to estimate the diffusion coefficient of both anionic and cationic chemical agents.

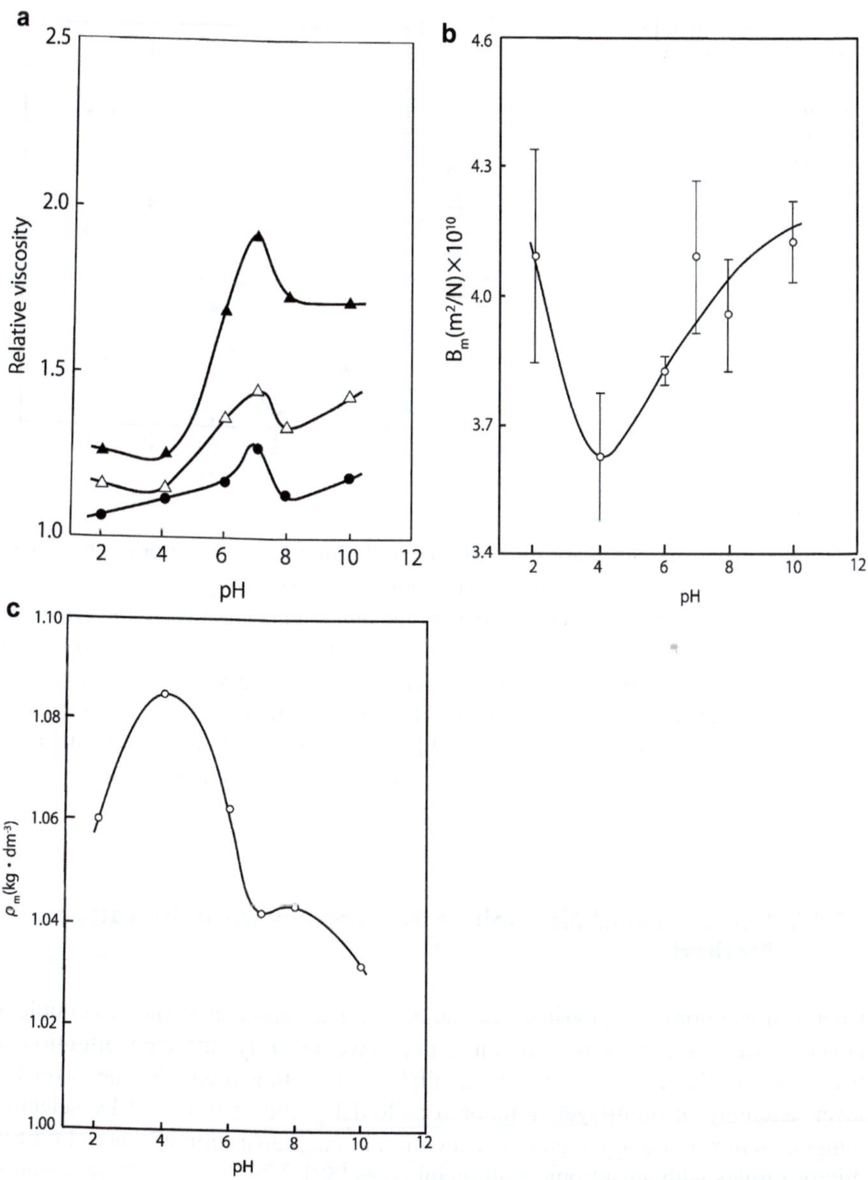

Fig. 10.7 (From Ref. [57], Figs. 2, 3, and 4) Effects of medium pH on (**a**) apparent relative viscosity of PPL microcapsule suspension, (**b**) adiabatic compressibility, and (**c**) density of PPL microcapsule membrane, at an ionic strength of 0.154

From the available data for the permeability coefficients of 5-sulfosalicylic acid as the anion and phenyltrimethylammonium chloride as the cation with different ionic strengths of the dispersing medium, the permeability coefficients were determined. The diffusion coefficients of the materials in the microcapsule membrane

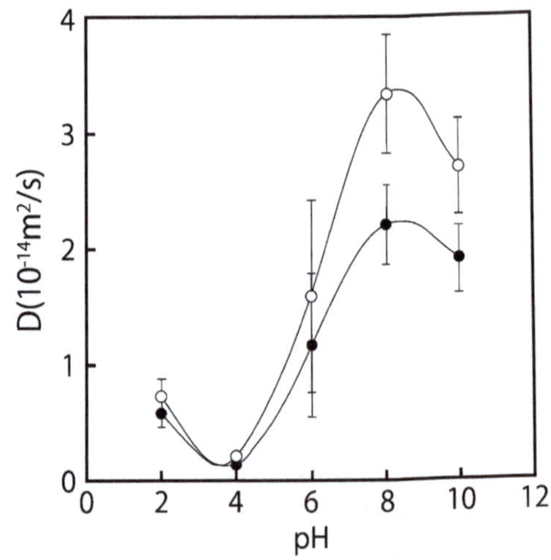

Fig. 10.8 (From Ref. [51], Fig. 8) Diffusion coefficients *D* of 5-sulfosalicylic acid (*closed circle*) and phenyltrimethylammonium chloride (*open circle*) in PPL microcapsule membrane

calculated from the permeability coefficients divided by the membrane thickness determined by single-particle light scattering are shown in Fig. 10.8 [51]. The pH dependence of the diffusion coefficients of the anion and cation is characterized by the same marked change at approximately pH 6 as the thickness and outer diameter or the volume of the microcapsule membrane. Thus, it is suggested that, at high pH, a swollen gel containing a large amount of dispersing medium in its microcapsule membrane has a large diffusion coefficient. The relatively small diffusion coefficients of the anion at high pH may be attributed to the electrostatic repulsion with COO^- in the microcapsule membrane.

10.3.2 Microcapsule/Microsphere Prepared by Insolubilization Method

Chemical reactions and physical associations of macromolecular species couple to generate various mesoscale structures that have recently attracted attention in nano-/microtechnology. Among the available fabrication methods, the layer-by-layer assembly of multilayer films onto colloidal particles followed by selective template removal has attracted much attention in the preparation of a novel type of microcapsules with anisotropic wall membranes [58].

Recently, another method of preparing anisotropic microcapsules has been studied, simple dialysis, which helps polymers form a unique ordered structure with a refractive index gradient. In the dialysis of Curdlan dissolved in aqueous sodium hydroxide into aqueous calcium chloride, the outflux of hydroxide anions changes the conformation of Curdlan molecules from a random coil to a triple helix owing to the pH change, and the influx of calcium cations cross-links helical Curdlan molecules intermolecularly, resulting in a cylindrical gel having a layer with significant birefringence and an amorphous layer with no birefringence in

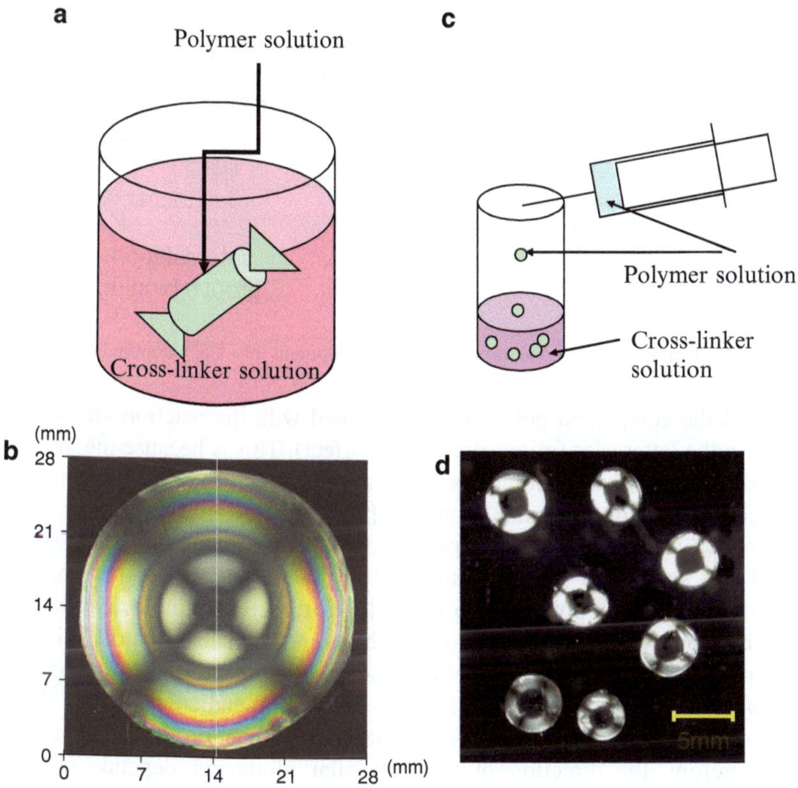

Fig. 10.9 (From Ref. [59], Figs. 1a and 2a and Ref. [60], Fig. 1b) (**a**) Schematic of dialysis-induced gelation of Curdlan solution, (**b**) top view of the slice of the Curdlan gel observed under crossed nicols, (**c**) schematic of preparation of Curdlan gel beads, and (**d**) cross-section of Curdlan gel beads observed under crossed nicols

alternating layers, as shown in Fig. 10.9a, b [59, 60]. This method can be used to prepare spherical particles by dripping an aliquot of Curdlan solution in calcium chloride aqueous solution, as shown in Fig. 10.9c, d. Although there is no separation between polymer solutions and the medium, a thin gel membrane produced with the first contact of the polymer solution and the medium at the contact surface can serve as a dialysis tube, resulting in dialysis-induced gelation [61]. This gelation for preparing anisotropic particles can be utilized for various combinations of both biological and synthetic rodlike or semiflexible polymer molecules and medium containing appropriate cross-linker molecules [62–65]. When polymer molecules move toward the periphery of the particle to form a gel membrane, while the central part becomes a dilute polymer solution, as often seen in the process, a core-shell structure is obtained, resulting in microcapsules.

The largest difference in condition between conventional interfacial polymerization explained in Sect. 10.3.1 and the present method is that a conformational

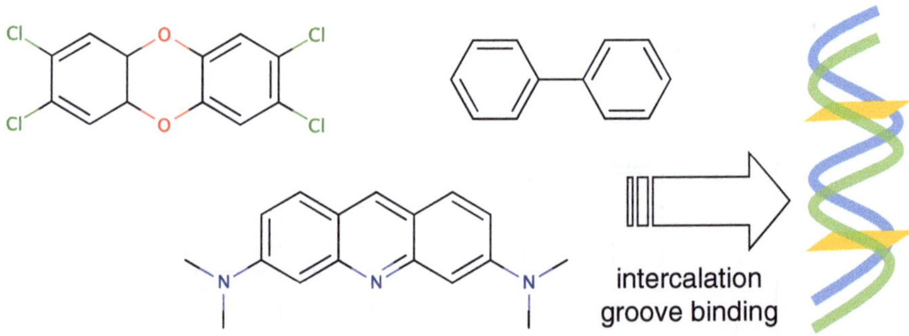

Fig. 10.10 Illustration of intercalation of a planar compound into DNA double helix

change of the component polymers is associated with the reaction on the contact surface in the latter case (micro-rheological effect). This is because the gradient of the chemical potential of polymers along the polymer chain anchored to the gel phase could result in molecular alignment [5, 6].

The structure of the anisotropic gel prepared using a dialysis tube or a rectangular cell was studied by X-ray scattering, light scattering, and visual observation under a pair of circular polarizers [6, 63–65]. The direction of molecular orientation determined by experiments was shown to be perpendicular to the diffusion direction in most cases [63, 64]. On the other hand, in some special cases, the orientation direction was partly circumferential and partly perpendicular to the diffusion direction, which could be attributed to another interaction force in gels [5, 6, 65]. Therefore, the direction of the molecular alignment depends on the gel formation mechanism in the system. Although the structure of the wall membrane of microcapsules prepared by this method has not yet been investigated, a large birefringence similar to that observed in the gels using a dialysis tube has been observed, suggesting an anisotropic structure similar to that of the latter gel.

The unique optical properties of a microcapsule/microsphere could be used as a probe of chemical agents in the environment. It is challenging to develop materials consisting of biomolecules without damaging their unique bio-functions, since materials toxic to them can selectively adsorb onto them. As a typical example, carcinogen and environmental hormones with planar aromatic groups intercalate into DNA double helices, thereby inhibiting the interpretation of hereditary codes, as shown in Fig. 10.10. This specific property of DNA is used for the most efficient adsorbents of such specific toxic agents selectively. For this application, DNA must be in the water-insoluble state [66, 67]. A water-insoluble anisotropic gel prepared from double-helical DNA by the simple dripping method is considered to be an adsorbent with which toxic agents can be extracted effectively (an advantageous gel property) and detected even when agents produce no color in the visible wavelength range (an advantageous birefringent property). A theoretical analysis of the adsorption of acridine orange, an intercalator (model carcinogen), showed that the adsorption dynamics was explained quantitatively [68].

The alternating structure of the above Curdlan microcapsule/microsphere could be utilized for a drug delivery carrier. Effective drugs often produce harmful side effects. In such a case, sustained releases of various types such as stepwise release and constant release are required; however, common exponential release is not preferred [69]. When a drug is contained in microcapsules with a structure gradient or the core is in two-phase equilibrium, the drug release profile could deviate from the conventional exponential behavior and tends toward constant release that could provide us with an efficient drug release. In the case that the core is in two-phase equilibrium, a typical release curve and a polynomial theoretical equation are shown for arginate microcapsules [70].

10.4 Recent Trends in Biological and Medical Application

Biomedical and pharmaceutical applications of microcapsules are one of the trends in recent decades, as shown in Fig. 10.11. In these applications, various nano-/microcapsules are used for all clinical processes of examination, diagnosis, medication, healing, and other biochemical support systems. That is, nano-/microcapsules are used for bio-imaging and biosensing in the examination and diagnosis processes for drug, gene, and cell carriers, and embolization materials in medication and healing processes, as well as for cell culture scaffolds, enzyme immobilization, and affinity chromatography in biochemical support systems [17, 71, 72]. One of the recently developed nano-/microcapsules has multifunctions such as a cell culture scaffold/cell carrier [73] and bio-imaging particles/embolization materials [74]. Furthermore, all-in-one systems such as those with functions of detecting malignant cells, real-time imaging, releasing drugs, and monitoring the effect of the therapy have been developed [75]. In the field of ultrasound in

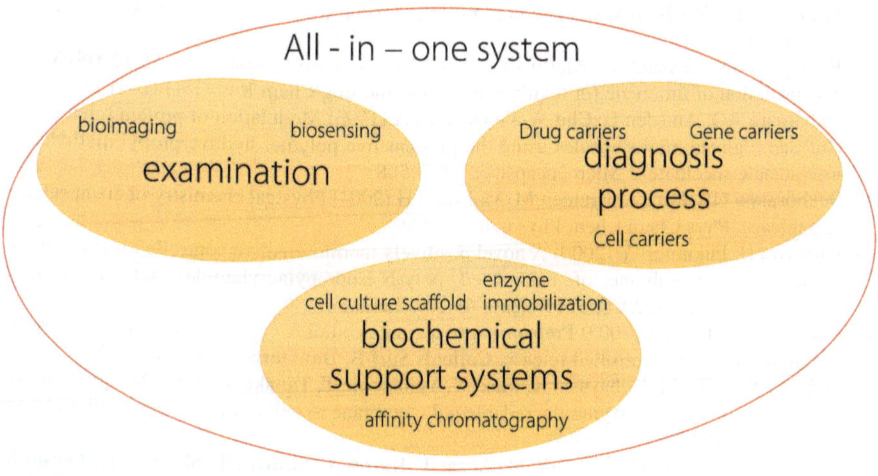

Fig. 10.11 Multifunctional microcapsules for various uses

medicine, microcapsules with a bubble core (microbubbles) are typical examples for illustrating multifunctional effects [76]. These microcapsules have acoustic impedance very different from that of aqueous media such as blood. Therefore, microcapsules can be used as an ultrasonic contrast medium without any adverse effects such as cardiovascular disorder. When an ultrasonic wave is applied to microbubbles, the difference in acoustic impedance between blood and air bubbles makes the bubbles vibrate in the same phase, resulting in the microcapsules being trapped, broken, and transported by ultrasonic waves, and thus they could be used not only as a drug delivery system but also for the removal of thrombosis and control of blood flow [77].

References

1. (a) Nixon JR (1976) Microencapsulation. Marcel Dekker, New York; (b) Kondo T (1979) Microcapsule processing and technology. Marcel Dekker, New York
2. (a) Kondo T (2001) Microcapsules: their science and technology Part I. Various preparation methods. J Oleo Sci 50:1–11; (b) Kondo T (2001) Microcapsules: their science and technology Part II. Properties. J Oleo Sci 50:81–95; (c) Kondo T (2001) Microcapsules: their science and technology Part III. Industrial, medical, and pharmaceutical applications. J Oleo Sci 50:143–152
3. Kondo T (1978) Microcapsules: their preparation and properties. In: Matijevic E (ed) Surface and colloid science, vol 10. Plenum, New York, pp 1–43
4. Dobashi T, Tomita N, Maki Y, Chang CP, Yamamoto T (2011) An analysis of anisotropic gel forming process of chitosan. Carbohydr Polym 84:709–712
5. Yamamoto T (2011) Rheology of polymer chain in the interface between sol and gel phases. Abstract for 59th rheology conference, Kiryu (Japan), pp 6–7 (in Japanese)
6. Maki Y, Furusawa K, Yasuraoka S, Okamura H, Hosoya N, Sunaga M, Dobashi T, Sugimoto Y, Wakabayashi K (2014) Universality and specificity in molecular orientation in anisotropic diffusion method. Carbohydr Polym 108:118–126
7. Yun J, Im JS, Lee YS, Bae TS, Lim YM, Kim HI (2010) pH and electro-responsive release behavior of MWCNT/PVA/PAAc composite microcapsules. Colloids Surf A Physicochem Eng Asp 368:23–30
8. Rodrigues SN, Fernandes I, Martins IM, Mata VG, Barreiro F, Rodrigues AE (2008) Microencapsulation of limonene for textile application. Ind Eng Chem Res 47:4142–4147
9. Okhamafe AO, Amsden B, Chu W, Goosen MFA (1996) Modulation of protein release from chitosan-alginate microcapsules using the pH-sensitive polymer hydroxypropyl methylcellulose acetate succinate. J Microencapsul 13:497–508
10. Sukhorukov GB, Fery A, Brumen M, Möhwald H (2004) Physical chemistry of encapsulation and release. Phys Chem Chem Phys 6:4078–4089
11. Ichikawa H, Fukumori Y (2000) A novel positively thermosensitive controlled release microcapsule with membrane of nano-sized poly(N-isopropylacrylamide) gel dispersed in ethylcellulose matrix. J Control Release 63:107–119
12. Chang CP, Dobashi T (2003) Preparation of alginate complex capsules containing eucalyptus essential oil and its controlled release. Colloids Surf B: Biointerfaces 32:257–262
13. Dobashi T, Koike M, Kobayashi K, Maki Y, Yamamoto T, Tanaka S (2009) An application of microcapsules having enzyme-degradable gel membrane to cell culture. Progr Colloid Polym Sci 136:149–153
14. Tanaka M, Motomura T, Kawada M, Anzai T, Kasori Y, Shiroya T, Shimura K, Onishi M, Mochizuki A (2000) Blood compatible aspects of poly(2-methoxyethylacrylate) (PMEA) relationship between protein adsorption and platelet adhesion on PMEA surface. Biomaterials 21:1471–1481

15. Konishi M, Fukuhara N, Kondo T (1969) Studies on microcapsules II Preparation of polyphthalamide microcapsules. Chem Pharm Bull 17:804–809
16. Shigeri Y, Konishi M, Kondo T, Shiba M, Tomioka S (1970) Studies on microcapsules VI Effect of variation in polymerization condition on microcapsule size. Can J Chem 48:2047–2051
17. Ma G, Su ZG (2013) Microspheres and microcapsules in biotechnology: design, preparation and applications. Pan Stanford Publishing, Singapore
18. Sato M, Takahashi H, Awazu M, Ohshima T (1999) Production of ultra-uniformly-sized silica particles by applying ac superimpose on dc voltage. J Electrostat 46:171–176
19. Narita T, Hosoya E, Yamamoto T, Dobashi T (2003) Gibbs free energy expression for the system polystyrene in methylcyclohexane and its application to microencapsulation. Langmuir 19:5240–5245
20. Jay AW, Edwards MA (1968) Mechanical properties of semipermeable microcapsules. Can J Physiol Pharmacol 46:731–737
21. Leblond P (1973) Red cell shape. Springer, Berlin, p 95
22. (a) Uyesaka N (1991) RBC deformability investigated with newly developed nickel mesh filter. J Jpn Soc Biorheol 5:172–181 (In Japanese); (b) Himura H, Noguchi CT, Uyesaka N, Hasegawa S, Blanchette-Mackie EJ, Schechter AN, Rodger GP (1995) Sickle cell rheology is determined by polymer fraction–not cell morphology. Am J Hematol 48:19–28
23. Matsukawa H, Watanabe A (1979) Fuji in situ process microcapsules. In: Kondo T (ed) Microencapsulation. Techno Books, Tokyo, pp 339–348
24. Sakanishi A, Mitaku S, Takano Y (1976) Linear viscoelasticity of suspensions of spherical shell structures and erythrocyte membrane. Thromb Res 8(Suppl II):35–44
25. (a) Dobashi T, Furukawa T, Narita T, Shimofure S, Ichikawa K, Chu B (2001) Determination of swelling ratio of poly(urea-urethane)microcapsules by single-particle light scattering. Langmuir 17:4525–4528; (b) Dobashi T, Furukawa T, Narita T, Ichikawa K (2002) Coupling of chemical cross-linking, swelling and phase separation in microencapsulation. Langmuir 18:6031–6034
26. Dobashi T, Ishimaru H, Sakanishi A, Ichikawa K (2003) An ultrasonic study of swollen microcapsule membrane. Langmuir 19:3071–3073
27. Dehoux T, Tsapis N, Audoin B (2012) Relaxation dynamics in single polymer microcapsules probed with laser-generated GHz acoustic waves. Soft Matter 8:2586–2589
28. Ichikawa K (1994) Dynamic mechanical properties of polyurethane-urea microcapsules on coated paper. Appl Polym Sci 54:1321–1327
29. (a) Timoshenko S, Woinowsky-Krieger S (1959) Theory of plates and shells, vol xiv. McGraw-Hill, New York; (b) Ugural AC (1999) Stresses in plates and shells, vol xx. WCB/McGraw Hill Boston; (c) Niordson FI (1985) Shell theory, vol xiv. Elsevier Science Pub Co., Amsterdam/New York; (d) Ramm E, Wall WA (2004) Shell structures—a sensitive interrelation between physics and numeric. Int J Numer Methods Eng 60(1):381–427
30. Fery A, Weinkamer R (2007) Mechanical properties of micro- and nanocapsules: single-capsule measurements. Polymer 48:7221–7235
31. Binnig G, Quate CF, Gerber C (1986) Atomic force microscope. Phys Rev Lett 56:930–933
32. Keyang X, David MH, Igor L, Taylor GW (1998) Atomic force microscopy used for the surface characterization of microcapsule immunoisolation devices. J Biomed Mater Res A 41:461–467
33. Fernandes PAL, Delcea M, Skirtach AG, Mohwald H, Fery A (2010) Quantification of release from microcapsules upon mechanical deformation with AFM. Soft Matter 6:1879–1883
34. Huang D, Swanson EA, Lin CP, Schuman JS, Stinson WG, Chang W, Hee MR, Flotte T et al (1991) Optical coherence tomography. Science 254(5035):1178–1181
35. Rogowska J, Patel NA, Fujimoto JG, Brezinski ME (2004) Optical coherence tomographic elastography technique for measuring deformation and strain of atherosclerotic tissues. Heart 90:556–562

36. Matsunami Y, Ichikawa K (2002) Characterization of the structure of poly(urea-urethane) microcapsules. Int J Pharm 242:147–153
37. Dobashi T, Chu B (1999) Light scattering studies of microcapsules in suspension. In: Surface characterization method surfactant science series 87. Marcel Dekker, New York, pp 249–268
38. (a) Dobashi T, Yeh F, Ying Q, Ichikawa K, Chu B (1995) An experimental investigation on the structure of microcapsules. Langmuir 11:4278–4283; (b) Dobashi T, Takenaka M, Yeh F, Wu G, Ichikawa K, Chu B (1996) Scattering from poly(urea-urethane) microcapsule suspension. J Colloid Interface Sci 179:640–642
39. Flory PJ (1953) Principles of polymer chemistry. Cornell University Press, Ithaca
40. Fedors RF (1974) A method for estimating both the solubility parameters and molar volumes of liquids. Polym Eng Sci 14:147–154
41. Miyamoto T, Matsunami Y, Ichikawa K, Hasegawa H (2004) Elemental Mapping in Poly (urea-urethane) Microcapsules by ESI-TEM-The Control of Physical Properties in Nanometer-scale. Fujifilm Res Dev 49:44–48
42. Furukawa T, Hung C, Terao K, Yamamoto T, Dobashi T, Ichikawa K (2003) Effect of affinity between core medium and wall membrane on permeability of dye through microcapsule membrane. Trans Mater Res Soc Jpn 28:989–992
43. Kondo T (1990) Preparation and permeability characteristics of microcapsule membranes. J Control Release 11:215–224
44. Sato T, Yamamoto T, Shibako S, Ichikawa K, Dobashi T (2003) Permeability of azo-dye through poly(urea-urethane) microcapsule membrane. J Membr Sci 213:25–31
45. (a) Williams G, Watts DC (1970) Non-symmetrical dielectric relaxation behaviour arising from a simple empirical decay function. Trans Faraday Soc 66:80; (b) Grassberger P, Procaccia I (1982) The long time properties of diffusion in a medium with static traps. J Chem Phys 77:6281; (c) Palmer RG, Stein DL, Abrahams E, Anderson PW (1984) Models of hierarchically constrained dynamics for glassy relaxation. Phys Rev Lett 53:958–961
46. Yamamoto T, Dobashi T, Kimura M, Chang CP (2002) An approach to analysis of pigment release from microcapsules with size distribution. Colloids Surf B 25:305–311
47. Chang CP, Yamamoto T, Kimura M, Sato T, Ichikawa K, Dobashi T (2003) Release characteristics of an azo dye from poly(ureaurethane) microcapsules. J Control Release 86:207–211
48. Wakata Y, Ichikawa K (1999) Thermoresponsive microcapsule heat sensitive recording material and multicolor heat sensitive recording material. US Patent 5,916,680, June 1999
49. Sakuma S, Ohshima H, Kondo T (1989) Charge distribution in poly(N, N-l-lysinediylterephthaloyl) microcapsule membranes. J Colloid Interface Sci 133:253–256
50. Miyauchi E, Togawa Y, Makino K, Ohshima H, Kondo T (1992) Dependence on pH of permeability towards electrolyte ions of poly(L-lysine-alt-terephthalic acid) microcapsule membranes. J Microencapsul 9:329–333
51. Dobashi T, Narita T, Masuda J, Makino K, Mogi T, Ohshima H, Takenaka M, Chu B (1998) Light scattering of single microcapsule with a hydrogel membrane. Langmuir 14:745–749
52. Makino K, Miyauchi E, Togawa Y, Ohshima H, Kondo T (1994) Chap 28: pH- and ionic-strength-dependent permeation through poly(L-lysine-alt-terephthalic acid) microcapsule membranes. In: Polymers of biological and biomedical significance, ACS symposium series vol 540, pp 314–323. American Chemical Society, Easton. doi:10.1021/bk-1994-0540
53. Kuhn W, Hargitay B, Katchalsky A, Eisenberg H (1950) Reversible dilation and contraction by changing the state of ionization of high-polymer acid networks. Nature 165:514–516
54. de Rossi D, Kajiwara K, Osada Y, Yamauchi A (1991) Polymer gels. Plenum Press, New York
55. Tanaka T, Fillmore DJ (1979) Kinetics of swelling of gels. J Chem Phys 70:1214–1218
56. Narita T, Yamamoto T, Suzuki D, Dobashi T (2003) Dynamics of the volume phase transition of a hydrogel membrane of a microcapsule. Langmuir 19:4051–4054
57. Nagase Y, Kita R, Dobashi T, Kondo T (1996) Viscoelastic properties of hydrogel microcapsule membranes. Polym Gels Networks 4:7–15

58. Tong W, Song X, Gao C (2012) Layer-by-layer assembly of microcapsules and their biomedical applications. Chem Soc Rev 41:6103–6124
59. Dobashi T, Nobe M, Yoshihara H, Yamamoto T, Konno A (2004) Liquid crystalline gel with refractive index gradient of Curdlan. Langmuir 20:6530–6534
60. Dobashi T, Yoshihara H, Nobe M, Koike M, Yamamoto T, Konno A (2005) Liquid crystalline gel beads of Curdlan. Langmuir 21:2–4
61. (a) Dobashi T, Furusawa K, Kita E, Minamisawa Y, Yamamoto T (2007) DNA liquid crystalline gel as adsorbent of carcinogenic agent. Langmuir 23:1303–1306; (b) Thiele H (1954) Ordered coagulation and gel formation. Discuss Faraday Soc 18:294–314; (c) Maki Y, Wakamatsu M, Ito K, Furusawa K, Yamamoto T, Dobashi T (2009) Optical anisotropy of calcium-induced alginate gels. J Biorheol 23:24–28
62. Maki Y, Ito K, Hosoya N, Yoneyama C, Furusawa K, Yamamoto T, Dobashi T, Sugimoto Y, Wakabayashi K (2011) Anisotropic structure of calcium-induced alginate gels by optical and small-angle X-ray scattering measurements. Biomacromolecules 12:2145–2152
63. Furusawa K, Sato S, Masumoto J, Hanazaki Y, Maki Y, Dobashi T, Yamamoto T, Fukui A, Sasaki N (2012) Studies on the formation mechanism and the structure of the anisotropic collagen gel prepared by dialysis-induced anisotropic gelation. Biomacromolecules 13:29–39
64. Wu ZL, Kurokawa T, Sawada D, Hu J, Furukawa H, Gong JP (2011) Anisotropic hydrogel from complexation-driven reorientation of semirigid polyanion at Ca^{2+} diffusion flux front. Macromolecules 44:3535–3541
65. Furusawa K, Minamisawa Y, Dobashi T, Yamamoto T (2007) Dynamics of liquid crystalline gelation of DNA. J Phys Chem 111:14423–14430
66. Iwata K, Sawadaishi T, Nishimura S, Tokura S, Nishi N (1996) Utilization of DNA as functional materials: preparation of filters containing DNA insolubilized with alginic acid gel. Int J Biol Macromol 18:149–150
67. Kitamura H, Matsuura E, Nagata A, Sakairi N, Tokura S, Nishi N (1997) DNA-alginate complex recognized by autoantibodies against DNA. Int J Biol Macromol 20:75–77
68. Furusawa K, Wakamatsu M, Dobashi T, Yamamoto T (2007) Adsorption kinetics of carcinogen to DNA liquid crystalline gel beads. Langmuir 23:10081–10087
69. Langer R (1990) New methods of drug delivery. Science 249:1527–1533
70. Yeh KW, Chang CP, Yamamoto T, Dobashi T (2011) Release model of alginate microcapsules containing volatile tea-tree oil. Colloids Surf A Physicochem Eng Asp 380:152–155
71. Benita S (2005) Microencapsulation. CRC Press, Boca Raton
72. Donbrow M (1991) Microcapsules and nanoparticles in medicine and pharmacy. CRC Press, Boca Raton
73. Newman KD, McBurney MW (2004) Poly(D, L lactic-co-glycolic acid) microspheres as biodegradable microcarriers for pluripotent stem cells. Biomaterials 25:5763–5771
74. Uehara M, Maki Y, Dobashi T (2011) Preparation of 12-hydroxystearic acid microsphere containing oil-based contrast medium. Trans Mater Res Soc Jpn 36:379–382
75. Lee JH, Lee K, Moon SH, Lee Y, Park TG, Cheon J (2009) All-in-one target-cell-specific magnetic nanoparticles for simultaneous molecular imaging and siRNA delivery. Angew Chem Int Ed 48:4174–4179
76. (a) Blomley MJK, Cooke JC, Unger EC, Monaghan MJ, Cosgrove DO (2001) Microbubble contrast agents: a new era in ultrasound. BMJ 322; (b)Unnikrishnan S; Klibanov AL (2012) Microbubbles as ultrasound contrast agents for molecular imaging: preparation and application. Am. J. Roentgenol 199:292–299
77. Yamakoshi Y, Ozawa Y, Ida M, Masuda N (2001) Effects of Bjerknes forces on gas-filled microbubble trapping by ultrasonic waves. Jpn J Appl Phys 40:3852

Biomedical Application of Soft Nano-/Microparticles

11

Jie Wu, Xi Zhao, Yue-Qi Wang, and Guang-Hui Ma

Abstract

In recent years, the soft hydrogel nano-/microparticles have been investigated intensively. Their great porosity, elasticity, and biocompatibility benefit their application, especially in the biomedical field as drug delivery system. Proteins, peptides, or other bioactive molecules can be entrapped inside of these particles with less bioactivity loss and released slowly through the pores of hydrogel. After modifying with targeting molecules or using stimuli-responsive materials, these particles can deliver entrapped drugs to targeting tissues or cells. Recently, hydrogel nano-/microparticles have also been proposed as vaccine adjuvant. The introduction of hydrogel particles in vaccine formulations not only protects the bioactivity of antigens but also increases the immune response. However, due to the high viscosity and water content and weak mechanical strength of hydrogel, there exist many difficulties in preparation and characterization of hydrogel particles. In order to conquer these problems, several novel techniques have been promoted, which are introduced in this review. After reviewing the development of preparation and characterization of hydrogel nano-/microparticles, the application study of these particles on drug delivery systems and on vaccine adjuvants is summarized.

J. Wu • G.-H. Ma (✉)
National Key Lab of Biochemical Engineering, Institute of Process Engineering, Chinese Academy of Sciences, Beijing 100190, People's Republic of China
e-mail: ghma@home.ipe.ac.cn

X. Zhao • Y.-Q. Wang
National Key Lab of Biochemical Engineering, Institute of Process Engineering, Chinese Academy of Sciences, Beijing 100190, People's Republic of China

University of Chinese Academy of Sciences, Beijing 100190, People's Republic of China

R. Kita, T. Dobashi (eds.), *Nano/Micro Science and Technology in Biorheology*, DOI 10.1007/978-4-431-54886-7_11

Keywords
Hydrogel nano-/microparticles • Biomedical application • Drug delivery system • Vaccine adjuvant

11.1 Introduction

Hydrogels, the soft materials with high percent of water inside, have potential applications in the biomedical field, because of their high hydrophilicity, elasticity, porous structure, and biocompatibility with biological tissues. They have been widely studied as tissue engineering materials [42, 46], wound dressing [65, 83, 84], and biosensors [26, 66]. However, the large size of conventional bulk hydrogels with dimensions above centimeters may bring some inconveniences. For example, in order to be used as long-effective drug delivery systems, the bulk hydrogels have to be implanted by surgery, which would reduce the patient compliance [72]. In order to remedy this disadvantage, several strategies were proposed. One way is developing injectable hydrogels with sol–gel transition responding to external stimuli, such as temperature [64], pH [124], or UV [29]. However, during the sol–gel transition process, high initial burst release happens and constant release is hard to achieve [34, 36]. Another strategy which attracted much attention in recent years is preparing hydrogel microparticles or nanoparticles. The hydrogel particles of micro- or nanoscale size have several advantages over bulk hydrogels. Firstly, the particles can be administered by various routes. Due to their small size, the particles are injectable, and they can penetrate through mucosa membranes, so they are also useful for oral, nasal, and other mucosal drug delivery. Secondly, since the diffusion time of molecules in hydrogel increases in proportion to the square of the size of the hydrogel, hydrogel particles with smaller size obtain better stimuli responsibility than bulk hydrogel [80]. Moreover, hydrogel particles could be modified or conjugated with various functional groups or molecules for special purpose, such as targeting drug delivery and stimuli-responsive drug delivery. The positive features of their excellent biocompatibility with versatile biological activities provide hydrogel nano-/microparticles numerous opportunities for further development.

11.2 Preparation

To prepare hydrogel nano-/microparticles, it is common to adopt emulsification method in the literatures [22, 62]. The water phase contains hydrophilic materials and is dispersed in oil phase by stirring, homogenizing, or spraying. After solidifying or cross-linking, the particles are obtained. And by adopting different materials or cross-linking agents, it is possible to prepare particles with various structures. By this way, Wei et al. prepared chitosan microparticles with different structures, including solid, porous in surface, and hollow structures [107]. Typically, the chitosan particles were cross-linked by glutaraldehyde (CG particles). In this process, the amino groups on chitosan chains reacted with the aldehyde groups of glutaraldehyde to solidify the particles with solid structure and smooth surface. In order to prepare chitosan particles with different structures, Wei developed a two-step cross-linking method. Firstly, p-phthaldehyde was used to cross-link chitosan, and then the particles were cross-linked again by glutaraldehyde. Because the Schiff base formed between p-phthaldehyde and chitosan was unstable in acetone and ethanol, by washing with these solvents, the chitosan particles with hollow structure (C-PG particles) were prepared. By adding different amounts of p-phthaldehyde and glutaraldehyde, the wall thickness of particles can be adjusted. Furthermore, Wei introduced the one- and two-step cross-linking methods to prepare composited particles composed of the mixture of chitosan and quaternized chitosan (in a 1:1 mass ratio). Due to the steric hindrance of quaternized groups to the cross-linking, the composite particles cross-linked with glutaraldehyde had hollow structure with pores on the surface (CH-G particles). And for those composite particles prepared by the two-step cross-linking method, the macroporous structures were observed (CH-PG particles). The confocal laser scanning microscope (CLSM) images of these particles are shown in Fig. 11.1.

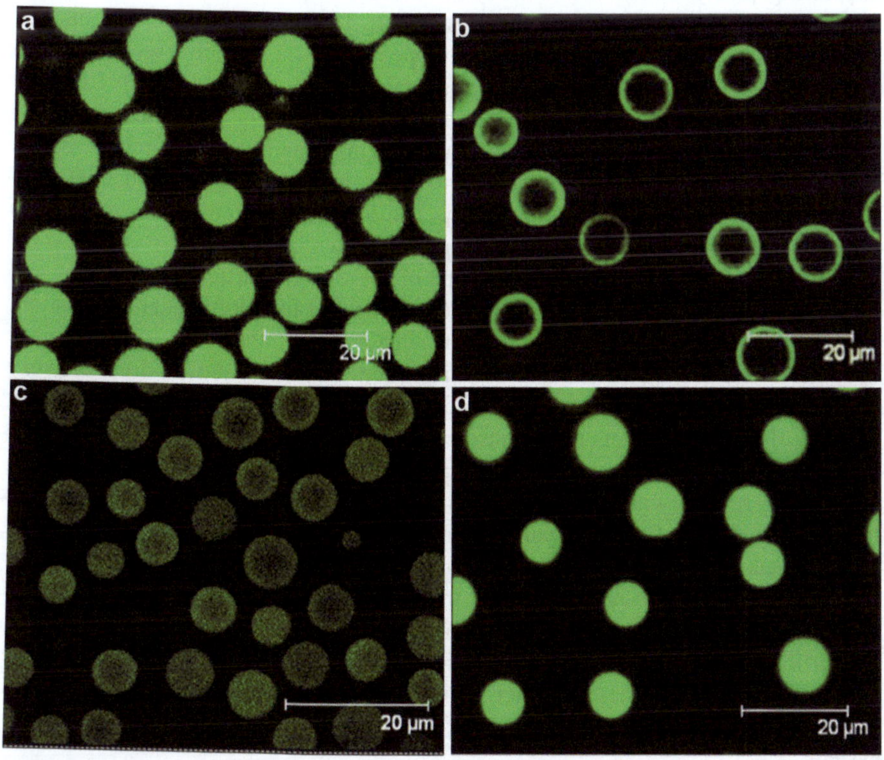

Fig. 11.1 CLSM images of CG (**a**), C-PG (**b**), CH-G (**c**), and CH-PG (**d**) particles (Modified from Wei et al. [107] with permission from John Wiley and Sons

Besides emulsification, the method of precipitation also has been used to prepare hydrogel particles, especially for those particles in nanoscale [8, 25, 44]. To induce the precipitation of polymer, there are mainly two methods, including adding non-solvent or the agent with opposite charges. Masotti et al. used both of the methods to prepare chitosan nanoparticles [53]. The methanol was used as non-solvent to precipitate chitosan, and the average diameter of formed nanoparticles was 45 ± 9 nm. DNA was added as anionic agent and formed nanoparticles with chitosan with an average size of 38 ± 4 nm. For preparing chitosan nanoparticles, tripolyphosphate (TPP) is the ionic cross-linking agent with the highest frequency of utilization [15, 56, 67]. By varying the composition and preparation conditions, the nanoparticles with different characteristics, such as size and surface charge, can be obtained [24].

Besides chitosan, alginate has also been frequently used to prepare hydrogel particles by precipitation. Anionic alginate tends to form “egg-box junctions” with multivalent cations, especially Ca^{2+} [86]. The formed Ca-alginate particles usually have an average diameter of nanoscale and been used for gene delivery and bone repair [88, 117]. But for some special applications, precipitation has also been used to prepare particles with bigger size. For example, in order to treat phenylketonuria (PKU) patients, which is an inherited metabolic disorder, *Lactococcus lactis*-expressing phenylalanine ammonia-lyase (LLEP) is proved effective. However, LLEP tends to lose its bioactivity in the stomach. Thus, in order to develop its oral formulation with proper bioactivity, entrapped LLEP in particles seems to be a good choice. The average size of LLEP is about 1 μm, and to entrap LLEP in particles, the particles should be in micrometer range. Zhang et al. used precipitation method to prepare LLEP entrapped alginate microparticles by combining with spray-solidification technique, which is to spray alginate solution that contained LLEP directly into $CaCl_2$ solution [121]. The formed Ca-alginate microparticles have an average diameter around 58–59 μm as shown in Fig. 11.2.

However, as Fig. 11.2 shows, the shape of these microparticles is irregular and their size distribution is broad. For particles prepared for biomedical application, their size and shape are very important factors [87]. Particles with different diameters tend to aim to different organs when they are used as drug delivery system, which decreases the bioavailability of the entrapped drug [21]. Moreover, the size of particles affects their transmembrane efficiency when the particles are used as oral drug delivery system. Wei et al. fed the Sprague–Dawley (SD) rats with chitosan particles with different sizes (2.1, 7.2, and 12.5 μm) [108]. The bioadhesive ability of chitosan particles in the gastrointestinal tract was determined by modified everted sac experiment combined with flow cytometry analysis. By counting the number of particles in per mg tissue, it was found that more particles with 2.1 μm adhered on the mucus due to their small size (Fig. 11.3). Furthermore, the particles with small size (2.1 μm) could penetrate the mucosal membrane and be absorbed in the gastrointestinal tract, while those bigger particles (12.5 μm) could not. This study proved the importance of size control of particles as drug delivery system.

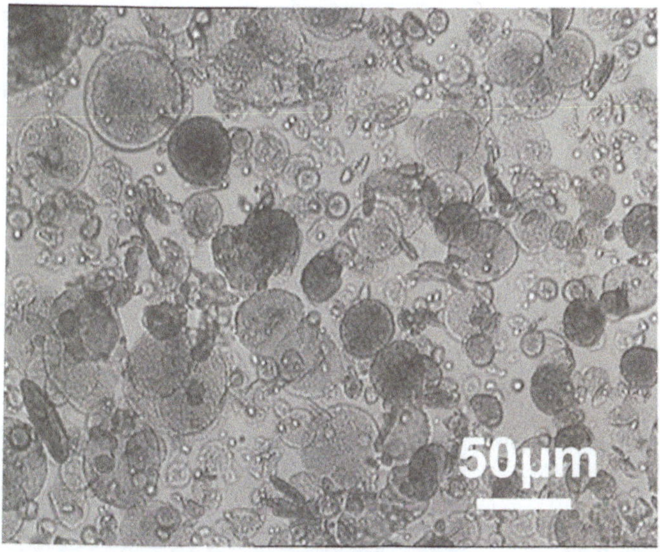

Fig. 11.2 Optical images of LLEP entrapped alginate microparticles (Reproduced from Zhang et al. [121] with permission from ACS Publications)

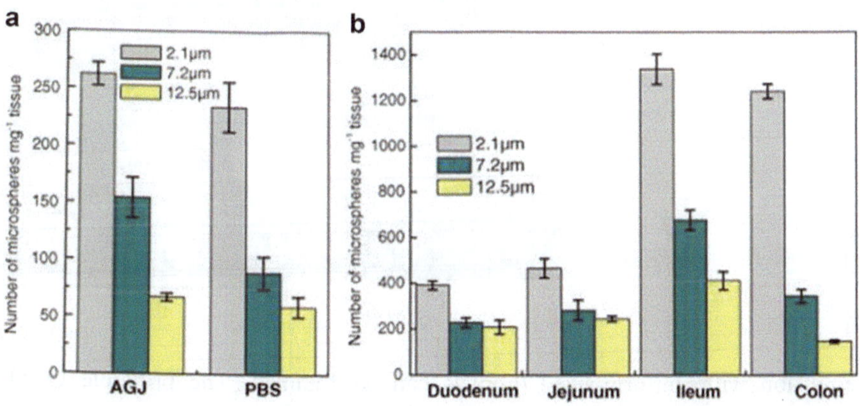

Fig. 11.3 Bioadhesion of chitosan microparticles with different diameters in the stomach (**a**) and the intestinal tract (**b**) (means ± SD, $n = 6$). AGJ means artificial gastric juice (hydrochloric acid solution, pH 1.2); PBS means phosphate-buffered saline (pH 7.2) (Reproduced from Wei et al. [108] with permission from Elsevier)

Thus, in order to obtain particles with narrow size distribution, several novel preparing methods were developed, and Shirasu porous glass (SPG) membrane emulsification technique is one representative of them. The SPG membrane is consisted of Al_2O_3–SiO_2 and has uniform-sized pores on its wall, so by pressing the dispersed phase through the SPG membrane into continuous phase, the

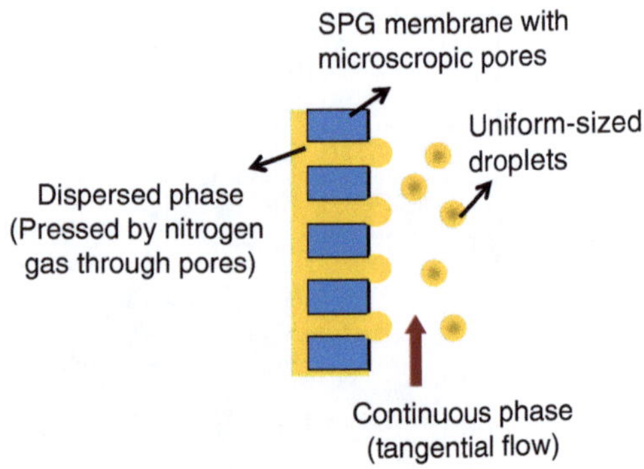

Fig. 11.4 Principle of SPG membrane emulsification

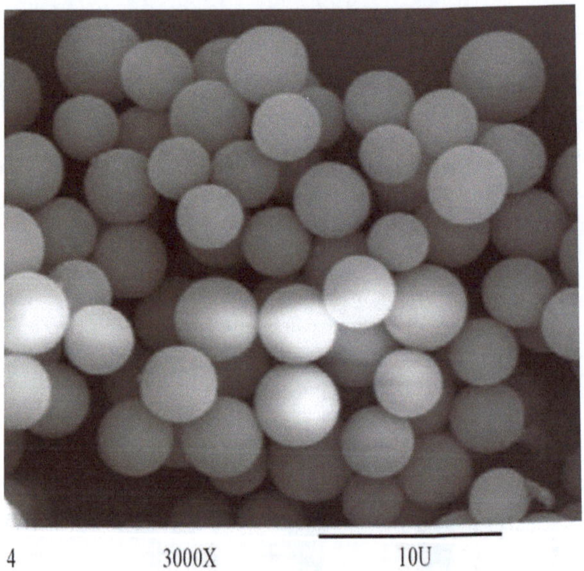

Fig. 11.5 SEM photographs of chitosan microspheres prepared by membrane emulsification technique (Modified from Wang et al. [99] with permission from Elsevier)

emulsion with uniform-sized droplets can be obtained. The principle of SPG membrane emulsification is illustrated in Fig. 11.4 [99].

Wang et al. used SPG membrane emulsification to prepare chitosan microparticles, and the number-average diameter of prepared particles is 13.8 μm with coefficient of variation (CV) value below 15 % [99]. The scanning electron microscopy (SEM) photograph of chitosan microparticles is shown in Fig. 11.5.

Besides chitosan particles, agarose [125, 126], alginate [122], and poly (N-isopropylacrylamide-co-acrylic acid) [P(NIPAM-co-AA)] hydrogel particles [101] prepared by SPG membrane emulsification technique were reported, and all of these formed particles have narrow size distribution.

However, by this SPG membrane emulsification technique, which is also called general SPG membrane emulsification technique, the diameter of formed droplets

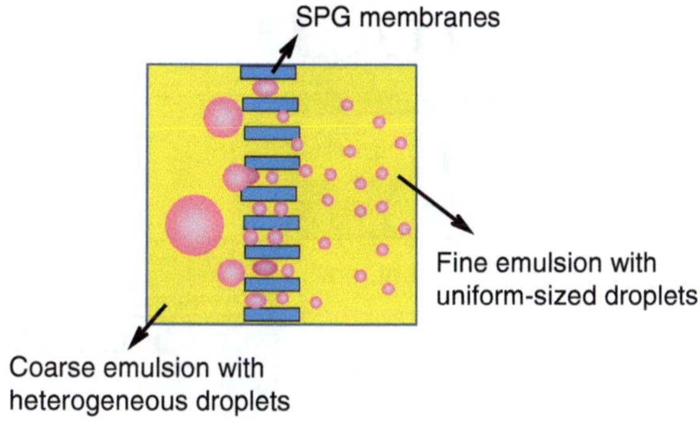

Fig. 11.6 Principle of premix SPG membrane emulsification

is several times bigger than the pore size of used membrane. So in order to prepare particles in nanoscale, the membrane with very tiny pores should be adopted, which would dramatically lower the productivity. And for some viscous systems, the preparation process would be very difficult if not impossible.

To remedy these disadvantages, a novel SPG membrane emulsification, premix SPG membrane emulsification, was developed. In this technique, the dispersed phase is firstly mixed with the continuous phase by stirring, homogenizing, or other methods, and the mixed solution is called coarse emulsion or pre-emulsion. Then, the coarse emulsion is pressed through SPG membrane one or several times by nitrogen gas to obtain droplets with narrow size distribution. The principle of premix SPG membrane emulsification is shown in Fig. 11.6.

By premix SPG membrane emulsification technique, it is probable to prepare uniform-sized particles with nanoscale or for systems with high viscosity. For example, agarose, the natural polysaccharide, has high viscosity and been widely used as chromatography medium. In order to separate biomolecules more quickly and efficiently, small agarose beads are desirable due to their big surface area. However, the reduction of bead size results in small spaces among the beads and high column back pressure. In order to resist the high pressure, one way is increasing the agarose content of beads. But, with the increase of agarose content, the water phase becomes very viscous, so that it is very difficult to prepare uniform-sized agarose beads with high agarose content and small size by stirring or other common techniques. The development of premix SPG membrane emulsification solved this problem. By this technique, agarose beads with high polymer content (14 wt.%) and small diameter (less than 10 μm) have been prepared with CV of 9.8 % [127, 128]. And for other systems with high viscosity, this technique still works. Lv et al. used this technique to prepare chitosan and N-((2-hydroxy-3-trimethylammonium) propyl) chitosan chloride (HTCC) nanoparticles with polydispersity index 0.029 and 0.034 separately [51].

Microchannel (MC) emulsification technique is another useful method to produce monodisperse droplets or particles [38, 69]. The main preparation process is pumping the dispersed phase into continuous phase through special silicon plate, stainless steel, or PMMA array with predesigned microgrooves or microchannels (Fig. 11.7), and the typical structure of silicon array is illustrated in Fig. 11.8.

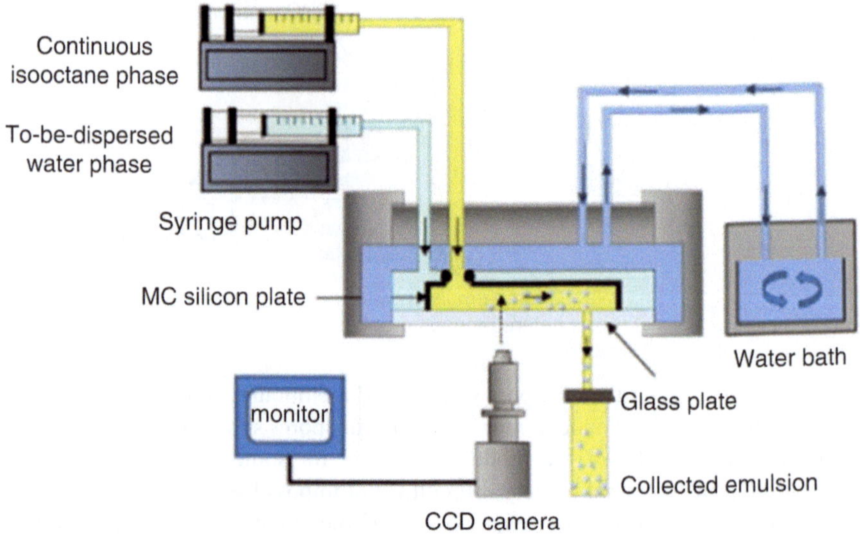

Fig. 11.7 Microchannel emulsification experimental setup based on silicon plate (Modified from Chuah et al. [13] with permission from Elsevier)

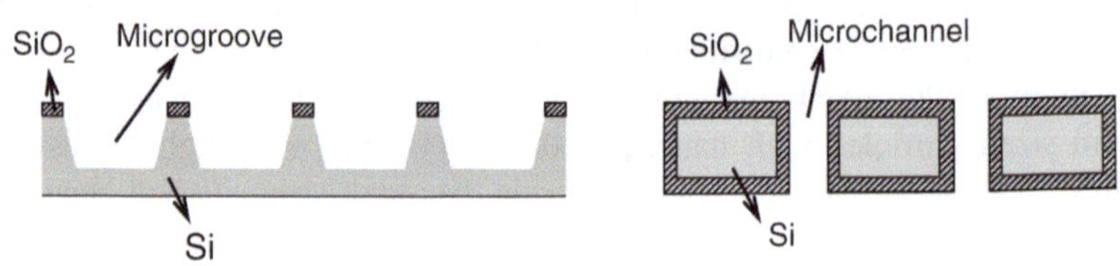

Fig. 11.8 Silicon MC array with microgrooves (*left*) or with straight – through microchannels – (*right*)

The size of produced droplets is mainly affected by the size of the microgrooves or microchannels, the flow rate of dispersed phase, and the compositions of both dispersed phase and continuous phase. By optimizing these factors, uniform-sized particles can be obtained. For instance, Chuah et al. prepared monodisperse alginate microspheres with an average size of 6.2 μm and CV of below 10 % by MC emulsification [13]. Besides alginate microspheres, gelatin microbeads [32], poly (N-isopropylacrylamide) (PNIPAM) microparticles [31], and chitosan microparticles [114] with uniform size have been fabricated successfully by MC technique.

Microfluidic (MF) technique is another important method that utilized microfabricated devices to control the size of formed droplets [16], which has been reported frequently in recent years. MF technique is divided into planar MF and three-dimensional axisymmetric MF. The main component of planar MF device is poly(methyl methacrylate) (PMMA) plate with microchannels on silicon

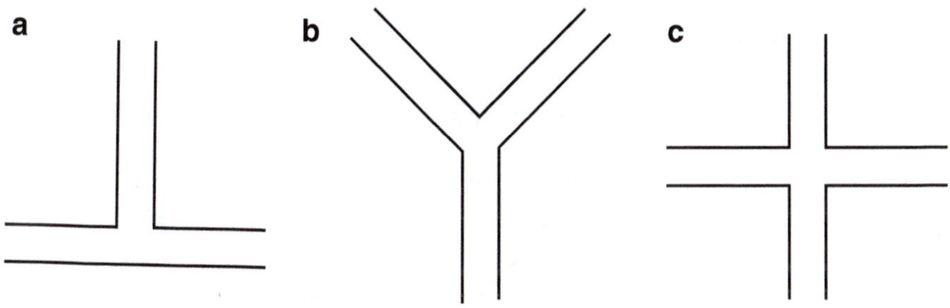

Fig. 11.9 Typical structures of microchannels of planar MF devices: (**a**) T junction, (**b**) Y junction, and (**c**) cross junction

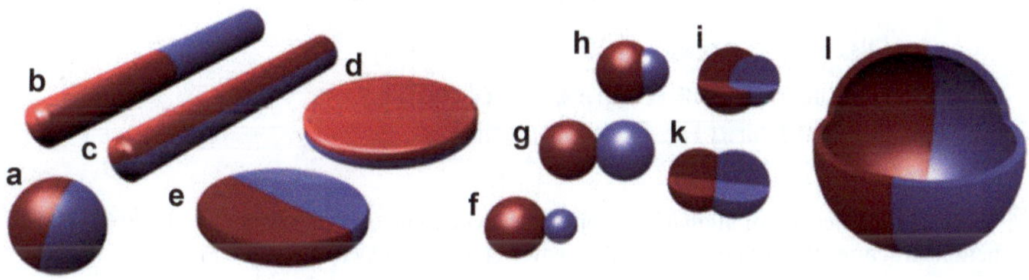

Fig. 11.10 Different types of Janus particles (JPs): *spherical* (**a**), two types of *cylindrical* (**b**, **c**), and *disk-shaped* (**d**, **e**) JPs. (**f–k**) Various kinds of *dumbbell-shaped* JPs with asymmetric or snowman character (**f**), symmetric appearance (**g**, **k**), attached nodes (**h**), and eccentric encapsulation (**i**). (**l**) Janus vesicles or capsules (Reproduced from Walther and Müller [97] with permission from ACS Publications)

mold. The structure of microchannels includes several types, such as T junction, Y junction, and cross junction (Fig. 11.9).

These special structures of microchannels provide the possibility to prepare particles with various structures or morphologies. For example, Hwang et al. used MF device with T junction structure and prepared nonspherical hydrogel microparticles with ɣ-Fe_2O_3 nanoparticles inside [30]. The dispersed phase (magnetic solution) and continuous phase (mineral oil) were independently pressed by air through the microchannels to form emulsion. And by adjusting the channel geometries, pressure for dispersed phase and continuous phase, microgels with different morphologies were prepared, including spheres, disks and plugs.

Another type of anisotropic microparticles, Janus particles, can also be prepared by MF technique. Janus particles have hemispherically distinct geometries as shown in Fig. 11.10 [97]. Due to their unique anisotropy, the preparation and application of Janus particles have been reported in many literatures [11, 59, 63].

For Janus particles, whether they have narrow size distribution is important for their biomedical application. By using MF device with Y junction structure, Aketagawa et al. prepared monodisperse Janus hydrogel beads with CV of less

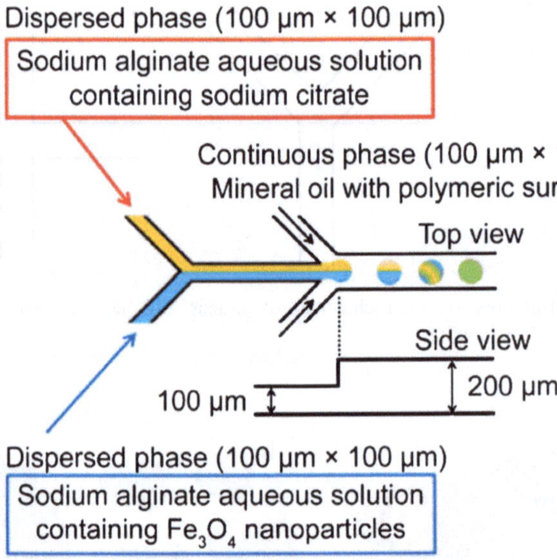

Fig. 11.11 Fabrication of alginate microparticles with magnetic anisotropy by MF device with Y junction structure (Reproduced from Aketagawa et al. [1]. (Open Access))

than 5 % [1]. Two different dispersed phases (sodium alginate solution containing sodium citrate and sodium alginate solution containing Fe_3O_4 nanoparticles) were separately injected through the two inlets of Y junction slowly to form a two-phase parallel stream (Fig. 11.11). After entering into a continuous phase, biphasic droplets were formed. At the presence of electrolyte, the Fe_3O_4 nanoparticles aggregated and precipitated at the bottom of the droplets. Then, the Janus alginate microparticles with magnetic anisotropy were prepared.

However, for these planar MF devices, the formed droplets have to contact the wall of the microchannels during formation and flow, this may damage the formed structure of some fragile particles [96]. In order to conquer this problem, a novel three-dimensional axisymmetric MF device was developed. A classic example of axisymmetric MF device is glass capillary device as shown in Fig. 11.12.

In glass capillary device, the dispersed phase is surrounded by the continuous phase, which avoids the contact between newly formed droplets and the wall of microchannels. Moreover, by designing glass capillary devices with different structures, the droplets and microparticles with various structures or morphologies can be prepared easily as shown in Fig. 11.13 [12, 14, 48, 77, 102, 104].

As three promising techniques, SPG membrane emulsification, MC, and MF technique, all of them show their advantages to provide droplets or particles with narrow size distribution over other common methods. Moreover, these techniques are energy efficient because they do not need high energy to disperse the emulsion. Comparing these three methods, MC and MF techniques are superior to SPG membrane emulsification on the size distribution, structure, and shape control of particles, while the latter shows its advantages on the large-scale production

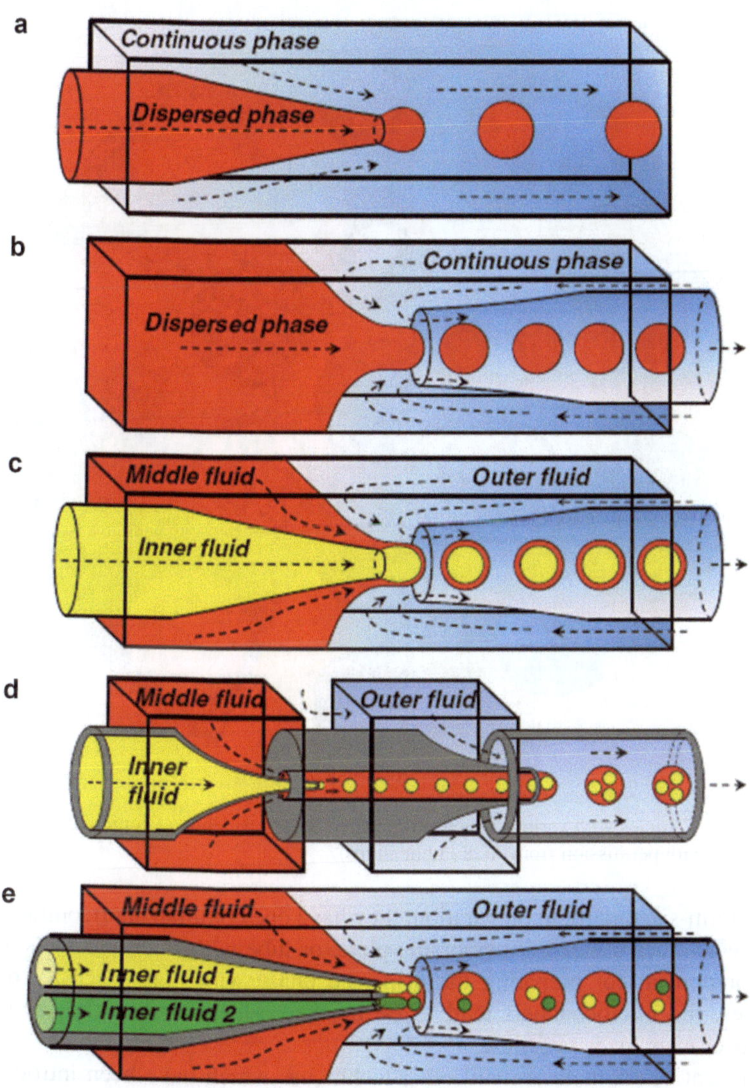

Fig. 11.12 Axisymmetric glass capillary devices: (**a**) co-flow of two immiscible fluids, (**b**) countercurrent flow of two immiscible fluids with flow focusing, (**c**) combination of co-flow and countercurrent flow of three immiscible fluids, (**d**) two sequential co-flow droplet generators, (**e**) injection of two distinct inner phases of double emulsions using a two-bore injection tube (Reproduced from Vladisavljević et al. [96] with permission from Springer)

[96]. Furthermore, SPG membrane emulsification technique can be utilized to prepare submicro- and nanoparticles. But for MC and MF techniques, due to the limit of the processing technology, the size of microchannels is at least several micrometers; thus it is hard to utilize MC and MF techniques to prepare nanoparticles. However, in spite of these techniques having been proved useful

Fig. 11.13 Particles with various structures prepared by MF technique (Reproduced from Wang et al. [102] with permission from ACS Publications)

for uniform-sized particle preparation, they have one common shortcoming. All of these techniques utilized a membrane, plate, or tube with microchannels as their key component; for preparing the dispersed phase with high viscosity or solid granules, the microchannels tend to be blocked which hinders the preparation process or even leads to failure.

In recent years, there are still other techniques which have been introduced to prepare uniform-sized hydrogel particles, such as ink-jet printing [19, 33], imprint lithography [9], and microfluidic electrospray technique [49]. Some of them show their advantages in size distribution and structure control of particles. With more intensive study, these techniques may play a greater role in the preparation of hydrogel particles.

11.3 Characterization

Like other polymer particles, the size and size distribution, surface characteristics, morphology, and interior structure of hydrogel particles are important factors for their applications. For accurate measurement, it needs to be taken in consideration that hydrogel particles have some special characteristics which would affect their characterization.

For example, most of the hydrogel particles have high water content, and some particles even contain more than 90 % (w/w) of water. Due to their high water content, the hydrogel particles have porous structure, which facilitate the adsorption and diffusion of biomacromolecules. But it is difficult to characterize the porous structure of hydrogel particles because of their weak mechanical strength. The commonly used method for pore structure characterization, mercury intrusion porosimetry, needs high pressure, which might crush or deform the particles [5]. Before using the scanning electron microscopy (SEM) to observe the morphology of particles, the samples have to be dried, and the drying process might damage the gel network as stated before [54]. In order to keep the original structure of hydrogel particles, Du et al. adopted critical point drying to dry agarose beads [20]. Although the drying process is very mild, the particles still shrink and the porous structure of the hydrogel could not be observed clearly due to dehydration.

To avoid the effect of the drying process, the techniques which can test the samples in liquid environment have been investigated for characterization of hydrogel particles. Atomic force microscopy (AFM) and confocal laser scanning microscopy (CLSM) are two representative techniques among them. The measurement conducted by AFM can achieve high spatial resolution of the surface morphology of samples [4, 41]. Seeber et al. utilized AFM to observe surface morphology change of the thermosensitive microgel below and above its lower critical solution temperature (LCST) [76]. The swell and collapse of microgel can be observed and measured by detecting the change of layer thickness.

However, in order to measure accurately, the particles need to be fixed onto a substrate. Usually electrostatic attraction between particles with substrate is utilized to fix the particles. But for some samples uncharged or with weak charges, chemical cross-linking has to be used, and the cross-linking may affect the structure of hydrogel particles. Moreover, for soft hydrogel particles, the force of AFM probe may cause the deformation of particles and the obtained results are not so reliable, especially for the measurement of height [9].

CLSM is another technique for obtaining optical images of hydrogel particles in liquid surroundings. The key feature of CLSM is its ability to observe the interior structure without the need of destroying the samples. Yang et al. prepared superporous agarose (SA) microspheres and used CLSM to prove the existence of pores (Fig. 11.14) [115]. It is clearly observed from Fig. 11.12c, d that inside of SA microspheres, there are some fluorescence signals, which proves fluorescein isothiocyanate (FITC)-labeled IgG (FITC-IgG) can enter the interior of microspheres through the superpores. While for those homogeneous agarose (HA) microspheres, without the existence of superporous structure, FITC-IgG cannot enter the interior of microspheres and only binds with protein A on the surface of microspheres (Fig. 11.12a, b).

CLSM can also be utilized to observe the diffusion of molecules in hydrogel particles. Wu et al. prepared chitosan hydrogel microparticles and entrapped FITC-labeled bovine serum albumin (BSA) as a model drug inside [111]. With the release

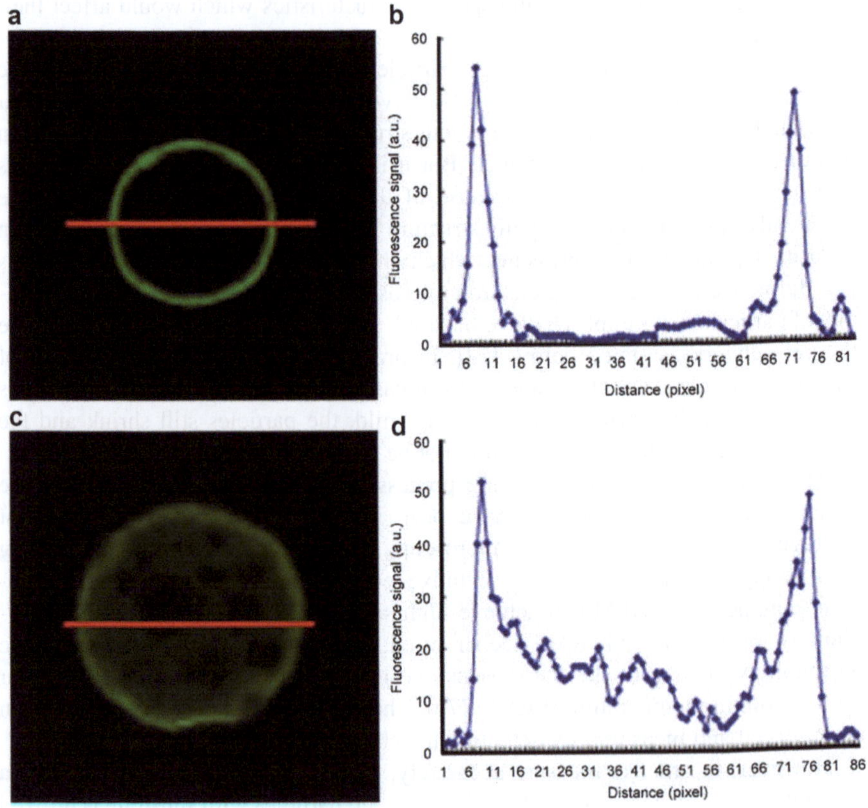

Fig. 11.14 Surface availability on HA and SA microspheres for binding with protein A. (**a**, **c**) Sectional images were obtained from the middle of HA (**a**) and SA (**c**) beads by CLSM. (**b**, **d**) Graphical representations of the fluorescence intensities were obtained from (**a**) and (**c**) images, respectively (Reproduced from Yang et al. [115] with permission from Elsevier)

of drug, the fluorescence intensity of microparticles decreased. However, in order to characterize by CLSM, the tested samples need to be fluorescence labeled, and the pretreating process is relatively complex and time-consuming.

Besides the porous structure, the mechanical strength is another important property of hydrogel particles. Conventional hydrogels have weak mechanical strength and they tend to deform or be broken under compression. But for some applications, whether the particles have essential high mechanical strength is important. For instance, the particles used as separation medium must have mechanical strength high enough to keep their integrity under operating pressure. Otherwise, these particles will collapse, and the deformed particles tend to block the gaps between particles, which lift the column back pressure rapidly and may lead to the failure of separation. In order to measure their mechanical strength, one method is measuring the pressure–flow rate curve of particles in the column. By this method, the tested particles need to be packed in the column and the flow rate of

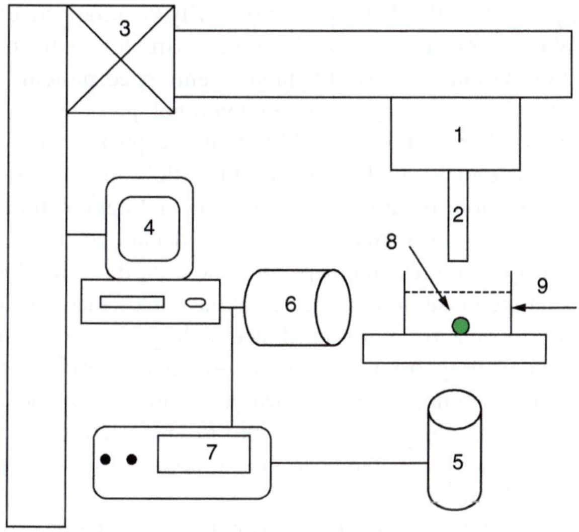

Fig. 11.15 Schematic diagram of the micromanipulation rig: (*1*) force transducer, (*2*) probe, (*3*) stepping motor, (*4*) computer with a data acquisition board, (*5*) lower-view microscope, (*6*) side-view microscope, (*7*) video recorder, (*8*) agarose microsphere in water, and (*9*) glass chamber (Reproduced from Yan et al. [113] with permission from Elsevier)

solution (usually water) though this column is measured constantly. With the increase of pressure up to a point, the particles deform or collapse and the column back pressure increases sharply. However, for this method, amount of samples needs to be used. In order to reduce the amount of samples, several techniques, which can detect the mechanical strength of individual particle, have been proposed, including AFM [6, 57, 116] and microelectromechanical systems (MEMS) [37]. Micromanipulation technique is another one which detects the mechanical strength of a particle in microscale. Yan et al. used this technique to test individual agarose microsphere [113]. The equipment is shown in Fig. 11.15. Single microsphere is compressed by a probe controlled by a force transducer. Then, by measuring the change of force and the diameter of the microsphere, the mechanical strength of the microsphere is calculated.

11.4 Application

11.4.1 As Drug Delivery System

Most of the literatures related to the application of hydrogel nano- and microparticles in biomedical field focus on drug delivery system. As other delivery systems are based on particles, the utilization of hydrogel particles helps to protect loaded drug from degradation of enzyme, acid, or other adverse surroundings in vivo. In addition, hydrogel particles have some special characteristics, which facilitate their biomedical application. Firstly, hydrogels contain high percent of water, and the presence of water improves the biocompatibility of hydrogels and makes them nonirritant for entrapped biomacromolecules. Thus, the introduction of hydrogels will help keep the bioactivity of entrapped drugs. For example, Wang and Wu combined agarose hydrogel particles with poly(lactic-co-glycolic acid) (PLGA) microspheres to stabilize entrapped protein drugs [98]. The model drug, insulin, was encapsulated in hydrogel particles and then dispersed in PLGA microspheres. For sole PLGA microspheres, the encapsulation process, the hydrophobicity of PLGA, and the acidic microenvironment due to the polymer degradation induced the degradation and aggregation of insulin, and about 18 % insulin in PLGA microspheres was denatured. But after entrapping drugs in the agarose particles and dispersing these particles in PLGA microspheres, only 1 % denatured insulin was detected in the release medium. The high retention efficiency of drug bioactivity was attributed to drug encapsulation by agarose particles, which isolated drug from the aqueous–organic interfaces and hydrophobic and acidic microenvironment. Schoubben et al. carried out another similar study [74]. They introduced alginate particles into PLGA microparticles and found the addition of alginate particles stabilized the entrapped insulin and prolonged the release period up to 4 months.

Secondly, although the main function of drug carriers based on particles is releasing drug slowly and constantly to prolong the half-life of drug in blood circulation, some of hydrogel particles have more functions than that. These hydrogel particles possess stimuli responsibility, which means that they change their shape, size, or other characteristics responding to external stimuli, such as temperature, pH, and so forth. Depending on this special property, these hydrogel particles have been developed as injectable drug delivery system, targeting drug delivery system, and stimuli-responsive delivery system [58, 71, 92, 122].

In order to obtain injectable systems, the thermosensitivity of hydrogels would work. For example, Hsiao et al. designed thermosensitive chitosan nanocapsules and developed injectable depot delivery system based on these nanocapsules [28]. They synthesized carboxymethyl-hexanoyl chitosan (CHC), firstly, and CHC would self-assemble to nanocapsules. With the adding of negatively charged sodium β-glycerophosphate (β-GP), these nanocapsules showed thermosensitivity, that is, they transformed from solution state into gel at body temperature. Antiepileptic drug ethosuximide was loaded as model drug in this system and the in vivo study proved its treating efficiency.

For developing targeting drug delivery systems, the hydrogel nano- and microparticles with pH sensitivity have been frequently investigated. Chitosan, the positively charged polysaccharide, and alginate, the negatively charged polysaccharide, are the most commonly used materials to prepare pH-sensitive particles for colon-targeting drug delivery or tumor-targeting drug delivery. Li et al. fabricated nanoparticles composed of quaternized chitosan and alginate [43]. These nanoparticles showed pH-dependent drug release profiles. BSA as a model drug was released slowly in simulated gastric fluid (pH 1.2) and rapidly in simulated intestinal fluid (pH 7.4).

The positive charge of chitosan not only helps the colon-targeting drug delivery but also promotes the drug absorption through the mucosal membrane. Wei et al. fabricated an oral delivery system for siRNA based on N-((2-hydroxy-3-trimethylammonium)propyl) chitosan chloride (HTCC) nanoparticles (HNP)

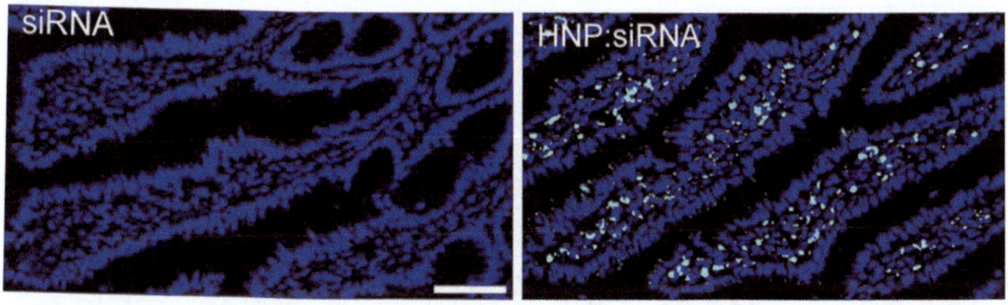

Fig. 11.16 Absorption of siRNA (*left*) and HNP:siRNA (*right*) in the small intestine. Scale bar: 100 μm (Modified from Wei et al. [110] with permission from Elsevier)

[110]. The positive charge of HNP facilitated the permeation of siRNA in the intestine tract as shown in Fig. 11.16. By co-encapsulating siRNA and paclitaxel (PTX) in HNP, the complex system could achieve effective tumor suppression effect.

Moreover, the stimuli responsibility of hydrogel particles can be utilized to realize special release profiles, like trigger release. Inspired by the natural plant, squirting cucumber, Liu et al. prepared thermosensitive hydrogel capsules and designed an interesting system based on these capsules for nanoparticle delivery with thermo-triggered squirting property [47]. The shell of hydrogel capsules is composed of PNIPAM, which would shrink dramatically at the temperature above its LCST. By the aim of microwave or infrared irradiation, when the capsules were heated above the LCST of PNIPAM, their shell would shrink rapidly and result in squirting delivery of entrapped nanoparticles.

Besides the stimuli of pH and temperature, biomolecules have also been used as stimulus to regulate the drug release rate from hydrogel particles. For example, to treat diabetes I, the patients need to inject insulin two or three times every day. In order to avoid the daily injection and keep the blood glucose level in normal range, glucose-responsive microcapsules have been promoted. Zhang et al. fabricated monodispersed glucose-responsive hydrogel microcapsules by UV-initiated polymerization of glucose-responsive 3-acrylamidophenylboronic acid (AAPBA) and thermo-responsive PNIPAM as the microcapsule shell [123]. These microcapsules exhibit reversible swelling/shrinking behavior and different drug release rate responding to glucose concentration changes.

Multi-stimuli-responsive hydrogel particles have also been reported. Sahiner et al. designed hydrogel nanoparticles with core–shell structure and magnetic Fe_3O_4 inside of them [70]. These particles are sensitive to pH change and magnetic field. Moreover, these particles could be further modified with other functional groups or combined with other metals to achieve more stimuli responsibility.

Thirdly, hydrogel particles usually have porous network structure and high permeability for drug, ions, or other water-soluble metabolites, which make them attractive candidates for use in the biomedical field. With the porous structure, hydrogel particles can load more drugs than solid polymer particles. Moreover, by

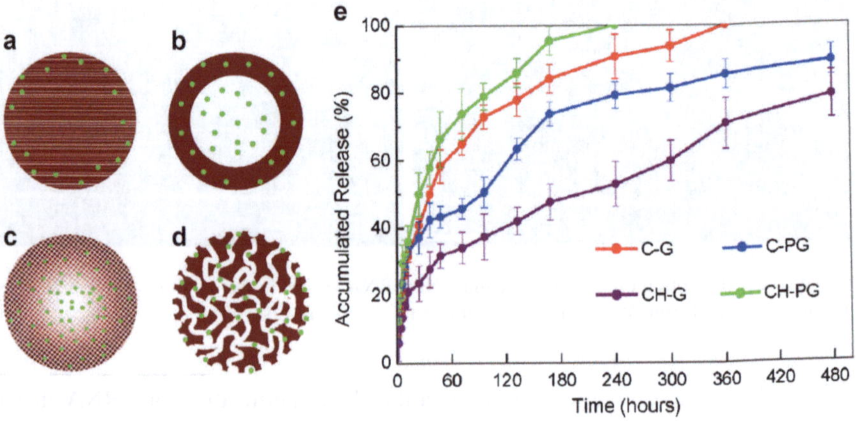

Fig. 11.17 Schematic representation of BSA (*green dots*) distribution in microspheres prepared by different methods: (**a**) CG microspheres, (**b**) C-PG microspheres, (**c**) CH-G microspheres, (**d**) CH-PG microspheres, and (**e**) their different release profiles (Modified from Wei et al. [107] with permission from John Wiley and Sons)

changing the porous structures, the different drug release profiles would be obtained to fulfill various treatment requirements. Wei et al. designed four types of chitosan microspheres with different structures and investigated their drug release profiles (Fig. 11.17) [107]. The solid particles (CG) released drug with high initial burst because most of the drug is dispersed near the particle surface. And C-PG microspheres with hollow structure showed triphase drug release profile. The diffusion of drug in the particle shell resulted in the first burst release, and then the drug in the particle cavity is diffused outside. After that the concentration gradient of drug in the microspheres dominated the last slow release. The particles with hollow structure and pores on the surface (CH-G) released drug with minimal initial burst and would be beneficial for controlled release of protein drugs. The particles with macroporous structure (CH-PG) showed a strong initial burst, and the authors suggested using these particles for pulsed therapy, for instance, vaccine administration.

Furthermore, the porous structure of hydrogel particles can also be utilized for some special purposes. Lv et al. prepared chitosan nanoparticles, and during the preparation process of nanoparticles with temperature-programmed solidification technique, the loaded hydrophobic drug, paclitaxel (PTX), formed nanocrystals in the hydrogel network (Fig. 11.18) [51]. The in situ crystallization of PTX not only helped the increase of drug loading content but also the decrease of burst release of the drug.

In order to achieve better treatment efficiency, Lv et al. modified their hydrogel particles by PEGylation and conjugated PEG terminal with cyclic Arg-Gly-Asp (RGD) peptide [52]. PEGylation of nanoparticles prolonged their circulation time in blood, and modification with cyclic RGD peptide gave them better affinity to the tumor cells. With the synergistic effect of PEGylation, cyclic RGD peptide

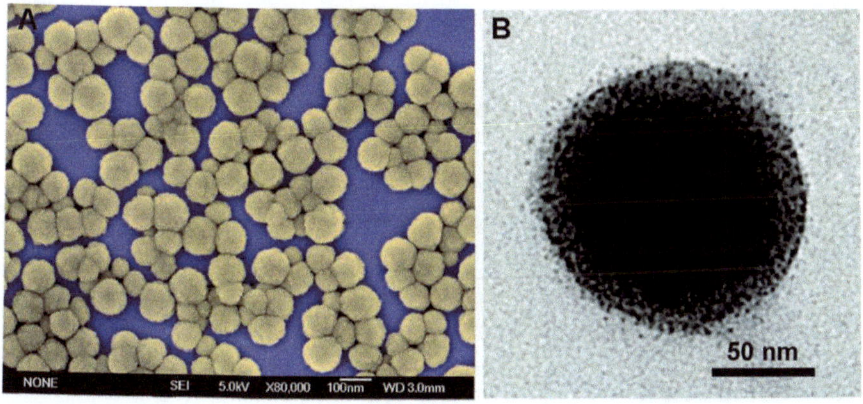

Fig. 11.18 SEM and transmission electron microscope (TEM) images of chitosan nanoparticles loaded with PTX nanocrystals (Reproduced from Lv et al. [51] with permission from ACS Publications)

modification, and chitosan nanoparticles, this system achieved high survival rate and low side effects in animal tests.

Besides the typical porous structure of hydrogel, with the introduction of novel preparation techniques, hydrogel particles with various special structures have been fabricated and utilized to achieve unique functions [17, 105].

Zhou et al. used layer-by-layer method on the preparation of microgels with core–membrane structure, which could realize pulsed drug release without external trigger [129]. They, respectively, coated polyelectrolyte sodium polystyrene sulfonate (PSS) and poly(allylamine hydrochloride) (PAH) on chitosan–alginate microgels, and the drug release profile of these microgels consisted of two phases: a little- or no-drug release period (lag phase) and a majority-of-drug release after lag time (II phase). The authors suggested this system would be useful as pulsed drug delivery system. However, because vigorous ultrasound was used to disperse the particles during polyelectrolyte coating and the particles were washed three times to remove superfluous polyelectrolyte after every coating, the entrapped drug diffused outside into external solution, which resulted in the sharp decrease of encapsulation efficiency with the increase of layer number.

Another method which attracted much attention is utilizing Pickering emulsions to prepare microspheres with nanoparticles as shell, and these microspheres with special structures are named as colloidosomes [2, 18, 93]. The physical properties of colloidosomes can be precisely adjusted by choosing different nanoparticles and changing the preparation conditions. Such versatility of colloidosomes promised their application in many fields. But only in recent years, hydrogel nanoparticles have been utilized to prepare colloidosomes [7, 78, 109] that used chitosan nanoparticles as emulsifier to prepare PLGA microcapsules (Fig. 11.19). The chitosan nanoparticles adhered on the surface of PLGA droplets and stabilized these droplets in water phase. Just by evaporating the oil solvent, the chitosan nanoparticles coated with PLGA microcapsules were formed.

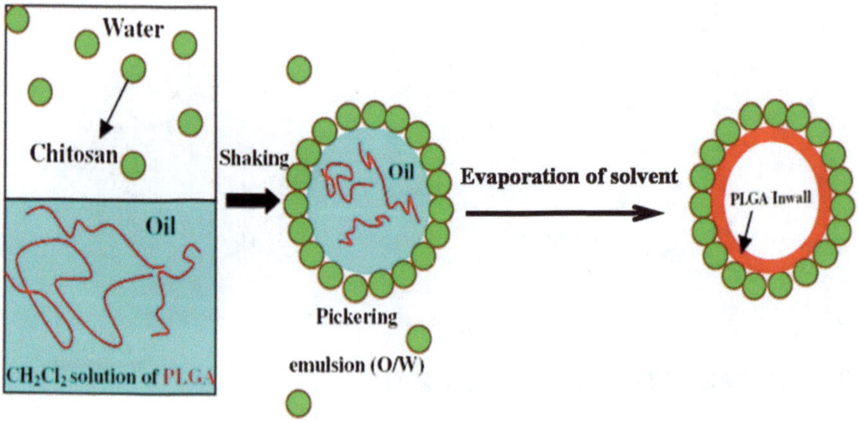

Fig. 11.19 Schematic of the preparation of chitosan nanoparticles coated with PLGA microcapsules by solvent evaporation based on Pickering emulsion droplets (Modified from Wei et al. [109] with permission from Elsevier)

You et al. also fabricated alginate particles which assembled colloidosomes, which showed high pH sensitivity [118]. Due to the small size of particles and big spaces between particles, colloidosomes behaved better and have rapid stimuli-responsive ability compared to general hydrogel microspheres. And the formed colloidosomes showed nonuniform deformation of morphology responding to the change of pH due to the nonuniformity dispersion of particles on the shell.

In order to prepare colloidosomes, Pickering emulsions should have high physical stability which could be assessed by measuring the rheological property of emulsions. The rheological properties of Pickering emulsions are affected by multiple factors, such as the type of used particle and oil, particle concentration, the volume ratio of oil and water, temperature, pH, and ionic strength [23, 82, 91]. Chen et al. evaluated the influence of particle type on rheological behavior of Pickering emulsions, and they found that Pickering emulsion stabilized by moderately hydrophobic particles exhibited relative large yield stresses and thixotropic flow behavior with high stability [10]. The group led by Kawaguchi carried out systematic study about the rheological behavior of Pickering emulsions composed of silicone oil and different particles, and their research provided valuable guidance for studying the stabilization mechanism of Pickering emulsions [35, 89, 90].

11.4.2 As Vaccine Adjuvant

In recent years, besides as drug delivery system, another important application of hydrogel particles, as vaccine adjuvant, has been investigated intensively. With the development of vaccine, the attenuated and inactivated antigens are gradually replaced by subunit vaccine, recombinant vaccine, nucleic vaccine, and other vaccines with better safety. However, those novel vaccines have low immunogenicity and need some strategies to enhance immune response. Using particles as delivery system or vaccine adjuvant is a promising method to solve this problem [39]. The antigens can be encapsulated, adsorbed or cross-linked in the particles, and targeted to the antigen-presenting cells (APCs). The adding of particles in vaccine formulation kept the antigens from degradation and increased the amount of antigens being taken by APCs. For example, Kovacsovics-Bankowski et al. found that antigens linked to beads were presented with MHC class I molecules by APCs up to 10^4-fold more efficiently than soluble antigens [40]. Moreover, some particles have been reported that they are not only delivery system but also work as adjuvant.

Liu et al. used two types of chitosan microparticles with recombinant anthrax protective antigens together [50]. One of them is called as CS-CL MPs and prepared by premix membrane emulsification method. The formed chitosan droplets without emulsion were cross-linked with glutaraldehyde. Another one is named as CS-NH_2 MPs and prepared by layer-by-layer technology with alginate and chitosan. To prepare CS-NH_2 MPs, the alginate microparticles were prepared, firstly. Then these particles were dispersed in chitosan acid solution to coat chitosan polysaccharide layer on the surface. After that, the coated particles were dispersed in alginate solution and chitosan solution again to prepare microparticles with chitosan bilayers. The main difference of these two chitosan microparticles is whether the microparticles remain plenty of amino groups on their surface as shown in Fig. 11.20.

They found that CS-NH_2 MPs with abundant amino groups cannot only stimulate more antigen-specific antibody titers and cytokine production but also activate the complement system than CS-CL MPs. It is partly attributed to the stabilization effect of free amino groups on CS-NH_2 MPs surface for C3b, otherwise C3b would be inactivated quickly.

Besides activating the complement system, other adjuvant effects of hydrogel particles have also been reported. Wang et al. developed another chitosan microparticle system as vaccine adjuvant with multiple mechanisms of immunity enhancement [103]. In their study, the chitosan microparticles were prepared by thermal-gelation method between quaternized chitosan and glycerophosphate, an

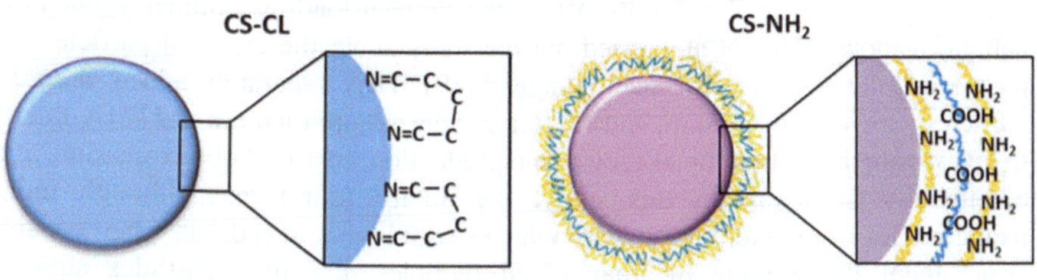

Fig. 11.20 Surface chemical structures of CS-CL MPs (*left*) and CS-NH_2 MPs (*right*) (Modified from Liu et al. [50] with permission from ACS Publications)

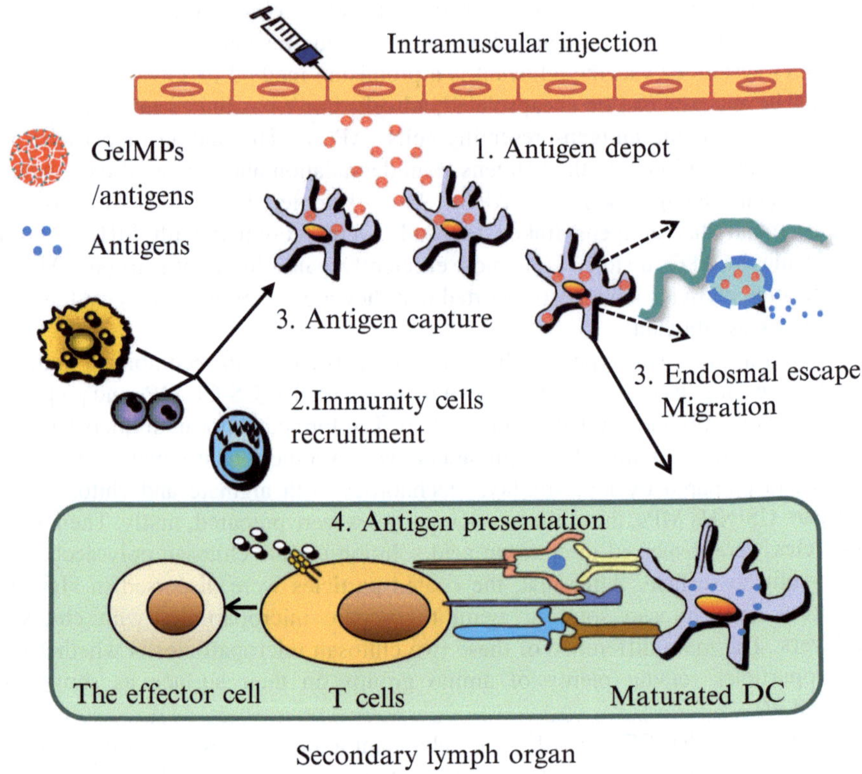

Fig. 11.21 Four main immunity enhancement mechanisms of chitosan microparticles

ionic cross-linking agent. Four main immunity enhancement mechanisms are proved active for this adjuvant system (Fig. 11.21), including: (1) antigen depot, (2) recruiting immunity cells to enhance antigen capture, (3) enabling endosomal escape due to the pH sensitivity of particles, and (4) activating dendritic cells (DCs).

For those hydrogel particles used as vaccine adjuvant, in order to induce high immune response, the optimization of their physical properties is essential. Their physical properties will not only affect the antigen-loading efficiency but also cellular response. Yue et al. carried out the study about the effect of particle size on the cellular responses in macrophages [119]. They separately added chitosan particles with 430 nm, 1.9 μm, and 4.8 μm into the culture medium of J774A.1 cells to investigate the cellular uptake, cytokine production, costimulator expression, and cytotoxicity in detail. In this study, the author found that although more nanoparticles were taken up by macrophages with higher speed, but when considering about the volume of internalized particles, the microparticles showed advantages (Fig. 11.22). Moreover, the size of particles would also affect their trafficking pathway. The author concluded that nanoparticles would be more

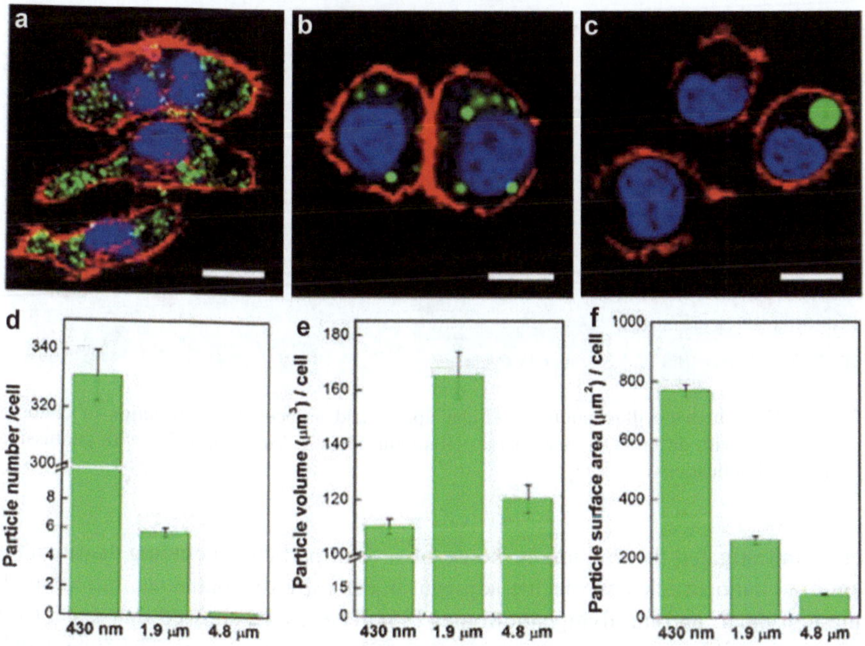

Fig. 11.22 Particle internalization of J774A.1 cells after a 24 h incubation with different sized particles. The CLSM images exhibiting the J774A.1 cells after incubation with 430 nm particles (**a**), 1.9 μm particles (**b**), and 4.8 μm particles (**c**). Particles are clearly distributed in the cell with strong *green* fluorescence. The actin segments bound with rhodamine–phalloidin appear *red* in color, whereas the nucleus dyed with 4,6-diamidino-2-phenylindole (DAPI) is *blue*. Scale bars: 10 μm. The data obtained by flow cytometry (FACS) were further expressed by three means of particle number (**d**), volume (**e**), and surface area (**f**). FACS data were acquired from 15,000 cells per sample and each experiment was performed in triplicate (Reproduced from Yue et al. [119] with permission from Elsevier)

beneficial as vaccine adjuvant which would be transported by both lysosomal and non-lysosomal trafficking, and microparticles would be the better choice as drug delivery system due to their non-lysosomal trafficking pathway.

Besides particle size, Yue et al. investigated the effect of the surface charge of particles on cellular response [120]. They prepared three types of chitosan nanoparticles by SPG membrane emulsification to ensure identical physicochemical properties among them except for the surface charge. The fabricated carboxymethyl chitosan nanoparticles (N-NPs) have a negative charge (-45.84 ± 2.18 mV), and those particles being coated with chitosan (M-NPs) or quaternized chitosan (P-NPs) have a neutral charge (0.51 ± 1.31 mV) and a positive charge (39.25 ± 2.68 mV), separately. They found that P-NPs could achieve high internalization rate and amount. Moreover, P-NPs could escape from lysosome, while N-NPs and M-NPs would highly colocalize with lysosome (Fig. 11.23).

For these vaccine formulations, not only the physical characteristics of particles but also the injection site needs to be considered before application. Mohanan

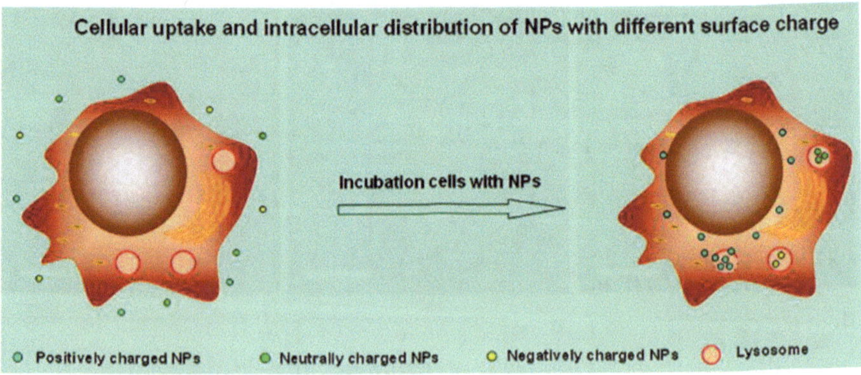

Fig. 11.23 Schematic illustration of cellular uptake and intracellular distribution of chitosan nanoparticles with different surface charges (Reproduced from Yue et al. [120] with permission from ACS Publications)

et al. investigated the immune response of several particles including N-trimethyl chitosan nanoparticles as vaccine adjuvant injected by subcutaneous, intradermal, intramuscular, and intralymphatic routes [55]. The results showed that except for the level of IgG, IgG2a immune responses and IFN-ɣ-producing Th1 type T-cell responses were affected strongly by the administration route.

Besides injection, hydrogel particles have been proved useful for other administration routes, such as mucosal vaccination, and those routes attracted much attention in recent years. The main advantage of mucosal vaccination is that high level of secretory IgA (sIgA) at mucosal site can be induced, and sIgA is important for mucosal protection against respiratory pathogens. Bal et al. investigated the immunogenicity of N-trimethyl chitosan (TMC) nanoparticles as vaccine adjuvant after nasal and intradermal administration, and ovalbumin as a model antigen was encapsulated in nanoparticles [3]. Although the mice which took an intradermal vaccination have higher antibody level in serum than the group of nasal vaccination, the latter could provoke higher sIgA titer.

For mucosal vaccination, materials with positive charge show their advantages. They can adhere on the mucosal membrane by electrostatic interaction to prolong the residence time and promote the penetration of antigens. Slütter et al. compared the adjuvanticity of nanoparticles made from chitosan, N-trimethyl chitosan chloride (TMC), and poly(lactic-co-glycolic acid) (PLGA) for oral vaccination [85]. TMC nanoparticles increased the M-cell-specific transport efficiency and facilitated the opening of tight junctions between the epithelial cells more efficiently than others due to their positive surface charge. Verheul et al. used another way to prove the importance of surface charge of particles for mucosal vaccination [94]. They modified TMC nanoparticles with polyethylene glycol (PEG) to shield their surface charge and used these particles both for nasal and intradermal vaccinations. The nanoparticles with PEGylation lost their ability as vaccine adjuvant compared to that of non-PEGylated one when used intranasally, while the PEGylated nanoparticles for intradermal vaccination could enhance immune responses as well as the non-PEGylated one.

Verheul et al. found that besides surface charge, the particle stability in the administration site would also affect the immune response [95]. They prepared TMC nanoparticles and reacetylated TMC (TMC-RA) nanoparticles for nasal vaccination. They found that TMC-RA would be degraded by lysozyme more extensively than TMC, and this difference would result in the inferior adjuvant effect of TMC-RA particles.

11.4.3 Other Applications

As the research of hydrogel nano- and microparticles moves along, their application is not only limited to drug delivery and vaccine adjuvant but also extended to other places. For example, agarose particles have been used widely for chromatographic separation. Their inherent neutral, hydrophilicity, porosity, and abundance of hydroxyl groups made them an ideal column packing material for biomacromolecule separation. Various types of agarose beads have been prepared as ion-exchange medium, gel filtration medium, and metal-chelating medium. To improve their separation efficiency, recent studies focused on the preparation of superporous agarose (SA) beads. Gottschalk et al. used double emulsification method to prepare SA microparticles to immobilize human red blood cells (RBCs) as affinity chromatography medium [27]. The human RBC glucose transporter GLUT1 in the membranes of the immobilized cells can be used to separate cytochalasin B. The porous structure of agarose particles provided superior cell stability than the solid ones, which would be useful to achieve stable separation efficiency. Shi et al. adopted another technique using solid granules as porogenic agent to prepare SA microparticles [79]. With the existence of wide pores of SA, low back pressure, high protein capacity, and separation efficiency were achieved, and all of these are favorable for high-speed protein separation. Superporous agarose hydrogel particles can also be utilized for other applications. Yang et al. prepared superporous agarose (SA) microparticles as solid support for microfluidic immunoassay [115]. They conjugated protein A onto SA particles and used them to detect goat IgG. The superporous structure of agarose particles facilitated the diffusion of IgG through particles and enhanced the detection sensitivity.

Although most of the applications about chitosan particles focused on the drug delivery, Wei et al. found an interesting property of chitosan particles which would broaden their application range [106]. Chitosan particles cross-linked with glutaraldehyde (CG) or formaldehyde (CF) possessed autofluorescence property as shown in Fig. 11.24. This found is important for biomedical application of chitosan particles. The distribution of particles in the body could be observed easily by in vivo image system without the need of pre-labeling with radio or fluorochromes. Wei et al. utilized this property to investigate the bioprocess of chitosan microparticles in rats following oral administration as shown in Fig. 11.25 [108]. These autofluorescent chitosan particles could be used as tracer agent.

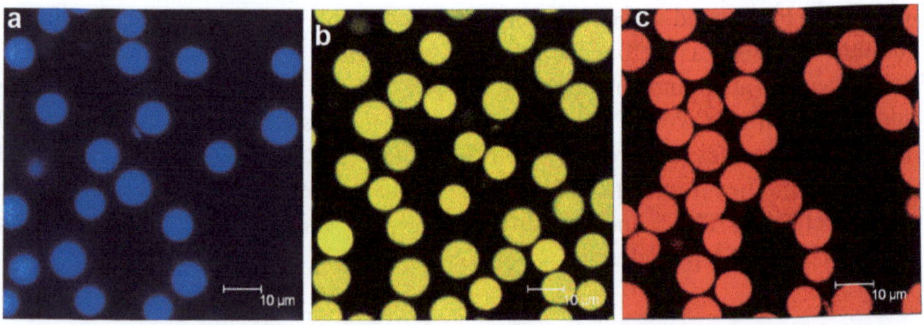

Fig. 11.24 CLSM images of (**a**) selectively reduced CG microspheres excited at 365 nm, (**b**) CG microspheres excited at 488 nm, and (**c**) CF microspheres excited at 543 nm. The scale bars represent 10 μm (Reproduced from Wei et al. [106] with permission from John Wiley and Sons)

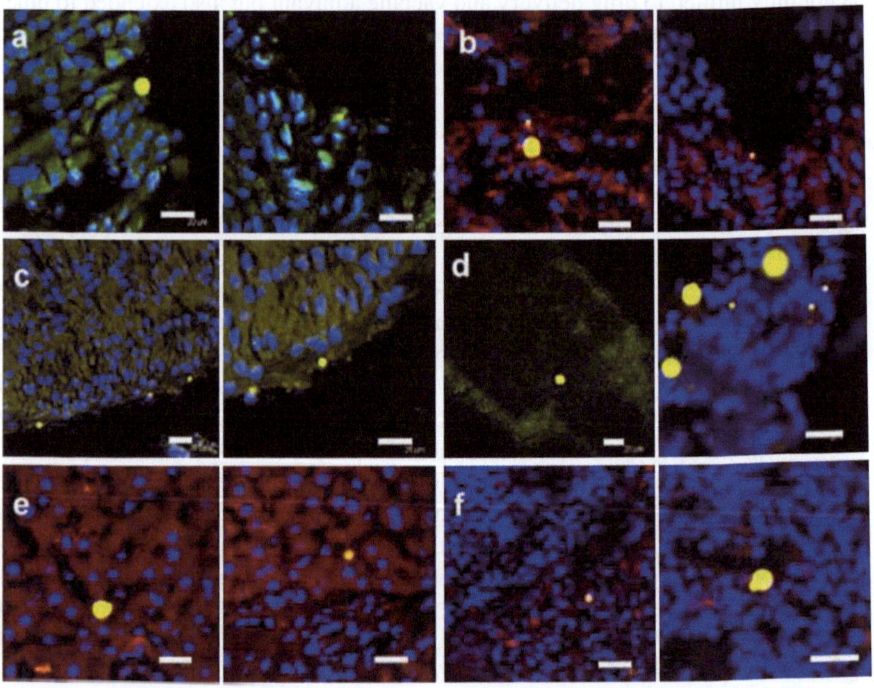

Fig. 11.25 Distribution of CG microspheres with different diameters in systemic tissues: the (**a**) stomach, (**b**) small intestine, (**c**) colon, (**d**) mesenteric vessel and lymph node, (**e**) liver, and (**f**) spleen. Scale bars represent 20 μm in each case (Reproduced from Wei et al. [108] with permission from Elsevier)

Hydrogel particles have also been reported for tissue engineering. They can be utilized to entrap drug or enzyme as bone regeneration templates. To achieve this purpose, calcium titanium phosphate–alginate (CTP–alginate) and hydroxyapatite–alginate (HAp–alginate) microspheres were developed and used to immobilize glucocerebrosidase, the therapeutic enzyme for Gaucher disease [68]. The enzymes, adsorbed onto the ceramic particles or encapsulated inside of the polymer particles, were released from the microspheres with different release profiles. Sakai et al. used jet technique and alginate microparticles as template to prepare agarose microcapsules with hollow-core structure [73]. Feline kidney cells (CRFK cells) were enclosed in microcapsules and grew fast to form embryoid body-like spherical tissues.

At present, hydrogel particles have also been applied as coating materials [75], magnetic resonance imaging (MRI) contrast agent [60], and wound dressing [81].

But in order to further broaden their application range, there are still some disadvantages of hydrogel particles that need to be conquered. For example, the softness and elasticity of hydrogels due to the presence of high water content are often considered to be their advantages because they reduced the frictional irritation between hydrogels and biotissues [61]. But the high water content also resulted in the weak mechanical strength of hydrogels, and the integrity of hydrogel particles might be destroyed during application. In order to increase the mechanical strength, several strategies have been proposed, such as increasing the cross-linking density and adding inorganic components [100]. However, both of these methods have their disadvantages. The high cross-link density would largely affect the size and structure of hydrogel and might make the hydrogel become brittle [61]. The introduction of inorganic components might reduce the biocompatibility of hydrogels. Xu et al. developed another strategy to prepare composite microgels with high mechanical strength [112]. The composite microgels consisted of two networks, which contributed to the lift of mechanical strength. The compressive strength of composite hydrogels was lifted 60–100 times than that of general hydrogel and was up to 17–30 MPa. After that, they also utilized these microgels to prepare bulk hydrogel with high mechanical strength [45]. The optimized hydrogels have compressive strength up to 21.94 MPa.

11.5 Conclusion

As stated above, the soft hydrogel nano- and microparticles have many advantages over other "hard" particles. The high water content, porous structure, elasticity, and stimuli responsibility of hydrogel particles promoted the studies of them in biomedical applications. However, most of these studies are still at the stage of lab research and small-scale experiments. In order to use those particles in clinical tests, there are several obstacles that need to be conquered:

1. Most of hydrogel particles consist of natural polymers, and it is hard to control the batch stability of polymers. The properties of natural polymers are affected by numerous external conditions, such as the sources, the storage conditions, and the treating process, which limited the clinical application of hydrogel particles. Because for those hydrogel particles used as drug delivery system and vaccine adjuvant, the batch stability and quality control are important. Thus, at present

the main biomedical applications of chitosan are limited in topical formulations, such as wound dressing and nasal spray.

2. The size control of natural polymer particles is comparatively more difficult than those of synthetic polymer due to the high viscosity of formers, especially for nanoparticles with high polymer content. The development of membrane emulsification and microfluidic method partly solved this problem, but in large-scale production, there are still many challenges. And due to the high water content, the storage of hydrogel particles is very difficult. Heat-drying and freeze-drying would destroy the shape of particles and induce the aggregate of particles, and the treated particles could not re-disperse in water. Storage of hydrogel particles in water is inconvenient for long-distance transport, and the entrapped drug or antigens will be released in water, which results in low treating efficiency. Adding freeze-drying protecting agents can improve the re-dispersity in water after lyophilization, but the side effects of additives need to be seriously considered. Adopting supercritical drying technique can keep the original shape and dispersity of hydrogel particles, but it is hard to realize in large-scale production.
3. Characterization of hydrogel particles is another question. Surface area, pore size distribution, and volume of particles are hard to be measured quantitatively by conventional equipments. Without definite quality measure methods, the choice and evaluation of particles would be difficult. In order to broaden the application of hydrogel particles, more studies should be carried out which concerned about the above points.

References

1. Aketagawa K, Hirama H, Torii T (2013) Hyper-miniaturisation of monodisperse Janus hydrogel beads with magnetic anisotropy based on coagulation of Fe_3O_4 nanoparticles. J Mater Sci Chem Eng 1:1–5
2. Ao Z, Yang Z, Wang JF et al (2009) Emulsion-templated liquid core-polymer shell microcapsule formation. Langmuir 25:2572–2574
3. Bal SM, Slütter B, Verheul R et al (2012) Adjuvanted, antigen loaded N-trimethyl chitosan nanoparticles for nasal and intradermal vaccination: adjuvant- and site-dependent immunogenicity in mice. Eur J Pharm Sci 45:475–481
4. Bardajee GR, Pourjavadi A, Soleyman R (2011) Novel nano-porous hydrogel as a carrier matrix for oral delivery of tetracycline hydrochloride. Colloids Surf A Physicochem Eng Asp 392:16–24
5. Barrande M, Beurroies I, Denoyel R et al (2009) Characterisation of porous materials for bioseparation. J Chromatogr A 1216(41):6906–6916
6. Best JP, Cui JW, Müllner M et al (2013) Tuning the mechanical properties of nanoporous hydrogel particles via polymer cross-linking. Langmuir 29(31):9824–9831
7. Binks BP, Murakami R, Armes SP et al (2006) Effects of pH and salt concentration on oil-in-water emulsions stabilized solely by nanocomposite microgel particles. Langmuir 22:2050–2057
8. Cai X, Tong X, Shen XY et al (2009) Preparation and characterization of homogeneous chitosan-polylactic acid/hydroxyapatite nanocomposite for bone tissue engineering and evaluation of its mechanical properties. Acta Biomater 5(7):2693–2703

9. Caldorera-Moore M, Kang MK, Moore Z et al (2011) Swelling behavior of nanoscale, shape- and size-specific, hydrogel particles fabricated using imprint lithography. Soft Matter 7:2879–2887
10. Chen J, Vogel R, Werner S et al (2011) Influence of the particle type on the rheological behavior of Pickering emulsions. Colloids Surf A Physicochem Eng Asp 382:238–245
11. Chen QH, Li QQ, Lin JH (2011) Synthesis of Janus composite particles by the template of dumbbell-like silica/polystyrene. Mater Chem Phys 128(3):377–382
12. Chu LY, Utada AS, Shah RK et al (2007) Controllable monodisperse multiple emulsions. Angew Chem Int Ed 119(47):9128–9132
13. Chuah AM, Kuroiwa T, Kobayashi I et al (2009) Preparation of uniformly sized alginate microspheres using the novel combined methods of microchannel emulsification and external gelation. Colloids Surf A Physicochem Eng Asp 351:9–17
14. Chung BG, Lee KH, Khademhosseini A et al (2012) Microfluidic fabrication of microengineered hydrogels and their application in tissue engineering. Lab Chip 12:45–59
15. Csaba N, Köping-Höggård M, Alonso MJ (2009) Ionically crosslinked chitosan/tripolyphosphate nanoparticles for oligonucleotide and plasmid DNA delivery. Int J Pharm 382:205–214
16. Desmarais SM, Haagsman HP, Barron AE (2012) Microfabricated devices for biomolecule encapsulation. Electrophoresis 33:2639–2649
17. Díez-Pascual AM, Wong JE (2010) Effect of layer-by-layer confinement of polypeptides and polysaccharides onto thermoresponsive microgels: A comparative study. J Colloid Interface Sci 347:79–89
18. Dinsmore AD, Hsu MF, Nikolaides MG et al (2002) Colloidosomes: selectively permeable capsules composed of colloidal particles. Science 298:1006–1009
19. Dohnal J, Štěpánek F (2010) Inkjet fabrication and characterization of calcium alginate microcapsules. Powder Technol 200:254–259
20. Du KF, Bai S, Dong XY et al (2010) Fabrication of superporous agarose beads for protein adsorption: effect of $CaCO_3$ granules content. J Chromatogr A 1217:5808–5816
21. Eldridge JH, Hammond CJ, Meulbroek JA (1990) Controlled vaccine release in the gut-associated lymphoid tissues: orally administered biodegradable microspheres target the Peyer's patches. J Control Release 11:205–214
22. Franco CL, Price J, West JL (2011) Development and optimization of a dual-photoinitiator, emulsion-based technique for rapid generation of cell-laden hydrogel microspheres. Acta Biomater 7(9):3267–3276
23. Frith WJ, Pichot R, Kirkland M et al (2008) Formation, stability, and rheology of particle stabilized emulsions: influence of multivalent cations. Ind Eng Chem Res 47(17):6434–6444
24. Gan Q, Wang T, Cochrane C et al (2005) Modulation of surface charge, particle size and morphological properties of chitosan–TPP nanoparticles intended for gene delivery. Colloids Surf B: Biointerfaces 44:65–73
25. Gan Q, Wang T (2007) Chitosan nanoparticle as protein delivery carrier – systematic examination of fabrication conditions for efficient loading and release. Colloids Surf B: Biointerfaces 59(1):24–34
26. Gao D, Liu JJ, Wei HB et al (2010) A microfluidic approach for anticancer drug analysis based on hydrogel encapsulated tumor cells. Anal Chim Acta 665(1):7–14
27. Gottschalk I, Gustavsson PE, Ersson B et al (2003) Improved lectin-mediated immobilization of human red blood cells in superporous agarose beads. J Chromatogr B 784:203–208
28. Hsiao MH, Larsson M, Larsson A et al (2012) Design and characterization of a novel amphiphilic chitosan nanocapsule-based thermo-gelling biogel with sustained in vivo release of the hydrophilic anti-epilepsy drug ethosuximide. J Control Release 161(2012):942–948
29. Hu XH, Zhou J, Zhang N et al (2008) Preparation and properties of an injectable scaffold of poly(lactic-co-glycolic acid) microparticles/chitosan hydrogel. J Mech Behav Biomed Mater 1(4):352–359

30. Hwang DK, Dendukuri D, Doyle PS (2008) Microfluidic-based synthesis of non-spherical magnetic hydrogel microparticles. Lab Chip 8:1640–1647
31. Ikkai F, Iwamoto S, Adachi E et al (2005) New method of producing mono-sized polymer gel particles using microchannel emulsification and UV irradiation. Colloid Polym Sci 283:1149–1153
32. Iwamoto S, Nakagawa K, Sugiura S et al (2002) Preparation of gelatin microbeads with a narrow size distribution using microchannel emulsification. AAPS Pharm Sci Technol 3(3):72–76
33. Iwanaga S, Saito N, Sanae H et al (2013) Facile fabrication of uniform size-controlled microparticles and potentiality for tandem drug delivery system of micro/nanoparticles. Colloids Surf B: Biointerfaces 109:301–306
34. Jiang GQ, Wang YJ, Ding FX (2013) Thermosensitive hydrogel incorporating microspheres for injectable implant delivery of naltrexone. Adv Mater Res 647:71–79
35. Kawazoe A, Kawaguchi M (2011) Characterization of silicone oil emulsions stabilized by TiO2 suspensions pre-adsorbed SDS. Colloids Surf A Physicochem Eng Asp 392:283–287
36. Khodaverdi E, Tafaghodi M, Ganji F et al (2012) *In vitro* insulin release from thermosensitive chitosan hydrogel. AAPS Pharm Sci Technol 13(2):460–466
37. Kim K, Cheng J, Liu Q et al (2009) Investigation of mechanical properties of soft hydrogel microcapsules in relation to protein delivery using a MEMS force sensor. J Biomed Mater Res A 92A(1):103–113
38. Kobayashi I, Mukataka S, Nakajima M (2005) Novel asymmetric through-hole array microfabricated on a silicon plate for formulating monodisperse emulsions. Langmuir 21:7629–7632
39. Koppolu B, Zaharoff DA (2013) The effect of antigen encapsulation in chitosan particles on uptake, activation and presentation by antigen presenting cells. Biomaterials 34:2359–2369
40. Kovacsovics-Bankowski M, Clark K, Benacerraf B et al (1993) Efficient major histocompatibility complex class I presentation of exogenous antigen upon phagocytosis by macrophages. Proc Natl Acad Sci U S A 90:4942–4946
41. Latterini L, Tarpani L (2012) AFM measurements to investigate particulates and their interactions with biological macromolecules. In: Frewin C (ed) Atomic force microscopy investigations into biology – from cell to protein. InTech, Rijeka, pp 87–98
42. Lau TT, Ho LW, Wang DA (2013) Hepatogenesis of murine induced pluripotent stem cells in 3D micro-cavitary hydrogel system for liver regeneration. Biomaterials 34(28):6659–6669
43. Li T, Shi XW, Du YM et al (2007) Quaternized chitosan/alginate nanoparticles for protein delivery. J Biomed Mater Res A 83(2):283–290
44. Li J, Huang QR (2012) Rheological properties of chitosan – tripolyphosphate complexes: from suspensions to microgels. Carbohydr Polym 87(2):1670–1677
45. Li PC, Xu K, Tan Y et al (2013) A novel fabrication method of temperature-responsive poly (acrylamide) composite hydrogel with high mechanical strength. Polymer 54:5830–5838
46. Liang YY, Jeong J, DeVolder RJ et al (2011) A cell-instructive hydrogel to regulate malignancy of 3D tumor spheroids with matrix rigidity. Biomaterials 32(35):9308–9315
47. Liu L, Wang W, Ju XJ et al (2010) Smart thermo-triggered squirting capsules for nanoparticle delivery. Soft Matter 6:3759–3763
48. Liu L, Wu F, Ju XJ et al (2013) Preparation of monodisperse calcium alginate microcapsules via internal gelation in microfluidic-generated double emulsions. J Colloid Interface Sci 404:85–90
49. Liu Z, Shum HC (2013) Fabrication of uniform multi-compartment particles using microfluidic electrospray technology for cell co-culture studies. Biomicrofluidics. doi:10.1063/1.4817769
50. Liu Y, Yin Y, Wang LY et al (2013) Engineering biomaterial-associated complement activation to improve vaccine efficacy. Biomacromolecules 14:3321–3328

51. Lv PP, Wei W, Yue H et al (2011) Porous quaternized chitosan nanoparticles containing paclitaxel nanocrystals improved therapeutic efficacy in non-small-cell lung cancer after oral administration. Biomacromolecules 12:4230–4239
52. Lv PP, Ma YF, Yu R et al (2012) Targeted delivery of insoluble cargo (paclitaxel) by PEGylated chitosan nanoparticles grafted with Arg-Gly-Asp (RGD). Mol Pharmaceutics 9:1736–1747
53. Masotti A, Marino F, Ortaggi G et al (2007) Fluorescence and scanning electron microscopy of chitosan/DNA nanoparticles for biological applications. In: Méndez-Vilas A, Díaz J (eds) Modern research and educational topics in microscopy, vol 2. Formatex, Badajoz, pp 690–696
54. Meng QQ, Wang JX, Ma GH et al (2009) Lyophilization of CNBr-activated agarose beads with lactose and PEG. Process Biochem 44:562–571
55. Mohanan D, Slütter B, Henriksenn-Lacey M et al (2010) Administration routes affect the quality of immune responses: a cross-sectional evaluation of particulate antigen-delivery systems. J Control Release 147:342–349
56. Morris GA, Castile J, Smith A et al (2011) The effect of prolonged storage at different temperatures on the particle size distribution of tripolyphosphate (TPP) – chitosan nanoparticles. Carbohydr Polym 84(4):1430–1434
57. Nakano K, Tozuka Y, Yamamoto H et al (2008) A novel method for measuring rigidity of submicron-size liposomes with atomic force microscopy. Int J Pharm 355:203–209
58. Nayak AK, Pal D (2011) Development of pH-sensitive tamarind seed polysaccharide-alginate composite beads for controlled diclofenac sodium delivery using response surface methodology. Int J Biol Macromol 49:784–793
59. Nie L, Liu SY, Shen WM et al (2007) One-pot synthesis of amphiphilic polymeric Janus particles and their self-assembly into supermicelles with a narrow size distribution. Angew Chem Int Ed 46:6321–6324
60. Oerlemans C, Seevinck PR, van de Maat GH et al (2013) Alginate–lanthanide microspheres for MRI-guided embolotherapy. Acta Biomater 9:4681–4687
61. Pal K, Banthia AK, Majumdar DK (2009) Polymeric hydrogels: characterization and biomedical applications – a mini review. Des Monomers Polym 12:197–220
62. Paques JP, van der Linden E, van Rijn CJM et al (2013) Alginate submicron beads prepared through w/o emulsification and gelation with CaCl2 nanoparticles. Food Hydrocoll 31:428–434
63. Park BJ, Brugarolas T, Lee D (2011) Janus particles at an oil-water interface. Soft Matter 7:6413–6417
64. Park MR, Chun CJ, Cho CS et al (2010) Enhancement of sustained and controlled protein release using polyelectrolyte complex-loaded injectable and thermosensitive hydrogel. Eur J Pharm Biopharm 76(2):179–188
65. Pereira R, Carvalho A, Vaz DC et al (2013) Development of novel alginate based hydrogel films for wound healing applications. Int J Biol Macromol 52:221–230
66. Quinn CAP, Connor RE, Heller A (1997) Biocompatible, glucose-permeable hydrogel for in situ coating of implantable biosensors. Biomaterials 18(24):1665–1670
67. Rampino A, Borgogna M, Blasi P et al (2013) Chitosan nanoparticles: preparation, size evolution and stability. Int J Pharm 455(1–2):219–228
68. Ribeiro CC, Barrias CC, Barbosa MA (2003) Calcium phosphate-alginate microspheres as enzyme delivery matrices. Biomaterials 25:4363–4373
69. Rondeau E, Cooper-White JJ (2008) Biopolymer microparticle and nanoparticle formation within a microfluidic device. Langmuir 24:6937–6945
70. Sahiner N, Butun S, Ilgin P (2011) Soft hydrogel particles with high functional value. Colloids Surf A Physicochem Eng Asp 381:74–84
71. Sajeesh S, Sharma CP (2006) Novel pH responsive polymethacrylic acid-chitosan-polyethylene glycol nanoparticles for oral peptide delivery. J Biomed Mater Res B Appl Biomater 76(2):298–305

72. Sakaguchi G, Tambara K, Sakakibara Y et al (2005) Control-released hepatocyte growth factor prevents the progression of heart failure in stroke-prone spontaneously hypertensive rats. Ann Thorac Surg 79(5):1627–1634
73. Sakai S, Hashimoto I, Kawakami K (2008) Production of cell-enclosing hollow-core agarose microcapsules via jetting in water-immiscible liquid paraffin and formation of embryoid body-like spherical tissues from mouse ES cells enclosed within these microcapsules. Biotechnol Bioeng 99(1):235–243
74. Schoubben A, Blasi P, Giovagnoli S et al (2009) Novel composite microparticles for protein stabilization and delivery. Eur J Pharm Sci 36:226–234
75. Scott EA, Nichols MD, Cordova LH et al (2008) Protein adsorption and cell adhesion on nanoscale bioactive coatings formed from poly(ethylene glycol) and albumin microgels. Biomaterials 29:4481–4493
76. Seeber M, Zdyrko B, Burtovvy R et al (2011) Surface grafting of thermoresponsive microgel nanoparticles. Soft Matter 7:9962–9971
77. Shah RK, Kim JW, Agresti JJ et al (2008) Fabrication of monodisperse thermosensitive microgels and gel capsules in microfluidic devices. Soft Matter 4:2303–2309
78. Shah RK, Kim JW, Weitz DA (2010) Monodisperse stimuli-responsive colloidosomes by self-assembly of microgels in droplets. Langmuir 26(3):1561–1565
79. Shi QH, Zhou X, Sun Y (2005) A novel superporous agarose medium for high-speed protein chromatography. Biotechnol Bioeng 92(5):643–651
80. Siegel RA, Gu YD, Lei M et al (2010) Hard and soft micro- and nanofabrication: an integrated approach to hydrogel-based biosensing and drug delivery. J Control Release 141:303–313
81. Silan C, Akcali A, Otkun MT et al (2012) Novel hydrogel particles and their IPN films as drug delivery systems with antibacterial properties. Colloids Surf B: Biointerfaces 89:248–253
82. Simon S, Theiler S, Knudsen A et al (2010) Rheological properties of particle-stabilized emulsions. J Dispers Sci Technol 31(5):632–640
83. Singh B, Pal L (2012) Sterculia crosslinked PVA and PVA-poly(AAm) hydrogel wound dressings for slow drug delivery: Mechanical, mucoadhesive, biocompatible and permeability properties. J Mech Behav Biomed Mater 9:9–21
84. Singh B, Sharma S, Dhiman A (2013) Design of antibiotic containing hydrogel wound dressings: biomedical properties and histological study of wound healing. Int J Pharm 457 (1):82–91
85. Slütter B, Plapied L, Fievez V et al (2009) Mechanistic study of the adjuvant effect of biodegradable nanoparticles in mucosal vaccination. J Control Release 138:113–121
86. Smidsrod O, Skjakbraek G (1990) Alginate as immobilization matrix for cells. Trends Biotechnol 8:71–78
87. Soane RJ, Frier M, Perkins AC (1999) Evaluation of the clearance characteristics of bioadhesive systems in humans. Int J Pharm 178(1):55–65
88. Son KD, Yang DJ, Kim MS (2012) Effect of alginate as polymer matrix on the characteristics of hydroxyapatite nanoparticles. Mater Chem Phys 132(2–3):1041–1047
89. Sugita N, Nomura S, Kawaguchi M (2008) Rheological and interfacial properties of silicone oil emulsions stabilized by silica particles. J Dispers Sci Technol 29(7):931–936
90. Suzuki T, Morishita C, Kawaguchi M (2010) Effects of surface properties on rheological and interfacial properties of Pickering emulsions prepared by fumed silica suspensions pre-adsorbed poly(N-Isopropylacrylamide). J Dispers Sci Technol 31(11):1479–1488
91. Tzoumaki MV, Moschakis T, Kiosseoglou V et al (2011) Oil-in-water emulsions stabilized by chitin nanocrystal particles. Food Hydrocoll 25(6):1521–1529
92. Vaghani SS, Patel MM, Satish CS (2012) Synthesis and characterization of pH-sensitive hydrogel composed of carboxymethyl chitosan for colon targeted delivery of ornidazole. Carbohydr Res 347:76–82
93. Velev OD, Furusawa K, Nagayama K (1996) Assembly of latex particles by using emulsion droplets as templates. 1. Microstructured hollow spheres. Langmuir 12:2374–2384

94. Verheul RJ, Slütter B, Bal SM et al (2011) Covalently stabilized trimethyl chitosan-hyaluronic acid nanoparticles for nasal and intradermal vaccination. J Control Release 156:46–52
95. Verheul RJ, Hagenaars N, van Es T et al (2012) A step-by-step approach to study the influence of N-acetylation on the adjuvanticity of N, N, N-trimethyl chitosan (TMC) in an intranasal nanoparticulate influenza virus vaccine. Eur J Pharm Sci 45:467–474
96. Vladisavljević GT, Kobayashi I, Nakajima M (2012) Production of uniform droplets using membrane, microchannel and microfluidic emulsification devices. Microfluid Nanofluid 13:151–178
97. Walther A, Müller AHE (2013) Janus particles: synthesis, self-assembly, physical properties, and applications. Chem Rev 113:5194–5261
98. Wang N, Wu XS (1998) A novel approach to stabilization of protein drugs in poly(lactic-co-glycolic acid) microspheres using agarose hydrogel. Int J Pharm 166:1–14
99. Wang LY, Ma GH, Su ZG (2005) Preparation of uniform sized chitosan microspheres by membrane emulsification technique and application as a carrier of protein drug. J Control Release 106:62–75
100. Wang Y, Chen DJ (2012) Preparation and characterization of a novel stimuli-responsive nanocomposite hydrogel with improved mechanical properties. J Colloid Interface Sci 372:245–251
101. Wang YX, Qin J, Wei Y et al (2013) Preparation strategies of thermo-sensitive P(NIPAM-co-AA) microspheres with narrow size distribution. Powder Technol 236:107–113
102. Wang W, Zhang MJ, Chu LY (2013) Functional polymeric microparticles engineered from controllable microfluidic emulsions. Acc Chem Res. doi:10.1021/ar4001263
103. Wang YQ, Wu J, Fan QZ et al (2013) Novel vaccine delivery system induces robust humoral and cellular immune responses based on multiple mechanisms. Adv Healthcare Mater. doi:10.1002/adhm.201300335
104. Wang W, Zhang MJ, Xie R et al (2013) Hole-shell microparticles from controllably evolved double emulsions. Angew Chem Int Ed 52:8084–8087
105. Wang JY, Hu YD, Deng RH et al (2013) Multiresponsive hydrogel photonic crystal microparticles with inverse-opal structure. Langmuir 29(28):8825–8834
106. Wei W, Wang LY, Yuan L et al (2007) Preparation and application of novel microspheres possessing autofluorescent properties. Adv Funct Mater 17:3153–3158
107. Wei W, Yuan L, Hu G et al (2008) Monodisperse chitosan microspheres with interesting structures for protein drug delivery. Adv Mater 9999:1–5
108. Wei W, Wang LY, Yuan L et al (2008) Bioprocess of uniform-sized crosslinked chitosan microspheres in rats following oral administration. Eur J Pharm Biopharm 69:878–886
109. Wei ZJ, Wang CY, Zou SW et al (2012) Chitosan nanoparticles as particular emulsifier for preparation of novel pH-responsive Pickering emulsions and PLGA microcapsules. Polymer 53:1229–1235
110. Wei W, Lv PP, Chen XM et al (2013) Codelivery of mTERT siRNA and paclitaxel by chitosan-based nanoparticles promoted synergistic tumor suppression. Biomaterials 34:3912–3923
111. Wu J, Wei W, Wang LY et al (2008) Preparation of uniform-sized pH-sensitive quaternized chitosan microsphere by combining membrane emulsification technique and thermal-gelation method. Colloids Surf B: Biointerfaces 63:164–175
112. Xu K, Tan Y, Chen Q et al (2010) A novel multi-responsive polyampholyte composite hydrogel with excellent mechanical strength and rapid shrinking rate. J Colloid Interface Sci 345(2):360–368
113. Yan Y, Zhang ZB, Stokes JR et al (2009) Mechanical characterization of agarose micro-particles with a narrow size distribution. Powder Technol 192:122–130
114. Yang CH, Huang KS, Chang JY (2007) Manufacturing monodisperse chitosan microparticles containing ampicillin using a microchannel chip. Biomed Microdevices 9(2):253–259

115. Yang Y, Nam SW, Lee NY et al (2008) Superporous agarose beads as a solid support for microfluidic immunoassay. Ultramicroscopy 108:1384–1389
116. Yao KD, Liu WG, Lin Z et al (1999) In situ atomic force microscopy measurement of the dynamic variation in the elastic modulus of swollen chitosan/gelatin hybrid polymer network gels in media of different pH. Polym Int 48:794–798
117. You JO, Peng CA (2005) Calcium-alginate nanoparticles formed by reverse microemulsion as gene Carriers. Macromol Symp 219(1):147–153
118. You JO, Rafat M, Auguste DT (2011) Cross-linked, heterogeneous colloidosomes exhibit pH-induced morphogenesis. Langmuir 27:11282–11286
119. Yue H, Wei W, Yue ZG et al (2010) Particle size affects the cellular response in macrophages. Eur J Pharm Sci 41:650–657
120. Yue ZG, Wei W, Lv PP et al (2011) Surface charge affects cellular uptake and intracellular trafficking of chitosan-based nanoparticles. Biomacromolecules 12:2440–2446
121. Zhang YL, Jia XY, Wang LY et al (2011) Preparation of Ca-alginate microparticles and its application for phenylketonuria oral therapy. Ind Eng Chem Res 50:4106–4112
122. Zhang YL, Wei W, Lv PP et al (2011) Preparation and evaluation of alginate-chitosan microspheres for oral delivery of insulin. Eur J Pharm Biopharm 77:11–19
123. Zhang MJ, Wang W, Xie R et al (2013) Microfluidic fabrication of monodisperse microcapsules for glucose-response at physiological temperature. Soft Matter 9:4150–4159
124. Zhao LL, Zhu LJ, Liu FY et al (2011) pH triggered injectable amphiphilic hydrogel containing doxorubicin and paclitaxel. Int J Pharm 410(1–2):83–91
125. Zhou QZ, Wang LY, Ma GH et al (2007) Preparation of uniform-sized agarose beads by microporous membrane emulsification technique. J Colloid Interface Sci 311:118–127
126. Zhou QZ, Liu XY, Liu SJ et al (2008) Preparation of uniformly sized agarose microcapsules by membrane emulsification for application in sorting bacteria. Ind Eng Chem Res 47(17):6386–6390
127. Zhou QZ, Wang LY, Ma GH et al (2008) Multi-stage premix membrane emulsification for preparation of agarose microbeads with uniform size. J Membr Sci 322:98–104
128. Zhou QZ, Ma GH, Su ZG (2009) Effect of membrane parameters on the size and uniformity in preparing agarose beads by premix membrane emulsification. J Membr Sci 326:694–700
129. Zhou GC, Lu Y, Zhang H et al (2013) A novel pulsed drug-delivery system: polyelectrolyte layer-by-layer coating of chitosan–alginate microgels. Int J Nanomedicine 8:877–887

Control of the Multi-scale Structure of Scaffolds and Its Application in Tissue Engineering

12

Kazuya Furusawa

Abstract

In the field of tissue engineering, various engineered tissues have been constructed by culturing cells with cell scaffolds that play the role of extracellular matrices. However, to construct engineered tissues that exhibit native biological functions has been challenging. This is because biological tissues feature complex hierarchical structures. At a microscopic scale, biological tissues such as muscle, bone, tendon, and cornea present highly ordered structures. Conversely, at a macroscopic scale, local structures, cell types, cell density, and the compositions of biomolecules within biological tissues are heterogeneously distributed. The heterogeneities of tissues are spatially regulated and are not completely random. Because the complex hierarchical structures contribute various biological functions, we must construct engineered tissues that feature biomimetic hierarchical structures to regenerate biological functions. Thus, fabricating cell scaffolds that exhibit biomimetic multi-scale structures are crucial because these scaffolds can serve as a template for use in constructing engineered tissues that possess biomimetic hierarchical structures. To fabricate biomimetic cell scaffolds, we must control the multi-scale structures of the cell scaffolds. In this chapter, numerous examples of methodologies used for controlling microscopic and macroscopic structures of scaffolds are described and their applications in the field of tissue engineering are introduced.

K. Furusawa (✉)
Faculty of Advanced Life Science, Hokkaido University, Kita-ku Kita 10 Nishi 8, Sapporo, Hokkaido, Japan
e-mail: kfurusawa@sci.hokudai.ac.jp

R. Kita, T. Dobashi (eds.), *Nano/Micro Science and Technology in Biorheology*,
DOI 10.1007/978-4-431-54886-7_12

Keywords
Multi-scale structure • Cell scaffold • Tissue engineering

12.1 Introduction

Langer and Vacanti defined "tissue engineering" as an interdisciplinary field developed for creating and repairing biological tissues [1]. The final goal in tissue engineering is to construct whole tissues that exhibit native biological functions. One of the strategies used to achieve this goal involves constructing an engineered tissue by culturing cells with cell scaffolds (Fig. 12.1). However, to construct

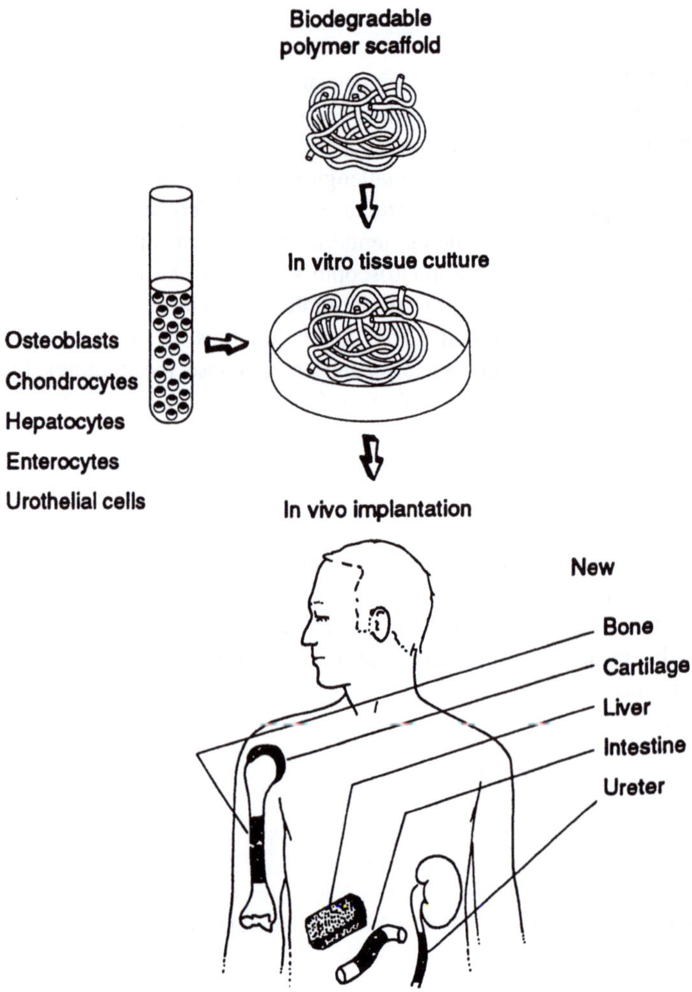

Fig. 12.1 A strategy for constructing engineered tissues (Adapted from Langer and Vacanti [1]). Copyright 1993 AAAS

engineered tissues that display native biological functions is challenging because native tissues possess complex 3D hierarchical structures. Because the 3D hierarchical structures contribute to the biological functions of tissues, we must construct engineered tissues that feature biomimetic 3D hierarchical structures to regenerate native biological functions. For this purpose, cell scaffolds that exhibit biomimetic multi-scale structures are well suited, because these scaffolds can be used as a template for developing the multi-scale structures of biological tissues. To fabricate such biomimetic cell scaffolds, we must control the multi-scale structures of cell scaffolds.

Various methods have been developed to fabricate cell scaffolds featuring multi-scale structures. To control the multi-scale structures of cell scaffolds, numerous rheological techniques such as the applications of external fields, microfluidic technologies, and self-organizations of biopolymer solutions have been used. In this chapter, several examples of the methods used to control the multi-scale structures of cell scaffolds are described. To design the multi-scale structures of cell scaffolds, understanding how the structure of cell scaffolds affects cellular behaviors is helpful. Therefore, in Sect. 12.2, the effects that microscopic and macroscopic structures of cell scaffolds exert on cellular behaviors are briefly discussed. In Sect. 12.3, we review, also briefly, the materials used to prepare cell scaffolds. In Sects. 12.4 and 12.5, several examples of methods used to control the microscopic and macroscopic structures of cell scaffolds and their applications in the field of tissue engineering are described. In Sect. 12.6, we introduce dialysis-induced anisotropic gelation, which can be used to control both the microscopic and macroscopic structures of cell scaffolds, and then discuss the application of this method in the field of tissue engineering.

12.2 Effects of the Structure and Properties of Cell Scaffolds on Cellular Behaviors

12.2.1 Cell-Extracellular Matrix Interaction

Biological tissues consist of cells and the extracellular matrix (ECM) [2]. The ECM consists of various proteins such as collagen, fibronectin, and laminin, as well as proteoglycans. Therefore, the ECM is regarded as a multicomponent material. Almost all the cells in our body adhere to the ECM, and they do so mainly by means of the interaction between integrin and ECM proteins. Integrin is a transmembrane protein present on the surface of cells that binds to ECM proteins such as collagen, fibronectin, and laminin [3]. The interaction between integrin and ECM proteins translates cues from the ECM into intracellular signal transduction. This adhesion-mediated signal transduction depends on the combinations of integrin and ECM proteins, demonstrating that cells can sense the compositions of the ECM. Therefore, cells behave distinctly depending on the compositions of the ECM. Moreover, cells sense the physical properties of the ECM, such as structural and mechanical properties, through adhesion-mediated signal transduction, and cells respond to these physical properties of the ECM [4–10]. The cell-ECM interaction affects various cellular behaviors such as proliferation, morphology change, adhesion, migration, and differentiation, and thus, the interaction directs the morphogenetic process. Therefore, designing the physical properties of cell scaffolds is crucial for constructing functional engineered tissues.

12.2.2 Effects of the Microscopic Structure of ECM on Cellular Behaviors

Highly ordered structures are often observed in the ECM of connective tissues such as bone, tendon, cornea, and blood vessel. For example, oriented collagen fiber arrays are present in the tendon tissue [11]. The highly ordered structures of the ECM affect not only mechanical properties but also various cellular behaviors in the connective tissues. Pioneering studies on the effects of the ordered structures of ECMs on cell behaviors were conducted by Weiss [4, 12], who showed that cells aligned parallel to the orientation of fibers in blood clots and fish scales (Fig. 12.2). This phenomenon is known as "contact guidance." Weiss and Taylor also reproduced contact guidance in vitro using glass that had been scored using a MicroLathe [12]: cells elongated along an orientation axis of the microgrooves of the scored glass. Since these seminal studies, the mechanism of contact guidance has been intensively investigated. Although contact guidance has been widely suggested to occur as a result of anisotropic force generation by lamellipodia and filopodia [13, 14], the mechanism of contact guidance remains unresolved. Conversely, contact guidance has been widely reported to induce directional cell migration [15–17]. Directional cell migration is well known to play a key role in tissue morphogenesis [2]. Therefore, the control of contact guidance is crucial when

Fig. 12.2 Contact guidance of connective tissue cells cultured on a sheet of parallel collagen fibers obtained from the interior of a fish scale (Adapted from Weiss [12]). Copyright 1959 AIP Publishing LLC

constructing engineered tissues that feature biomimetic hierarchical structures, and hence, methods are required to produce cell scaffolds exhibiting anisotropic structures (anisotropic cell scaffolds) and to control the anisotropic structures of the scaffolds. As introduced in Sect. 12.4, various techniques have been developed to prepare anisotropic cell scaffolds and these techniques have been applied in the field of tissue engineering.

12.2.3 Influence of Macroscopic Structure on the Functions of Tissues

At a macroscopic scale, cell density and morphology and the structure, properties, and compositions of the ECM of native tissues are heterogeneously distributed. The heterogeneities observed in native tissues are spatially regulated and are not completely random. For example, the bone density and the hierarchical structure present at the peripheral parts of long bones are considerably distinct from those at inner parts. Long bones contain cortical bone at the periphery and cancellous bone in the interior. The cortical bone exhibits high bone density and a highly ordered hierarchical structure, whereas the cancellous bone exhibits low bone density and a spongelike structure composed of trabecular bones. The macroscopic inhomogeneity contributes the mechanical properties of long bones [18–20]. By contrast, the confined compression modulus of joint cartilage increases with increasing distance from the articular surface [21]. The compression modulus gradient contributes the functions of cartilage, such as lubrication and shock absorption. Furthermore, the vascular networks present in native tissues consist of vessels of various diameters and lengths and feature self-similar structures [22, 23]. The hierarchical structure of vascular networks results in the efficient transportation of oxygen and nutrients. Thus, mimicking the heterogeneities of tissues is critical for the construction of highly functional engineered tissues.

Tissue morphogenetic processes are controlled through the spatial regulation of cellular functions. Therefore, spatially regulating cellular behavior is crucial for obtaining engineered tissues that mimic the heterogeneities of native tissues. As mentioned in Sect. 12.2.1, cellular behaviors depend on local compositions, structures, and mechanical properties of the ECM. Therefore, controlling the spatial distributions of the compositions, structures, and mechanical properties of cell scaffolds can facilitate the spatial regulation of cellular behavior. For example, culturing cells on a cell scaffold exhibiting an elastic modulus gradient generates a cell density gradient (Fig. 12.3) [24–26]. The formation of a cell density gradient can be attributed to the directional migration of cells induced by the physical cue provided by the elastic modulus gradient of the cell scaffold. This type of directional migration is known as "durotaxis."

In recent years, various methodologies have been developed to control the spatial distribution of compositions, structures, and mechanical properties of cell scaffolds. In Sect. 12.5, several examples of the methods used to regulate the spatial distributions of composition, structure, and mechanical properties of cell scaffolds

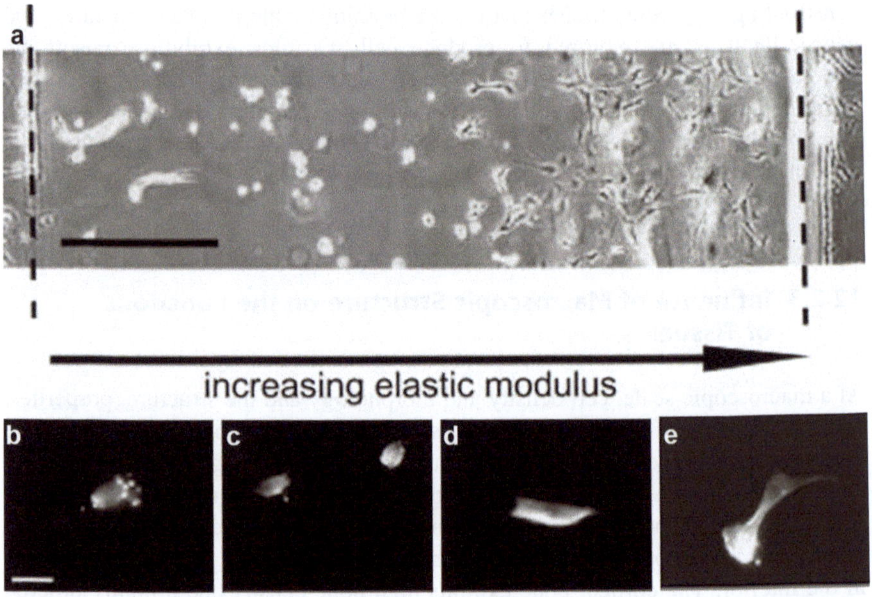

Fig. 12.3 (**a**) Durotactic behavior of vascular smooth muscle cells cultured on a polyacrylamide hydrogel featuring an elastic modulus gradient. Fluorescent images of F-actin of cells located on the softer region (**b**), on the intermediate regions (**c**–**d**), and on the stiffer region (**e**) (Adapted from Zaari et al. [26]). Copyright 2004 John Wiley & Sons Inc

are presented and the applications of these methods in the field of tissue engineering are described.

12.3 Materials Used for Cell Scaffolds

Cell scaffolds have been produced using various materials such as glass substrates, polymeric solids, and polymer hydrogels. Because using these materials has advantages and disadvantages, we must select the most appropriate materials required to achieve a specified purpose. Here, we briefly review the materials that are commonly used in tissue engineering for the purpose of fabricating cell scaffolds.

12.3.1 Glass Substrates

Glass has been traditionally used as a substrate for culturing cells. Because cells cannot directly adhere to the surface of glass substrates, we must coat the surface of glass substrates with ECM proteins before seeding cells. However, the surface chemical properties of glass substrates can be readily modified using a variety of silane coupling agents. The modified surface of glass substrates can adsorb or bind various biomolecules, allowing the adhesive interactions between cells and the glass substrates to be controlled and the differentiation behaviors of cells to be guided [27].

However, the Young's modulus of conventional glass substrates is approximately 70 GPa, which is substantially higher than that of bone or teeth, the hardest tissues in our body. Therefore, glass substrates cannot provide biomimetic mechanical microenvironments to cultured cells. Furthermore, glass substrates are not biodegradable and therefore cannot be implanted in the body. Thus, to achieve the goal of tissue engineering, cell scaffolds are required that present highly physiological cultivation conditions and can be implanted in the body.

12.3.2 Polymeric Solids

Polymeric solids have been used most widely as common substrate materials because they can be readily and inexpensively cast into diverse shapes. Surface topographic morphologies of polymeric solids can be modified using various microfabrication techniques such as photolithographic techniques and photoablation [28]. Numerous microfluidic devices have been developed using these microfabrication techniques and used to investigate the microrheological properties of cells and biopolymers [29, 30]. The surface chemical properties of polymeric solids can also be modified using various microfabrication techniques [31, 32]. Recently, several methods to fabricate micro-patterned cell scaffolds have been developed using combinations of modification techniques devised for controlling the surface topographic morphologies and chemical properties of polymeric solids. The topographic effects of cell scaffolds on cellular behaviors such as morphology change and differentiation have been studied by culturing cells on scaffolds that feature micro-patterned surfaces. For example, the mechanism of contact guidance has been intensively investigated using substrates featuring modified surface topographic morphologies and chemistries [7, 13, 14].

Polymeric scaffolds that present macroporous structures, such as sponges and foams, have been prepared using various methods; these scaffolds can be used to culture cells in a 3D environment. Specifically, because macroporous materials are composed of biodegradable polyesters such as polycaprolactone (PCL), polylactic acid (PLA), and poly-3-hydroxybutyrate (P3HB), can be degraded in vivo, and are highly biocompatible, these materials have been used as cell scaffolds to fabricate implantable engineered tissues [33, 34]. The most famous engineered tissue is the earlike engineered tissue that was fabricated by culturing chondrocytes in biodegradable cell scaffolds and then implanting these into the dorsal region of nude mice [35]. Biopolymers such as nucleic acids, proteins, polysaccharides, and proteoglycans can be used as scaffold materials. However, several biopolymers dissolve in water, and biopolymers used as materials for preparing cell scaffolds must be insoluble in water. To make biopolymers insoluble in water, gelation of biopolymer solutions has been often used.

12.3.3 Polymer Hydrogels

A polymer hydrogel consists of a 3D polymer network and water. The polymer hydrogel swells considerably in water and thus exhibits a biomimetic composition of the body. The mechanical properties of polymer hydrogels are similar to those of ECMs in biological tissues, and these properties can be controlled by regulating the cross-linking density of 3D polymer networks. Polymer hydrogels of diverse shapes can be prepared, including spheres, fibers, and disks. As described in this section, various technologies are available for preparing hydrogels featuring anisotropic structures that mimic the ordered structures observed in biological tissues. The characteristics and advantages of polymer hydrogels allow us to investigate various cellular behaviors under biomimetic conditions and to fabricate biomimetic engineered tissues. Numerous polymer hydrogels have been used as cell scaffolds in the fields of cell biology and tissue engineering. Hydrogels composed of biopolymers are especially useful, because these hydrogels are highly biodegradable and biocompatible and they can provide microenvironments to cultured cells that are more physiological than those provided by other types of hydrogels. Furthermore, the toxicities of the degradation products of biopolymer hydrogels, such as oligopeptides and oligosaccharides, are low.

Collagen hydrogels are widely used in the fields of cell biology and tissue engineering because collagen is a major component of the ECM. The collagen molecule can be enzymatically extracted from animal tissues such as bone, tendon, and dermis. The collagen extracted enzymatically is known as atelocollagen, in which the telopeptides at the two terminal regions of native collagen are digested. Because the major antigenic site of collagen is located in the telopeptide, atelocollagen has low antigenicity. Atelocollagen can be dissolved in acidic aqueous solutions such as hydrochloric acid and acetic acid solutions, and atelocollagen hydrogels can be prepared by neutralizing the pH of these solutions. Atelocollagen hydrogels have been traditionally used as a cell scaffold to investigate diverse cellular behaviors. Furthermore, cells can be readily cultured in atelocollagen hydrogels. The behaviors of cells cultured in 3D environments are highly distinct from those of cells cultured in atelocollagen hydrogels [5, 36, 37]. However, atelocollagen hydrogels prepared in accordance with manufactures' protocols are extremely soft and fragile and exhibit an isotropic structure. Therefore, the handling of atelocollagen hydrogels has been limited, and isotropic atelocollagen hydrogels do not serve as a major template in the construction of engineered tissues featuring biomimetic hierarchical structures.

The composition of Matrigel (or EHS-gel) mimics that of ECMs in native tissues. Matrigel mainly consists of laminin (60 %), Type IV collagen (30 %), and entactin (8 %). The behaviors of cells grown on Matrigel have been determined to be unlike those of cells grown on collagen gels. For example, functions of primary hepatocyte grown on Matrigel, such as albumin secretion, were maintained long term; by comparison, these functions of cells grown on collagen gels declined rapidly [38, 39].

Hydrogels composed of polysaccharides have also been used as cell scaffolds because these are inexpensive to obtain. Alginate is a polysaccharide extracted

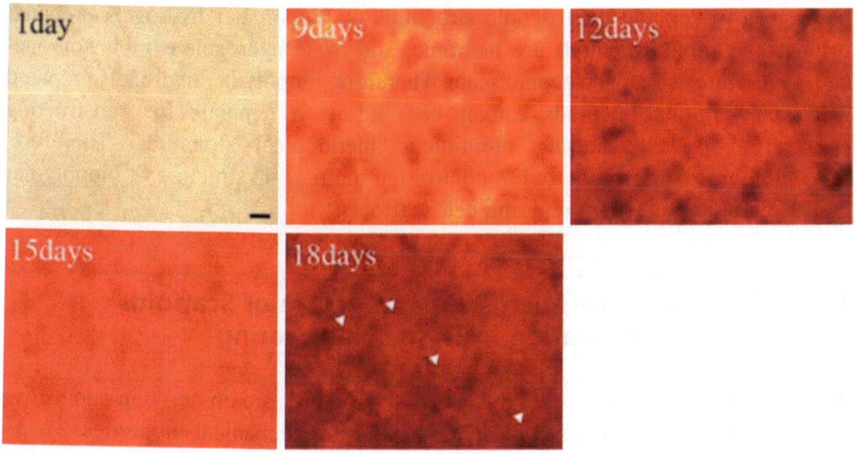

Fig. 12.4 Engineered bone tissues constructed by culturing MC3T3-E1 cells in agarose hydrogels. The specimens were stained using Alizarin Red S. Arrowheads indicate stained cells in the agarose hydrogel (Adapted from Hanazaki et al. [44])

from brown algae that dissolves readily in water. Alginate hydrogels can be prepared by adding multivalent metal cations such as Ca^{2+}, Mg^{2+}, and Cu^{2+}. Cells cannot directly adhere to the surface of alginate hydrogels. Therefore, biochemical modifications of alginate molecules are required to enable cell-alginate adhesive interactions. Conjugation of the arginine-glycine-aspartic acid (RGD) sequence to alginate molecules is often the biochemical modification performed to mediate cell-alginate adhesive interactions [40–44]. RGD-modified alginate hydrogels have been used to investigate various cellular behaviors in 2D and 3D environments. For example, the adipogenic differentiation of adipose-derived stem cells (ASCs) was enhanced when the cells were cultured in RGD-modified alginate hydrogels [41]. The enhancement of adipogenic differentiation can be attributable to the adhesive interaction between ASCs and the RGD sequence conjugated to the alginate. Moreover, RGD-modified alginate hydrogels can be used as carriers to deliver stem cells [43].

Agarose is a polysaccharide extracted from algae such as *Gelidiaceae* and *Rhodophyta*. Agarose dissolves in hot water, and agarose hydrogels are prepared by cooling concentrated agarose solutions. Agarose hydrogels have been frequently used to culture cells in 3D environments. For example, Hanazaki et al. constructed artificial bone tissue by culturing osteoblastic cells in an agarose gel (Fig. 12.4) [44]. Furthermore, agarose hydrogels have been used for investigating the effects of mechanical stimuli on chondrocyte behaviors in 3D environments and for studying the effect of the ECM deposited by chondrocytes on the mechanical properties of agarose hydrogels [45–47].

Biopolymer hydrogels offer several advantages when used for constructing engineered tissues. Specifically, biopolymer hydrogels composed of ECM proteins can mimic cell-ECM interactions in vitro. Therefore, these polymer hydrogels can be regarded as biomimetic cell scaffolds suitable for use in constructing engineered tissues. However, the microscopic structures of biopolymer hydrogels prepared using conventional methods are isotropic, and these hydrogels exhibit homogeneous structures at a macroscopic scale. Therefore, biopolymer hydrogels prepared using conventional methods cannot be used as a template for constructing engineered tissues featuring biomimetic hierarchical structures. Therefore, methodologies that can be used to control the multi-scale structure of biopolymer hydrogel are required for preparing cell scaffolds.

12.4 Control of the Microscopic Structure of Scaffolds and Its Application in Tissue Engineering

As mentioned in Sect. 12.2.2, highly ordered structures are observed in connective tissues, and these structures contribute anisotropic mechanical properties. Therefore, reconstructing the anisotropic structures of native connective tissues is critical for regenerating engineered tissues that possess highly biomimetic structures. Cell scaffolds featuring anisotropic structures (anisotropic cell scaffolds) can be used as a template for constructing engineered tissues that mimic the anisotropic structures of connective tissues. To prepare anisotropic cell scaffolds, controlling the molecular orientation of cell scaffolds is critical. Various techniques have been developed to control the molecular orientation of cell scaffolds. In this section, the methods used to prepare anisotropic cell scaffolds are briefly reviewed.

12.4.1 Isotropic-to-Anisotropic Transitions of Rigid and Semiflexible Biopolymer Solutions

Polymers featuring rigid or semiflexible structures exhibit isotropic-to-anisotropic phase transition. Several biopolymers such as collagen and DNA are regarded as main-chain liquid crystal polymers because these biopolymers possess rigid helical structures. Concentrated solutions of liquid crystal biopolymers exhibit diverse liquid crystalline textures [48–52]. For example, slow evaporation of concentrated DNA solutions generates a liquid crystalline phase that presents a columnar hexagonal structure. These liquid crystalline textures depend on the concentration of DNA molecules and the molecular weight of DNA and on the types of salts added. Similarly, highly concentrated Type I collagen solutions exhibit liquid crystalline phases that mimic the ordered structures observed in the collagen matrix of connective tissues (Fig. 12.5) [51–54]. The morphologies of liquid crystal phases change from a spherulite phase to a dense cholesteric phase when the concentration of collagen is increased [53, 54]. Giraud-Guille and coworkers obtained a dense collagen matrix displaying highly ordered structures by neutralizing the pH of highly concentrated collagen solutions [55] and then used the dense collagen matrix to investigate various properties of fibroblast, such as migration, density, and expression of matrix metalloproteinases, on collagen hydrogels [56]. The fibroblasts seeded on the surface of collagen hydrogels infiltrated the hydrogels to reach deep parts; however, the fibroblasts infiltrated the dense collagen hydrogel

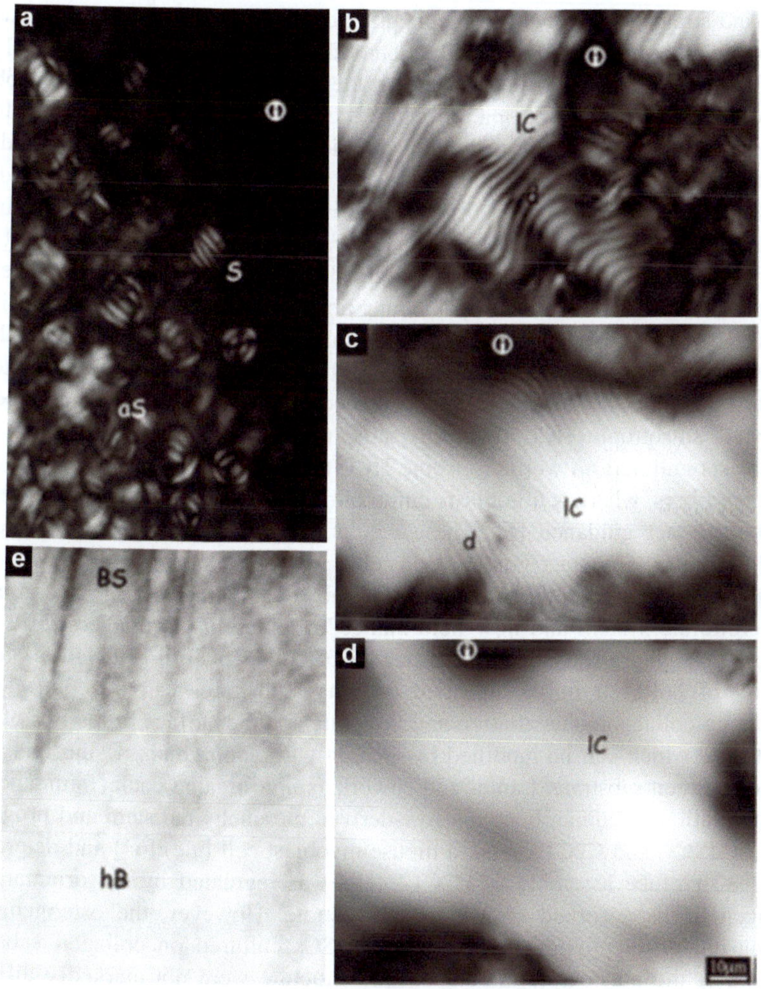

Fig. 12.5 Liquid crystalline phases observed in dense collagen matrices (Adapted from Mosser et al. [53]). Copyright 2006 Elsevier

considerably more slowly than they infiltrated dilute collagen hydrogels. These results indicate that the dense collagen matrix reduces the infiltration of cells from the surface of the gel to the interior of the gel. By contrast, another group reported that a cholesteric collagen film cast from a dense collagen solution can induce the contact guidance of human fibroblasts [57].

12.4.2 Flow Field-Induced Orientation of Polymers

External fields such as flow and magnetic fields are tools that can be used to manipulate molecular alignment in scaffolds. When flow fields such as shear flow field and stretched flow field are applied to polymer solutions, polymer chains are deformed and oriented along the direction of the flow field [58–62]. If we apply the flow fields to polymer solutions during the gelation processes, we can obtain hydrogels featuring anisotropic structures. Elsdale and Bard prepared hydrated collagen lattices (HCLs) featuring anisotropic structures by applying a unidirectional drainage flow field to neutralized collagen solutions at semi-dilute concentrations (approximately 0.1 wt%) during the gelation processes [5]. In the anisotropic HCLs, collagen fibers aligned parallel to the drainage direction. Using anisotropic HCLs, the investigators examined the morphology, motility, adhesion, and proliferation of human embryonic lung fibroblasts under 2D and 3D conditions and observed that the fibroblasts aligned parallel to the orientation of the collagen fibers in the HCLs; the orientation of the fibroblasts here could be attributed to contact guidance as described in the preceding paragraph. Dunn and Ebendal prepared anisotropic HCL by using same method and found that chick-heart fibroblasts cultured on wet anisotropic HCL aligned parallel to the orientation of collagen fibers, whereas fibroblasts cultured on air-dried anisotropic HCL did not exhibit contact guidance [6]. These results suggest a key role of 3D culture conditions in the contact guidance of cells. Recently, Lanfer et al. developed a method to finely control the orientation and density of collagen fibrils by combining a microfluidic system and flow field-induced orientation of collagen fibrils [63]; these investigators showed that the degree of collagen fibril orientation and the density of collagen fibrils adsorbed on glass substrates can be controlled by regulating collagen concentration and flow rate and the surface properties of glass substrates, which can be modified by coating with copolymers. Using this anisotropic collagen substrate, Lanfer and coworkers investigated contact guidance and differentiation of human bone marrow-derived mesenchymal stem and progenitor cells (hMSCs) and C2C12 cells (a mouse myoblast cell line) [64] and determined that the myotube assembly of C2C12 cells was regulated by the orientation of collagen fibrils adsorbed on the glass substrate. However, the osteogenic and adipogenic differentiation behaviors of hMSCs cultured on collagen substrates that did or did not feature the anisotropic structure were not markedly different, although contact guidance of the hMSCs was induced on the collagen substrate featuring the anisotropic structure.

12.4.3 Magnetic Field-Induced Orientation of Polymers

Magnetic field can also be used to manipulate the molecular alignment in cell scaffolds. Torbet et al. prepared anisotropic cell scaffolds composed of various fibrous proteins by using a strong magnetic field [65–67]. These investigators showed that the birefringence of the fibrous protein scaffolds, which is related to the degree of fibril orientation, can be controlled by the magnitude of the applied magnetic field. For example, anisotropic collagen hydrogels can be prepared by applying magnetic fields stronger than 1.9 tesla [66]. The contact guidance behaviors of various cells embedded in magnetically aligned collagen scaffolds have been investigated. Guido and Tranquillo described a method to systematically and quantitatively investigate contact guidance behaviors in anisotropic collagen gels prepared by applying magnetic fields [68] and demonstrated that a cell orientation parameter that can be used to quantify cell orientation distribution was linearly proportional to the birefringence intensity of the anisotropic collagen gels. A modified magnetic molecular alignment method was used to prepare an engineered corneal stroma composed of the multilayered collagen gel, in which the orientation of the collagen fibrils changed orthogonally in each layer [69]; keratocytes seeded on the engineered corneal stroma infiltrated the scaffold and the cells aligned parallel to the orientation of the collagen fibrils in each layer.

Anisotropic cell scaffolds prepared using isotropic-to-anisotropic phase transitions exhibit highly ordered microscopic structures that may mimic the microscopic structure of the ECM in connective tissues. Furthermore, because external fields can be applied to finely control molecular alignment in cell scaffolds, the cell scaffolds prepared using flow fields feature well-defined anisotropic structures. These anisotropic cell scaffolds can be used for investigating cellular behaviors under highly physiological conditions and can be used as the template for reconstructing the local anisotropic structures of native tissues. However, because many of the anisotropic cell scaffolds introduced above are homogeneous at a macroscopic scale, they cannot serve as templates that mimic the macroscopic structure of native tissues. Therefore, controlling the macroscopic heterogeneities of cell scaffolds is also crucial.

12.5 Control of the Macroscopic Heterogeneities of Scaffolds and Its Application in Tissue Engineering

As mentioned in the preceding section, at a macroscopic scale, heterogeneous structures are observed in native tissues, and the distribution of these structures is spatially regulated and is not completely random. Such heterogeneous structures could be generated by spatially regulated cellular functions and could contribute the macroscopic functions of native tissues. Therefore, we must regenerate the macroscopic structure of native tissues to obtain highly functional engineered tissues. The spatial distribution of cell density must be controlled and cellular functions must be spatially regulated and, to achieve this, cell scaffolds that feature spatially regulated macroscopic structures are required. To date, numerous methods have been developed to control the macroscopic structures of cell scaffolds, and these methods are briefly reviewed in this section.

12.5.1 Cell Scaffolds Featuring Gradient Properties

As mentioned in Sect. 12.2.2, directional cell migration plays a key role in tissue morphogenesis. The mechanisms that control the direction of cell migration have been intensively investigated in the field of developmental biology. Many of the directional cell migrations observed during development have been widely reported to be regulated by concentration gradients of ligand molecules such as fibroblast growth factors (FGFs), bone morphogenetic proteins, and stromal cell-derived factors. These ligands are known as "chemokines," and the directional cell migration induced by a concentration gradient of chemokines is called "chemotaxis." If a concentration gradient of chemokines can be generated in vitro, the morphogenetic processes that occur in vivo could be reproduced. Therefore, controlling chemotaxis in vitro can be considered a critical part of the methodology used to construct engineered tissues featuring biomimetic hierarchical structures. However, because several chemokines are water-soluble biomolecules, maintaining a concentration gradient of chemokines long term is challenging. Hence, techniques are required to immobilize chemokines on cell scaffolds. A polyethylene glycol (PEG) hydrogel featuring a concentration gradient of covalently immobilized FGF was developed using a gradient maker and photopolymerization [70]. On the PEG hydrogel featuring the FGF concentration gradient, vascular smooth muscle cells (VSMCs) aligned parallel to the direction of the concentration gradient and migrated from a region of low FGF concentration to a region of high FGF concentration. Therefore, PEG hydrogels can be used for investigating chemotaxis in vitro, and the technique of immobilizing FGF to generate the concentration gradient can be used to control chemotaxis and cell orientation.

Cells not only sense chemical stimuli by means of specific ligand-receptor interactions but also detect physical stimuli through the adhesive interactions that occur between cells and the ECM. As mentioned in Sect. 12.2.3, cells seeded on cell scaffolds featuring an elastic modulus gradient migrate from soft regions to rigid regions, and this directional cell migration is called "durotaxis." Therefore, by regulating the spatial distribution of various properties of cell scaffolds, directional cell migration is controlled to construct engineered tissues that exhibit highly biomimetic structures. Numerous cell scaffolds featuring gradients of mechanical properties have been developed to investigate cellular behaviors under physiological conditions in vitro and to test the potential utility of such cell scaffolds in the field of tissue engineering. Microfluidic devices have been frequently used to fabricate cell scaffolds that exhibit gradients of mechanical properties [25, 26]. Collagen-coated polyacrylamide hydrogels that exhibit an elastic modulus gradient have been prepared by photopolymerizing precursor solutions under a cross-linker gradient generated using a gradient generator, which was a microfluidic device fabricated using poly(dimethylsiloxane) [71]. VMSCs seeded on this gradient polyacrylamide hydrogel migrated from the soft side to the rigid side. The durotaxis of VMSCs was evaluated using the tactic index developed for a biased persistent random walk, and the durotaxis was shown to increase when the magnitude of the elastic modulus gradient was increased. Moreover, the degree of cell orientation was also reported to increase when the magnitude of the elastic modulus gradient was increased. Thus, cell scaffolds featuring an elastic modulus gradient can control not only the spatial distribution of cells but also cell morphology. Gradient biomaterials composed of ECM proteins are useful for investigating the effects of gradient properties on cellular behaviors under physiological conditions. A 3D collagen gel featuring an elastic modulus gradient was developed using an H-shaped microfluidic device [72] and used to investigate the effect of the elastic modulus gradient on neurite growth; neurite growth was directed and enhanced by the elastic modulus gradient, demonstrating that the 3D collage gel can be used as a substrate to guide the directional regrowth of axons in various nervous systems.

12.5.2 Cell Scaffolds Featuring Multichannel Structures

To fabricate large engineered tissues is challenging because cells in inner parts of the large engineered tissues lack pathways that transport oxygen, nutrients, and waste products; thus, the cells die because of a shortage of oxygen and nutrients and the accumulation of waste products. To avoid this problem, we must introduce transport pathways into engineered tissues. Moreover, transport pathways in native tissues consist of tubular structures of various diameters and lengths, and they possess self-similar structures. The hierarchical structure of transport pathways enables efficient transportation. Therefore, a methodology to introduce biomimetic transport pathways is critical for obtaining highly functional engineered tissues.

Cell scaffolds featuring multichannel structures can be used as a template to introduce transport pathways into engineered tissues. Unidirectional freeze-drying of polymer hydrogels generates a cell scaffold featuring multichannel structures [73–79]. A gelatin hydrogel exhibiting aligned porous structures was used as a cell scaffold for culturing human umbilical vein endothelial cells (HUVECs) [75]. In the gelatin hydrogel, HUVECs formed aligned cellular entities along tubular pore walls. Moreover, cell scaffolds featuring aligned multichannel structures can also be used as a template to construct other tissues such as peripheral nervous systems [76, 78], cartilage [77], and skeletal muscle [79].

12.6 Control of Multi-scale Structures of Scaffolds and Its Application in Tissue Engineering

The methods reviewed in the previous sections can be used to control either the microscopic structure or the macroscopic heterogeneities of cell scaffolds, but they cannot be used to control multi-scale structures. To obtain engineered tissues exhibiting biomimetic hierarchical structures, cell scaffolds featuring multi-scale structures are required. In this final section, a methodology to control the multi-scale structure of cell scaffolds is described.

12.6.1 Dialysis-Induced Anisotropic Gelation

Dialysis of concentrated biopolymer solutions in monovalent or multivalent ionic solutions generates hydrogels that exhibit anisotropic structures [80–88]. The phenomenon was discovered during investigations on the gelation of sodium alginate

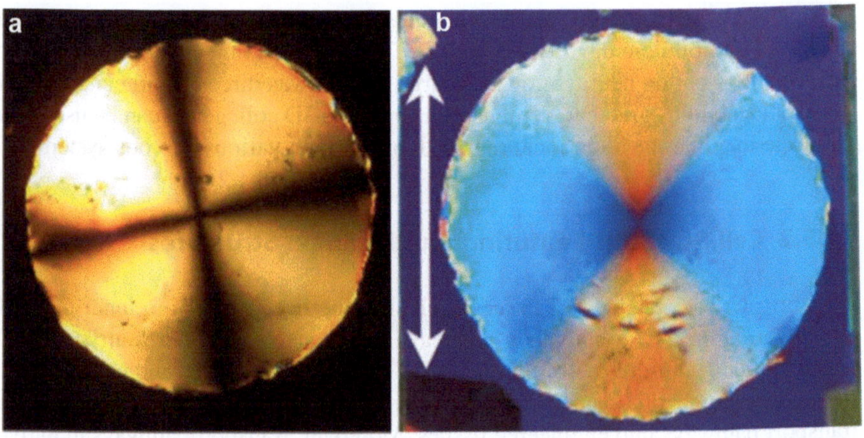

Fig. 12.6 Anisotropic alginate hydrogels observed under polarized light (**a**) and circular polarized light (**b**). The *double arrow* indicates the slow axis of the circular polarizer (Adapted from Maki et al. [87]). Copyright 2011 America Chemical Society

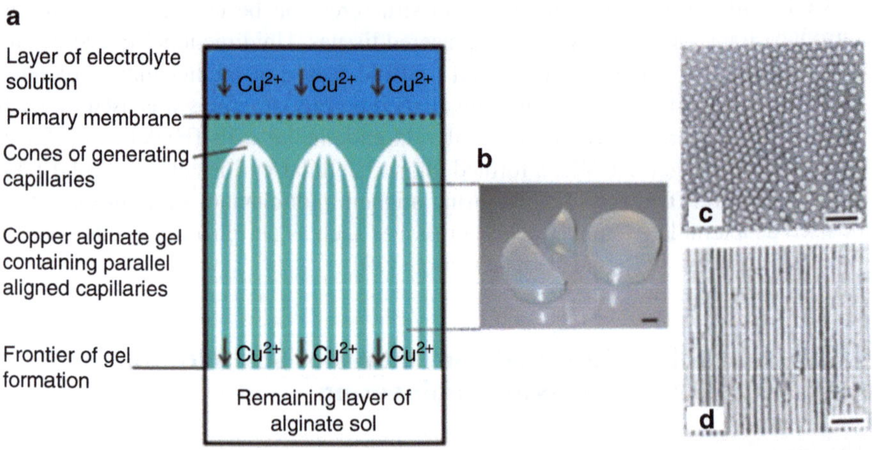

Fig. 12.7 Multichannel structure of ACG (Adapted from Prang et al. [90]). Copyright 2006 Elsevier

solution induced by calcium ions (Ca^{2+}) [80, 89]. The anisotropic alginate gel has a birefringence gradient (Fig. 12.6). Furthermore, anisotropic alginate hydrogels featuring a multichannel structure can be prepared under appropriate preparation conditions (Fig. 12.7). Therefore, these anisotropic alginate gels exhibit an anisotropic structure at microscopic scales and gradient properties and a multichannel structure at macroscopic scales, suggesting that anisotropic alginate gels possess a multi-scale structure and can be used as a template for constructing engineered tissues featuring multi-scale hierarchical structures. The anisotropic alginate gel that features a multichannel structure is known as an alginate capillary gel (ACG), and ACGs have been used as a cell scaffold in the field of tissue engineering [90–93]. The applications of ACG in the field of tissue engineering are introduced in Sect. 12.6.3.

Dialysis-induced anisotropic gelation was observed in systems in which various combinations of concentrated biopolymer solutions and monovalent or multivalent ionic solutions were used. Dialysis of concentrated curdlan solution into calcium chloride solution generated a cylindrical hydrogel that displayed a concentric turbid ring pattern in a cross section perpendicular to the long axis of the gel [81]. An inner part of concentric ring pattern was easily separated upon pushing, which suggests that structural transition occurred at the interface between the concentric turbid ring pattern and the inner part of gel. Dobashi and Yamamoto reported that the curdlan hydrogel displayed a unique birefringence pattern and that the birefringence intensity changed with position relative to the center of gel. Therefore, the curdlan hydrogel exhibited an anisotropic structure and gradient properties. Furthermore, numerous anisotropic hydrogels composed of biopolymers such as DNA, carboxymethyl cellulose (CMC), κ-carrageenan, sacran, polylysine, and polyglutamic acid have been developed using dialysis-induced anisotropic gelation [82–86]. Specifically, anisotropic CMC hydrogels that were prepared by dialyzing concentrated CMC solutions at basic pH values in calcium chloride solutions exhibit not only gradient properties but also multichannel structures. Moreover, anisotropic hydrogels composed of a synthetic liquid crystalline polyelectrolyte, poly (2,2′-disulfonyl-4,4′-benzidine terephthalamide) (PBDT), was prepared using a similar methodology [94, 95].

Hydrogels composed of acid-soluble biopolymers such as collagen and chitosan can be prepared by neutralizing the pH. Therefore, dialysis-induced anisotropic gelation can be applied to mixtures of acid-soluble biopolymer solutions and neutral pH buffer solutions or other alkaline salt solutions. Dialyzing aqueous, acidic chitosan solutions in NaOH solutions generates anisotropic chitosan hydrogels [96]. Similarly, dialyzing acid-soluble atelocollagen solutions in neutral phosphate buffer solutions generates collagen gels that possess anisotropic and multichannel structures and gradient properties (multichannel collagen gel: MCCG) (Fig. 12.8) [84, 88]. MCCGs of various shapes can be prepared by modifying the dialysis-induced anisotropic gelation process. MCCG beads can be prepared readily by adding the atelocollagen solution dropwise into the phosphate buffer solution, whereas MCCG needles can be generated by injecting the atelocollagen solution into the phosphate buffer solution through a syringe needle. MCCG films can be prepared using the following procedure: First, the atelocollagen solution is sandwiched between a petri dish and a circular cover glass. Next, the phosphate buffer solution is poured into the petri dish. The interface between the atelocollagen solution and the phosphate buffer solution gels immediately, and the resulting gel membrane functions as a dialysis membrane. The MCCG layer then grows from the gel membrane to the center of the cover glass, and, finally, we obtain circular MCCG films. Photographs of MCCG films are presented in Fig. 12.8a–c. The number of channels increases when the distance from the center of the gel is increased, whereas the diameter of channels decreases when the distance from the center of the gel is increased (Fig. 12.8d, e). The morphological features of MCCGs effectively mimic the vascular networks observed in native tissues. As discussed in Sect. 12.2.1, collagen is a major

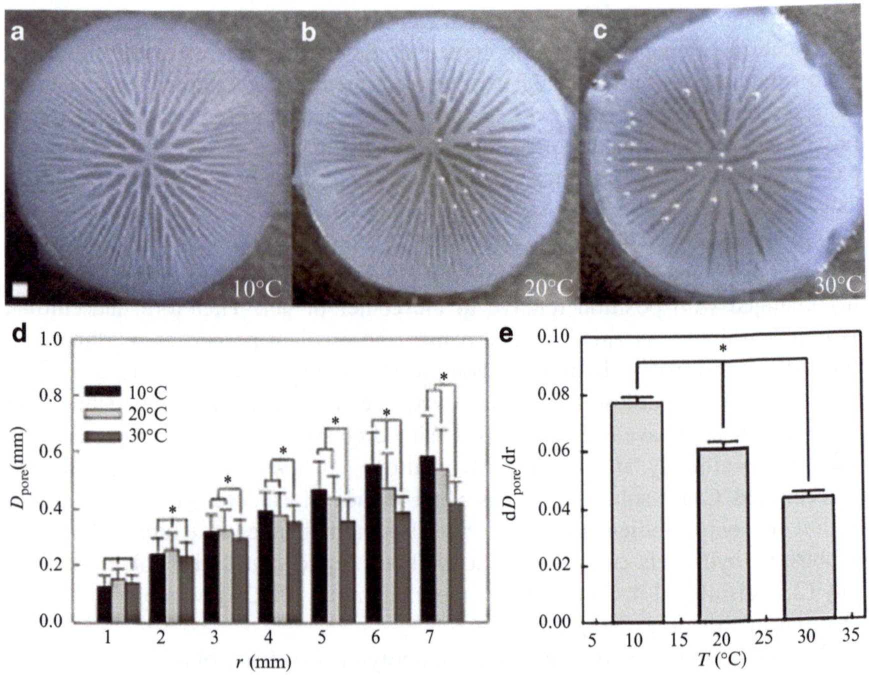

Fig. 12.8 (**a–c**) Photographs of MCCGs prepared at various temperatures. (**d–e**): Effect of preparation temperature on channel diameter (D_{pore}) and on the gradient of channel diameter (dD_{pore}/dr) (Adapted from the supporting information in Hanazaki et al. [100]). Copyright 2013 America Chemical Society

component of ECMs, and the ECM in connective tissues exhibits an anisotropic structure and heterogeneities of structure and properties. MCCGs closely mimic numerous features of the ECM in connective tissues, and, furthermore, MCCGs are biocompatible because they are prepared without using any toxic chemicals. Therefore, the MCCG is a biomimetic cell scaffold extremely well suited for constructing engineered tissues that feature biomimetic hierarchical structures.

12.6.2 Mechanism of Formation of Anisotropic Hydrogels Prepared Using Dialysis-Induced Anisotropic Gelation

The anisotropic structure of anisotropic alginate gels lacking a multichannel structure was investigated by characterizing the optical anisotropy and measuring the small-angle X-ray scattering (SAXS) pattern [87]. Observation of the birefringence sign indicated that the alginate molecules aligned perpendicular to the diffusion direction of Ca^{2+}. The 1D SAXS profiles showed that MCCGs were composed of rodlike fibrils, and the alignment factor (A_f), which was estimated using the azimuthal SAXS intensity profile, suggested that the rodlike fibrils were oriented perpendicular to the diffusion direction of Ca^{2+}. Similarly, the microscopic structure of MCCGs was studied by measuring the small-angle light scattering (SALS) pattern and using confocal laser scanning microscopy (CLSM) [88]. The SALS pattern suggested that the rodlike fibrils align perpendicular to the diffusion direction of ions in the phosphate buffer solution. A real-space microscopic structure observed using CLSM clearly showed that the collage fibrils aligned parallel to the interface between the tubular pore and the gel matrix. Moreover, a similar molecular alignment was observed in the case of anisotropic PBDT hydrogels [95]. Therefore, dialysis-induced anisotropic gelation probably aligns polymers or fibrils perpendicular to the direction of ionic flow from the extradialytic solution to the intradialytic solution. However, the detailed mechanism of the molecular orientation that occurs during dialysis-induced anisotropic gelation remains unclear. In the case of the anisotropic PBDT hydrogel, the PBDT molecules were suggested to be oriented as a result of the sharp concentration gradient of Ca^{2+} at the diffusion front. By contrast, collagen fibrils in MCCGs were oriented parallel to the interface between the tubular pore region and the gel matrix region, indicating that the formation of the multichannel structure affects the orientation of collagen fibrils. Furthermore, no birefringence was observed in the case of atelocollagen hydrogels lacking a multichannel structure, which can be prepared by dialyzing atelocollagen solutions in phosphate solutions containing sodium chloride at high concentrations. As discussed below, experimental results demonstrated that multichannel structures form because of a liquid-liquid phase separation of collagen solutions. Therefore, the orientation of collagen fibrils in MCCGs can be attributed to the liquid-liquid phase separation. In both cases, the formation of unstable interfaces induces molecular orientation. However, experimental evidence supporting the proposed mechanisms has not been obtained. Therefore, the detailed mechanism of the molecular orientation that occurs during dialysis-induced anisotropic gelation must be further systematically investigated.

Dialyzing alginate solutions into multivalent metal cation solutions such as calcium chloride and copper sulfate solutions generates an alginate gel that possesses a multichannel structure (ACG). The morphological characteristics of the gel, such as the number of channels and the average channel diameter, can be controlled by regulating the molecular weight of alginate, the concentration of alginate, the concentration of calcium chloride, and the pH of the alginate solution [89]. The mechanism of ACG formation was intensively investigated by Kohler et al. [97–99], who proposed that the multichannel structure of ACG formed because of a hydrodynamic convectional flow generated at the front of gel formation. The hydrodynamic convectional flow was attributed to the contraction of the alginate chains caused by the cross-linking reaction. In their series of papers, Kohler and coworkers theoretically and experimentally showed that the multichannel structure was formed above a critical velocity of contraction of the alginate chain. However, several assumptions of the theory are inconsistent with the morphologies of the gel and the phenomena observed during the gelation process; in particular, the mechanism of the coarsening of channels that is observed near the initial gel membrane formed during the onset of the gelation process cannot be explained by the theory.

A similar multichannel structure was observed in MCCGs. The mechanism by which the multichannel structure of MCCGs forms was investigated by tracing the growth of the channel diameter in a cross section perpendicular to the diffusion direction of ions in the phosphate buffer solution. The growth process of channel diameter can be expressed using a power law function of gelation time with an exponent of 1/3, suggesting that the multichannel structure of MCCGs forms because of the spinodal decomposition of the collagen solution (Fig. 12.9) [88]. Conversely, hydrodynamic convectional flow was also observed at the front of gel formation, as in the case of ACG; these results indicate that the formation of the multichannel structure of MCCGs can be attributed to the coupling between the liquid-liquid phase separation and the hydrodynamic convectional flow.

Although several open questions related to the mechanism of dialysis-induced anisotropic gelation remain, the anisotropic hydrogels produced using this approach mimic the complex hierarchical structures of biological tissues extremely well. Therefore, anisotropic hydrogels have been used as a template cell scaffold to construct engineered tissues featuring biomimetic hierarchical structures.

12.6.3 Application of Anisotropic Gels Prepared Using Dialysis-Induced Anisotropic Gelation in Tissue Engineering

ACG has been used as a scaffold for regenerating various tissues in vitro and in vivo. As mentioned in Sect. 12.5.2, the vascularization is required for constructing large engineered tissues. ACG can be used as a template for constructing vascular networks in engineered tissues. HUVECs and human smooth muscle cells (HuSMCs) were cocultured using a modified ACG [93]. To facilitate cell adhesion, fibronectin was coated on the surface of the multichannel structure of the modified ACG. In this modified ACG, endothelial lumen-like structures composed of HUVECs and HuSMCs were formed. However, the cell density was low in the endothelial lumen-like structures, and a polarization of the epithelial cells, which is characteristic of the cells that form the epithelial lumen, was not observed. This might be attributed to the disparities in compositions and densities of the ECM proteins on the channel surface, because cell adhesion and epithelial cell polarity depend strongly on the compositions of ECM proteins. Therefore, anisotropic hydrogels that exhibit multichannel structures and are composed of ECM proteins are required for constructing biomimetic epithelial lumens. In another case, calcified ACG was prepared from mixtures of alginate sol and hydroxyapatite powder [92]. The calcified ACG was used as a cell scaffold for culturing human bone marrow stromal cells. Furthermore, ACG was also used as a guide to promote unidirectional axonal regrowth in the injured spinal cord in vivo [90].

MCCG mimics the hierarchical structure of native tissues at multiple scales. Therefore, MCCG can be used as a biomimetic cell scaffold. Figure 12.10 shows mouse calvarial osteoblast-like cells (MC3T3-E1 cells) cultured on MCCG films. The MC3T3-E1 cells were seeded on the surface parallel to the orientation of the

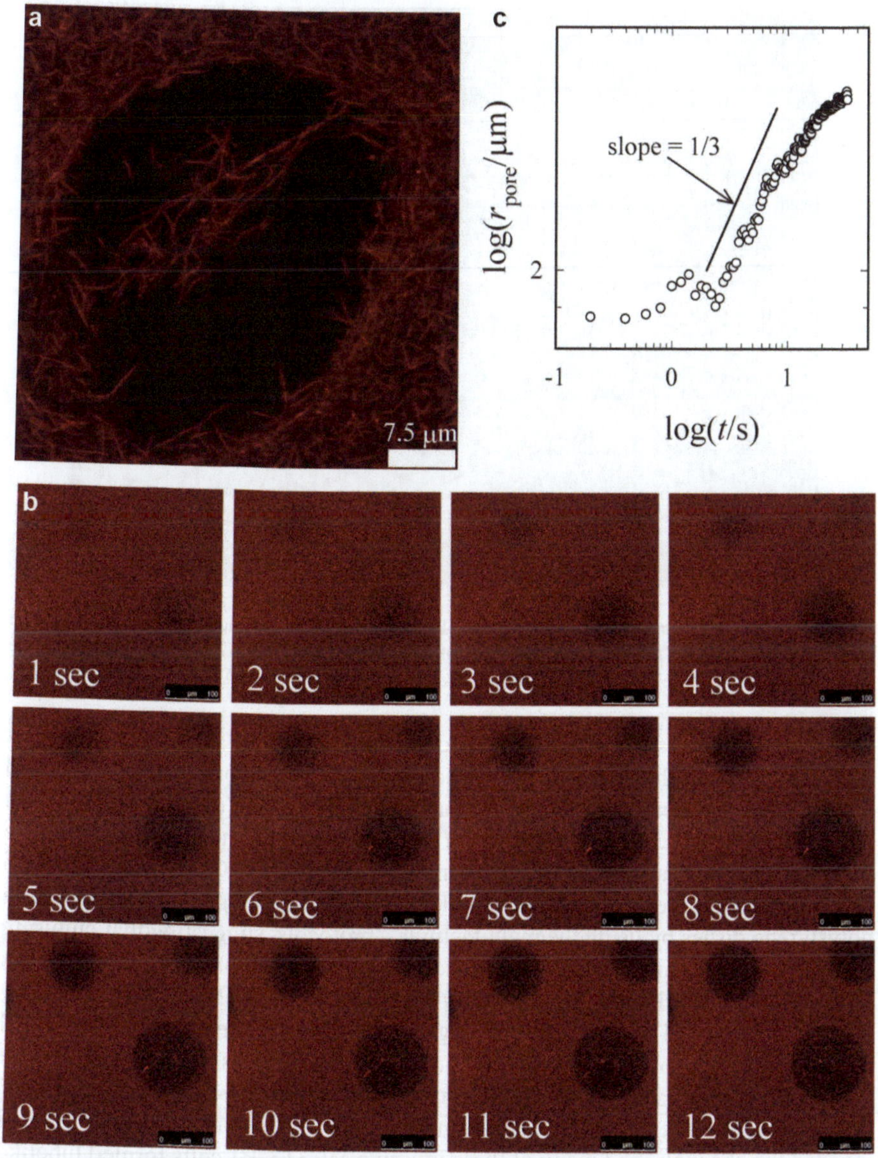

Fig. 12.9 (**a**): High-resolution CLSM image of the channel structure of an anisotropic collagen gel. The magnification is 292× (including a digital zoom factor of 4.64). (**b**): Time-lapse images showing the phase separation of the collagen solution during dialysis. (**c**): Time course of the change in channel radius r_{pore} during the phase separation of the collagen solution

multichannel structure. After 5 days of cultivation, several MC3T3-E1 cells were found to have penetrated the gel surface and invaded the gel matrix of the MCCG film. The invading MC3T3-E1 cells were aligned parallel to the orientation of the

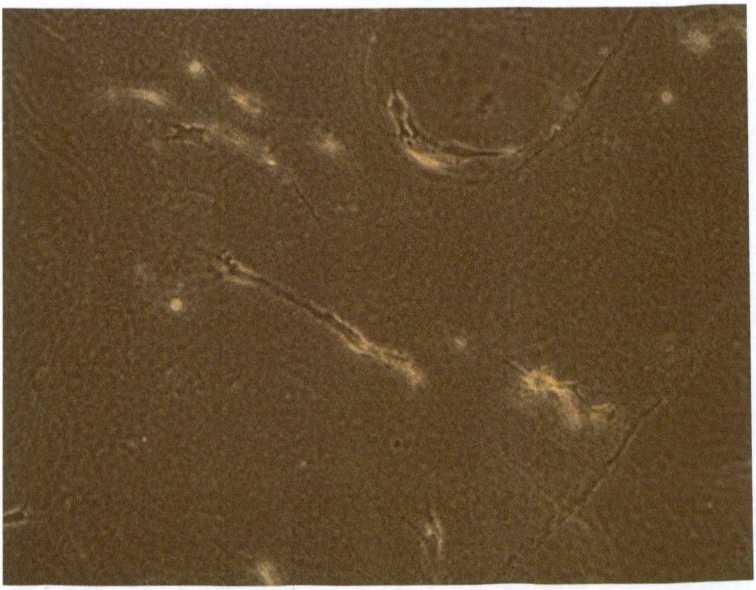

Fig. 12.10 Phase-contrast micrograph of MC3T3-E1 cells that had invaded into the MCCG. The *double arrow* indicates the orientation of the collagen fibrils of the MCCG. Arrowheads indicate the cells aligned parallel to the orientation of the collagen fibrils of the MCCG

collagen fibrils. This alignment of cells could have occurred as a result of contact guidance. Because the degree of orientation of the collagen fibrils can be controlled by regulating pH, ionic strength, and preparation temperature (unpublished), we can control the contact guidance behavior of cells by culturing them on MCCGs. The formation of mineralized tissue in a sample cultured on MCCGs was shown to occur faster than the formation in samples cultured on collagen gels prepared on polystyrene dishes by using the manufacturer's protocols, suggesting that culturing on MCCGs enhances the osteogenic differentiation of MC3T3-E1 cells [100]. Moreover, changes in the multi-scale structure of MCCGs induced by culturing MC3T3-E1 cells were investigated using a combination of CLSM and synchrotron radiation SAXS. In this study, MC3T3-E1 cells were seeded on the surface perpendicular to the orientation of the multichannel structure. The CLSM image of the engineered tissue is shown in Fig. 12.11. Almost all the osteoblasts had invaded the channels, and many of these invasive MC3T3-E1 cells formed tubelike structures. These results clearly showed that engineered tissues exhibiting a heterogeneous distribution of cell density were constructed. The heterogeneity of engineered tissues can be controlled by changing the multichannel structure of MCCGs, which can be also altered by regulating preparation conditions such as temperature and the pH and composition of the phosphate buffer solution. The SAXS intensity profiles obtained for the engineered tissues constructed using MCCGs showed that the average diameter of bundles of collagen fibrils increased up to 7 days in culture and then decreased slightly over time, whereas the average

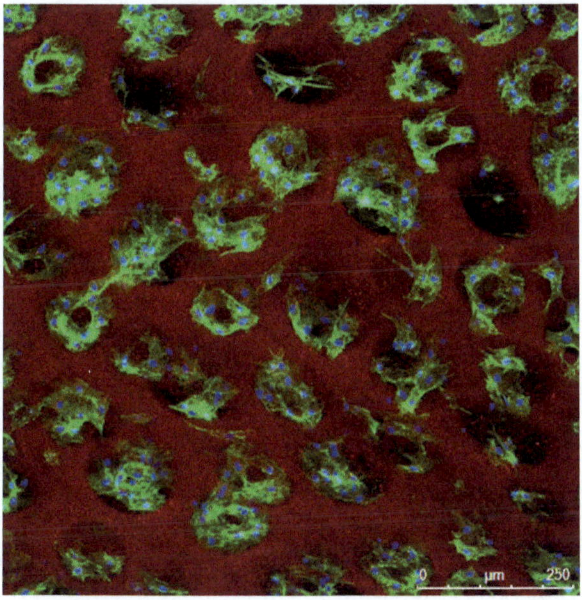

Fig. 12.11 CLSM image of the engineered tissue constructed by culturing MC3T3-E1 cells on MCCGs. *Red, collagen*; *green, actin*; and *blue*, nucleus (Adapted from Hanazaki et al. [100]). Copyright 2013 America Chemical Society

diameter of the collagen fibrils did not change noticeably. These results showed that the MC3T3-E1 cells modified the diameter of collagen fibril bundles, but not that of collagen fibrils. Moreover, an epithelial lumen-like structure was constructed by culturing Madin-Darby canine kidney (MDCK) cells on MCCGs (unpublished). F-actin was localized at the inner surface of the epithelial lumen-like structure, suggesting that the MDCK cells were correctly polarized. The polarization of MDCK cells can be attributed to an adhesive interaction between the MDCK cells and the matrix proteins present at the inner wall of the channels. Preliminary results have shown that MCCGs serve as an excellent template for constructing epithelial lumens such as vascular networks, lymphatic networks, and kidney tubules in the engineered tissues, leading to the regeneration of engineered tissues featuring large volumes and biomimetic hierarchical structures.

12.7 Conclusion

During the past decade, stem cell technologies have been investigated extensively. In the near future, various functional cells generated from pluripotent stem cells such as embryonic stem cells and induced pluripotent stem cells may become available. However, constructing engineered tissues that exhibit the biological functions of native tissues by only culturing stem cells on plastic dishes is challenging; this is because the biological functions of native tissues are attributed to their complex 3D hierarchical structures. Therefore, diverse techniques have been developed to construct engineered tissues that possess the complex 3D hierarchical structures of native tissues. To construct biomimetic engineered tissues, cell scaffolds featuring biomimetic structures can be used as a template. Although numerous methods to prepare biomimetic cell scaffolds are available, many of these methods can only mimic the structure of native tissues in a limited size range. Therefore, controlling the multi-scale structures of cell scaffolds is crucial for generating biomimetic engineered tissues.

In this chapter, dialysis-induced anisotropic gelation was introduced as an example of a method used to control the multi-scale structure of cell scaffolds. Specifically, multichannel collagen gels mimic various characteristics of the ECM of native tissues. Therefore, multichannel collagen gels could be used as a template for constructing engineered tissues that feature the complex 3D hierarchical structures of native tissues. Conversely, numerous microfabrication techniques have been recently used to construct biomimetic engineered tissues. Microfluidic devices can be used to precisely regulate the shape and size of cell scaffolds and to construct engineered tissues in which the arrangement of cells and cell scaffolds is controlled [101]. Cell printing technologies represent candidate methods that could be used to manipulate the hierarchical structures of engineered tissues [102, 103]. We expect that engineered tissues and organs exhibiting native biological functions will be constructed by combining various microfabrication techniques with the methods introduced in this chapter that are used to control the structures of cell scaffolds.

References

1. Langer R, Vacanti JP (1993) Tissue engineering. Science 260:920–926
2. Alberts B, Johnson A, Lewis J, Raff M, Roberts K, Walter P (2008) Molecular biology of the cell, 5th edn. Garland Science, New York
3. Hynes RO (1992) Integrin: versatility, modulation, and signaling in cell adhesion. Cell 69:11–25
4. Weiss P, Garber B (1952) Shape and movement of mesenchyme cells as functions of the physical structure of the medium contributions to a quantitative morphology. Proc Natl Acad Sci U S A 38:264–280
5. Elsdale T, Bard J (1972) Collagen substrata for studies on cell behavior. J Cell Biol 54:626–637
6. Dunn GA, Ebendal T (1977) Contact guidance on oriented collagen gels. Exp Cell Res 111:475–479
7. Clark P, Connolly P, Curtis ASG, Dow JAT, Wilkinson CDW (1991) Cell guidance by ultrafine topography in vitro. J Cell Sci 99:73–77
8. Engler AJ, Sen S, Sweeney HL, Discher DE (2006) Matrix elasticity directs stem cell lineage specification. Cell 126:677–689
9. Ren K, Crouzier T, Roy C, Picart C (2008) Polyelectrolyte multilayer films of controlled stiffness modulate myoblast cell differentiation. Adv Funct Mater 18:1378–1389
10. Reilly CG, Engler AJ (2010) Intrinsic extracellular matrix properties regulate stem cell differentiation. J Biomech 43:55–62
11. Sliver FH, Freeman JW, Seehra GP (2003) Collagen self-assembly and the development of tendon mechanical properties. J Biomech 36:1529–1553
12. Weiss P (1959) Cellular dynamics. Rev Mod Phys 31:11–20

13. Dalton BA, Walboomers XF, Dziegielewski M, Evans MDM, Taylor S, Jansen JA, Steele JG (2001) Modulation of epithelial tissue and cell migration by microgrooves. J Biomed Mater Res 56:195–207
14. Teixeira AI, Abrams GA, Bertics PJ, Murphy CJ, Nealey PF (2003) Epithelial contact guidance on well-defined micro- and nanostructured substrates. J Cell Sci 116:1881–1892
15. Dickinson RB, Guido S, Tranquillo RT (1994) Biased cell migration of fibroblasts exhibiting contact guidance in oriented collagen gels. Ann Biomed Eng 22:342–356
16. Provenzano PP, Inman DR, Eliceiri KW, Trier SM, Keely PJ (2008) Contact guidance mediated three-dimensional cell migration is regulated by Rho/ROCK-dependent matrix reorganization. Biophys J 95:5374–5384
17. Doyle AD, Wang FW, Matsumoto K, Yamada KM (2009) One-dimensional topography underlies three-dimensional fibrillar cell migration. J Cell Biol 184:481–490
18. Rho JY, Kuhn-Spearing L, Zioupos P (1998) Mechanical properties and the hierarchical structure of bone. Med Eng Phys 20:92–102
19. Gibson LJ (1985) The mechanical behavior of cancellous bone. J Biomech 18:317–328
20. Keller TS (1994) Predicting the compressive mechanical behavior of bone. J Biomech 27:1159–1168
21. Schinagl RM, Gurskis D, Chen AC, Sah RL (1997) Depth-dependent confined compression modulus of full-thickness bovine articular cartilage. J Orthop Res 15:499–506
22. Goldberger AL, West BJ (1987) Fractals in physiology and medicine. Yale J Biol Med 60:421–435
23. Masters BR (2004) Fractal analysis of the vascular tree in the human retina. Annu Rev Biomed Eng 6:427–452
24. Lo CM, Wang HB, Dembo M, Wang YL (2000) Cell movement is guided by the rigidity of the substrate. Biophys J 79:144–152
25. Wong JY, Valasco A, Rajagopalan P, Pham Q (2003) Directed movement of vascular smooth muscle cells on gradient-compliant hydrogels. Langmuir 19:1908–1913
26. Zaari N, Rajagopalan P, Kim SK, Engler AJ, Wong JY (2004) Photopolymerization in microfluidic gradient generators: microscale control of substrate compliance to manipulate cell response. Adv Mater 16:2133–2137
27. Curran JM, Chen R, Hunt JA (2006) The guidance of human mesenchymal stem cell differentiation in vitro by controlled modifications to the cell substrate. Biomaterials 27:4783–4793
28. Becker H, Locascio LE (2002) Polymer microfluidic devices. Talanta 56:267–287
29. Domachuk P, Tsioris K, Omenetto FG, Kaplan DL (2010) Bio-microfluidics: biomaterials and biomimetic designs. Adv Mater 22:249–260
30. Lautenschläger F, Piel M (2013) Microfabricated devices for cell biology: All for one and one for all. Curr Opin Cell Biol 25:116–124
31. Falconnet D, Csucs G, Grandin HM, Textor M (2006) Surface engineering approaches to micropattern surfaces for cell-based assays. Biomaterials 27:3044–3063
32. Chen G, Ito Y (2001) Gradient micropattern immobilization of EGF to investigate the effect of artificial juxtacrine stimulation. Biomaterials 22:2453–2457
33. Mikos AG, Thorsen AJ, Czerwonka LA, Bao Y, Langer R, Winslow DN, Vacanti JP (1994) Preparation and characterization of poly(L-lactic acid) foams. Polymer 35:1068–1077
34. Tanaka Y, Yamamoka H, Nishizawa S, Nagata S, Ogasawara T, Asawa Y, Fujihara Y, Takato T, Hoshi K (2010) The optimization of porous polymeric scaffolds for chondrocyte/atelocollagen based tissue-engineered cartilage. Biomaterials 31:4506–4516
35. Puelacher WC, Mooney D, Langer R, Upton J, Vacanti JP, Vacanti GA (1994) Design of nasoseptal cartilage replacements synthesized from biodegradable polymers and chondrocytes. Biomaterials 15:774–778
36. Cukierman E, Pankov R, Yamada KM (2002) Cell interactions with three-dimensional matrices. Curr Opin Cell Biol 14:633–639

37. Grinnell F, Petroll WM (2010) Cell motility and mechanics in three-dimensional collagen matrices. Annu Rev Cell Dev Biol 26:335–361
38. Bissell DM, Arenson DM, Maher JJ, Roll FJ (1987) Support of cultured hepatocytes by a laminin-rich gel. J Clin Investig 79:801–812
39. Oda H, Kozawa K, Hitomi Y, Kakinuma A (1995) Laminin-rich matrix maintains high level of hepatocyte nuclear factor 4 in rat hepatocyte culture. Biochem Biophys Res Commun 212:800–805
40. Rowley JA, Madlambayan G, Mooney DJ (1999) Alginate hydrogels as synthetic extracellular matrix materials. Biomaterials 20:45–53
41. Kang SW, Cha BH, Park H, Park KS, Lee KY, Lee SH (2011) The effect of conjugating RGD into 3D alginate hydrogels on adipogenic differentiation of human adipose-derived stromal cells. Macromol Biosci 11:673–679
42. Connelly JT, Garcia AJ, Levenston ME (2007) Inhibition of in vitro chondrogenesis in RGD-modified three-dimensional alginate gel. Biomaterials 28:1071–1083
43. Yu J, Du KT, Fang Q, Gu Y, Mihardja SS, Sievers RE, Wu JC, Lee RJ (2010) The use of human mesenchymal stem cells encapsulated in RGD modified alginate microspheres in the repair of myocardial infarction in the rat. Biomaterials 31:7012–7020
44. Hanazaki Y, Ito D, Furusawa K, Fukui A, Sasaki N (2013) Change in the viscoelastic properties of agarose gel by HAp precipitation by osteoblasts cultured in an agarose gel matrix. J Biorheol 26:21–28
45. Buschmann MD, Gluzband YA, Grodzinsky AJ, Hunziker EB (1995) Mechanical compression modulates matrix biosynthesis in chondrocyte/agarose culture. J Cell Sci 108:1497–1508
46. Mauck RL, Seyhan SL, Ateshian GA, Hung CT (2002) Influence of seeding density and dynamic deformational loading on the developing structure/function relationships of chondrocyte-seeded agarose hydrogels. Ann Biomed Eng 30:1046–1056
47. Sasaki N, Imai T, Hashimoto A, Yasuda H (2009) Effect of pericellular matrix formation by chondrocytes cultured in agarose gel on the viscoelastic properties of agarose gel matrix. J Biorheol 23:95–101
48. Strzelecka TE, Davidson MW, Rill RL (1988) Multiple liquid crystal phases of DNA at high concentrations. Nature 331:457–460
49. Livolant F, Levelut AM, Doucet J, Benoit JP (1989) The highly concentrated liquid-crystalline phase of DNA is columnar hexagonal. Nature 339:724–726
50. Livolant F, Leforestier A (1996) Condensed phase of DNA: structures and phase transitions. Prog Polym Sci 21:1115–1164
51. Giraud-Guille MM (1989) Liquid crystalline phases of sonicated type I collagen. Biol Cell 67:97–101
52. Giraud-Guille MM (1992) Liquid crystallinity in condensed type I collagen solutions a clue to the packing of collagen in extracellular matrices. J Mol Biol 224:861–873
53. Mosser G, Anglo A, Helary C, Bouligand Y, Giraud-Guille MM (2006) Dense tissue-like collagen matrices formed in cell-free conditions. Matrix Biol 25:3–13
54. Giraud-Guille MM, Mosser G, Belamie E (2008) Liquid crystallinity in collagen systems in vitro and in vivo. Curr Opin Colloid Interface Sci 13:303–313
55. Besseau L, Giraud-Guille MM (1995) Stabilization of fluid cholesteric phases of collagen to ordered gelated matrices. J Mol Biol 251:197–202
56. Helary C, Foucault-Bertaud A, Godeau G, Coulomb B, Giraud-Guille MM (2005) Fibroblast populated dense collagen matrices: cell migration, cell density, and metalloproteinases expression. Biomaterials 26:1533–1543
57. Kirkwood JE, Fuller GG (2009) Liquid crystalline collagen: a self-assembled morphology for the orientation of mammalian cells. Langmuir 25:3200–3206
58. Cerf R, Scheraga HA (1952) Flow birefringence in solutions of macromolecules. Chem Rev 51:185–261
59. LeDuc P, Haber C, Bao G, Wirtz D (1999) Dynamics of individual flexible polymers in a shear flow. Nature 399:564–566

60. Perkins TT, Smith DE, Chu S (1997) Single polymer dynamics in an elongational flow. Science 276:2016–2021
61. Sasaki N, Atkins EDT, Fulton WS (1991) Elongational flow studies on the molecular properties of collagen and its thermal denaturation. J Appl Polym Sci 42:2975–2985
62. Sasaki N, Hayakawa I, Hikichi K, Atkins EDT (1996) Deformation of λ-phage DNA molecules in an elongational flow field. J Appl Polym Sci 59:1389–1394
63. Lanfer B, Feudenberg U, Zimmermann R, Stamov D, Körber V, Werner C (2008) Aligned fibrillar collagen matrices obtained by shear flow deposition. Biomaterials 29:3888–3895
64. Lanfer B, Seib FP, Freudenberg U, Stamov D, Bley T, Bornhäuser M, Werner C (2009) The growth and differentiation of mesenchymal stem and progenitor cells cultured on aligned collagen matrices. Biomaterials 30:5950–5958
65. Torbet J, Freyssinet JM, Hudry-Clergeon G (1981) Oriented fibrin gels formed by polymerization in strong magnetic fields. Nature 289:91–93
66. Torbet J, Ronzière MC (1984) Magnetic alignment of collagen during self-assembly. Biochem J 219:1057–1059
67. Torbet J, Dickens MJ (1984) Orientation of skeletal muscle actin in strong magnetic field. Fed Eur Biochem Soc J 173:403–406
68. Guido S, Tranquillo RT (1993) A methodology for the systematic and quantitative study of cell contact guidance in oriented collagen gels correlation of fibroblast orientation and gel birefringence. J Cell Sci 105:317–331
69. Torbet J, Malbouyres M, Builles N, Justin V, Roulet M, Damour O, Oldberg Å, Ruggiero F, Hulmes DJS (2007) Orthogonal scaffold of magnetically aligned collagen lamellae for corneal stroma reconstruction. Biomaterials 28:4268–4276
70. DeLong SA, Moon JJ, West JL (2005) Covalently immobilized gradient of bFGF on hydrogel scaffold for directed cell migration. Biomaterials 26:3227–3234
71. Isenberg BC, Dimilla PA, Walker M, Kim S, Wong JY (2009) Vascular smooth muscle cell durotaxis depends on substrate stiffness gradient strength. Biophys J 97:1313–1322
72. Sundararaghavan HG, Monteiro GA, Firestein BL, Shreiber DI (2008) Neurite growth in 3D collagen gels with gradients of mechanical properties. Biotechnol Bioeng 102:632–643
73. Kang HW, Tabata Y, Ikada Y (1999) Fabrication of porous gelatin scaffolds for tissue engineering. Biomaterials 20:1339–1344
74. Vlierberghe SV, Cnudde V, Dubruel P, Masschaele B, Cosijns A, Daepe LD, Jacobs PJS, Hoorebek LV, Remon JP, Schacht E (2007) Porous gelatin hydrogels: 1. Cryogenic formation and structure analysis. Biomacromolecules 8:331–337
75. Dubruel P, Unger R, Vlierberghe SV, Cnudde V, Jacobs PJS, Schacht E, Kirkpatrick CJ (2007) Porous gelatin hydrogels: 2. In vitro cell interaction study. Biomacromolecules 8:338–344
76. Bozkurt A, Brook GA, Moellers S, Lassner FL, Sellhaus B, Weis J, Woeltje M, Tank J, Beckmann C, Fuchs P, Damink J, Schügner F, Heschel I, Pallua N (2007) In vitro assessment of axonal growth using dorsal root ganglia explants in a novel three-dimensional collagen matrix. Tissue Eng 13:2971–2979
77. Wu X, Liu Y, Li X, Wen P, Zhang Y, Long Y, Wang X, Guo Y, Xing F, Gao J (2010) Preparation of aligned porous gelatin scaffolds by unidirectional freeze-drying method. Acta Biomater 6:1167–1177
78. Saglam A, Perets A, Canver AC, Li HL, Kollins K, Cohen G, Fischer I, Lazarovici P, Lelkes PI (2013) Angioneural crosstalk in scaffolds with oriented microchannels for regenerative spinal cord injury repair. J Mol Neurosci 49:334–346
79. Kroehne V, Heschel I, Schügner LD, Bartsch JW, Jockusch H (2008) Use of a novel collagen matrix with oriented pore structure for muscle cell differentiation in cell culture and in grafts. J Cell Mol Med 12:1640–1648
80. Thiele H (1954) Ordered coagulation and gel formation. Discuss Faraday Soc 18:294–301
81. Dobashi T, Nobe M, Yoshihara H, Yamamoto T, Konno A (2004) Liquid crystalline gel with refractive index gradient of curdlan. Langmuir 20:6530–6534
82. Narita T, Tokita M (2006) Liesegang pattern formation in κ-carrageenan gel. Langmuir 22:349–352

83. Dobashi T, Furusawa K, Kita E, Minamisawa Y, Yamamoto T (2007) DNA liquid-crystalline gel as adsorbent of carcinogenic agent. Langmuir 23:1303–1306
84. Furusawa K, Saito H, Tsugueda A, Narazaki Y, Yamamoto T, Dobashi T (2008) Liquid crystalline gelation of aqueous solutions of structural proteins. Trans Mater Res Soc Jpn 33:467–469
85. Okajima MK, Miyazato S, Kaneko T (2009) Cyanobacterial megamolecule sacran efficiently forms LC gels with very heavy metal ions. Langmuir 25:8526–8531
86. Lin SC, Minamisawa Y, Furusawa K, Maki Y, Takeno H, Yamamoto T, Dobashi T (2010) Phase relationship and dynamics of anisotropic gelation of carboxymethyl cellulose aqueous solution. Colloid Polym Sci 288:695–701
87. Maki Y, Ito K, Hosoya N, Yoneyama C, Furusawa K, Yamamoto T, Dobashi T, Sugimoto Y, Wakabayashi K (2011) Anisotropic structure of calcium-induced alginate gels by optical and small-angle X-ray scattering measurements. Biomacromolecules 12:2145–2152
88. Furusawa K, Sato S, Masumoto J, Hanazaki Y, Maki Y, Dobashi T, Yamamoto T, Fukui A, Sasaki N (2012) Studies on the formation mechanism and the structure of the anisotropic collagen gel prepared by dialysis-induced anisotropic gelation. Biomacromolecules 13:29–39
89. Despang F, Dittrich R, Gelinsky M (2011) Novel biomaterials with parallel aligned pore channels by directed ionotropic gelation of alginate: mimicking the anisotropic structure of bone tissue. In: Cavrak M (ed) Advances in biomimetics. InTech, Europe, Rijeka, pp 349–372
90. Prang P, Müller R, Eljaouhari A, Heckmann K, Kunz W, Weber T, Faber C, Vroemen M, Bogdahn U, Wedner N (2006) The promotion of oriented axonal regrowth in the injured spinal cord by alginate-based anisotropic capillary hydrogels. Biomaterials 27:3560–3569
91. Dittrich R, Tomandl G, Despang F, Bernhardt A, Hanke T, Pompe W, Gelinsky M (2007) Scaffolds for hard tissue engineering by ionotropic gelation of alginate-influence of selected preparation parameters. J Am Ceram Soc 90:1703–1708
92. Bernhardt A, Despang F, Lode A, Demmler T, Hanke T, Gelinsky M (2009) Proliferation and osteogenic differentiation of human bone marrow stromal cells on alginate-gelatin-hydroxy-apatite scaffolds with anisotropic pore structure. J Tissue Eng Regen Med 3:54–62
93. Yamamoto M, James D, Li H, Butler J, Rafii S, Rabbany S (2010) Generation of stable co-cultures of vascular cells in a honeycomb alginate scaffold. Tissue Eng A 16:299–308
94. Yang W, Furukawa H, Gong JP (2008) Highly extensible double-network gels with self-assembling anisotropic structure. Adv Mater 20:4499–4503
95. Wu ZL, Kurokawa T, Sawada D, Hu J, Furukawa H, Gong JP (2011) Anisotropic hydrogel from complexation-driven reorientation of semirigid polyanion at Ca^{2+} diffusion flux front. Macromolecules 44:3535–3541
96. Dobashi T, Tomita N, Maki Y, Chang CP, Yamamoto T (2011) An analysis of anisotropic gel forming process of chitosan. Carbohydr Polym 84:709–712
97. Thumbs J, Kohler HH (1996) Capillaries in alginate gel as an example of dissipative structure formation. Chem Phys 208:9–24
98. Treml H, Kohler HH (2000) Coupling of diffusion and reaction in the process of capillary formation in alginate gel. Chem Phys 252:199–208
99. Treml H, Woelki S, Kohler HH (2003) Theory of capillary formation in alginate gels. Chem Phys 293:341–353
100. Hanazaki Y, Masumoto J, Sato S, Furusawa K, Fukui A, Sasaki N (2013) Multiscale analysis of changes in an anisotropic collagen gel structure by culturing osteoblasts. Am Chem Soc Appl Mater Interfaces 5:5937–5946
101. Onoe H, Okitsu T, Itou A, Kato-Negishi M, Gojo R, Kiriya D, Sato K, Miura S, Iwanaga S, Kuribayashi-Shigetomi K, Matsunaga YT, Shimoyama Y, Takeuchi S (2013) Metre-long cell-laden microfibers exhibit tissue morphologies and functions. Nat Mater 12:584–590
102. Mironov V, Boland T, Trusk T, Forgacs G, Markwald RR (2003) Organ printing: computer-aided jet-based 3D tissue engineering. Trends Biotechnol 21:157–161
103. Cui X, Boland T (2009) Human microvasculature fabrication using thermal inkjet printing technology. Biomaterials 30:6221–6227

Sensing of Biomolecules and Cells by Semiconductor Device

13

Toshiya Sakata

Abstract

In this chapter, a principle of semiconductor device based on field effect is introduced as a novel biosensing method, which allows to detect ionic charges for biological phenomena in a direct, label-free, real-time, and noninvasive manner, because most of biomolecules have intrinsic molecular charges such as DNA and cell–cell communication is closely related to ionic behaviors through ion channel proteins at the cell membrane. The platform based on the semiconductor principle is suitable for an applicable detection system in the field of clinical diagnosis, pharmaceutical discovery, tissue engineering, environmental investigation, and food control.

Keywords

Biosensing • Semiconductor principle • Biomolecular charge

13.1 Backgrounds

In 1956, John Bardeen, Walter H. Brattain, and William B. Shockley were jointly awarded the Nobel Prize in Physics "for their researches on semiconductors and their discovery of the transistor effect." The semiconductor technology has been developed in the field of electronic device industry according to Moore's law [1]. The size of the gate has reached to nm scale (~ a few 10 nm) in the present day, and the reducing of gate size has been reported to be difficult in the case of

T. Sakata (✉)
Department of Materials Engineering, Graduate School of Engineering,
The University of Tokyo, 7-3-1 Hongo, Bunkyo-ku, Tokyo 113-8656, Japan
e-mail: sakata@biofet.t.u-tokyo.ac.jp

R. Kita, T. Dobashi (eds.), *Nano/Micro Science and Technology in Biorheology*,
DOI 10.1007/978-4-431-54886-7_13

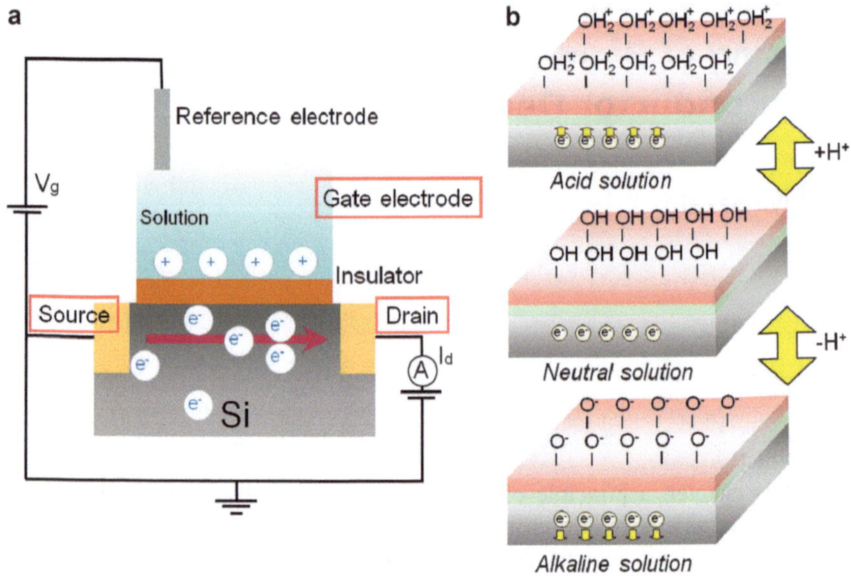

Fig. 13.1 Schematic illustration of IS-FET

silicon-based semiconductor technology. The downsizing of the gate will be realized by the use of carbon nanotube- or graphene-based nano-materials in the future.

Biological phenomena are found at nanoscale, as molecular behaviors such as DNA, antigen–antibody reaction, or ligand–protein receptor interaction at the cell membrane. On the other hand, microelectromechanical system (MEMS) technology is useful to produce various micro- or nanoscale devices by top-down approaches [2]. The developed devices are composed of a regular and functional structure, on which biomolecules are arrayed and detected in a multiplex system. Functional biomolecules are controlled on the multi-arrayed devices and in the micro-fluidic system, which are detected with an optical or electrochemical method.

In 1970, Bergverd et al. showed the electrical detection method of pH variation based on the change of positive charges of hydrogen ions using a field-effect transistor (FET) on the basis of the semiconductor principle [3]. This is called the ion-sensitive field-effect transistor (IS-FET), as shown in Fig. 13.1. The semiconductor material is separated with a solution across the gate insulator, of which the thickness is not more than a few hundred nm. The gate insulator is usually composed of oxides such as SiO_2, Ta_2O_5, Al_2O_3, and so on or nitrides such as Si_3N_4 [4]. The hydroxyl groups are formed at the surface between the solution and gate insulators and are so sensitive to hydrogen ions (Fig. 13.1b). These positive charges at the gate surface interact electrostatically with electrons at the channel in a silicon crystal. The field effect caused by charge density changes at the gate induces the change of the drain-source current and threshold voltage (V_T). This electrical response to hydrogen ions of IS-FET shows a Nernstian response, about 60 mV/pH around room temperature.

Recently, FET biosensors are being studied and developed to apply for clinical diagnosis, drug discovery, tissue engineering, and so on. Since FET biosensors can detect molecular recognition events accompanied by charge density changes without labeled materials and be easily arrayed by the use of the conventional semiconductor process which enables to measure multi-samples, the platform based on FET chips is suitable for a simple and cost-effective system for chip-based diagnosis. The downsizing of a system is significant for personalized medicine at home. Moreover, the electrical signals based on FET devices result in direct and quantitative analyses of bio-samples. One positive or negative charge of an ion or ionic molecule interacts electrostatically with one electron charge in a semiconductor device. Therefore, ion behaviors based on biological phenomena can be directly detected using semiconductor devices. Most of biological phenomena in vivo are closely related to electrical behaviors of charged media, for example, DNA molecules with negative charges based on phosphate groups, ions (potassium ions, sodium ions, and so on) through ion channels at the cell membrane keeping homeostasis in a body, and so on.

13.2 Concept of Semiconductor-Based Biosensing Devices

13.2.1 Ion-Sensitive Field-Effect Transistor (IS-FET)

As a reason of a bad tooth, three elements of *Streptococcus mutans*, "quality of tooth," and "saccharinity" affect on a bad tooth in the course of time. *Streptococcus mutans* induces acidification by dissolving saccharides in food and drink. As a result, acidification of dental plaque is in progress on a tooth. That is to say, the enamel of the tooth begins to dissolve to less than pH 5.5 resulting in a bad tooth. Therefore, it is important to control meals considering pH variation in a mouth in order to prevent a bad tooth. Thus, pH measurement is needed even for health care in daily life and can be accomplished by engineering such as semiconductor technology.

The principle of IS-FET is based on potentiometric detection of charge density changes induced at a gate insulator/solution interface accompanied by pH variation. Hydrogen ions with positive charges at the gate insulator electrostatically interact with electrons in a silicon crystal across the thin gate insulator resulting in the V_T change.

Typical drain voltage (V_{DS})–drain current (I_D) characteristic of the FET is shown in Fig. 13.2. It is found that the FET can be operated correctly. Since the fabricated FET is a depletion type as can be seen in Fig. 13.2, the reference electrode is usually connected to the ground for the measurement of the interfacial potential between the gate insulator and solution using the circuit shown in Fig. 13.3. The pH-response characteristics of the FET with a Si_3N_4 gate are shown in Fig. 13.4. The time course of the interfacial potential was measured during calibration and is

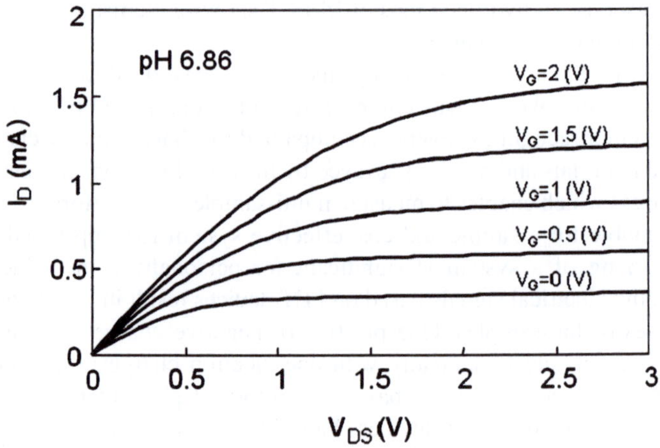

Fig. 13.2 V_{DS}–I_D characteristic of IS-FET

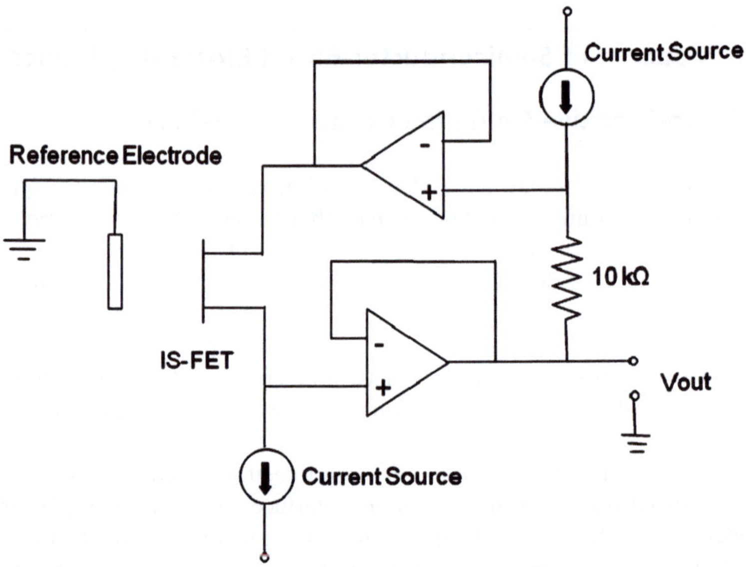

Fig. 13.3 Measurement circuit using IS-FET

shown in Fig. 13.4a. The arrows indicate the timing to change the buffer solutions. The interfacial potential changed rapidly after changing the buffer solution and became stable within 1 min. The calibration curve for the Si_3N_4 gate FET is shown in Fig. 13.4b. The relationship between pH and the output voltage is linear in the range from pH 1.68 to pH 9.18 with a correlation coefficient of 0.9999. The slope of

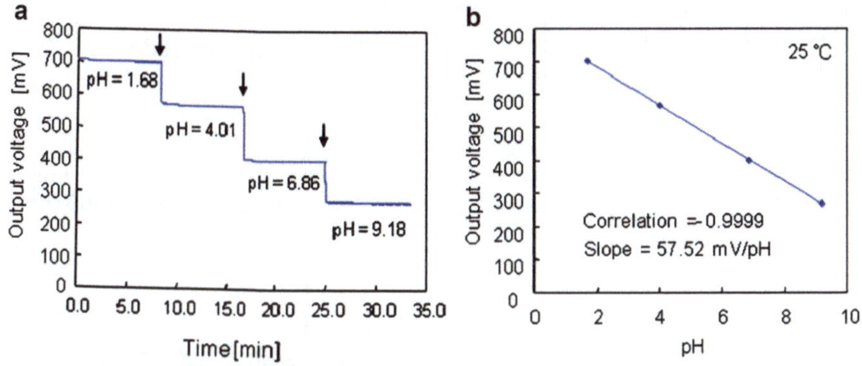

Fig. 13.4 (a) Change of output voltage for each pH and incubation time. (b) Correlation between output voltage and pH. Each output voltage for pH was averaged in the last 10 s before changing buffer solution. All of the measurements were performed at 25 °C

the calibration curve was 57.52 mV/pH, which is close to the theoretical slope at 25 °C. On the basis of these results, the operation of the Si_3N_4 gate FET was considered to be stable and no leakage through the gate insulator and no defect of the encapsulation could be observed.

13.2.2 Device Structures

An insulated-gate field-effect transistor (IG-FET) is composed of electrolyte–insulator–semiconductor. The structure is utilized as a pH sensor of IS-FET. The electrical characteristics of IG-FET are the same as the ones of IS-FET shown in Sect 13.2.1.

On the other hand, an extended-gate field-effect transistor (EG-FET) is shown in Fig. 13.5. The gate electrode is separated and extended from the metal oxide semiconductor (MOS)-FET. Using the EG-FET, materials and structures of the gate electrode can be varied and gate sensing membranes can be easily arrayed by a sputtering method, and MOS-FETs are not replaced and can be reused. Various kinds of gate materials can be designed as follows:

- Gold (Au) gate electrode

It is simple to immobilize biomolecules on the Au electrode, because spacer molecules with -SH (thiol group) are easily tethered. DNA or antibody molecules are immobilized by binding with spacer molecules on Au. However, the Au surface is polarized in solutions and a constant potential based on ion charges is not determined. Therefore, it is effective to tether the self-assembled monolayer (SAM) with both of the functional groups (−COOH, $-NH_2$, and so on) and -SH group at each end, when the Au gate electrode is utilized as the sensing surface of

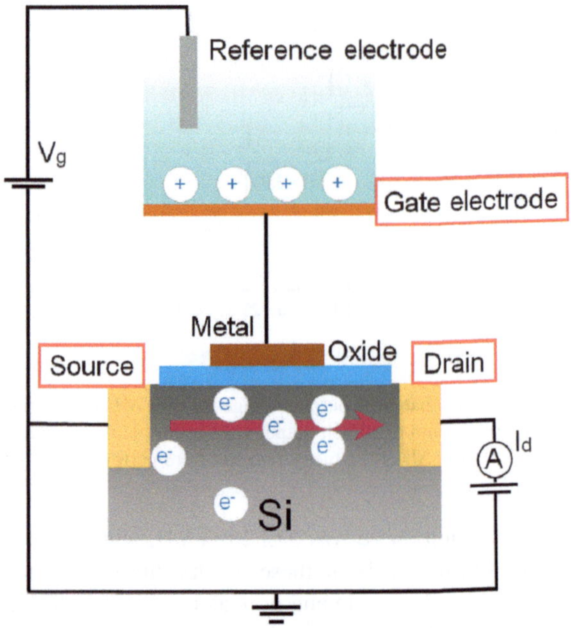

Fig. 13.5 Extended-gate field-effect transistor for biosensing

the EG-FET for biosensing, because their functional groups are bonded with various biomolecules.

– Indium–tin–oxide (ITO)-gate electrode

Transparent electrode is available for microscopic observation of cultured cells as if electrical monitoring is performed using a conventional culture dish. The ITO is basically oxide so that its surface is covered by hydroxyl groups in a buffer solution. The potential of the EG-FET with the ITO gate shows a near-Nernstian response to pH variation [5]. However, it is very important to control the making process of the ITO surface.

– Scaffold gate electrode

Using the EG-FET, the structure of the gate itself can be varied corresponding to applications. In the field of tissue engineering, the safety and functionality of artificial organs are required to transplant to a living body actually. This is why 3D culture of cells is required for making artificial organs. In order to perform 3D culture on the electrode, a nano- or microfiber scaffold is able to be developed as a gate electrode of EG-FET. Scaffold architecture affects cell binding and spreading. Cells binding to scaffolds with microscale architectures flatten and spread as if cultured on flat surfaces, while the scaffolds with nanoscale architectures have larger surface areas to adsorb proteins, presenting many binding sites to cell

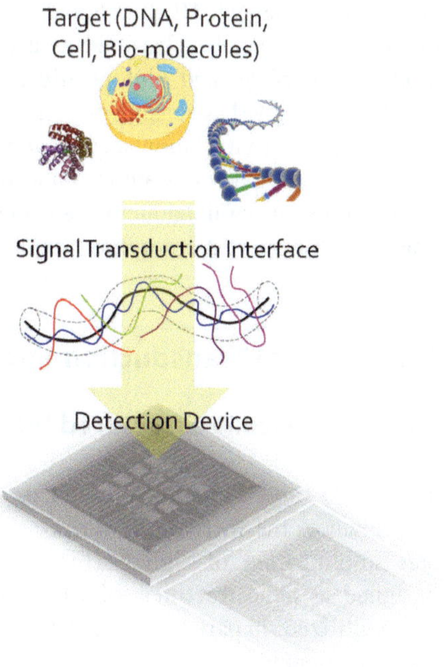

Fig. 13.6 Device structure for biosensing

membrane receptors. The adsorbed proteins may also change conformation, exposing additional cryptic binding sites [6].

– Cantilever gate electrode

A biosensing technique to measure the molecular charge distribution at the cell membrane is proposed using a principle of semiconductor. A cantilever of atomic force microscopy (AFM) is utilized in order to be in contact with the cell membrane and to move the electrode at nanoscale, and a "gate lever" with a platinum tip is studied to detect charges at the cell membrane using AFM. The gate lever as an electrode is extended from the MOS-FET as one of the EG-FET. Actually, negative charges based on sialic acids and so on at the cell membrane can be detected using the gate lever-FET. As a result, the distribution of the membrane protein with some charges at the cell membrane can be mapped at nanoscale using the proposed system.

The device structure for biosensing is composed of three functional elements such as the biological target, signal transduction interface, and detection device, as shown Fig. 13.6. In developing biosensing devices, a specific biological target should be designed for application. In this case, such a detection target is found in the field of medicine, biology, pharmaceutical discovery, tissue engineering, clinical diagnosis, food safety, and so on. In order to detect biological targets selectively, a signal transduction interface should be designed between the target and detection device. The functional interfaces are formed between solid and liquid phases, which can be classified as a physically structured interface, chemically modified interface, and biologically induced interface. Lastly, detection devices should be designed as sensing principles that enable to detect mass, viscoelasticity, charges, permittivity, fluorescence, and so on. In this chapter, the semiconductor-based biosensing devices are introduced as the detection device, but various kinds of materials of semiconductors are designed such as silicon, carbon nanotube, graphene, transparent amorphous oxide, and so on.

13.3 Signal Transduction Interface

13.3.1 Physically Structured Interface

The micro- and nanostructures on the gate electrode of semiconductor devices contribute to the increase of surface area for immobilization of probe molecules and amplification of signals (Fig. 13.7). Nanoparticles or nano-pillars can be designed as nanostructures, of which gold, silica, and carbon nanotubes, and so on are utilized. In particular, the detection limit of FET devices is closely related to the Debye length at the interface between the gate surface and solution. When the nano-pillar structure is made within the Debye length, longer or larger biomolecules such as DNA or protein can be detected within the Debye length. In the case of DNA sensing, particularly, extended DNA molecules can be detected parallel to the gate surface resulting in label-free DNA sequencing based on intrinsic molecular charges.

13.3.2 Chemically Synthesized Interface

Functional interfaces synthesized chemically are proposed as a monolayer or polymer membrane. A functional membrane as a monolayer sensitive to K^+ at the

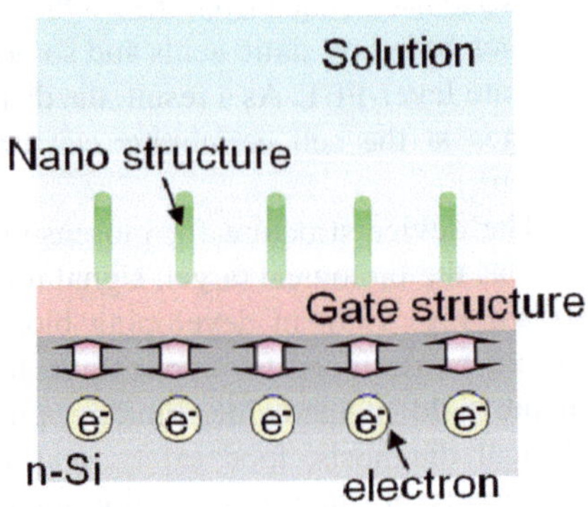

Fig. 13.7 Nanostructure-based FET

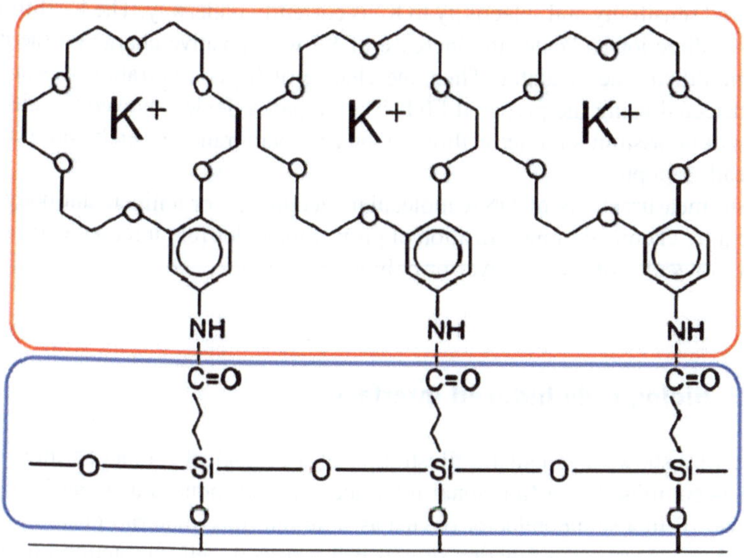

Fig. 13.8 Crown ether-modified gate transistor for K^+ detection

gate insulator is prepared using an 18-crown-6 ether derivative, as shown in Fig. 13.8. First, carboxyethylsilanetriol was used as silane coupling agent. The sensor was immersed in carboxyethylsilanetriol sodium salt 25 wt% in water for 24 h, then rinsed with Milli-Q, and dried in a vacuum at 120 °C for 1 h. Second, 4-aminobenzo-18-crown-6 (AB18C6) was used as an ionophore of K^+, and 1-ethyl-3-(3-dimethylaminopropyl) carbodiimide (EDC) was used as a condensation agent. The sensor was immersed in AB18C6 (1 mM) and EDC (10 mM) in N,N-dimethylformamide (DMF) for 24 h and rinsed. Finally, the chemical structure shown in Fig. 13.8 was synthesized. This monolayer does not contain a plasticizer, and the ionophore is chemically incorporated to the gate insulator, so there is no expected cytotoxicity.

In order to investigate the electrical characteristics of FET, a semiconductor parameter analyzer was used. Measurement was performed in 2[4-(2-hydroxyethyl)-1-piperazinyl] ethanesulfonic acid (HEPES) buffer solution, and the pH of the sample solution was adjusted to 7.4 with $Ca(OH)_2$ ($[Ca(OH)_2] =$ 1.7 mM). And K^+ concentration was adjusted to 10^{-4} M, 10^{-3} M, 10^{-2} M, and 10^{-1} M at pH 7.4. The prepared FET was immersed in the sample solution, and then the V_T change was measured. The effect of K^+ concentration on the electrical signal of prepared FET was investigated. The electrical response of prepared FET to K^+ concentration was 4.2 mV/decade. The slope was relatively small compared to that of the Nernstian response. This could be because the density of ionophore immobilized onto the FET gate surface is low. Further investigation toward improved sensitivity and selectivity to K^+ is currently underway. The V_T shift to the negative direction indicates the increase of K^+ with positive charge on the crown ether-modified gate insulator. Thus, the change of K^+ concentration was successfully detected using the prepared FET. The prepared device showed the electrical signals of potassium ion release through the cell membrane caused by programmed cell death "apoptosis."

Also, measurements of DNA molecular recognition or antigen–antibody reaction require chemical immobilization of probe molecules for target molecules using spacer molecules of a monolayer or polymer membrane.

13.3.3 Biologically Induced Interface

Figure 13.9 shows a schematic illustration of a biologically induced interface. A cell A is composed as a functional interface cell and induces a signal based on a target cell B to a semiconductor device as if an interface material plays its role. A cultured cell A causes a function of cell B due to cell–cell interaction on the gate sensing surface. Here, the vascular endothelial cell and tumor cell correspond to cell A and B, respectively, for example. The noninvasive and real-time measurement of invasion of cancer cells to the vascular endothelial cell layer and basement membranes was performed using the vascular endothelial cell-based field-effect transistor (VEC-FET), which is based on potentiometric detection of molecular recognition on the gate insulator. The shift of the V_T caused by charge density changes on the gate insulator can be monitored during the invasion process of tumor cells. The negative shift of the V_T has been successfully detected around 4 h after addition of invasive tumor cells (HeLa cells). This result indicates that the positive charges were induced on the surface of the gate insulator as a result of degradation of a negative-charged basement membrane on the gate insulator by secretion of enzyme from tumor cells. Thus, we can design in vivo-like gate FET in order to detect and reproduce biological events in vivo such as cell–cell interaction on the gate sensing surface in vitro. The platform based on the biologically induced gate FET is suitable for an in vivo-like sensing.

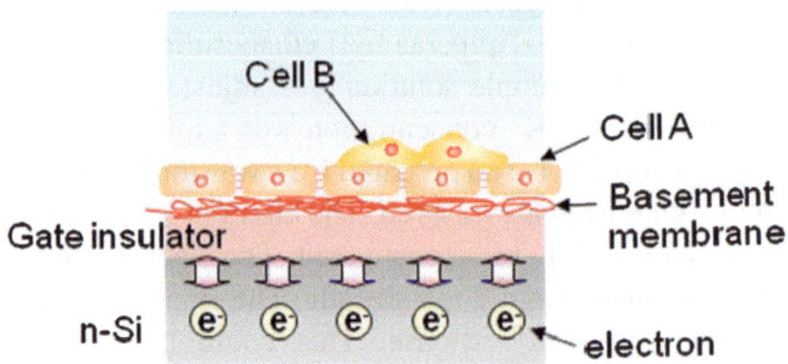

Fig. 13.9 Biologically induced interface

13.4 In Vitro Cell Sensing with Semiconductor Platform [7–16]

In vitro cell sensing results in the significant information of cell functions conveniently. When artificial cells such as an induced pluripotent stem (iPS) cell are transplanted into a body after control of differentiation induction, their functions contribute to regeneration of tissues or organs. In this case, the regenerated cells are required to be safe and show good quality in a body. So, the evaluation of cellular activities is needed to check the quality of living cells before transplantation. On the other hand, the effect of drugs on living cells is being investigated including a side effect in the field of pharmaceutical discovery. In this case, dying cells, such as programmed cell death, apoptosis, are detected by cell sensing methods. Thus, in vitro cell sensing plays important roles for quality check of transplanted cells or evaluation of degradation of cells based on drug effects in the field of regenerative medicine, clinical diagnosis, drug screening, and so on. Here, some case studies of in vitro cell sensing are introduced as follows.

13.4.1 In Vitro Monitoring of Cellular Respiration Activity

Glucose-stimulated electrical activity in pancreatic β-cells was provided as evidence of cell electrical activity [17], although muscle and nerve cells were believed to be specialized cells with electrical excitability. After the investigation, numerous works on the electrophysiology of β-cells were carried out and it has been elucidated [18–20]. The patch-clamp technique makes it possible to demonstrate the channels that contribute to the generation of electrical activity. The adenosine triphosphate (ATP)-regulated K^+ channels (K_{ATP} channels) provide the crucial link between metabolic and electrophysiological effects of glucose. K^+ flows out of the cell through spontaneously active K_{ATP} channels at low glucose and ATP concentrations. The K^+ gradients existing over the β-cell membrane result in a negative membrane potential of approximately −70 mV. Increasing glucose concentrations subsequently elevates intracellular ATP concentrations and closes K_{ATP} channels. Moreover, the generation of electrical activity is a key step in glucose-induced insulin release, and it is widely thought to represent the primary mechanism by which Ca^{2+} is imported into the β-cell cytosol, causing an increase in intracellular Ca^{2+} concentration and triggering several processes that ultimately result in insulin exocytosis [21].

Basically, the semiconductor-based biosensing device that was used in this study is IS-FET, which has been commonly used as a pH sensor [3]. The principle of IS-FET is based on the direct detection of hydrogen ion concentration based on the site-binding model [22] at the gate insulator, as also described above. On the other hand, the glucose-induced insulin secretion process includes the activation of ATP generation, which is closely related to respiration activity in the mitochondrion. The induction of respiration activity would increase hydrogen ion concentration at the cell/gate interface because of CO_2 release in solutions [23]. As a result of activation of respiration, the increase of hydrogen ion concentration at the cell/gate interface

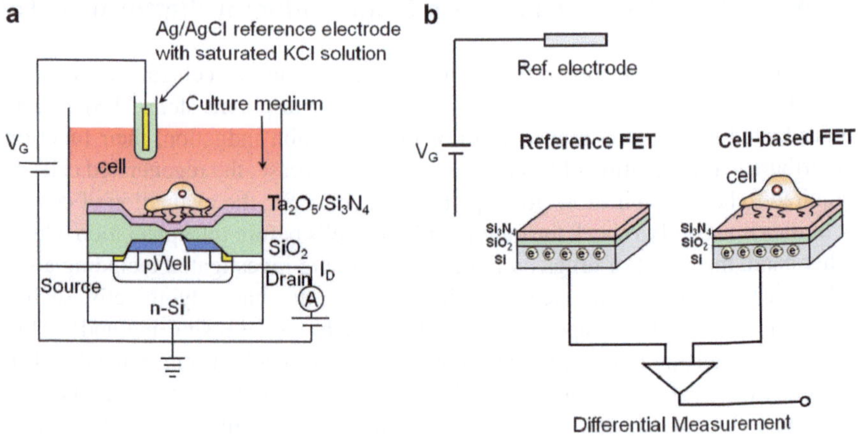

Fig. 13.10 Schematic illustration of cell monitoring. (**a**) Structure of cell-based FET. (**b**) Differential measurement between reference and cell-based FETs

will be directly detected as pH variation using IS-FET. In this section, we introduce the real-time, label-free, and noninvasive monitoring of electrical activity of rat pancreatic β-cells using a cell-based FET, focusing on the correlation between the respiration activity accompanied by the insulin secretion process due to glucose induction and electrical activity of pancreatic β cells.

Rat pancreatic β-cells (RIN-5 F) were used for the electrical detection of glucose response using the FETs in the present study. The β-cells were introduced to the sensing areas of the FET chip in a RPMI1640 culture medium, pH 7.4 (Invitrogen), including 2 mg/ml glucose supplemented with 10 % fetal bovine serum (FBS; JRH Biosciences) and 1 % penicillin/streptomycin (Invitrogen) at 37 °C in an atmosphere of 5.0 % CO_2 for 3 days after being maintained on a culture dish with the controlled RPMI1640 medium at 37 °C in an atmosphere of 5.0 % CO_2 for 1 week. In the electrical measurements, 400 μl of the controlled medium without glucose was placed in the FET chamber, and 10 mg/ml glucose was added to the culture medium at 10 μl. Image analyses were carried out using an optical microscope in parallel with electrical detection.

Figure 13.10 shows the concept of the cell-based FET. The diagram for the measurement of electrical signals is shown in Fig. 13.10a. The pancreatic β-cells were cultured on the gate surface of the FET, which is composed of the Ta_2O_5/Si_3N_4/SiO_2 layers. The Ta_2O_5 gate surface was not chemically modified with functional molecules. The potential of a measurement solution is controlled using a Ag/AgCl reference electrode with saturated KCl solution. Two types of FETs were prepared in the present study; one is the cell-based FET with pancreatic β-cells cultured on the gate surface, and the other is the reference FET without cells (Fig. 13.10b). Using these FETs, differential measurements were performed in order to eliminate the common background noises such as temperature changes and changes in ion concentration.

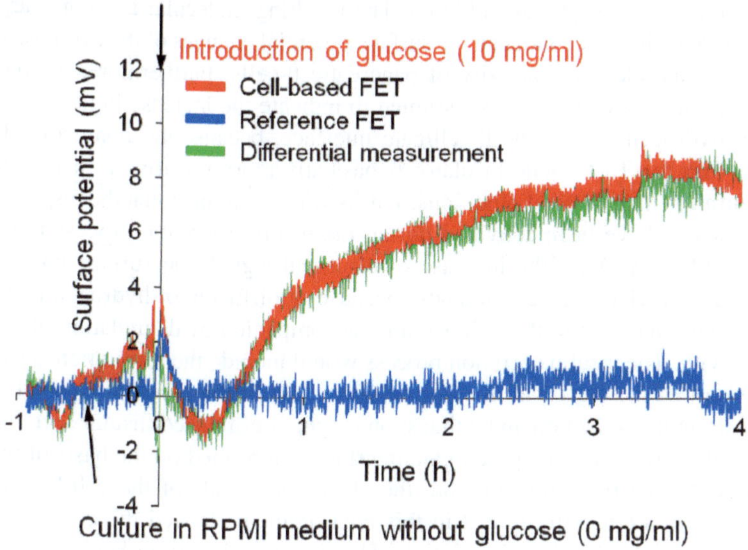

Fig. 13.11 Electrical monitoring of pancreatic β-cell activity

Figure 13.11 shows the shift of the surface potential of gate sensing area after introduction of glucose at a concentration of 10 mg/ml using the pancreatic β-cell-based FET. The β-cells aggregated on the Ta_2O_5 gate surface in the culture medium, because it was difficult for β-cells to adhere to the nonmodified gate but easy for them to aggregate like a spheroid. As shown in Fig. 13.11, the surface potential of the β-cell-based FET shifted gradually in the positive direction at about 8.5 mV for about 3 h. The average of surface potential changes due to the introduction of glucose was 10.3 mV for about 3 h using five-cell-based FETs. On the other hand, almost no surface potential changes of the reference FET was found, although a smaller shift was detected owing to the background noises such as temperature changes when glucose was injected to the gate. The electrical signal of the β-cell-based FET indicates the increase in positive charges or the decrease in negative charges based on ions or charged molecules through β-cells at the β-cell/gate interface following the introduction of glucose into the β-cell-based FET. About five β-cells were cultured on the gate sensing area, of which the gate width was 10 μm and the gate length was 340 μm. The surface potential changes were based on the charge density changes of the glucose response of about five β-cells. The uptake of glucose into β-cells causes the increase in ATP/ADP concentration, the depolarization of K_{ATP} channel, the influx of Ca^{2+} through ion channel, and the exocytosis of insulin and C-peptide together with Zn^{2+} to form microcrystals within secretory granules [24]. In this study, insulin and C-peptide secretion has been detected by the use of enzyme-linked immunosorbent assay (ELISA) method based on the reaction with an anti-insulin monoclonal antibody, of which the concentration was about

32 ng/ml at 2 h after glucose addition. The resulting molecular and ion charges at the β-cell/gate interface induced the surface potential changes of the cell-based FET that prove the electrical activity of pancreatic β-cells. Furthermore, the positive shift of surface potential can be assumed to indicate the increase in the concentration of hydrogen ions at the β-cell/gate interface, because the nonmodified FET device with the Ta_2O_5 gate insulator is basically more sensitive to hydrogen ion concentration than those of other ions. This is why we assume that the respiration of β-cells would have been activated during the insulin secretion process following glucose addition. After the increase of electrical signal, the surface potential of β-cell-based FET decreased gradually because of diffusion of hydrogen ions from the cell/gate interface to the bulk solution accompanied by degradation of respiration activity. The insulin secretion process would include the respiration activity of β-cells so that the electrical monitoring by the use of the β-cell-based FET might demonstrate the reduction and exhaustion of glucose-induced insulin secretion.

The pH variation at the β-cell/gate interface determined on the basis of respiration activity can be estimated from the electrical signals of the cell-based FET. Basically, the FET chip utilized in this measurement showed the change of gate voltage 51.4 mV/pH for pH variation. The average change in surface potentials after the addition of glucose was about 10.3 mV with a standard error of ± 2.2 mV for five-cell-based FETs. Therefore, the respiration activity triggered by glucose caused the change of about pH 0.2 at the interface between the pancreatic β-cells and the gate surface. Strictly, the shift of pH from 7.4 to 7.2 was detected at the β-cell/gate interface by the cell-based FETs. The amount of eliminated CO_2 was calculated to be about 1.2×10^{-8} M on the basis of principle that changes in hydrogen ion concentration corresponds to pH variation, according to the equilibrium of CO_2 in solutions. In the calculation, O_2 consumption, ATP synthesis, and so on in the citric acid cycle inside the mitochondrion could be estimated from the amount of eliminated CO_2, which was determined from the electrical signals of the cell-based FET. However, the effect of a buffer solution on pH and the diffusion of CO_2 at the β-cell/gate interface have to be considered in order to estimate accurately the consumption and the generation based on the respiration activity. Also, the electrical signals of the β-cell-based FET seem to include the effect of ion or molecular charges in the insulin secretion process other than the pH variation based on respiration activity. In a previous work [13], the noninvasive monitoring of transporter function was shown using oocyte-based FETs. The size of an oocyte was about 1 mm and larger than that of a somatic cell (about 10 μm). Therefore, the amplification of electrical signals based on the transporter function was expected by the use of the oocyte-based FETs. However, the electrical detection of β-cell activity obtained in this study has demonstrated the possibility of detection of the somatic cell function using β-cell-based FETs.

13.4.2 Evaluation of In Vitro Fertilization (IVF)-Based Embryo with Semiconductor Principle

Recently, assisted reproductive technology (ART) has been expected to be one of the therapeutic methods of sterility. Engineers other than obstetricians have been required for assured success of ART programs. For in vitro fertilization (IVF) of one of ART programs, how to evaluate embryo quality and select embryo in condition are significant. Morphological evaluation has been widely used to rank embryo quality because microscopic analysis is noninvasive and useful in predicting pregnancy rates [25, 26]. However, the standard of classification for embryo quality seems to be ambiguous among operators because it is a subjective method. Moreover, elective single-embryo transfer (eSET) will be recommended in the future in order to prevent multiple pregnancies [27]. Therefore, the novel principle to evaluate the quality of a single embryo quantitatively and noninvasively in a real-time manner is being required for practical use in ART.

In our previous work, the respiration activity of fertilized ova of a sea urchin was monitored noninvasively, quantitatively, and continuously as the change of pH by the use of the principle of semiconductor-based IS-FET [28]. The detection principle of IS-FET is based on the potentiometric detection of charge density changes at the gate insulator and is applied for a biosensing circuit [29]. Since the gate insulator usually consists of Si_3N_4 or Ta_2O_5 with a hydroxyl group at the surface in solutions, furthermore, the IS-FET is sensitive to the concentration of hydrogen ion with a positive charge and should be utilized as the pH sensor. Therefore, pH variation based on the respiration activity of an embryo will be monitored quantitatively and noninvasively in a real-time manner using a single-embryo-coupled gate FET for eSET (eSET-FET), because pH at the interface between an embryo and the gate membrane of FET will change sensitively according to the dissolving of carbon dioxide into a medium generated by the metabolism and respiration activity in an embryo. Thus, the platform based on eSET-FET sensor will be valuable for the development of an evaluation system to select a single embryo with good quality for eSET in the future.

On the other hand, oxygen consumption has been considered to be the parameter that provides the best indication of overall metabolic activity of a single embryo [30–34], although embryo metabolism has previously been assessed by measurement of nutrient consumption, such as glucose, pyruvate, and amino acids [35–38]. As one of the detection methods for the evaluation of embryo quality, the electrochemical system is being developed. Shiku et al. reported previously the detection concept of oxygen consumption based on the respiration activity of an embryo [39]. In this method, the oxygen reduction current was detected near the surface of a single embryo using the cyclic voltammetry technique. However, this method is unsuitable for real-time measurement for a long term such as cleavage of mammalian embryo.

In this case study, we introduce a semiconductor-based embryo sensing device for eSET (eSET-FET) to monitor single mouse embryo activity based on cellular respiration in a real-time, quantitative, and noninvasive manner. Additionally, we report to have developed the simultaneous analysis system composed of microscopic observation and electrical measurement of eSET-FET under the adequate embryo culture condition in an incubator.

Insulated gate FETs as semiconductor devices were fabricated using the standard integrated circuit technology except for deposition of the gate electrode. The N-channel depletion mode FET was designed in a 2 mm × 5 mm chip (ISFETCOM Co. Ltd.). The gate insulator was composed of Ta_2O_5, Si_3N_4, and SiO_2 layers, of which each thickness was 50, 100, and 100 nm, respectively. The gate surface contacting directly with the embryo and culture medium was Ta_2O_5 layer, of which the safety was confirmed by estimating the rate of transplantation of cultured embryos on the surface to recipient mice. The fabricated FET chip was mounted on a flexible polyimide film with patterned copper electrodes and bonded. The FET chip was encapsulated with a glass ring except for the sensing area and was utilized in a culture medium as the eSET-FET sensor. The eSET-FET sensor was immersed in a culture medium (KSOM; Arc Resources JAPAN) with a platinum Pt reference electrode. Polydimethylsiloxane (PDMS) was utilized for the mounting of a glass ring and Pt electrode on the device because of nontoxicity for an embryo culture. A single mouse embryo was put on one gate sensing area of the eSET-FET (Fig. 13.12a). The embryologist put a single mouse embryo on a gate area by using a micropipet.

A mammalian embryo is very sensitive to external environments such as the ion strength of culture medium, temperature, and so on. This is why the total system for evaluation of embryo based on IVF should be prepared appropriately. Figure 13.12b shows the simultaneous analysis system of embryo activity by microscopic observation and electrical measurement using the eSET-FET sensor in the incubator. Using the simultaneous analysis system, the subjective and objective evaluation of an embryo under the adequate condition of an embryo culture can be performed in a real-time and noninvasive manner. Moreover, the sensing area was covered by 20 μl of droplet culture medium, where a single mouse embryo was kept on the gate of the semiconductor, as shown in Fig. 13.12c. The mineral oil of 1 ml was introduced onto a dropped culture medium of 20 μl to prevent drying. The droplet culture is useful to control the position of a single embryo on the gate and detect ion concentration change due to low volume.

The electrical characteristics of the eSET-FET sensor such as the gate voltage (V_G)–drain current (I_D) characteristics and the surface potential at the gate surface were measured in a culture medium using a semiconductor parameter analyzer (B1500A, Agilent) and a custom-made potentiometric analyzer (Radiance Ware Inc.), respectively. As the basic electrical characteristic, the threshold voltage shift ΔV_T was determined after an embryo was placed on the gate. The ΔV_T was defined as a difference of the V_G–I_D characteristics at a constant drain current of 700 μA. The time course of the surface potential at the gate surface was monitored using a circuit with which the potential change at the interface between an aqueous solution and the gate insulator can be read out directly at a constant drain current. In the present study, the gate voltage and the drain current were set to be 1 V and 700 μA, respectively.

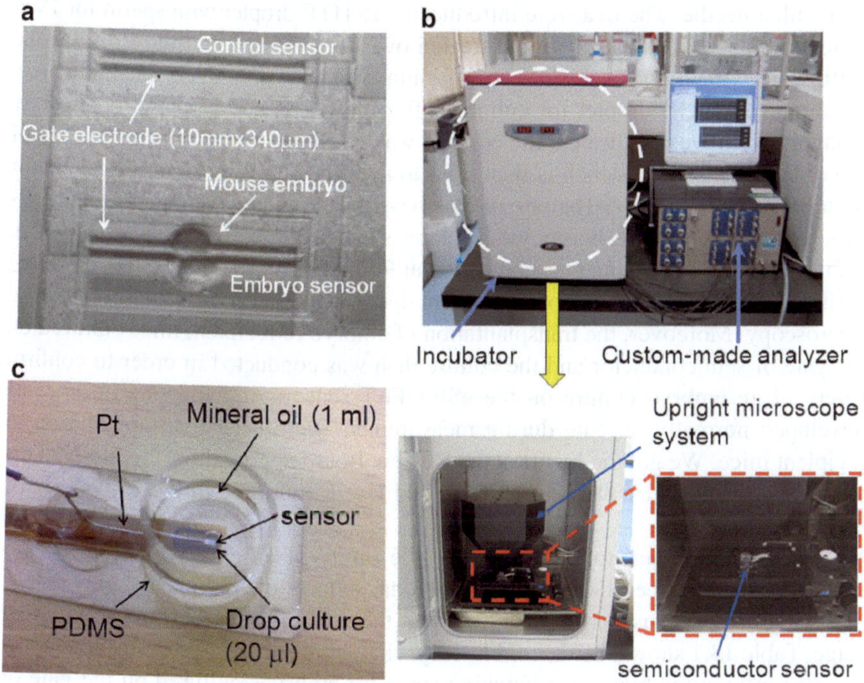

Fig. 13.12 Measurement device for embryo monitoring. (**a**) Photograph of a single mouse embryo on the gate of eSET-FET sensor. (**b**) Simultaneous embryo analysis system. Electrical measurement using the eSET-FET sensor can be continuously performed in various intervals of second or minute to ~ a few weeks or more in the incubator controlled at 37 °C and 5 % CO_2. The upright microscopy was set up in the incubator. (**c**) Sensing structure. A droplet of 20 μl was covered by mineral oil so that a long-term embryo culture can be conducted. Mineral oil is nontoxic and has the permeability of gases such as CO_2 and O_2

All the nutrient media to be used for IVF were prepared on the day before experiments, and a balanced gas in droplet was controlled through overnight under the adequate condition (5 % CO_2 and 37 °C). Male ICR mice (9–10 weeks old; Charles River JAPAN Inc.) were sacrificed by cervical dislocation. Sperms collected from caudal epididymides of ICR male mouse were incubated for 1.5 h at 5 % CO_2 and 37 °C. In this case, sperms of 2×10^5 cells/ml were seeded in HTF (human tubal fluid; Arc Resources JAPAN) droplet of 200 μl and subsequently cultured again in fresh HTF covered by mineral oil (Mineral Oil Light; REPROLINE) at 5 % CO_2 and 37 °C. Female B6D2F1 mice (9–10 weeks old; Charles River JAPAN Inc.) were kept for 48 h from intraperitoneal injection of 5 IU PMSG (pregnant mare's serum gonadotropin; ASKA Pharmaceutical JAPAN) and subsequently sacrificed by cervical dislocation after 15–16 h from intraperitoneal injection of 10 IU hCG (human chorionic gonadotropin; ASKA pharmaceutical JAPAN) for ovarian hyperstimulation. Oviducts were excised and drew to take out ova with a needle. The ova were introduced into HTF droplet with sperm for IVF. After about 5 h from insemination, treated ova were washed and moved from the fertilization medium to KSOM (potassium simplex optimized medium; Arc Resources JAPAN) droplet for culture. After the release of a second polar body from an ovum was confirmed, the embryo was set up on a gate sensing surface of the eSET-FET sensor, which is then used to measure the embryo in the incubator system with microscopy. The rate of blastocyst hatching and average of total cells in single embryo at the blastocyst-hatching stage were investigated as shown in Table 13.1. Each counting was conducted at 90 h after measurements (IVF). Total cells were counted by the use of fluorescent dye (DAPI) and confocal laser scanning microscopy. Moreover, the transplantation of embryo to recipient mice cultured on the gate of semiconductor and the culture dish was conducted in order to confirm safety of an embryo culture on the eSET-FET sensor. Actually, the blastocysts developed normally on gate during measurement were used for implantation to recipient mice. We got the Institutional Review Board approval for this study. This study was overseen by the Animal Care and Use Committee in the University of Tokyo.

It is very important to confirm firstly the safety of an embryo culture on the semiconductor devices like on a conventional culture dish. Actually, the mouse embryo after IVF was normally cultured on the gate until the blastocyst-hatching stage. Table 13.1 shows the rate of blastocyst hatching and the average of total cells in an embryo at the blastocyst-hatching stage for embryos cultured on the gate or the culture dish as control, respectively. The rates of blastocyst hatching were around 90 % for both conditions. The average of total cells for embryos cultured on the gate was also similar with that for embryos cultured on the culture dish. The culture of embryo on the gate was found to be of no difference with the one on the culture dish. Furthermore, the safety of an embryo culture on the gate was investigated by transplantation to the recipient mouse (Table 13.2). The rates of

Table 13.1 Rate of blastocyst hatching and average number of cells in a single mouse embryo at blastocyst-hatching stage for embryos cultured on the gate of semiconductor and the culture dish

	Gate of semiconductor	Culture dish
Rate of blastocyst hatching (%)	88.1	96.8
Average number of cells in an embryo (%)	71.5	72.4

Number of embryo: 59 on gate, 424 on culture dish

Table 13.2 Result of transplantation of embryo to recipient mouse cultured on the gate of the semiconductor and the culture

	Number of transferred embryo	Number of offspring	Rate of offspring (%)
Culture on gate of semiconductor	20	10 (♂6, ♀4)	50
Culture on dish	20	8 (♂5, ♀3)	40

All offspring grew normally. Deformed babies and abnormal behaviors were not found in this study

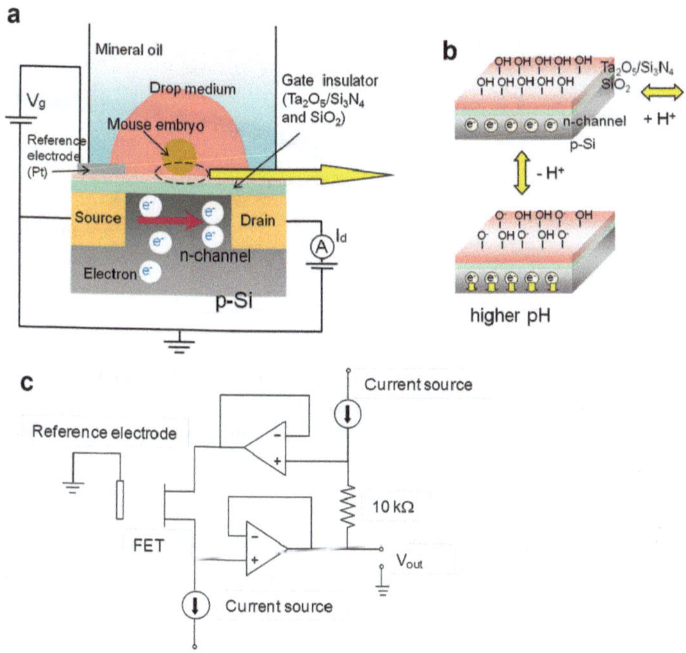

Fig. 13.13 (**a**) Schematic illustration for measurement of surface potential of eSET-FET sensor. (**b**) The Si_3N_4 or Ta_2O_5 gate surface with a hydroxyl group in solutions is sensitive to pH based on the concentration of hydrogen ions. (**c**) Circuit for measuring surface potential of eSET-FET sensor. The output voltage was continuously monitored as the surface potential change of eSET-FET sensor using this circuit, according to the conceptual structure of eSET-FET sensor shown in (**a**)

offspring were compared for IVF embryo transfer (ET) of embryos cultured on the gate or culture dish, respectively. Even embryos cultured on the gate of eSET-FET sensor obtained about the same offspring rate for IVF-ET with embryos cultured on the conventional dish. These data mean that the semiconductor device itself never show toxicity for mouse embryo under the culture condition. Thus, the embryo sensing by the use of the eSET-FET sensor was confirmed to be safe for the evaluation of embryo quality after IVF.

The detection principle of the eSET-FET sensor is based on the potentiometric detection of charge density changes at the gate insulator, on which specific binding between target and probe molecules is made for molecular recognition. Figure 13.13 shows the conceptual structure of the eSET-FET sensor for detection of embryo activity after IVF. A single mouse embryo was set on the gate sensing surface. The gate surface of the eSET-FET sensor is immersed in a measurement solution together with a platinum (Pt) reference electrode (Fig. 13.12c). Basically, ionic or molecular charges at the gate interact electrostatically with electrons in the silicon crystal through the thin gate insulator and induce electrical signals by the field effect, resulting in the source and drain current (I_D) change at the channel

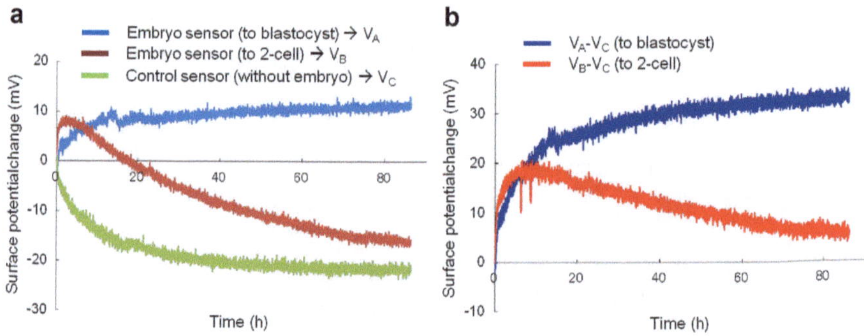

Fig. 13.14 Real-time and noninvasive monitoring of surface potential of eSET-FET sensor. (**a**) The eSET-FET sensor with embryo and the control sensor without embryo were measured to analyze embryo activity based on respiration after IVF. (**b**) Differential measurements of eSET-FET sensor with embryo against control sensor

(Fig. 13.13a). Eventually, the output voltage is monitored as the shift of surface potential using a circuit (Fig. 13.13c) with which the potential change at the interface between an aqueous solution and the gate insulator can be read out directly at a constant I_D. The charge density changes based on hydrogen ions can be detected as the shift of surface potential of eSET-FET sensor, because a hydroxyl group reacts equivalently with a hydrogen ion with positive charge according to pH (Fig. 13.13b). Here, this measurement principle is found to be noninvasive for an embryo culture. The details of the eSET-FET sensor and the fabrication process have been reported previously by our group. Furthermore, two types of eSET-FET sensors were prepared for differential measurements in this study: one was the eSET-FET sensor on which a single mouse embryo was kept and the other was the control FET sensor without an embryo (Fig. 13.12a). Using them, differential measurements were performed in order to eliminate the common unexpected signals such as temperature change, nonspecific adsorption, and change in ion concentration.

The change of surface potential at the gate surface of eSET-FET sensor was monitored after IVF (Fig. 13.14). The surface potential of eSET-FET sensor with single embryo to blastocyst increased gradually as shown in Fig. 13.14a. The increase of surface potential at the beginning of cleavage indicates the increase of hydrogen ions with positive charges based on the dissolving of carbon dioxide generated by cellular respiration. Since little surface potential change was measured before fertilization, when virginal ova were kept on the gate surface, the shift of surface potential based on the increase of hydrogen ion would reveal the change of metabolism due to fertilization. Mitochondria play an important role in respiration inside cells. Pyruvate molecules produced by glycolysis are transported into the mitochondrion matrix through the inner membrane, where they are oxidized and combined with coenzyme A to form carbon dioxide, acetyl-CoA, and NADH. The mouse embryo after fertilization would become activated accompanied by cleavage

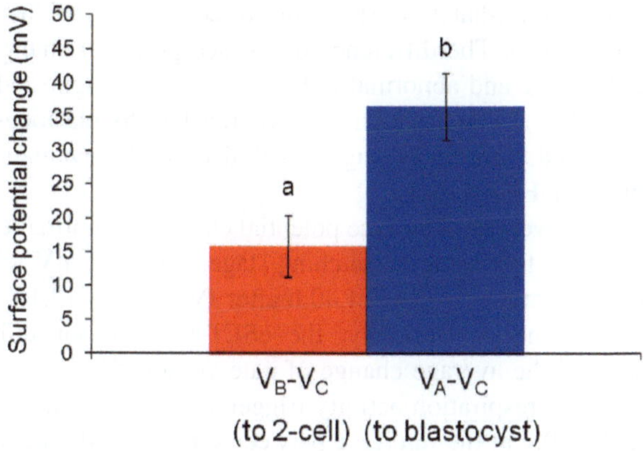

Fig. 13.15 Surface potential change of eSET-FET sensor with embryo against control sensor without embryo at 40 h. Eight-cell stage was often observed at around 40 h after IVF for normal eSET-FET. Different letters indicate statistical significance ($P < 0.01$)

resulting in the change of shape and amount of mitochondria [40], which produce ATP and are closely related to the metabolism of embryo. On the other hand, the surface potential change of eSET-FET sensor with single embryo accompanied by 2-cell block decreased gradually, which was similar with that of control sensor without embryo, although it increased drastically only for several hours before 2-cell stage. This electrical signal indicates an abnormal behavior of embryo before 2-cell block and that hydrogen ions with positive charges based on respiration activity increased drastically at the interface between embryo and sensor surface. This may be because metabolism in single embryo was activated accompanied by autophagy before cell death [41], because proteins would be degraded to amino acids in autolysosome, which would be utilized for cellular respiration. Lastly, the control sensor without embryo showed actually the pH variation of culture medium for 90 h under the embryo culture condition. The surface potential change of control sensor reached to about −20 mV at around 40 h, which could be calculated as pH 7.4–7.7 because the semiconductor sensor had the detection ability of about 60 mV/pH. Moreover, the differential electrical signals were calculated as shown in Fig. 13.14b. Basically, the electrical signal of sample sensor against control sensor should be evaluated considering unexpected signals such as temperature change, pH variation of culture medium, and so on. The electrical signal of eSET-FET sensor with normal embryo against control FET (normal eSET-FET) was clearly different from that of eSET-FET sensor with abnormal embryo against control FET (abnormal eSET-FET) after around 2-cell stage. Interestingly, the surface potential for normal eSET-FET increased gradually after IVF but its inclination changed significantly after around 2-cell stage of 10–15 h. This might show the change of metabolism during cleavage at around 2-cell stage. Figure 13.15 shows the differential signal at 40 h after IVF in Fig. 13.14b, which corresponds to around 8-cell stage for normal eSET-FET particularly. All the signals were calculated based on the differential measurements between sample sensor with embryo and control sensor without embryo. Twenty-one embryos were cultured at 37 °C on the gate in monitoring surface potential. Eighteen embryos of them were normally cultured to blastocyst-hatching stage, but three embryos of them stopped cleavage around 2-cell stage. The difference of surface potential changes at 40 h between normal eSET-FET and abnormal eSET-FET was distinguished as a significant difference ($p < 0.01$). Thus, which embryo reaches to blastocyst normally or not can be predicted at the early stage of cell division by evaluating surface potential change of eSET-FET sensor.

The average of surface potential change for normal eSET-FETs, which occurred cleavage to blastocyst-hatching stage, was 36.7 mV with standard error ±4.9 mV ($n = 18$) at 8-cell stage of 40 h after IVF in Fig. 13.15, considering the differential measurements. Basically, the eSET-FET sensor utilized in this measurement showed the average change of gate voltage 55.6 mV/pH for pH variation. Therefore, the respiration activity triggered by fertilization caused the change of about $\Delta pH = 0.7$ at the interface between embryo and gate surface. Strictly, the shift of pH from 7.4 to 6.7 was detected at the interface by the eSET-FET sensor. The eliminated carbon dioxide was calculated as about 1.6×10^{-7} M based on the concentration change of hydrogen ion corresponds to pH variation, according to the equilibrium of carbon dioxide in solutions ($CO_2 + H_2O \rightleftarrows H^+ + HCO_3^-$). In the calculation, the oxygen consumption, the ATP synthesis, and so on in the citric acid cycle inside the mitochondrion could be estimated from the amount of eliminated carbon dioxide, which is worked out from the electrical signal of the eSET-FET sensor. However, the diffusion of carbon dioxide at the interface between an embryo and the gate surface or from an embryo surface apart from the gate has to be considered in order to estimate accurately its generation. The volume of culture medium might affect actually on signals of pH variation based on embryos. In this case, however, the volume of 20 μL used in this study was large enough for the size of embryo (about 100 μm), and actually, we could not find electrical signals when the embryo did not contact with gate area and placed far from the gate in the same volume. Therefore, we need to put directly embryo on the gate surface for measurement. Therefore, the respiration activity of a single mouse embryo obtained by the electrical signals of eSET-FET sensor should be significant for convenient and noninvasive evaluation of embryo quality in the future.

13.4.3 Apoptosis Sensing with Transparent Device

Cells can be directly and noninvasively observed by the use of microscopy and imaging of fluorescent functional molecules as well as morphological changes. In vitro microscopic observation gives much information of cell functions to researchers. The observation by the use of inverted microscopy allows high-resolution imaging of cells on a conventional cell culture dish due to its transparency. However, the quantitative analysis of cell activities such as ion behaviors through an ion channel at the cell membrane would be difficult using conventional microscopy.

Most of biological phenomena in vivo are closely related to exchanges of charged ions such as potassium ion, sodium ion, and so on through an ion channel at the cell membrane in order to keep homeostasis. The direct detection of ions enables easily to analyze various biomolecular recognition events. The principle of semiconductor devices based on field effect can be utilized in order to realize to detect directly ionic mediums in solutions [42]. Recently, some original manners for electrical measurement of cell functions have been performed using the semiconductor devices [43–45]. The principle of semiconductor-based biosensing devices is based on the potentiometric detection of charge density changes induced at a gate insulator/solution interface accompanied by specific biomolecular recognition events. Ionic charges at the gate insulator electrostatically interact with electrons in silicon crystal across the thin gate insulator resulting in the threshold voltage change. The platform based on the semiconductor device has some great advantages in terms of real-time, noninvasive, and label-free measurements, ease of downsizing, and integration by the conventional semiconductor microfabrication process.

Thus, an electrical measurement during a microscopic observation contributes to simultaneous cell analysis based on subjective and objective evaluation of living or dying cells, deepening one's understanding of cell functions. Living cell behavior is closely related to respiration activity, resulting in change of pH in solutions. On the other hand, dying cells are known to release ions through ion channels at cell membrane (programmed cell death (apoptosis)) [46]. However, these experimental results have not been shown in a direct and real-time manner yet.

In this case study, we introduce a direct and real-time monitoring of programmed cell death, apoptosis, in the early stage. The simultaneous monitoring of cell function is demonstrated by use the semiconductor-based electrical monitoring and the microscopic observation. Here, we propose the cell-based transparent-gate transistor (cell-based TGT) with indium–tin–oxide (ITO) gate electrode because microscopic observation can be easily performed in a conventional way because of its transparency.

We have designed and fabricated an EG-FET. The gate electrode was extended from metal electrode of metal oxide semiconductor (MOS) [47]. Indium–tin–oxide (In_2O_3-SnO_2; ITO) was used as the material of gate electrode. Several developments of ITO electrode have been performed because of its transparency [5, 48]. The EG-FET prepared in this study is a FET with an extended ITO gate electrode of which sensing area is separated from the channel area of the FET (Fig. 13.16). The ITO thin film was deposited by sputtering. The sputtering time and electric power were 2 min and 100 W, respectively. Multi sensing device with the extended ITO gate electrodes was prepared. The ITO gate was encapsulated with an epoxy resin (ZC-203; Nippon Pelnox) except for the sensing area, of which the diameter was 5 mm.

In order to investigate the electrical characteristics of FET, a semiconductor parameter analyzer and a custom-made real-time analyzer were used. The electrical measurement was performed in HEPES buffer solution, and pH of the sample solution was adjusted to 7.4 with $Ca(OH)_2$ ($[Ca(OH)_2] = 1.7$ mM). K^+ concentration was adjusted to 1×10^{-4}, 1×10^{-3}, 1×10^{-2}, and 1×10^{-1} M with KCl. The prepared ITO gate was immersed in the sample solution, and then the threshold

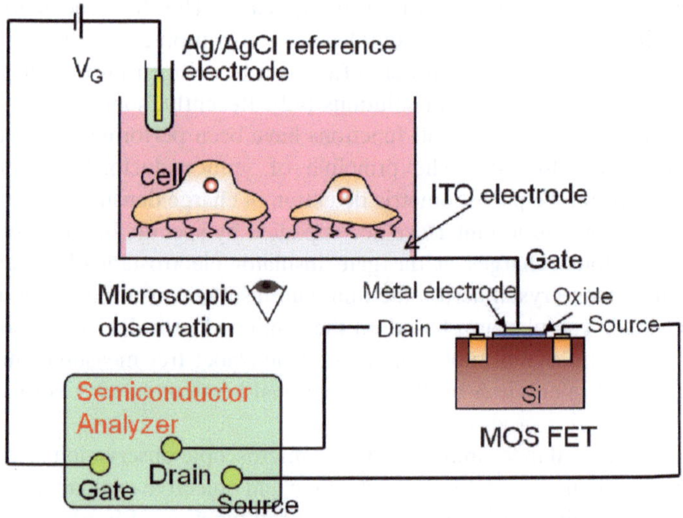

Fig. 13.16 Structure for measurements of electrical characteristics using cell-based transparent-gate transistor (TGT) with indium–tin–oxide (ITO) gate. A metal oxide semiconductor (MOS) was used as a transducer, of which the gate electrode was extended from the metal surface. Cells were cultured on the ITO gate electrode in a cell culture medium at 37 °C and 5 % CO_2 in the incubator system. A Ag/AgCl electrode with a saturated KCl solution was used as a reference electrode

voltage (V_T) and surface potential changes were measured. The ΔV_T is detected in the gate voltage (V_G)–drain current (I_D) electrical characteristic of semiconductors, which shifts in the negative direction with increasing K^+ concentration with positive charges. Moreover, real-time monitoring of ΔV_T was performed by the use of the custom-made real-time analyzer, because a circuit for measuring is designed accordingly [29].

To investigate the electrical characteristics of a cell-based transparent-gate transistor (TGT) under the steady condition, we have cultured HeLa cells on the ITO gate at 37 °C and 5 % CO_2 in the incubator system. Dulbecco's Modified Eagle Medium (DMEM, Invitrogen) (+1 % FBS) was used as culture medium. The number of seeded cells was about 1×10^5 cells/ml. The ITO gate was placed on the stage of microscopy in an incubator system. The electrical measurement of cell-based TGT was performed under the adequate condition of cell culture in the incubator. Simultaneously, the images of HeLa cells on the ITO gate were taken in the interval of 10 min in the incubator system. After confirming that cells were confluent, the change of threshold voltage was monitored. The apoptosis was initiated by injecting TNF-related apoptosis-inducing ligand (TRAIL, 500 ng/ml). The actual electrical signals for target was evaluated by differential measurements, which were performed in order to eliminate common background noises such as temperature changes, nonspecific adsorption, changes in ion concentration, and so on. Two kinds of cell-based TGTs with HeLa cells on ITO gate were prepared for

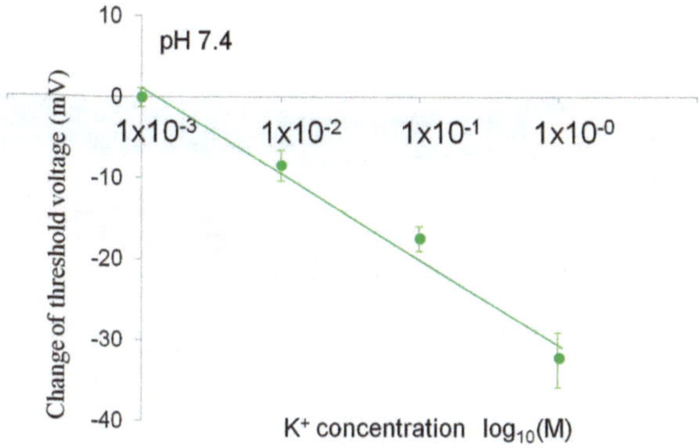

Fig. 13.17 Change of threshold voltage of indium–tin–oxide (ITO) gate field-effect transistor (FET) for K^+ concentration. The effect of K^+ concentration on the change of threshold voltage of ITO gate was investigated. The K^+ concentration was varied from 1×10^{-3} to 1×10^{0}, which was controlled at pH 7.4. The threshold voltage change was determined at a constant drain current of 700 μA and drain voltage of 1 V

differential measurements: one was the cell-based TGT stimulated with TRAIL, while another cell-based TGT was measured without adding TRAIL.

When programmed cell death, apoptosis, is induced by ligand proteins such as TRAIL, volume reduction of cells is observed accompanied by DNA fragmentation, chromatin concentration, and so on in them resulting in fragmentation of cells. In particular, the previous work has shown the possibility of H_2O, K^+, and Cl^- release through cell membrane before shrinkage of cells in the early stage of apoptosis [46]. However, the real-time monitoring of ion release has not been actually performed, but each ion behavior was suggested using transfected cells with downregulated potassium ion channels. In this research, the semiconductor-based biosensing devices have the ability to detect directly change of ion concentration. Therefore, the ion release will be detected directly in the early stage of apoptosis by the principle of field effect of semiconductor-based biosensing devices. The detection of apoptosis in the early stage may contribute to a high-throughput drug screening using arrayed devices.

The concept of cell-based TGT with cells on the ITO gate electrode is shown in Fig. 13.16. The ITO gate electrode is extended from metal electrode of conventional MOS-FET. Due to transparency of the ITO gate electrode, cultured cells can be easily observed by conventional inverted microscopy. When cells are cultured on the ITO gate electrode, we can monitor various cell functions in a real-time manner using the cell-based TGT with ITO gate. The effect of K^+ concentration on the electrical signal of ITO gate FET is shown in Fig. 13.17. The gate voltage for K^+ concentration of 1×10^{-3} M measured by the semiconductor parameter analyzer was standardized as the start point of measurement. The electrical response of ITO gate FET to K^+ concentration was about 11 mV/decade. The gate voltage of ITO gate responded to the change of K^+ concentration, although it was small compared

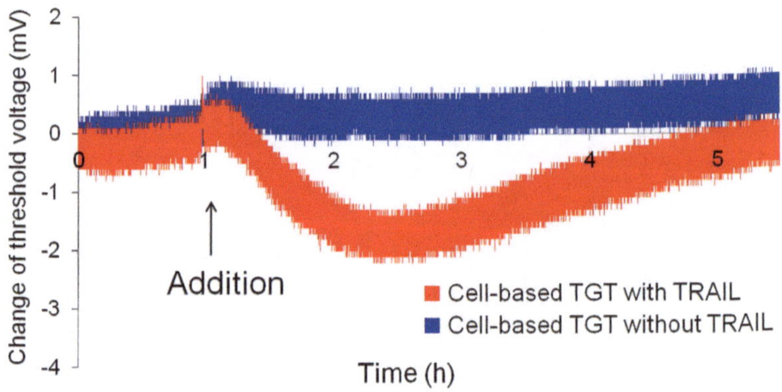

Fig. 13.18 Real-time monitoring of change of threshold voltage based on apoptosis using a cell-based transparent-gate transistor (TGT) with indium–tin–oxide (ITO) gate

with the potential slope of the Nernstian response. The electrical characteristics of ITO gate might be changed according to making process conditions such as oxygen concentration in sputtering ITO. The adequate response to K^+ concentration would demonstrate the detection of K^+ release based on apoptosis of cells. The gate voltage shift (corresponding to ΔV_T) to the negative direction indicates the increase of positive charges of K^+ at the ITO gate. The ITO gate FET prepared in this study was sensitive to K^+ concentration but not to pH, although the graph was not shown here. This is very effective to detect selectively ion behaviors based on cell functions without considering pH variation. The difference of potential response between different ions might be resulted from the correlation between each ion size and number of hydroxyl groups at ITO surface in solutions. However, this reason is complicated, and it should be very important to control the making process of the ITO surface.

Considering these electrical characteristics of cell-based TGT with ITO gate, a direct and real-time monitoring of apoptosis was investigated by the use of the device. Figure 13.18 shows the threshold voltage change of cell-based TGT with ITO gate before and after addition of TRAIL for apoptosis induction. The shift of threshold voltage indicates the ΔV_T. The surface potential was kept constant before the addition of TRAIL, but the threshold voltage shifted to the negative direction until about 2 h after the introduction of TRAIL, while no signal was found for control cell-based TGT, in which HeLa cells were cultured on the ITO gate but a culture medium without TRAIL was added. According to the electrical characteristics of ITO gate for K^+ response, the negative shift would be caused by K^+ release through ion channel in the early stage of apoptosis. Basically, K^+ and Cl^- release based on apoptosis was reported by the previous work [46], where they were not monitored in a real-time and direct manner. Therefore, we have clarified directly K^+ release due to apoptosis in the early stage by the use of the principle of semiconductor-based biosensing devices. Actually, the microscopic observation was performed simultaneously during electrical measurement. The snap shots of

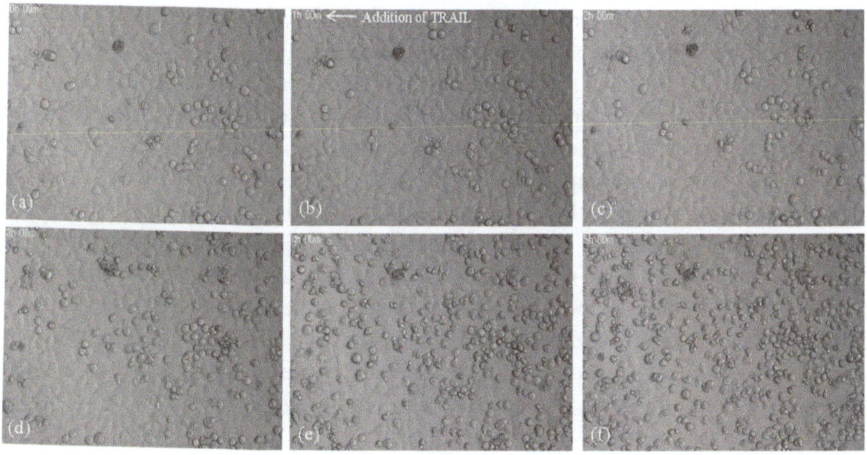

Fig. 13.19 Microscopic observation of apoptotic process of HeLa cells for incubation time. TRAIL was introduced into ITO gate with cultured HeLa cells at 1 h. (a) 0 h, (b) 1 h, (c) 2 h, (d) 3 h, (e) 4 h, (f) 5 h

cultured cells at the interval of 10 min were taken by the use of inverted microscopy in the incubation system. By the use of microscopic observation of shrinkage of HeLa cells accompanied by H_2O, K^+, and Cl^- release, apoptosis of HeLa cells on ITO gate was clearly observed around 2–3 h after introduction of TRAIL, as shown in Fig. 13.19. The morphological change has almost corresponded to the change of threshold voltage of cell-based TGT with ITO gate. The peak of threshold voltage shift based on K^+ release seems to be found a little earlier than the morphological change (Figs. 13.18 and 13.19). The detection of a few mV potential changes has corresponded to the K^+ concentration change of less than about a decade considering the electrical characteristic shown in Fig. 13.17. The concentration of K^+ inside cells is usually about 140 mM higher than that of about 5 mM outside cells. Strictly, the K^+ concentration in DMEM was originally 5.33 mM so that the K^+ concentration at the interface between cells and gate surface increased up to about 10 mM by apoptosis, according to the electrical characteristic of ITO gate shown in Figs. 13.17 and 13.18. The change of K^+ concentration in the range of intra- and extracellular concentrations has been detected by the use of the cell-based TGT with ITO gate. The control cell-based TGT has hardly shown the potential response after the introduction of culture medium without TRAIL, so that a few mV difference of electrical signals between the target and control devices was significant for K^+ release resulted from apoptosis. On the other hand, the increase of electrical signal of cell-based TGT was found after the peak potential based on apoptosis. The signal shift to the original base line should show the diffusion of K^+ from the interface between cells and gate surface, because almost all the cells occurred the shrinkage based on apoptosis and their morphological changes were hardly observed after the peak potential. Actually, the concentration change of potassium ion could be confirmed from about 5 mM to 6 mM in the whole culture medium as well as at the solution/transistor interface, after almost all of the HeLa cells occurred with apoptosis at 10 h after the addition of TRAIL.

In conclusion, we have shown the direct, nonlabel, and real-time monitoring of apoptosis using the semiconductor-based biosensing devices. As a result of the electrical monitoring on the basis of the principle of field effect, we could clarify directly K^+ release accompanied by Cl^- and H_2O release in the early stage of apoptosis. This is because the ITO gate electrode used in this study was sensitive to the concentration change of K^+, compared with pH variation. Moreover, the morphological change due to apoptosis was observed during the electrical monitoring of semiconductor device. The platform based on the semiconductor device with transparent electrode realizes subjective and objective evaluation of cell functions in a direct, noninvasive, and real-time manner and is useful as the measurement tool in the field of cell biology, clinical diagnosis, and drug discovery.

13.5 Molecular Recognition Based on Intrinsic Molecular Charges

13.5.1 Genetic Analysis [49–55]

The principle of genetic FET is based on the detection of charge density change on the gate surface which is induced by the specific binding of DNA molecules. Oligonucleotide probes are immobilized on the surface of the gate insulator. The genetic FET is immersed in a measurement solution together with a Ag/AgCl reference electrode with saturated KCl solution. The potential of a measurement solution is controlled and fixed by the gate voltage (V_G) through the reference electrode. When complementary DNA molecules are contained in a sample solution, hybridization occurs at the surface of the gate area. Since DNA molecules are negatively charged in an aqueous solution, a hybridization event can be detected by measuring a shift of the V_T. When an n-channel FET is used, the V_T shifts in the positive direction in response to DNA molecules. When intercalators are introduced into the double-stranded DNA after hybridization, the V_T of the genetic FET shifts in the negative direction, because the intercalators are positively charged. Since the intercalators react specifically with double-stranded DNA, an undesirable background noise caused by nonspecific adsorption of single-stranded target DNA can be eliminated and a more precise and reliable detection of a hybridization event can be realized.

A specific binding of charged biomolecules at the gate surface can be detected as a shift of the V_T, which can be determined in the gate voltage V_G and the drain current I_D (V_G–I_D) characteristics of the genetic FET (Fig. 13.20). The V_G–I_D characteristics of the genetic FET shifted along the gate voltage V_G axis in the positive direction after immobilization of oligonucleotide probes. Oligonucleotide probes immobilized were a normal type for R353Q locus of factor VII gene (probe: 5′-amino group-CCACTACCGGGGCACGT-3′ (17mer), target: 5′-ACGTGCCCCGGTAGTGG-3′ (17mer), melting temperature: 60 °C) [56]. In order to evaluate the V_T shift in more

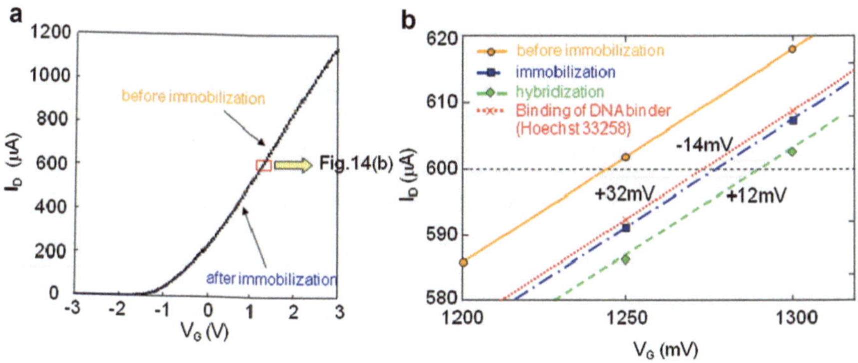

Fig. 13.20 V_G–I_D electrical characteristic of DNA-based FET

detail, the local area shown in Fig. 13.20b (surrounded area) was magnified. The V_T shifts after hybridization and specific binding of DNA binder are also shown in Fig. 13.20. When oligonucleotide probes were immobilized on the gate surface, the V_T shifted along the V_G axis by the amount of 32 mV. The positive shift is due to negative charges induced at the gate surface after immobilization process including cleaning, silanization, glutaraldehyde treatment, immobilization of oligonucleotide probes, and blocking with glycine. Immobilization of oligonucleotide probes and glycine blocking are considered to contribute to the V_T shift to a large extent. When the complementary target DNA was introduced to the gate surface and hybridized with oligonucleotide probes, the V_T shifted in the positive direction by the amount of 12 mV. This is due to increase of negative charges of the target DNA by hybridization. After hybridization, a DNA binder, Hoechst 33258 was introduced to the gate surface. The V_T shifted in the negative direction by the amount of 14 mV. The negative shift of the V_T indicates increase of positive charges at the gate surface and is due to specific binding of Hoechst 33258 to the double-stranded DNA. This is in contrast to the positive change of the V_T due to negatively charged DNA molecules. Thus, the charge density change at the gate surface after each molecular recognition event can be successfully detected using genetic FET.

DNA recognition events such as primer extension reaction can be also directly detected as electrical signal by the use of the genetic FET. The 11-base oligonucleotide probes on the genetic FET were hybridized with the 21-base target DNA at first. In order to extract small changes of the output voltages of the genetic FET after extension reaction, the output voltages before the introduction of DNA polymerase for both genetic and reference FETs were initialized and adjusted to zero as shown in Fig. 13.21 by adding or subtracting offset voltages, although the absolute output voltages are not zero and different between active and reference FETs. After washing, the genetic FETs were immersed in a reaction mixture and thermostable DNA polymerase was introduced into the gate surface. The V_T of the FET changed during primer extension reaction as shown in Fig. 13.21. Differential measurement was performed using a pair of FETs: one is the genetic FET with

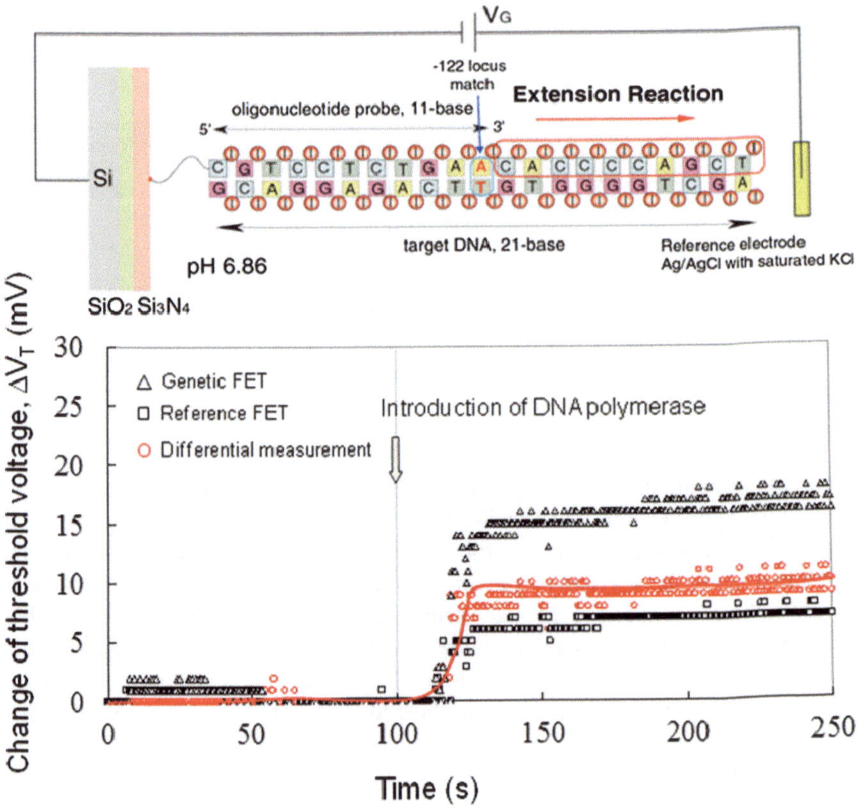

Fig. 13.21 Direct detection of extension reaction

immobilized oligonucleotides probes and the other is the reference FET without oligonucleotide probes. The common changes of the V_T due to ambient salt and pH changes and temperature difference in the buffer solution can be canceled out using the differential measurement. The difference of the ΔV_T between the genetic FET and the reference FET during extension reaction, ΔV_{Tdiff}, increased drastically up to about 10 mV. This positive change in the ΔV_{Tdiff} is rightly due to negative charges of polynucleotide extended by primer extension reaction. The ΔV_T of the reference FET is considered to be mainly due to the temperature change. From Fig. 13.21, we could demonstrate that the primer extension event on the gate surface was transduced directly into the electrical signal by the use of the genetic FET.

The effect of base length of the target DNA on the ΔV_{Tdiff} was investigated (Fig. 13.22 and Table 13.3 [56]). The linear relationship between the base length and the ΔV_{Tdiff} was obtained up to 41 bases. The ΔV_{Tdiff} after extension reaction increased to 24 mV, when target DNA with 41 bases was used. This is because the number of charges on the gate surface increased after primer extension with

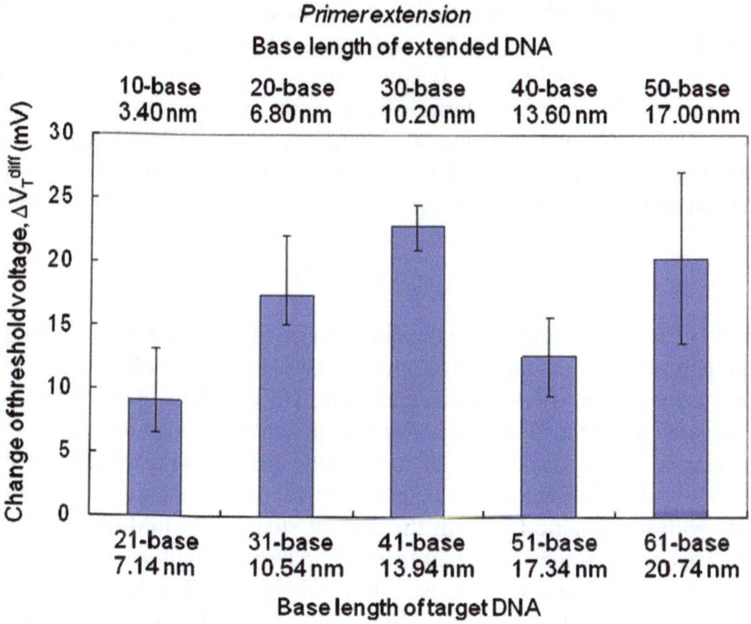

Fig. 13.22 Electrical signal for extended base length

Table 13.3 Sequence of oligonucleotides

Locus factor VII	Function	Sequence
Gene	−122-normal	
(−122)		
	Probe	5′-amino group-CGTCCTCTGAA-3′ (11mer)
	Target	5′- AGCTGGGGTGTTCAGAGGACG-3′ (21mer)
		5′-TGCAGCTCTCAGCTG GGGTGTTCAGAGGACG-3′ (31mer)
		5′-GGCGGCCAGGTGCA GCTCTCAGCTGGGGTGTTCAGAGGACG-3′ (41mer)
		5′-CATGGCCACTGGCGGCC AGGTGCAGCTCTCAGCTGGGGTGTTCAGAGGACG-3′ (51mer)
		5′-GCAGGGGATGCATGG CCACTGGCGGCCAGGTGCAGCTCTCAGCTGGGG TGTTCAGAGGACG-3′ (61mer)

increasing the template base length. However, the $\Delta V_{\mathrm{Tdiff}}$ did not follow the linear relationship, when the target DNA samples with 51 bases and 61 bases were used. The reason for this nonlinearity is considered to be related to the width of the electrical double layer at the interface between the gate insulator and an aqueous solution. The width of the electrical double layer, the Debye length, which is expressed in Eq. (13.1), is about 10 nm in the diluted salt solution (approximately 1 mM) and about 1 nm in the physiological solution (approximately 100 mM):

$$\delta = \left(\varepsilon\varepsilon_0 kT/2z^2q^2I\right)^{1/2} \tag{13.1}$$

Where δ is the Debye length, ε is the permittivity of the electrolyte solution, ε_0 is vacuum permittivity, k is the Boltzmann constant, T is the absolute temperature in Kelvin, z is the valency of the ions in the electrolyte, q is elementary charge, and I is the ionic strength of the electrolyte. The charge density change induced within the Debye length can be detected with the genetic FETs, while the charge density change induced outside the Debye length is shielded by counter ions and cannot be detected with the genetic FETs. In the present study, a 25 mM phosphate buffer solution was used for the V_T measurement. The Debye length at the gate insulator surface is therefore considered to be a few nanometers. The length of the target DNA with 41 bases is 13.94 nm, when it is straight. But oligonucleotide probes and the target DNA are flexible in the aqueous solution, and oligonucleotide probes are not always perpendicular to the surface of the gate insulator. It is therefore reasonable to ascribe saturation of the linear relationship between the V_T shift and the base length over 51 bases to the Debye length. Since the Debye length is dependent on the ionic strength of the aqueous solution as shown in Eq. (13.1), it is important to optimize the buffer concentration used for the measurement of the V_T shift.

The molecular recognition events such as hybridization and specific binding of DNA binder could be directly transduced into electrical signal using the genetic FETs. The change in the surface charge density could be detected as a shift of the V_T of the genetic FETs. The V_T shift after hybridization shown in Fig. 13.20, ΔV_T, can be expressed in Eq. (13.2), where $Q_{\text{ds-DNA}}$ is the charge per unit area of the double-stranded DNA after hybridization, $Q_{\text{ss-DNA}}$ is the charge per unit area of the single-stranded oligonucleotide probes, ΔQ_{DNA} is the charge difference per unit area after hybridization, and Ci is the gate capacitance per unit area:

$$\Delta V_T = \left(Q_{\text{ds-DNA}} - Q_{\text{ss-DNA}}\right)/Ci = \Delta Q_{\text{DNA}}/Ci \tag{13.2}$$

Since $\Delta V_T = 12$ mV and $\mathrm{Ci} = 4.3 \times 10^{-4}$ F/m^2 for the genetic FET, the amount of charges increased after hybridization is calculated to be 5.1×10^{-6} C/m^2. The base lengths of the oligonucleotide probe and the target DNA used in the experiment are both 17 bases, which correspond to 5.78 nm in length. Negative charges derived from phosphate groups are distributed along the double-stranded DNA from the gate surface to the bulk of the sample solution. We assume that these negative charges along the DNA molecules contributed to the V_T shift equally and that all the oligonucleotide probes were hybridized with the target DNA. Under these

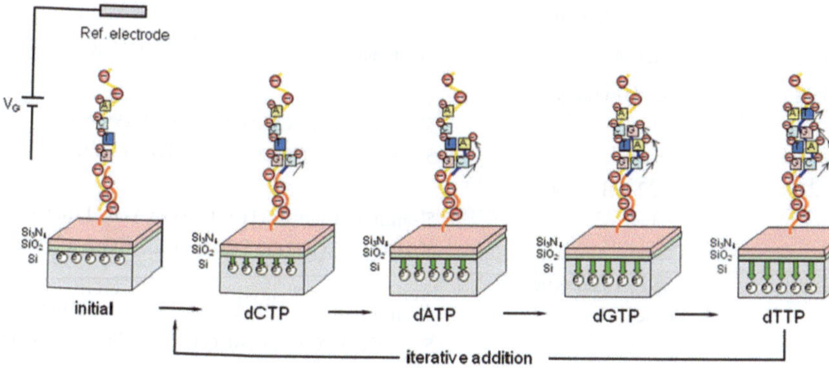

Fig. 13.23 Principle of label-free DNA sequencing based on intrinsic molecular charges

assumptions, the number of oligonucleotide probes on the channel region can be calculated to be 2.3×10^4, which corresponds to 1.9×10^8 /cm^2. The surface density of oligonucleotide probes immobilized on glass, silicon dioxide, and gold has been reported to be in the order of 10^9 to 10^{13}/cm^2, which was determined by different methods [57–62]. Since the density of the oligonucleotide probes is strongly dependent on the method and materials used for a substrate and immobilization, the number of oligonucleotide probes immobilized on silicon nitride could be increased by optimizing the immobilization method. It is noted that hybridization with 2.3×10^4 target DNA molecules resulted in the V_T shift of 12 mV. Therefore, detection of DNA molecules by the use of the genetic FET can be very sensitive, if hybridization is carried out sufficiently.

Although a number of methods for SNP analysis have been developed as described in the previous works, DNA sequencing techniques are still required to be improved in terms of cost, simplicity, and high throughput in order to analyze not only SNPs but also genomic variations such as insertion/deletion, short tandem repeat, etc. A new method for DNA sequencing is introduced in this section, which is based on detection of intrinsic charges of DNA molecules using the field effect.

Oligonucleotide probes are immobilized on the Si_3N_4 gate surface. The complementary target DNA is hybridized with the oligonucleotide probes on the gate surface. The hybridization events are followed by the introduction of DNA polymerase and one of each deoxynucleotide (dCTP, dATP, dGTP, or dTTP). DNA polymerase extends the immobilized oligonucleotide probes in a template-dependent manner (Fig. 13.23). As a result of extension reaction, negative charges increase at the gate surface of the FETs, because of intrinsic negative charges of incorporated molecules. This charge density change can be detected as a shift of the V_T of the FETs. Thus, iterative addition of each deoxynucleotide and measurement of V_T allow a direct, simple and nonlabeled DNA sequencing.

The base sequences of factor VII gene including two SNP sites and that of hereditary hemochromatosis (Table 13.4 [56]) were used to demonstrate the

Table 13.4 Sequence of oligonucleotides

Locus	Function	Sequence
R353Q	R353Q-wild type	
	Probe	5′-amino group-CCACTACCG-3′ (9mer)
	Target	5′-ACGTGCCCCGGTAGTGG-3′ (17mer)
−122	−122-wild type	
	Probe	5′-amino group-CGTCCTCTGAA-3′ (11mer)
	Target	5′AGCTGGGGTGTTCAGAGGACG-3′ (21mer)
C282Y	C282Y-wild type	
	Probe	5′-amino group-AGATATACGTG-3′ (11mer)
	Target	5′-CTCCACCTGGCACGTATATCT-3′ (21mer)

principle of DNA sequencing based on the FETs. We have paid special attention to the buffer concentration to be used for measuring charge density change at the gate surface. The potential change induced by adsorption of proteins at the gate surface was reported to be dependent on the electrolyte concentration [63]. It is therefore important to optimize the Debye length at the gate insulator/solution interface. In the present study, a 0.025 M phosphate buffer solution was used for measuring charge density change at the gate surface, while the conventional reaction mixture was used for single-base extension reaction.

The 11-base oligonucleotide probes were immobilized on the gate surface and hybridized with the 21-base target DNA for the base sequence of −122 (Table 13.4 [56]). We evaluated the FETs in combination with single-base extension for DNA sequencing. We prepared four kinds of buffer solution containing both DNA polymerase and one of dCTP, dATP, dGTP, or dTTP, respectively. The FETs hybridized with target DNA were immersed into the conventional reaction solutions for single-base extension reaction and the V_T shift was measured in a 0.025 M phosphate buffer solution after washing the FETs. The cycle of single-base extension and measurement of the V_T was repeated iteratively to determine the base sequence of the target DNA. When the base sequence of R353Q region of the factor VII gene was used as a target DNA, the V_T shifted in the positive direction only after single-base extension with the specific deoxynucleotides which were complementary to the base sequence of the target DNA (Fig. 13.24a). The V_T change for three-base extension, GGG, was 6.9 mV, which was bigger than that for one-base extension, but was not three times as big as that expected from the number of intrinsic charges. Although the linear relationship between the base length synthesized by the extension reaction and the V_T shift was obtained in the range from 0 to 30 bases (Fig. 13.22), it is important to detect single-base extension quantitatively, in order to reduce base call error especially for continuous sequence of the same base. The density and orientation of the immobilized oligonucleotide probes have to be controlled during a series of extension reactions at 72 °C. Further improvement of precision of the base call is also expected by automation of extension reaction and V_T measurements.

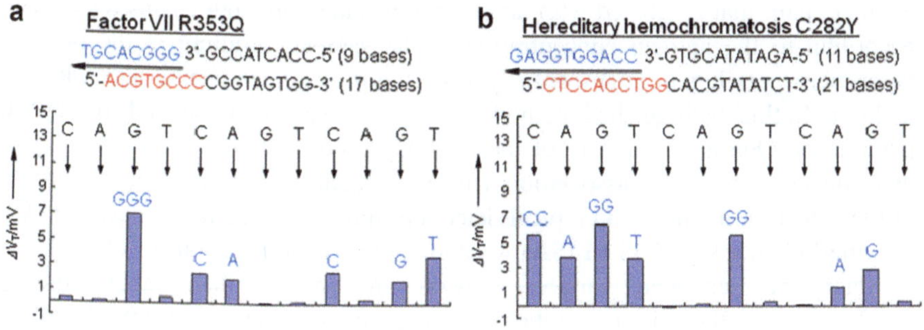

Fig. 13.24 Label-free DNA sequencing based on intrinsic molecular charges

The C282Y region of hereditary hemochromatosis gene was used as another example of DNA sequencing using the FET and single-base extension (Fig. 13.24b). The positive V_T shifts could be detected in accordance with the base sequence of the target DNA. In this case, the average V_T shifts for two-base incorporation was 5.8 mV with the standard deviation of 0.4 mV, while the average V_T shift for single-base extension was 3.2 mV. The V_T shifts for two-base extension was approximately twice as big as that for single-base extension. Thus, the results of iterative extension reaction and detection of the V_T indicated the ability of a direct, simple, and potentially precise DNA sequencing analysis using the FETs. The number of bases which can be analyzed by the proposed method is about ten bases at present. The V_T shift for single-base extension became gradually smaller as the number of bases increased more than ten. One of the reasons for this limitation would be the Debye length at the gate insulator/solution interface. Any charge density change induced outside the Debye length cannot be detected with the FETs. Lateral extension reaction in which DNA probes are extended in parallel with the gate surface would be effective for DNA sequencing with long bases. Another reason for the limitation would be peeling off the immobilized oligonucleotide probes from the surface of the gate insulator as the temperature stress of the extension reaction at 72 °C is applied repeatedly. The stronger immobilization method for oligonucleotide probes on the Si_3N_4 surface has to be adopted to analyze longer base sequence.

13.5.2 Detection of Low-Molecular-Weight Antigen [64]

About a decade ago, a novel immunoassay principle called open sandwich immunoassay (OS-IA) was proposed for the detection of mainly low-molecular-weight antigens in a noncompetitive mode [65–69]. This method is based on the antigen-dependent interchain interaction of separated V_L and V_H chains from a single antibody variable region. To perform OS-IA, for example, one of the separated chains is immobilized on the plate. When the sample containing the antigen and another chain that is labeled with an enzyme is added on, both chains reassociate according to the antigen concentration in the sample, resulting in the positive colorimetric or chemiluminescent signal on the plate after washing and substrate addition. In this distinguished method, the detection sensitivity depends on several parameters. One is the affinity of the antibody used, and another is the antigen-dependency of the V_H/V_L association. However, equally important is the sensitivity of the detection method, and typical detection limits obtained in the described OS enzyme-linked assays (OS-ELISA) were in the few nanomolar range [65–69].

Since the separated variable regions are smaller than (approximately $2 \times 2 \times 3.5$ nm) the original antibody to be captured within the Debye length at the gate surface, the detection principle of bio-FET will overcome the detection limit of OS-ELISA. Compared with the OS-ELISA, the OS-FET has the advantages of label-free and quantitative detection. Because of the label-free measurement, we can omit the experimental process of labeling. Moreover, we can estimate quantitatively the number of target molecules with charges and directly transduce the charge density changes based on the capture of target on the gate into the electrical signals using the OS-FET. In the present study, we propose a new detection method, open sandwich-based immunofield-effect transistor (OS-FET), for a label-free and highly sensitive detection of low-molecular-weight antigen, which is based on the detection of intrinsic molecular charges of one of the separated chains using the field effect. Here, we report on the direct transduction of open sandwich immunoassay at the gate surface into an electrical signal using the FET.

In our setup, the V_H chain is chemically immobilized on the Si_3N_4 gate surface. Then, the small antigen is sandwiched with the free V_L chain tethered with a negatively charged protein (MBP), and the V_H chain is immobilized on the gate surface (Fig. 13.25). In this study, bisphenol A (BPA) is utilized as a model antigen with a molecular mass of 228 and negligible charge at neutral pH [70]. The compound is a widely used small monomer in the manufacture of polycarbonate plastics and epoxy resins and has been reported to have estrogenic effects. While many methods were devised to detect BPA, the detection limit attained was more

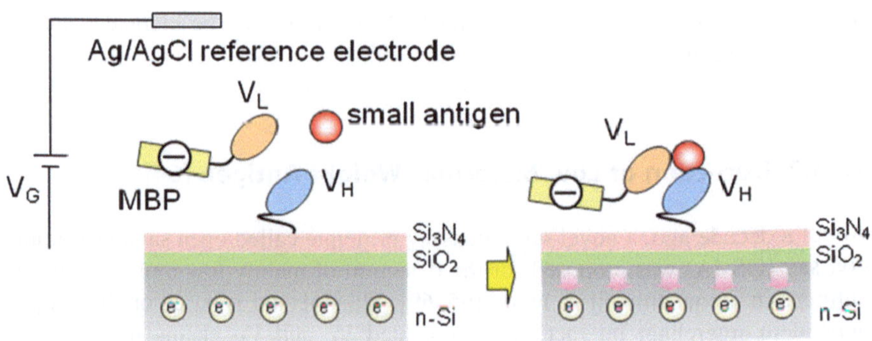

Fig. 13.25 Concept of OS-FET

than the nanomolar range [71–76]. However, recent research suggested that the exposure of BPA at the picomolar range changed some cell functions [76]. Therefore, the development of a more sensitive method to detect BPA is urgently needed. As shown in Fig. 13.25, the free V_L chain used is tethered with MBP with 13 negative charges per molecule (pI = 5.2), which is one of the routinely used fusion tags for recombinant protein production in *E. coli*, partly because it acts as an intramolecular chaperone and increases the expression yields of the fused proteins. Due to the large negative charge of MBP, the reassociation of the two chains is expected to induce the charge density changes at the gate surface. This charge density change can be detected as a shift of the V_T of the FET. Thus, the measurement of V_T allows a direct, simple, and nonlabeled OS-IA.

The FET chip is immersed in a measurement solution together with a Ag/AgCl reference electrode with saturated KCl solution. The potential of a measurement solution is controlled and fixed by the gate voltage (V_G) through the reference electrode. We have paid special attention to the buffer concentration used for measuring charge density change at the gate surface. The potential change induced by adsorption of proteins at the gate surface was reported to be dependent on the electrolyte concentration. It is, therefore, important to optimize the Debye length at the gate insulator/ solution interface. In the present study, a 0.025 M phosphate buffer solution was used for measuring charge density change at the gate surface, while the conventional reaction mixture was used during OS-IA reaction.

The shift of the gate voltage (V_G)–drain current (I_D) characteristic, the V_T shift, was measured after reacting BPA with MBP-linked V_L (MBP-V_L) and V_H chains (Fig. 13.26). The initial concentration of BPA added to the OS-FET chip was 1 μM. The resultant V_T shifted in the positive direction by the amount of 13.1 mV due to the negative charge of MBP (Fig. 13.26a). On the other hand, the effect of nonspecific adsorption of MBP-VL on the electrical signal was investigated by the use of the control FET (Fig. 13.26b). When the control FET chip with the immobilized V_H chains was reacted with the MBP-VL solution without BPA, the V_T shift was 4.5 mV to the positive direction. The electrical signal of the control FET implies not only the antigen-independent basal binding of MBP-V_L but also the presence of unexpected BPA, which might be included in the conventional disposal tube used in the experiments. Thus, for the highly sensitive detection of small uncharged antigens using the OS-FET, it is important to perform the differential measurement by the use of the control FET and the OS-FET.

The above results demonstrate that the reassociation of MBP V_L and V_H chains with a small antigen can be directly transduced into electrical signal using the FETs and that the OS-IA can be realized essentially on the basis of intrinsic molecular charges without any labels. From the point of view of practical application, it is required to detect BPA at the possible lowest adverse effect concentration. According to a report of the National Toxicology Program in 2008 (http://ntp.niehs.nih.gov/), the lowest, but controversial, dose levels that can induce a variety of adverse effects are 0.0024–0.01 μg/kg/day. Assuming that the amounts of BPA are taken from beverages, the concentration is estimated to be 10–60 ng/mL (43–260 nM), which is completely within the working range of OS-FET.

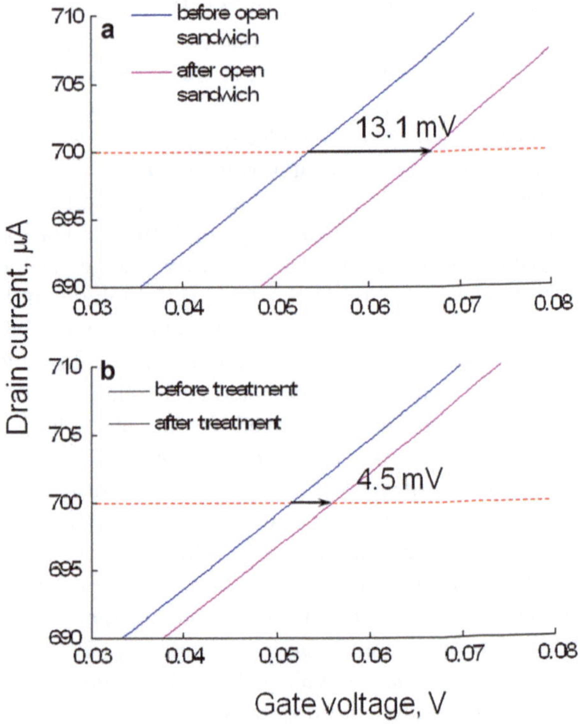

Fig. 13.26 V_G–I_D characteristic of OS-FET

It is possible to integrate multiple FETs and signal processing circuits in a single chip using advanced semiconductor technology. Simultaneous analyses of various small antigens in the environment or in our body can be realized on the basis of the FETs. Since the output of the FET is an electrical signal, standardization of the results obtained is easier than those with the chemiluminescence-based analyses. Therefore, the platform based on the FETs is suitable for a label-free, highly sensitive, and quantitative detection system for small antigen analysis in environmental, food, and clinical research.

Acknowledgments The authors wish to thank Professor K. Kataoka of the University of Tokyo in Japan and Professor Y. Miyahara of the Tokyo Medical and Dental University in Japan for their help and useful discussion. These works were partly supported by the Center for NanoBio Integration (CNBI) of the University of Tokyo and Sentan–Keisoku project and Core Research for Evolutional Science and Technology (CREST) project of the Japan Science and Technology Agency (JST).

References

1. Moore GE (1998) Proc IEEE 86:82–85
2. Rondelez Y, Tresset G, Nakashima T, Kato-Yamada Y, Fujita H, Takeuchi S, Noji H (2005) Nature 433:773–777
3. Bergveld P (1970) IEEE Trans Biomed Eng 17:70
4. McBride PT, Janata J, Comte PA, Moss SD, Johnson CC (1979) Anal Chim Acta 108:161–167
5. Lin J-L, Hsu H-Y (2010) Sensors 10:1798–1809
6. Abbott A (2003) Nature 424:870–872
7. Sakata T, Sugimoto H (2011) Jpn J Appl Phys 50:020216-1
8. Sakata T, Makino I, Kita S (2011) Eur Biophys J 40:699–704
9. Murakami T, Sakata T, Matsumoto A, Takai M, Ishihara K, Miyahara Y (2010) Trans Mater Res Soc Jpn 35:255–258
10. Matsumoto A, Sato N, Sakata T, Kataoka K, Miyahara Y (2009) Adv Mater 21:4372–4378
11. Sakata T, Miyahara Y (2009) IEEJ Trans Sensors Micromach 129:242–244
12. Sakata T, Makino I, Kita S, Miyahara Y (2008) Microelectron Eng 85:1337–1340
13. Sakata T, Miyahara Y (2008) Anal Chem 80:1493–1496
14. Matsumoto A, Sato N, Sakata T, Kataoka K, Miyahara Y (2009) J Solid State Electrochem 13:165–170
15. Sakata T, Miyahara Y (2008) Jpn J Appl Phys 47:368–370
16. Sakata T, Ueda A, Miyahara Y (2007) IEEJ Trans Electr Electron Eng 2:295–300
17. Dean PM, Matthews EK (1968) Nature 219:389–390
18. Henquin JC, Meissner HP (1984) Experientia 40:1043–1052
19. Ashcroft FM, Rorsman P (1989) Prog Biophys Mol Biol 54:87–143
20. Cook DL, Satin LS, Hopkins WF (1991) Trends Neurosci 14:411–414
21. Wollheim CB, Lang J, Regazzi R (1996) Diabetes Rev 4:276–297
22. Siu WM, Cobbold RSC (1979) IEEE Trans Electron Devices ED-26:1805
23. Hellerström C, Andersson A, Welsh M (1980) Horm Metab Res Suppl 10:37–43
24. Orci L, Ravazzola M, Amherdt M, Madsen O, Perrelet A, Vassalli JD, Anderson RG (1986) J Cell Biol 103:2273–2281
25. Veek LL (1991) Atlas of the human oocyte and early conceptus, vol 2. Williams & Wilkins, Baltimore, pp 151–153
26. Gardner DK, Lane M, Stevens J, Schlenker T, Schoolcraft WB (2000) Fertil Steril 73:1155–1158
27. Practical Committee of the American Society for Reproductive Medicine, Fertility and Sterility 2012, 97:835–842
28. Bergveld P (1971) IEEE Trans Bio-Med Eng BME-17:70–71
29. Sakata T, Kamahori M, Miyahara Y (2005) Jpn J Appl Phys 44(4B):2854
30. Fridhandler L, Hafez ES, Pincus G (1957) Exp Cell Res 13:132–139
31. Magnusson C, Hillensjo T, Tsafriri A, Hultborn R, Ahren K (1977) Biol Reprod 17:9–15
32. Overstrom EW, Duby RT, Dobrinsky J, Roche JF, Boland MP (1992) Theriogenology 37:269
33. Houghton FD, Thompson JG, Kennedy CJ, Leese H (1996) Mol Reprod Dev 44:476–485
34. Lopes AS, Larsen LH, Ramsing N, Lovendahl P, Raty M, Peippo J, Greve T, Callesen H (2005) Reproduction 130:669–679
35. Wales RG (1986) Reproduction 76:717–725
36. Javed MH, Wright RW Jr (1991) Theriogenology 35:1029–1037
37. Rieger D, Loskutoff NM, Betteridge KJ (1992) Reprod Fertil Dev 4:547–557
38. Khurana NK, Nieman H (2000) Biol Reprod 62:847–856
39. Shiku H, Shiraishi T, Ohya H, Matsue T, Abe H, Hoshi H, Kobayashi M (2001) Anal Chem 73:3751–3758
40. Mills RM, Brinster RL (1967) Exp Cell Res 47:337–344
41. Kroemer G, Levine B (2008) Nat Rev Mol Cell Biol 12:1004–1010
42. Bergveld P (2003) Sensors Actuators B 81:88

43. Straub B, Meyer E, Fromherz P (2001) Nat Biotechnol 19:121
44. Brittinger M, Fromherz P (2005) Appl Phys A 81:439
45. Wrobel G, Seifert R, Ingebrandt S, Enderlein J, Ecken H, Baumann A, Kaupp UB, Offenhausser A (2005) Biophys J 89:3628
46. Okada Y, Maeno E, Shimizu T, Dezaki K, Wang J, Morishima S (2001) J Physiol 532:3
47. Sakata T, Matsumoto S, Nakajima Y, Miyahara Y (2005) Jpn J Appl Phys 44(4B):2860
48. Chiou PY, Ohta AT, Wu MC (2005) Nature 436:370
49. Sakata T, Miyahara Y (2009) Curr Appl Phys 9:e210–e213
50. Sakata T, Maruyama S, Ueda A, Otsuka H, Miyahara Y (2007) Langmuir 23:2269–2272
51. Sakata T, Miyahara Y (2007) Biosens Bioelectron 22:1311–1316
52. Sakata T, Miyahara Y (2006) Angew Chem Int Ed 45:2225–2228
53. Sakata T, Miyahara Y (2005) Biosens Bioelectron 21:827–832
54. Sakata T, Miyahara Y (2005) ChemBioChem 6:703–710
55. Sakata T, Kamahori M, Miyahara Y (2004) Mater Sci Eng C 24:827–832
56. Kajiyama T, Miyahara Y, Kricka LJ, Wilding P, Graves DJ, Surrey S, Fortina P (2003) Genome Res 13:467–475
57. Chrisey LA, Lee GU, O'Ferrall CE (1996) Nucleic Acids Res 24:3031–3039
58. Steel AB, Herne TM, Tarlov MJ (1998) Anal Chem 70:4670–4677
59. Kumar A, Larsson O, Parodi D, Liang Z (2000) Nucleic Acids Res 28:e71
60. Peterson AW, Heaton RJ, Georgiadis RM (2001) Nucleic Acids Res 29:5163–5168
61. Huang E, Satjapipat M, Han S, Zhou F (2001) Langmuir 17:1215–1224
62. Guo Z, Gatterman MS, Hood L, Hansen JA, Petersdorf EW (2001) Genome Res 12:447–457
63. Nakajima H, Esashi M, Matsuo T (1980) Nippon Kagaku Kaishi 10:1499–1508
64. Sakata T, Ihara M, Makino I, Miyahara Y, Ueda H (2009) Anal Chem 81:7532–7537
65. Ueda H, Tsumoto K, Kubota K, Suzuki E, Nagamune T, Nishimura H, Schueler PA, Winter G, Kumagai I, Mahoney WC (1996) Nat Biotechnol 14:1714–1718
66. Suzuki C, Ueda H, Tsumoto K, Mahoney W, Kumagai I, Nagamune T (1999) J Immunol Methods 224:171–184
67. Aburatani T, Sakamoto K, Masuda K, Nishi K, Ohkawa H, Nagamune T, Ueda H (2003) Anal Chem 75:4057–4064
68. Suzuki T, Munakata Y, Morita K, Shinoda T, Ueda H (2007) Anal Sci 23:65–70
69. Lim S-L, Ichinose H, Shinoda T, Ueda H (2007) Anal Chem 79:6193–6200
70. Staples CA, Dorn PB, Klecka GM, O'Block ST, Harris LR (1998) Chemosphere 36:2149–2173
71. Matsunaga T, Ueki F, Obata K, Tajima H, Tanaka T, Takeyama H, Goda Y, Fujimoto S (2002) Anal Chim Acta 475:75–83
72. Shin H-S, Park C-H, Park S-J, Pyo H (2001) J Chromatogr A 912:119–125
73. Ohmura N, Lackie SJ, Saiki H (2001) Anal Chem 73:3392–3399
74. Goda Y, Kobayashi A, Fukuda K, Fujimoto S, Ike M, Fujita M (2000) Water Sci Technol 42:81–88
75. Inui H, Takehara A, Doi F, Nishi K, Takai M, Miyake S, Ohkawa H (2009) J Agric Food Chem 57:353–358
76. Welshons WV, Nagel SC, vom Saal FS (2006) Endocrinology 147:s56–s69

From Single-Molecule DNA Imaging to Development of a Gene Delivery System

14

Yuichi Yamasaki

Abstract

The fundamental properties of single DNA molecules clarified by single-molecule DNA imaging using fluorescence microscopy and related applications are briefly introduced in this chapter. Firstly, the basis of fluorescence video microscopy for single-molecule DNA imaging is discussed. The physical quantities determined from image analyses are in accordance with those obtained by conventional methods that measure the ensemble averages of physical properties, although the dynamic behavior of DNA molecules observed in fluorescence video microscopy seems to be somewhat different from that found by conventional characterization. DNA solution properties as a basis of polymer physics are also described in relation to optical tweezers. An essential concept of polymer physics, the "reptation model," is proven by single-molecule imaging and provides useful information about the molecular mechanics of DNA gel electrophoresis, which is a widely used tool in biology and biochemistry. A mechanism of DNA condensation that is closely related to the traditional problem in polymer physics known as the "coil-globule transition" is also clarified using single-molecule DNA imaging. This transition has recently attracted considerable attention as an elemental technology for gene therapy. A promising nonviral gene delivery system is also introduced along with the newly developed advanced imaging technique intravital real-time confocal laser scanning microscopy.

Keywords

Fluorescence microscopy • Single-molecule DNA imaging • Electrophoresis • DNA condensation • Nonviral gene delivery

Y. Yamasaki (✉)
Department of Materials Engineering, School of Engineering, The University of Tokyo, 7-3-1 Hongo, Bunkyo, Tokyo 113-8656, Japan
e-mail: yamasaki@bmw.t.u-tokyo.ac.jp

R. Kita, T. Dobashi (eds.), *Nano/Micro Science and Technology in Biorheology*,
DOI 10.1007/978-4-431-54886-7_14

14.1 Fluorescence Video Microscopy for Single-Molecule DNA Imaging

Real-time, direct observation of single DNA molecules in solution by means of fluorescence microscopy was originally proposed by Yanagida et al. [1]. In this section, their method is summarized briefly; for an elaborate discussion on this method, refer to Bustamante's review [2]. Yanagida and colleagues developed a DNA staining method with the commonly used nuclear staining fluorescent dye 4′,6-diamidino-2-phenylindole (DAPI) [3] to visualize single DNA molecules [4]. They demonstrated that several physical quantities can be deduced, including the radius of gyration and persistent length from the fluctuation of DNA conformation, and the elastic (entropic) constant from the contraction rate of stretched DNA molecules [1]. Other groups reported that the diffusion constant of a single DNA molecule could be deduced from Brownian motion analysis [5, 6]. This technical breakthrough has enabled us to analyze many properties of single DNA molecules including polymer physics, transport phenomena, electrostatic properties, and enzyme reactions.

The experimental apparatus for this method is composed of an epi-illuminated fluorescence microscope, a highly sensitive video camera, and an image processing system. An optical system for epi-illumination is depicted schematically in Fig. 14.1. In this optical system, a microscope objective lens that collects fluorescence also acts as a condenser for excitation light. Because the amount of backscattered excitation light is minimized in this optical system, one can observe high-contrast fluorescence images against a dark background.

While the principles of fluorescence microscopy have already been established, both fluorescent dyes and video cameras are still being developed. DAPI [2], the dimer TOTO of thiazole orange (TO) [7], and the dimer YOYO of oxazole yellow (YO) [8] are examples of dyes that are widely used for DNA imaging by fluorescence microscopy. Several analogues of these cyanine dyes with different emission properties and novel chromophores constructed from different molecular structures have also become commercially available [9]. Intercalating dyes such as TOTO and YOYO are known to extend the DNA contour length by 30 % [10, 11]. When these dyes bind to DNA by intercalation, they unwind the double helix of DNA markedly [12].

To obtain DNA fluorescence images, primitive camera tubes have been replaced by charge-coupled device (CCD) cameras equipped with various types of internal image intensifiers. The improvement of computer processing speed means that we can now easily handle normal video signals (30 frames per second) that used to be recorded on videotape. Therefore, this method has now become a relatively common technique in physical, chemical, and biological research fields. Several applications based on single-molecule DNA observation are described in the following sections.

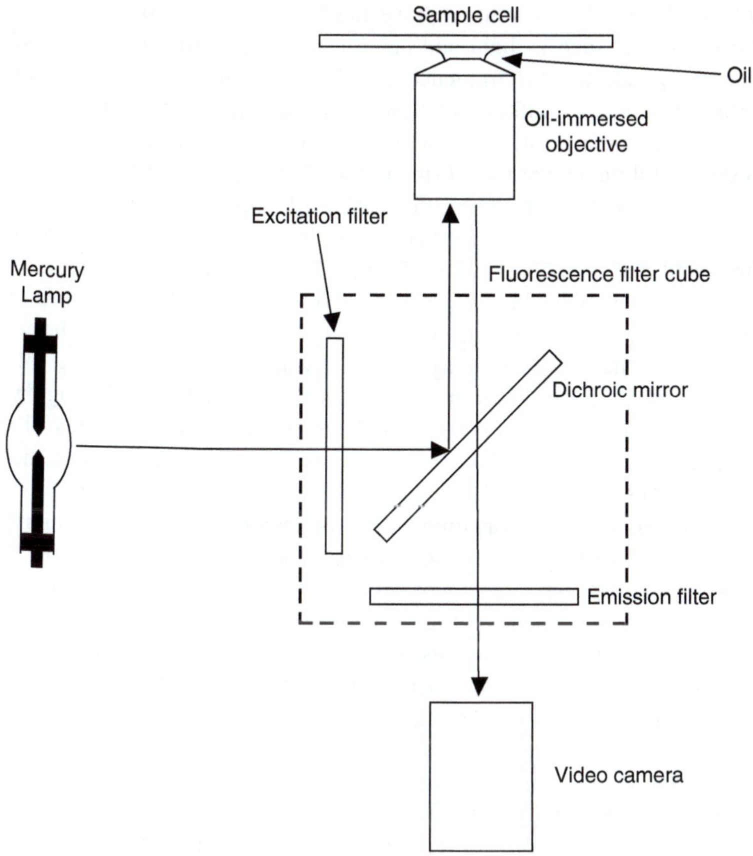

Fig. 14.1 Schematic diagram of an optical system of an epi-illuminated fluorescence microscope

14.2 Fundamentals of Polymer Physics

In this section, the fundamental properties of single DNA chains related to polymer physics are introduced. For experiments in polymer physics, monodisperse polymers should be used to obtain precise measurements. Several polymerization methods, known as living polymerizations, can provide monodisperse polymers with a very narrow distribution of molecular weight or degree of polymerization (DP), ideally a Poisson distribution. A DP distribution cannot be eliminated from usual polymer syntheses. In contrast, genetically prepared DNA molecules from viruses or bacteria have uniform length for each species, so DNA molecules in solution are suitable for polymer physics experiments. In the first half of this section, the fundamental properties of single DNA chains determined from simple direct imaging are described. Some applications using optical tweezers are provided in Sect. 14.3.

To reveal the dynamics of single polymer chains in solution, the diffusion coefficient D, radius of gyration R_g, and persistent length P are essential physical quantities. These polymer solution properties have been studied mainly by ensemble measurements such as light and neutron scattering, viscoelasticity, dielectric relaxation, and electric birefringence. As described in Chap. 2, the rheological properties of polymer solutions are closely related to the conformation of polymer chains. The enormous effort spent on these characterization techniques has evolved into a successful molecular model (the Rouse-Zimm model) and scaling concepts. However, realistic images of the fluctuation of polymer chains have not been readily accessible to most researchers. Therefore, it was worthwhile to attempt the direct and real-time observation of single DNA molecules in solution.

D of a polymer molecule can be accurately determined from the mean square displacement against time (diffusivity), which is analyzed by tracking the center of mass of DNA fluorescence images. The hydrodynamic radius R_H of DNA molecules can be deduced from the Stokes-Einstein relation. Yanagida et al. compared their results with a theoretical model of wormlike chains [13], and P was determined to be 47–70 nm, in good agreement with the value measured by light scattering (~50 nm) [1].

The situation for the estimation of R_g is somewhat different. Fluorescence images of long DNA molecules such as bacteriophage T4 and λ show a thick filament with an apparent length of less than 10 μm without any disturbances (flow, electric field, osmotic pressure) because of the shortening of end-to-end distance reflected by random walk statistics. During fluctuation, the apparent length, shape, and local segment density (reflected by fluorescence intensity) of DNA fluorescence images change very quickly. R_g can be estimated by measuring both long- and short-axis lengths of sequential fluctuating DNA images. As a result, this estimation includes a substantial margin of error.

Matsumoto et al. analyzed the Brownian motion of T4 DNA and reported that R_g and rotational relaxation showed qualitative agreement with the Zimm model [5, 14]. However, because of the walls of the thin sample chambers, D deduced from diffusivity were somewhat lower than those predicted by the Zimm model. In 1996, Smith et al. clearly demonstrated that the power-law scaling of D with DNA length L is in good agreement with the Zimm model ($D \sim L^{-3/5}$) using deeper chambers (depth of 75 μm) [6]. A scaling exponent ν of 0.611 ± 0.016 was reported. In a further detailed study, Robertson et al. reported a topology-independent scaling law of $D \sim L^{-\nu}$ where $\nu_L = 0.571 \pm 0.014$, $\nu_C = 0.589 \pm 0.018$, and $\nu_S = 0.571 \pm 0.057$ for linear, relaxed circular, and supercoiled DNA, respectively [15]. The renormalization group theory that takes into account the effect of excluded volume predicts $\nu = 0.588$ [16].

14.3 Optical Tweezers for DNA Manipulation

Ashkin first used optical tweezers for trapping micron-sized particles by laser radiation pressure [17]. Ashkin and his colleagues then expanded this technique to laser cooling to trap atoms. A prototype of current typical optical tweezers based on a single focused laser beam was also reported by Ashkin et al. [18]. They

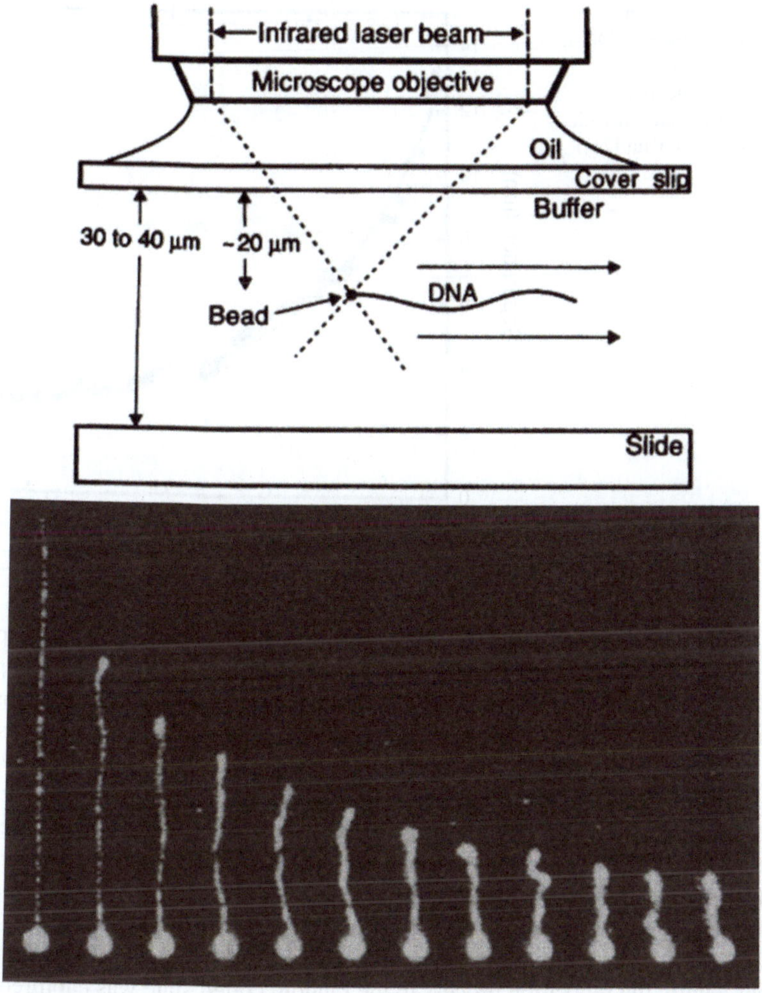

Fig. 14.2 Manipulation and visualization of a single DNA molecule using a fluorescence microscope equipped with optical tweezers. *Top*: Schematic diagram of the measurement chamber. *Bottom*: Sequential images showing the relaxation of a 39-μm-long DNA molecule after being stretched out in a fluid flow (4.5-s interval) (Reproduced with permission from Perkins et al. [23])

achieved optical trapping of 20-nm particles in water at room temperature and also manipulated viruses, bacteria, and organelles of protozoa by optical trapping [19, 20]. As an extension of these works, Chu and colleagues manipulated individual DNA chains using optical tweezers [21, 22].

The principle of optical tweezers to manipulate single DNA chains is illustrated schematically in Fig. 14.2 [23]. In typical setups using a single laser, an infrared laser beam is focused by an objective lens to trap beads attached to the end of DNA chains. Objectives with high numerical aperture (1.3 or higher for oil immersion)

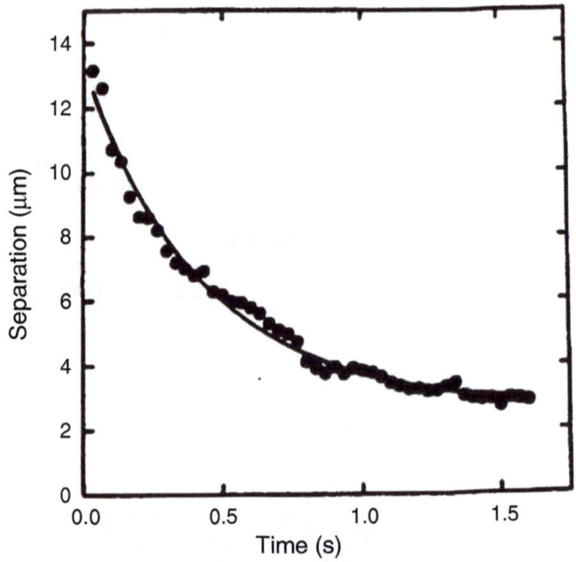

Fig. 14.3 Relaxation process of extended λ-DNA. The *solid line* shows the expected exponential behavior (Reproduced with permission from Chu [21])

used for high-resolution fluorescence imaging can also generate a strong trapping force. The Nd:YAG laser emitting at 1,064 nm is the most widely used example at present. To manipulate a single DNA chain in the simple setup, in which a laser is focused in a fixed position, microfluidic chambers are combined with a microscope stage or a microscope stage is mechanically moved with submicron accuracy.

In early studies such as that of Matsumoto et al. [24], shear force was generated by slowly moving a cover slip to extend DNA chains to nearly their contour length. Optical tweezers are a sophisticated method to carry out similar experiments. Chu first demonstrated the elastic properties of single DNA chains using a pair of controllable optical tweezers to manipulate two beads attached at both ends of a DNA molecule via biotin-avidin bridges [21]. After the extension of λ-DNA to its contour length, the relaxation process to the random coiled state was captured after the release of one bead. The distance between the two beads decreased exponentially like a simple Hookean spring (Fig. 14.3).

Perkins et al. subsequently performed relaxation experiments of fully extended DNA molecules using a setup with a single laser and stopped-flow chamber [23]. Fitting these relaxation data with an inverse Laplace transform showed sharp spectral peaks, indicating that the relaxation process was a sum of exponentials (internal relaxation modes). The longest relaxation time scaled as $L^{1.65}$, in agreement with the dynamic scaling prediction of the Zimm model. A detailed discussion of these findings is provided in Shaqfeh's review [25].

Optical tweezers opened up a new research field called single-molecule DNA mechanics. Video microscopes equipped with dual-controllable optical tweezers or a single laser with a micropipette allows us to measure the force during DNA stretching. Similar experiments are widely carried out to investigate protein folding

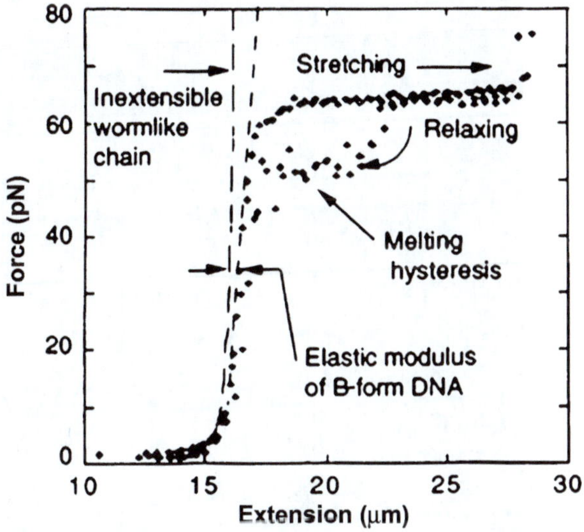

Fig. 14.4 Typical force-extension curve of λ-DNA (Reproduced with permission from Smith et al. [83])

using atomic force microscopy (AFM) by AFM fishing. In contrast to AFM fishing for proteins, double-stranded DNA does not produce characteristic structures in its force-extension curves for full extension of the B form (normal double helix form) (Fig. 14.4), [83]. In 1992, Smith et al. reported the first quantitative force (F)-extension (x) curves for λ-DNA at various NaCl concentrations using magnetic bead trapping to measure the elastic responses of B-form DNA [10]. Bustamante et al. pointed out that fitting these data with the wormlike chain model gave better agreement than that with the freely jointed chain model [26]. Marko and Siggia suggested an analytical formula for F-x curves in terms of L and P:

$$\frac{FP}{k_{\mathrm{B}}T} = \frac{1}{4(1-x/L)^2} - \frac{1}{4} + \frac{x}{L}, \tag{14.1}$$

where k_{B} is the Boltzmann constant and T is the absolute temperature [27]. Fitting F-x curves to this equation gives an intrinsic elastic persistent length of 50.8 nm. The overstretching region in F-x curves where normal double helix structure (B-form) is no longer retained is not a subject of this chapter, but modulations, unzipping, and melting of double helix structure are introduced in the review by Bustamante et al. [28].

Apart from dilute polymer solutions, the properties of concentrated polymer solutions are very complicated, but can be treated by the reptation model developed by de Gennes [29, 30] and Doi and Edwards [16]. The basic concept of this model is that a polymer chain in the entangled state is topologically constrained by the surrounding chains and relaxation processes are achieved by chain movement along its contour or inside a hypothetical tube. Perkins et al. directly observed the

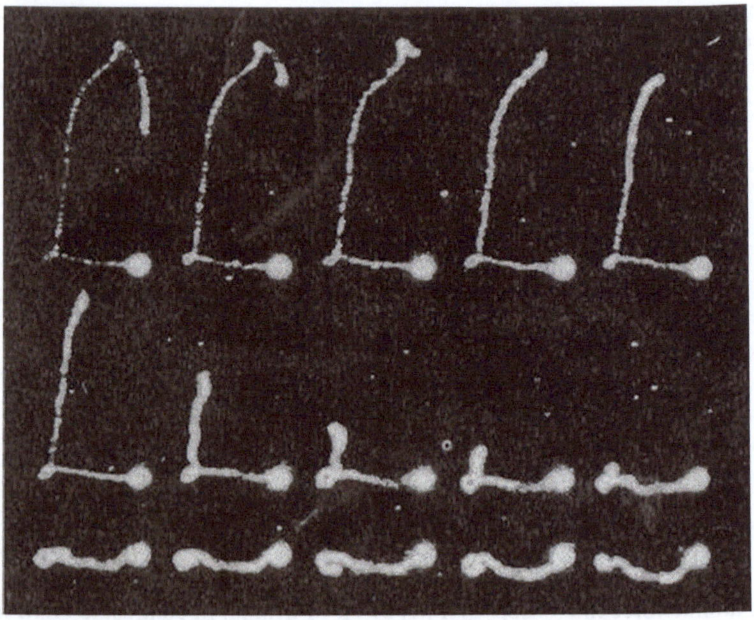

Fig. 14.5 Sequential images showing the relaxation of an 80-μm-long DNA molecule stretched along an indirect path drawn by a bead manipulated with optical tweezers. The times are 0, 2.3, 4.7, 7.0, and 9.3 s (*top row*); 11.6, 24.6, 37.7, 50.7, and 63.7 s (*middle row*); and 76.7, 89.7, 103, 115, and 128 s (*bottom row*) (Reprinted with permission from Perkins et al. [22])

reptation movement of single DNA chains [22]. In their report, a fluorescently stained DNA chain embedded in background of unstained DNA chains was first extended by manipulating the bead attached to the end of the stained DNA molecule using optical tweezers, and then the extended DNA chain was retracted along the path previously drawn by the bead from the unfixed end and relaxed to its equilibrium state. The top and middle rows of Fig. 14.5 show that a small loop made by the dragged bead stayed fixed for long time (~120 s), indicating the constraint from the surrounding unstained chains was hardly modulated because of their slow diffusion. These observations provide direct evidence for the fundamental properties of the reptation model.

As a further detailed study, Smith et al. [31] determined the self-diffusion coefficient D_{rep} of λ-DNA in a concentrated region of a similar sample to that used in their previous work [22]. Their obtained scaling exponent α of 1.8 ± 0.1 for λ-DNA at high concentration was close to the scaling prediction for $D_{rep} \sim L^{-\alpha}$ with $\alpha = 2$. Reptational diffusion of λ-DNA was observed at 0.63 mg/mL, which is 16 times higher than the overlap concentration, but not at 0.40 mg/mL. In addition, shorter DNA with a concentration of 0.63 mg/mL did not follow the reptation model. These observations suggest that both chain length and concentration are factors that control the applicability of the reptation model to semidilute polymer solutions.

14.4 DNA Gel Electrophoresis

DNA gel electrophoresis is a widely used tool in biology and biochemistry. However, DNA migration modes in gels were not clarified for a long time. Before using single-molecule DNA imaging, linear dichroism, fluorescence polarization, and birefringence were employed to investigate DNA orientation in gels. However, the results obtained from these methods were difficult to interpret. Direct observation of fluorescently stained DNA migrating in gels can provide easily understandable descriptions concerning the molecular motion of DNA chains in gels. The behavior of DNA inside gels resembles that of entangled polymer solutions. First, the fundamental principles of DNA gel electrophoresis are briefly introduced.

In buffer solutions, DNA molecules migrate toward the positive electrode under a steady electric field. The migration velocity of DNA can be modulated by changing the ionic strength or applied electric field. However, the velocity is independent of DNA length when electrophoresis is performed in homogeneous buffer solution without a supporting gel [32]. Electrophoretic mobility μ, which is the ratio of migration velocity to applied electric field, can be written as the ratio of DNA effective charge Q to friction coefficient η. Both of these quantities are proportional to DNA length in homogeneous buffer solution, indicating that μ is independent of DNA length.

As described in Sect. 14.3, the relaxation of a single polymer chain in entangled polymer solution can be modeled by reptation movement. In the case of DNA electrophoresis, DNA chains are dragged inside gels by the electrophoretic force. The movement of DNA chains in gels under steady electric field can be explained by the biased reptation model, which takes the electrophoretic force into account. When DNA chains migrate inside a gel, the dependence of μ on DNA length can be divided into three regimes that are characterized by the relationship between gel concentration, electric field strength, and DNA length [33].

The first regime is the Ogston regime [34] that shows weak length dependency of μ on DNA length (low separation capability), because the gel pore size is larger than that of shorter DNA fragments. The second regime is called the reptation regime without DNA chain stretching (gels with small pores and weak electric field). In the second regime, μ clearly depends on DNA length. Numerous mathematical descriptions for the reptation mode in DNA electrophoresis have been developed [35–38].

The third regime is called the reptation regime with DNA chain stretching (gels with small pores, strong electric field, and longer DNA chains). In this regime, μ of longer DNA chains is almost independent of DNA length because of the strong entanglement between the gel network and DNA chains. Therefore, agarose gels possessing large pores are suitable to separate long DNA chains, whereas polyacrylamide gels with small pores can be used only to separate short DNA fragments (<1 kbp). These three regimes are also observed in electrophoresis of SDS-solubilized proteins [39] and synthetic polyelectrolytes [40]. Therefore, these phenomena are the general properties of linear polyelectrolytes.

According to the biased reptation model [37] applied to DNA electrophoresis, the velocity v of DNA inside a hypothetical tube can be expressed by the following equation:

$$v = \frac{QhE}{L\xi}, \tag{14.2}$$

where h is the length of projection of a mean end-to-end distance of DNA chain to the direction of the electric field, E is the applied electric field, and ξ is the frictional coefficient between DNA and the hypothetical tube [35]. This equation can be transformed to μ:

$$\mu = \frac{Qh^2}{L^2\xi}. \tag{14.3}$$

If the mean end-to-end distance of DNA under a weak electric field is same to that of coiled DNA without an electric field, indicating the square of h is proportional to L, μ is proportional to the inverse of L. This behavior is observed in the second regime of DNA electrophoresis in gels. Under a strong electric field, as long DNA chains are stretched along the electric field, h is proportional to L. Then, the dependence of μ on DNA length decreases again in the third regime. DNA chains longer than 50 kbp cannot be separated under common electrophoresis conditions with a constant electric field [41].

The separation of long DNA chains (>50 kbp) using pulsed-field gel electrophoresis (PFGE) [42, 43] was the outstanding breakthrough in the third regime of DNA gel electrophoresis. In a PFGE chamber, the electric field is periodically altered between two electrode pairs with a fixed angle from 90° to 120°. PFGE allows DNA chains up to 10 Mbp in length to be separated. A molecular sieving effect is thought be generated because the switching time required for the direction change of DNA migration depends on DNA length.

To assess the above models, Smith et al. employed fluorescence video microscopy to observe the migration of DNA on gels under both steady and pulsed electric fields [44]. Under a steady electric field, DNA chains in the gel migrated in an indirect manner with repeated extension and contraction movements. With increasing electric field strength, DNA chains became stretched and better aligned with the electric field. Another configuration, "U-shaped" DNA chains that caught on obstacles in the gel, was often observed. After stretching to its contour length, the shorter arm of a DNA chain slipped into the tube made by the longer arm, indicating the DNA chain became free from the obstacle. Schwartz and Koval also reported similar observations [45].

In PFGE, the slipping motion of DNA chains is modulated in different directions by placing two electrode pairs with a fixed angle from 90° to 120° between them. Smith et al. reported that tugged DNA segments (tails) changed to leading segments (head) corresponding to changes in the electric field direction with an electrode angle of 120°, just like a train running on a switchback track [44]. These

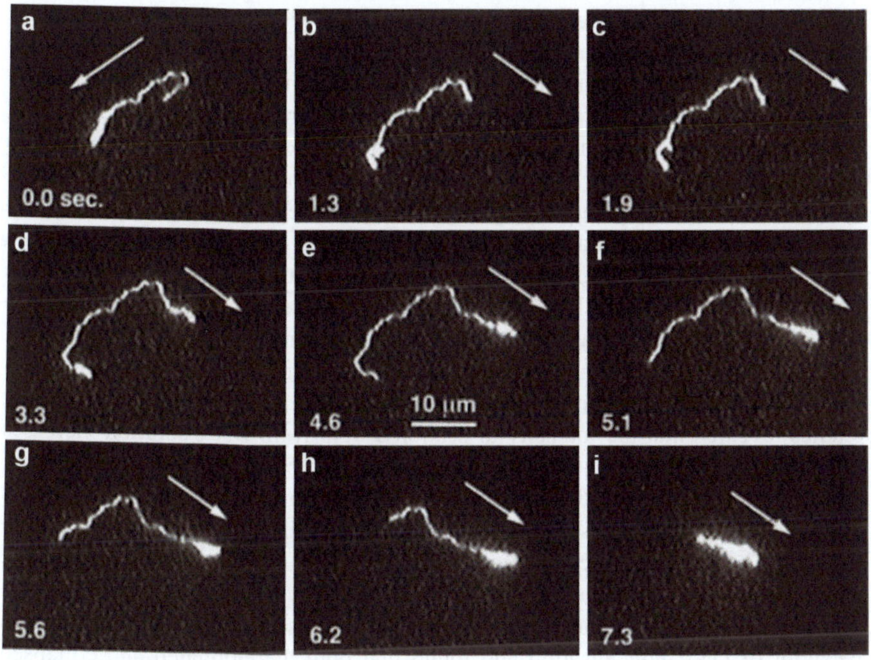

Fig. 14.6 Sequential images of T2 DNA migrating in PFGE in switchback mode with an electrode angle of 120° (Reproduced with permission from Gurrieri et al. [46])

reorientation dynamics are clearly shown in Fig. 14.6 [46]. Their computer simulations demonstrated that shorter DNA chains change direction more quickly and migrate faster than longer chains. Longer chains take a longer time to backtrack on their previous paths than shorter ones. When a 90° angle between electrodes is used in PFGE experiments, both ends can become the leading end, resulting in poor size separation.

14.5 DNA Condensation

The term DNA condensation is reserved for the conformational change of elongated DNA molecules into collapsed states with ordered morphologies, typically toroidal and rodlike states [47]. DNA condensation is a characteristic phenomenon related to a traditional problem of polymer physics, the coil-globule transition. Transmission electron microscopy, light scattering, and sedimentation have been used frequently to characterize DNA condensation. However, using these techniques, researchers could not easily determine whether DNA condensation is induced at single-molecule level or involves aggregation of multiple chains. Therefore, the definition of DNA condensation does not discriminate between single-molecule collapse and aggregation, although the former was predicted theoretically [48]. In concert with this prediction, Widom and Baldwin concluded that (1) the monomolecular condensation of λ-DNA induced by cobalt hexammine is not a two-state reaction and (2) the transition for monomolecular condensation is diffuse [49]. As described above, a consensus view for DNA condensation has not yet been provided. Thus, the direct, real-time observation of single DNA molecules may allow better characterization of DNA condensation.

DNA condensation is induced by neutralizing the negative charges of phosphate groups under various conditions. Multivalent cations (charge of 3+ or greater in normal aqueous solution), cationic surfactants, cationic lipids, and cationic polymers are well-known DNA condensing agents. Neutral polymers that exert a crowding effect through an excluded volume mechanism can also induce DNA condensation in the presence of adequate concentrations of salts (typically psi-condensation using poly(ethylene glycol) (PEG) and Na^{+}) [50]. Light scattering studies revealed that the intensity of scattered light from DNA solution increases continuously with the addition of condensing agents [51]. However, the scattering curves of DNA condensation cannot be used to determine how DNA molecules collapse into condensed states, particularly if it is in a continuous or discrete manner. Therefore, physicochemical characterization techniques that clarify the detailed condensation behavior of DNA are desired.

Yoshikawa and colleagues used fluorescence microscopy to reveal the transition behavior of DNA condensation and found that both the elongated and collapsed states coexisted during the transition region of DNA condensation induced by polyarginine [52]. Intensive study by the Yoshikawa group elucidated that this biphasic behavior of DNA condensation is a common feature independent of condensing agents. Mel'nikov et al. reported the relationship between the biphasic distribution of T4 DNA conformation and cooperative binding in DNA condensation induced by the cationic surfactant cetyltrimethylammonium bromide (CTAB) [53]. In this system, the binding of CTA ions onto DNA was quite similar to the aggregation of surfactant molecules observed in micelle formation in the closed association model. Until DNA condensation was complete, the bound fraction of CTA ions increased linearly with the concentration of CTAB, while the concentration of unbound CTA ions remained consistently lower. Takahashi et al. reported that the width of the coexistence region narrowed as the valence of polyamines increased [54]. Increasing the valence of the polyamine by one corresponded to a decrease of the transition point of DNA condensation, expressed as the concentration of added condensing agent required for the transition, by one order of magnitude. This experimental trend agreed with a theoretical prediction made using the modified Flory-Huggins theory taking into account the electrostatic interaction between phosphate groups of DNA and polyamines.

Yoshikawa and Matsuzawa used single-molecule DNA observation to visualize how T4 DNA chains collapse into condensed states in the case of psi-condensation using PEG and NaCl [55]. Because the PEG concentration required for DNA condensation increases the viscosity of the sample solution [50], the fluctuation of DNA chains slowed as the DNA molecules became entangled in gel matrices, which can easily be observed by a fluorescence microscope. During the condensation process of this system, nucleation occurred most frequently at the end of the DNA chains and the remaining elongated part successively wound around the nucleation center within 10 s, i.e., the nucleation and growth processes were confirmed as shown in Fig. 14.7 [55]. These studies reveal that the continuous increase of scattering intensity observed in light scattering experiments of DNA condensation can be explained by the gradual change in the distributions of both elongated and collapsed states in the transition region of DNA condensation that is characterized by the first-order phase transition of single DNA chains. In addition, Yamasaki and Yoshikawa demonstrated that DNA condensates collapsed by Fe^{3+} ions unfolded to the elongated state upon addition of ascorbic acid solution, which reduced Fe^{3+} to Fe^{2+} [56]. This indicates that the collapsed and elongated conformations of DNA can be controlled by the electrochemical redox reactions of condensing agents.

Optical tweezers have another advantage to characterize DNA condensates. The elastic responses of single DNA molecules measured by optical tweezers can reveal the binding force between DNA segments and the binding strength of condensing agents to DNA segments. Baumann et al. reported that under condensing conditions with spermidine, the elongation phase of plasmid-length DNA (1.3 μm) showed abrupt increases in binding force of 1–10 pN (stick-release pattern) (Fig. 14.8 (left)), which resembled the results of AFM fishing for proteins [57]. The authors suggested that DNA condensates unfold by releasing DNA fragments in turn or condensates are constructed from several independent folded segments aligned in series.

Another feature observed in *F*-*x* curves is baseline shifts at the region where λ-DNA is extended over 60 % of its contour (Fig. 14.8 (right)). The higher retractile force (~4 pN) corresponds to cobalt hexammine that efficiently induces DNA condensation and the lower (~1 pN) to spermidine. Assuming that this force is associated with DNA condensation, the authors pointed out the constant retractile force represents the force F_r required to convert condensed regions of the DNA chain to an extended form in the equilibrium of the two stable states. The work W done during a change in DNA extension Δx from x_1 to x_2 (0.44 μm) is

$$W = \int_{x_1}^{x_2} F\mathrm{d}x = F_r \Delta x. \tag{14.4}$$

Equation 14.4 gives W ~0.33 kT/bp for cobalt hexammine and W ~0.083 kT/bp for spermidine. These results were also reported by Murayama et al. [58].

To better understand the transition behavior of DNA condensation, not only conformation but also the electrostatic properties of the phosphate groups of DNA should be able to shed light. One of the most useful methods to investigate the electrostatic properties of polyelectrolytes is electrophoretic light scattering, which measures the frequency shift between the incident beam and scattered light from the electrophoresis analytes. The electrophoretic mobility reflecting the charge density of analytes is detected through the frequency shift. Raspaud et al. reported that aggregates of DNA fragments ranging from 50 to 1,000 bp showed

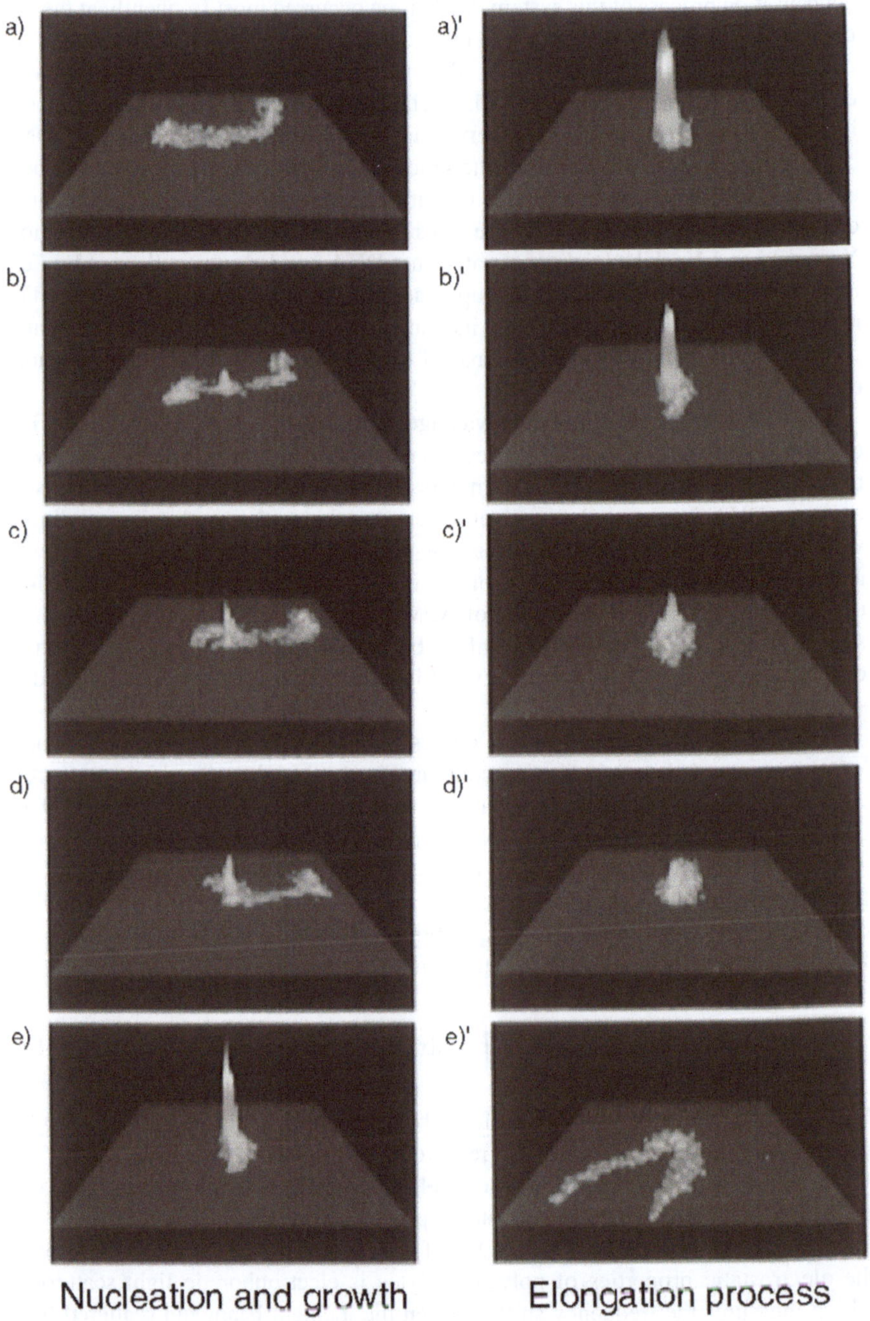

Fig. 14.7 (*Left*) Nucleation and growth processes during psi-condensation. (*Right*) Elongation of a single-molecule DNA condensate (pseudo 3D images) (Reprinted with permission from Yoshikawa and Matsuzawa [55])

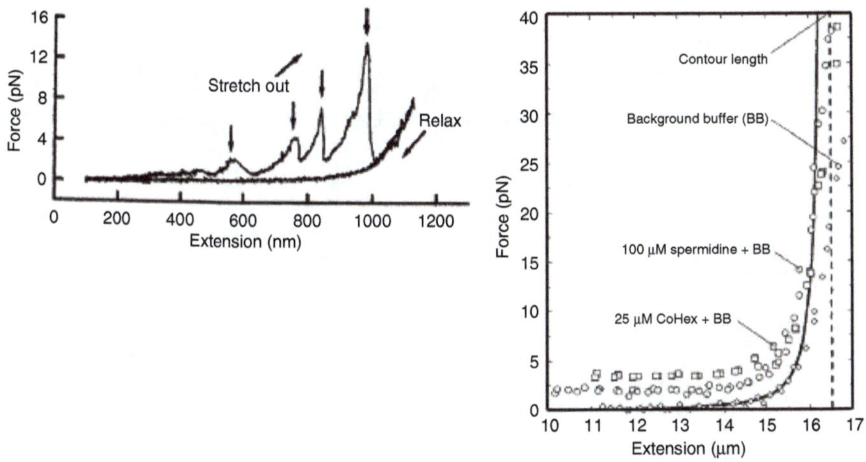

Fig. 14.8 Stretching of single DNA molecules under condensing conditions using optical tweezers. (*Left*) A stick-release pattern observed during stretching of plasmid-length DNA. (*Right*) *F-x* curves of λ-DNA molecules under condensing conditions with cobalt hexammine and spermidine (Reproduced with permission from Baumann et al. [57])

electroneutrality above a given spermine (4+) concentration and the onset of aggregation does not require the charge neutralization of DNA phosphates [59]. Yamasaki et al. also reported spermidine (3+)-condensed T4 DNA (166 kbp) showed charge neutralization [60].

When DNA condensation occurs at high DNA concentration, precipitation of DNA occurs after secondary aggregation. This process is commonly used to purify DNA. Therefore, precipitation experiments are another method to assess the electrostatic interaction between DNA and cationic compounds. In early precipitation experiments, Bungenberg de Jong showed that various polyelectrolytes precipitate upon addition of multivalent cations and resolubilize with further addition of cations, suggesting a reentrant transition [61]. Livolant and colleagues reported a similar phenomenon that aggregates of DNA fragments redissolved in an excess of multivalent cations (Fig. 14.9), [62, 63].

Concerning the reentrant transition, the plateau force of *F-x* curves for DNA reported by Baumann et al. [57] was reexamined by Murayama et al. [58]. The average plateau force in lower and higher spermidine concentration regions ([SPD] $\leq$1 mM, [SPD] $\geq$50 mM) ranged from 0.95 to 1.15 pN, whereas that in the moderate concentration region (2 mM $\leq$ [SPD] $\leq$ 10 mM) was around 1.7 pN. Changes in *F-x* curves observed in lower and moderate spermidine concentration regions are essentially the same in both studies. Murayama et al. pointed out that in the higher spermidine concentration region, no hysteresis was observed and *P* deduced from Eq. (14.1) was somewhat smaller than that without spermidine, suggesting these phenomena correspond to the elongation of DNA condensates, i.e., the reentrant transition (Fig. 14.10) [58].

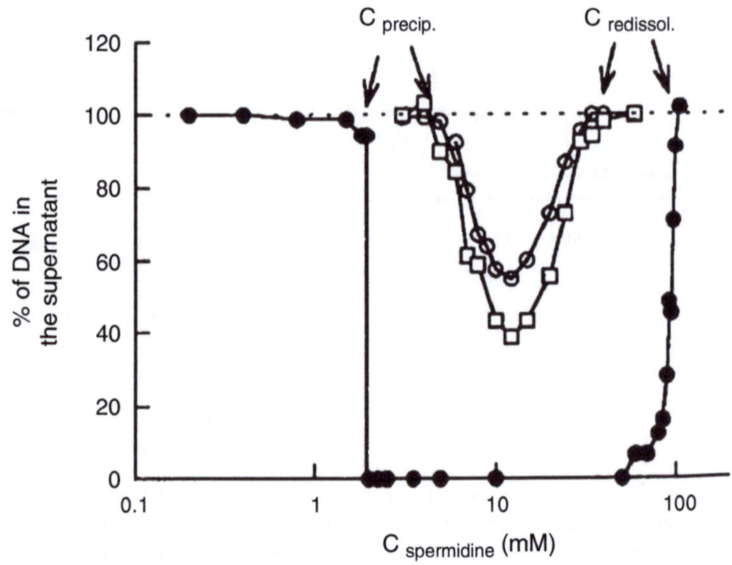

Fig. 14.9 Percentage of DNA in the supernatant after centrifugation, revealing that DNA precipitations can subsequently redissolve (Reprinted with permission from Raspaud et al. [63])

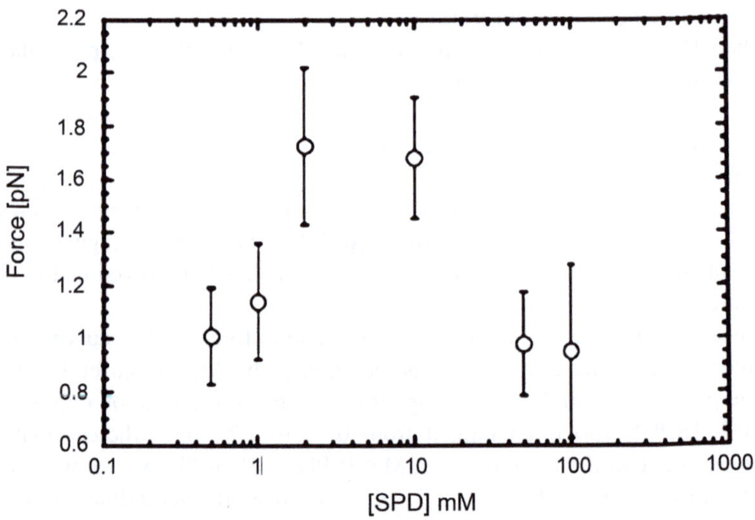

Fig. 14.10 Average force plateau values for the elastic responses of single DNA molecules at various spermidine concentrations corresponding to a reentrant transition (Reprinted with permission from Murayama et al. [58])

14.6 Application of DNA Condensation to Gene Delivery

In recent years, DNA condensation has received renewed interest in relation to advanced medical treatment such as gene therapy. This is because DNA condensation is the first important step to collapse therapeutic DNA to an adequate size for intravascular administration (~100 nm). Since the concept of gene therapy was first proposed [64], tremendous effort has been devoted to searching and designing gene delivery vectors. Despite numerous attempts, because of the serious problems associated with viral vectors, including a patient death during adenoviral treatment [65] and an adverse effect generated after retroviral gene therapy [66], only 1996-related clinical trials had been conducted globally between 1989 and 2013 (J Gene Med Clinical Trial site) [67]. Therefore, a development of nonviral gene delivery systems is required to provide safe treatment options.

Among the condensing agents described in Sect. 14.5, cationic polymers that interact strongly with DNA have great advantages for this application. DNA complexes with cationic polymers are usually regarded as polyplexes. There have been numerous efforts to modify the chemical structures of polymers over the last two decades [68–70]. In the wake of the successful example of PEGylation, which can improve blood retention time and decrease both immunogenicity and antigenicity and has been used to produce PEG-interferon and PEG-asparaginase [71], block copolymers of PEG with polycations have been widely investigated as nonviral carriers for therapeutic genes [72, 73]. Complexation of PEGylated polycations with DNA typically produces core-shell architectures, in which a polyplex core composed of polycations and DNA is surrounded by PEG.

As a cationic polymer, poly(L-lysine) (PLys) has been widely used since the 1970s as a condensation agent to model the structure of chromatin [74]. PEGylated PLys (PEG-PLys) is a potential candidate as a nonviral carrier system. PEG-PLys is synthesized by ring-opening polymerization of the N-carboxylic anhydride of ε-protected lysine (protected Lys-NCA) from an ω-amino group of α-methoxy-ω-amino PEG (MeO-PEG-NH_2) [75, 76]. Several derivatives of PEGylated cationic block copolymers were synthesized using the same method, and improved transfection efficiency was obtained by controlling the average DP of cationic segments and mixing the ratio of cationic polymer to DNA in these systems [77, 78]. Another strategy to synthesize similar block copolymers is covalent conjugation of PEG to oligolysine peptides prepared by solid-phase peptide synthesis [79].

Kanayama et al. found that a protecting group (benzyl ester) of poly(β-benzyl L-aspartate) (PBLA) underwent a quantitative aminolysis reaction with a variety of amine compounds under mild conditions [80]. Through a series of transfection and cytotoxicity assays using the PEGylated block copolymers of N-substituted poly(aspartamide) (PAsp), a highly promising candidate PEGylated poly [N-[N-(2-aminoethyl)-2-aminoethyl]aspartamide] (PEG-PAsp(DET)) exhibiting high efficiency and minimal toxicity was identified.

Recently, Nomoto et al. tested the colloidal stability of polyplexes in the blood stream using intravital real-time confocal laser scanning microscopy [81]. Snapshots obtained from observation of the blood flow in the blood vessels of

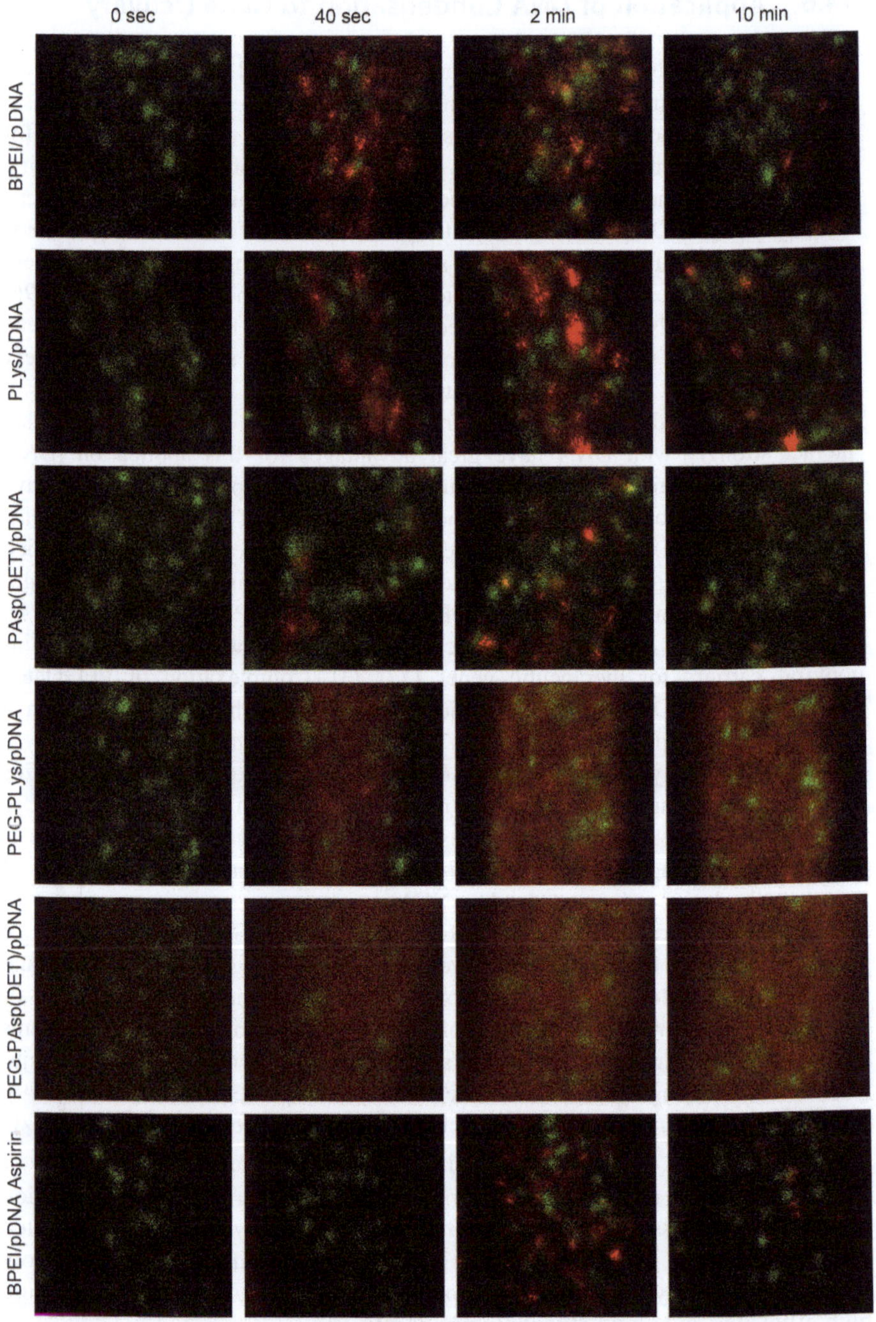

Fig. 14.11 Snapshots obtained from intravital real-time confocal laser scanning microscopy observation of blood flow in the blood vessels of a mouse earlobe. Polyplexes incorporating Cy5-labeled pDNA and DyLight 488-labeled platelets are shown in *red* and *green*, respectively (Reprinted with permission from Nomoto et al. [81])

mouse earlobe (Fig. 14.11) clearly indicated that agglomerations of Cy5-labeled plasmid DNA (red) several micrometers in size were formed immediately after intravenous injection of non-PEGylated polyplexes, plasmid DNA with branched polyethylenimine (BPEI), PLys, and PAsp(DET). The fluorescence intensity of Cy5-labeled plasmid DNA in these polyplexes decreased markedly by 10 min after injection, suggesting degradation of plasmid DNA. In contrast, PEGylated polyplexes formed with PEG-PLys and PEG-PAsp (DET) maintained high fluorescence intensity of over 70 % 10 min after injection because of their prolonged circulation in the blood.

Active targeting by introducing a cyclic RGD (arginine-glycine-aspartic acid) peptide (cRGD) into PEGylated polyplexes was recently reported as further development of the above study [82]. cRGD acts as a ligand for $\alpha_v\beta_3$ and $\alpha_v\beta_5$ integrin receptors that are overexpressed on the surface of tumor angiogenic endothelial cells. In this study, a polyplex composed of cRGD-PEG-PAsp(DET) and plasmid DNA encoding a soluble Fms-like tyrosine kinase-1 (sFlt-1) that reduces the free circulating level of vascular endothelial growth factor was intravenously administered into tumor-bearing mice subcutaneously inoculated with BxPC3 cells derived from human pancreatic adenocarcinoma. Successful sFlt-1 gene expression was expected to inhibit both neovasculature and tumor growth in this system. The antitumor efficacy observed in tumor growth curves is in good agreement with the intensity of sFlt-1 expression (Fig. 14.12), indicating that gene expression at the targeted site is successfully achieved by the ligand-conjugated PEGylated polyplex.

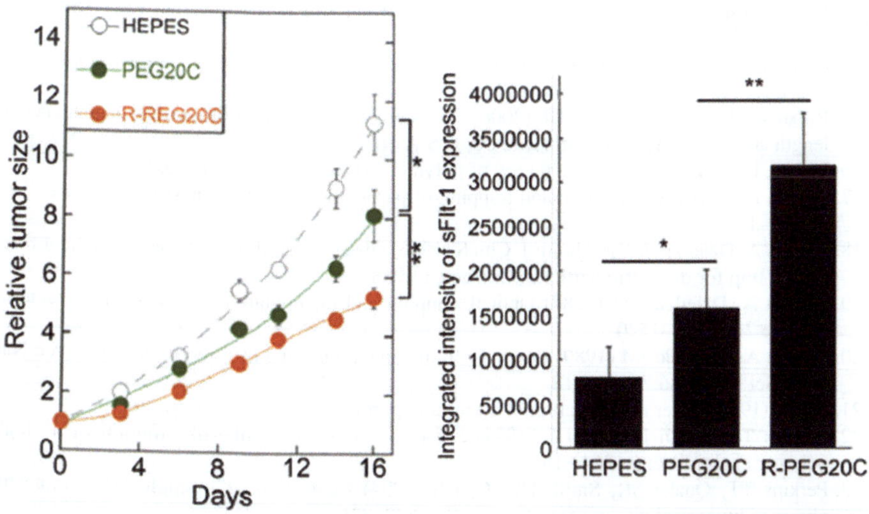

Fig. 14.12 In vivo evaluation after systemic administration of PEGylated polyplexes. HEPES: control, PEG20C: PEG-PAsp(DET), R-PEG20C: cRGD-PEG-PAsp (DET). (*Left*) Antitumor activity in terms of tumor size (mean ± SEM, $n = 3$). The PEGylated polyplexes were intravenously administered through the tail vein three times on days 0, 4, and 8. (*Right*) Expression of sFlt-1 protein in tumor tissue (Reproduced with permission from Ge et al. [82])

References

1. Yanagida M, Hiraoka Y, Katsura I (1982) Dynamic behaviors of DNA molecules in solution studied by fluorescence microscopy. Cold Spring Harb Symp Quant Biol 47:177–187
2. Bustamante C (1991) Direct observation and manipulation of single DNA molecules using fluorescence microscopy. Annu Rev Biophys Biophys Chem 20:415–446
3. Kapuscinski J (1995) DAPI: a DNA-specific fluorescent probe. Biotech Histochem 70:220–233
4. Morikawa K, Yanagida M (1981) Visualization of individual DNA-molecules in solution by light-microscopy – DAPI staining method. J Biochem 89:693–696
5. Matsumoto M, Sakaguchi T, Kimura H, Doi M, Minagawa K, Matsuzawa Y, Yoshikawa K (1992) Direct observation of Brownian motion of macromolecules by fluorescence microscope. J Polym Sci Polym Phys Ed 30:779–783
6. Smith DE, Perkins TT, Chu S (1996) Dynamical scaling of DNA diffusion coefficients. Macromolecules 29:1372–1373
7. Hirons GT, Fawcett JJ, Crissman HA (1994) TOTO and YOYO: new very bright fluorochromes for DNA content analyses by flow cytometry. Cytometry 15:129–140
8. Gurrieri S, Wells KS, Johnson ID, Bustamante C (1997) Direct visualization of individual DNA molecules by fluorescence microscopy: characterization of the factors affecting signal/background and optimization of imaging conditions using YOYO. Anal Biochem 249:44–53
9. Haugland RP (2004) Handbook of fluorescent probes and research chemicals. Molecular Probes, Eugene
10. Smith SB, Finzi L, Bustamante C (1992) Direct mechanical measurements of the elasticity of single DNA molecules by using magnetic beads. Science 258:1122–1126
11. Bakajin OB, Duke TAJ, Chou CF, Chan SS, Austin RH, Cox EC (1998) Electrohydrodynamic stretching of DNA in confined environments. Phys Rev Lett 80:2737–2740
12. Larsson A, Carlsson C, Jonsson M, Albinsson B (1994) Characterization of the binding of the fluorescent dyes YO and YOYO to DNA by polarized-light spectroscopy. J Am Chem Soc 116:8459–8465
13. Kratky O, Porod G (1949) Röntgenuntersuchung gelöster fadenmoleküle. Recl Trav Chim 68:1106–1115
14. Zimm BH (1956) Dynamics of polymer molecules in dilute solution: viscoelasticity, flow birefringence and dielectric loss. J Chem Phys 24:269–278
15. Robertson RM, Laib S, Smith DE (2006) Diffusion of isolated DNA molecules: dependence on length and topology. Proc Natl Acad Sci U S A 103:7310–7314
16. Doi M, Edwards S (1986) The theory of polymer dynamics. Clarendon, Oxford
17. Ashkin A (1970) Acceleration and trapping of particles by radiation pressure. Phys Rev Lett 24:156–159
18. Ashkin A, Dziedzic J, Bjorkholm J, Chu S (1986) Observation of a single-beam gradient force optical trap for dielectric particles. Opt Lett 11:288–290
19. Ashkin A, Dziedzic JM (1987) Optical trapping and manipulation of viruses and bacteria. Science 235:1517–1520
20. Ashkin A, Dziedzic JM (1989) Internal cell manipulation using infrared laser traps. Proc Natl Acad Sci U S A 86:7914–7918
21. Chu S (1991) Laser manipulation of atoms and particles. Science 253:861–866
22. Perkins TT, Smith DE, Chu S (1994) Direct observation of tube-like motion of a single polymer chain. Science 264:819–822
23. Perkins TT, Quake SR, Smith DE, Chu S (1994) Relaxation of a single DNA molecule observed by optical microscopy. Science 264:822–826
24. Matsumoto S, Morikawa K, Yanagida M (1981) Light microscopic structure of DNA in solution studied by the 4′,6-diamidino-2-phenylindole staining method. J Mol Biol 152:501–516

25. Shaqfeh ESG (2005) The dynamics of single-molecule DNA in flow. J Non-Newtonian Fluid Mech 130:1–28
26. Bustamante C, Marko JF, Siggia ED, Smith S (1994) Entropic elasticity of lambda-phage DNA. Science 265:1599–1600
27. Marko JF, Siggia ED (1995) Stretching DNA. Macromolecules 28:8759–8770
28. Bustamante C, Smith SB, Liphardt J, Smith D (2000) Single-molecule studies of DNA mechanics. Curr Opin Struct Biol 10:279–285
29. Gennes D (1971) Reptation of a polymer chain in the presence of fixed obstacles. J Chem Phys 55:572–579
30. Gennes D (1979) Scaling concept in polymer physics. Cornell Univ Press, Ithaca
31. Smith DE, Perkins TT, Chu S (1995) Self-diffusion of an entangled DNA molecule by reptation. Phys Rev Lett 75:4146–4149
32. Olivera BM, Baine P, Davidson N (1964) Electrophoresis of the nucleic acids. Biopolymers 2:245–254
33. Slater GW, Rousseau J, Noolandi J, Turmel C, Lalande M (1988) Quantitative analysis of the three regimes of DNA electrophoresis in agarose gels. Biopolymers 27:509–524
34. Ogston AG (1958) The space in a uniform random suspension of fibres. Trans Faraday Soc 54:1754–1757
35. Lumpkin OJ, Zimm BH (1982) Mobility of DNA in gel electrophoresis. Biopolymers 21:2315–2316
36. Stellwagen NC (1983) Accurate molecular weight determinations of deoxyribonucleic acid restriction fragments on agarose gels. Biochemistry 22:6180–6185
37. Slater GW, Noolandi J (1985) New biased-reptation model for charged polymers. Phys Rev Lett 55:1579–1582
38. Lalande M, Noolandi J, Turmel C, Rousseau J, Slater GW (1987) Pulsed-field electrophoresis: application of a computer model to the separation of large DNA molecules. Proc Natl Acad Sci U S A 84:8011–8015
39. Weber K, Osborn M (1969) The reliability of molecular weight determinations of dodecyl sulfate-polyacrylamide gel electrophoresis. J Biol Chem 224:4406–4412
40. Smisek DL, Hoagland DA (1989) Agarose gel electrophoresis of high molecular weight, synthetic polyelectrolytes. Macromolecules 22:2270–2277
41. Fangman WL (1978) Separation of very large DNA molecules by gel electrophoresis. Nucleic Acids Res 5:635–665
42. Carle GF, Olson MV (1984) Separation of chromosomal DNA molecules from yeast by orthogonal-field-alternation gel-electrophoresis. Nucleic Acids Res 12:5647–5664
43. Schwartz DC, Cantor CR (1984) Separation of yeast chromosome-sized DNAs by pulsed field gradient gel-electrophoresis. Cell 37:67–75
44. Smith SB, Aldridge PK, Callis JB (1989) Observation of individual DNA molecules undergoing gel electrophoresis. Science 243:203–206
45. Schwartz DC, Koval M (1989) Conformational dynamics of individual DNA molecules during gel electrophoresis. Nature 338:520–522
46. Gurrieri S, Smith SB, Wells KS, Johnson ID, Bustamante C (1996) Real-time imaging of the reorientation mechanisms of YOYO-labelled DNA molecules during 90° and 120° pulsed field gel electrophoresis. Nucleic Acids Res 24:4773–4781
47. Bloomfield VA (1996) DNA condensation. Curr Opin Struct Biol 6:334–341
48. Post CB, Zimm BH (1982) Theory of DNA condensation: collapse versus aggregation. Biopolymers 21:2123–2137
49. Widom J, Baldwin RL (1983) Monomolecular condensation of λ-DNA induced by cobalt hexammine. Biopolymers 22:1595–1620
50. Lerman LS (1971) A transition to a compact form of DNA in polymer solutions. Proc Natl Acad Sci U S A 68:1886–1890
51. Deng H, Bloomfield VA (1999) Structural effects of cobalt-amine compounds on DNA condensation. Biophys J 77:1556–1561

52. Minagawa K, Matsuzawa Y, Yoshikawa K, Matsumoto M, Doi M (1991) Direct observation of the biphasic conformational change of DNA induced by cationic polymers. FEBS Lett 295:67–69
53. Mel'nikov SM, Sergeyev VG, Yoshikawa K (1995) Transition of double-stranded DNA chains between random coil and compact globule states induced by cooperative binding of cationic surfactant. J Am Chem Soc 117:9951–9956
54. Takahashi M, Yoshikawa K, Vasilevskaya VV, Khokhlov AR (1997) Discrete coil-globule transition of single duplex DNAs induced by polyamines. J Phys Chem B 101:9396–9401
55. Yoshikawa K, Matsuzawa Y (1996) Nucleation and growth in single DNA molecules. J Am Chem Soc 118:929–930
56. Yamasaki Y, Yoshikawa K (1997) Higher order structure of DNA controlled by the redox state of Fe^{2+}/Fe^{3+}. J Am Chem Soc 119:10573–10578
57. Baumann CG, Bloomfield VA, Smith SB, Bustamante C, Wang MD, Block SM (2000) Stretching of single collapsed DNA molecules. Biophys J 78:1965–1978
58. Murayama Y, Sakamaki Y, Sano M (2003) Elastic response of single DNA molecules exhibits a reentrant collapsing transition. Phys Rev Lett 90:018102
59. Raspaud E, Chaperon I, Leforestier A, Livolant F (1999) Spermine-induced aggregation of DNA, nucleosome, and chromatin. Biophys J 77:1547–1555
60. Yamasaki Y, Teramoto Y, Yoshikawa K (2001) Disappearance of the negative charge in giant DNA with a folding transition. Biophys J 80:2823–2832
61. de Bungenberg Jong HG (1949) Complex colloid systems. In: Kruyt HR (ed) Colloid science, vol 2. Elsevier, New York, pp 335–432
62. Pelta J, Livolant F, Sikorav JL (1996) DNA aggregation induced by polyamines and cobalthexamine. J Biol Chem 271:5656–5662
63. Raspaud E, Olvera de la Cruz M, Sikorav JL, Livolant F (1998) Precipitation of DNA by polyamines: a polyelectrolyte behavior. Biophys J 74:381–393
64. Friedmann T, Roblin R (1972) Gene therapy for human genetic disease? Science 175:949–955
65. Marshall E (1999) Clinical trials – Gene therapy death prompts review of adenovirus vector. Science 286:2244–2245
66. Hacein-Bey-Abina S, von Kalle C, Schmidt M, Le Deist F, Wulffraat N, McIntyre E, Radford I, Villeval JL, Fraser CC, Cavazzana-Calvo M, Fischer A (2003) A serious adverse event after successful gene therapy for x-linked severe combined immunodeficiency. N Engl J Med 348:255–256
67. J Gene Med Clinical Trial site (2014) Gene therapy clinical trials worldwide http://www.abedia.com/wiley/index.html. Accessed 2 Apr 2014
68. Vijayanathan V, Thomas T, Thomas TJ (2002) DNA nanoparticles and development of DNA delivery vehicles for gene therapy. Biochemistry 41:14085–14094
69. Duncan R (2003) The dawning era of polymer therapeutics. Nat Rev Drug Discov 2:347–360
70. De Ilarduya CT, Sun Y, Düzgünes N (2010) Gene delivery by lipoplexes and polyplexes. Eur J Pharm Sci 40:159–170
71. Molineux G (2002) Pegylation: engineering improved pharmaceuticals for enhanced therapy. Cancer Treat Rev 28:13–16
72. Wolfert MA, Schacht EH, Toncheva V, Ulbrich K, Nazarova O, Seymour LW (1996) Characterization of vectors for gene therapy formed by self-assembly of DNA with synthetic block copolymers. Hum Gene Ther 7:2123–2133
73. Yang J, Wang HY, Yi WJ, Gong YH, Zhou X, Zhuo RX, Zhang XZ (2013) PEGylated peptide based reductive polycations as efficient nonviral gene vectors. Adv Healthcare Mater 2:481–489
74. Laemmli UK (1975) Characterization of DNA condensates induced by poly(ethylene oxide) and polylysine. Proc Natl Acad Sci U S A 14:4288–4292
75. Harada A, Kataoka K (1995) Formation of polyion complex micelles in an aqueous milieu from a pair of oppositely-charged block copolymers with poly(ethylene glycol) segments. Macromolecules 28:5294–5299

76. Katayose S, Kataoka K (1997) Water-soluble polyion complex associates of DNA and poly (ethylene glycol)-poly(L-lysine) block copolymer. Bioconjug Chem 8:702–707
77. Harada-Shiba M, Yamauchi K, Harada A, Takamisawa I, Shimokado K, Kataoka K (2002) Polyion complex micelles as vectors in gene therapy – pharmacokinetics and in vivo gene transfer. Gene Ther 9:407–414
78. Miyata K, Kakizawa Y, Nishiyama N, Harada A, Yamasaki Y, Koyama H, Kataoka K (2004) Block catiomer polyplexes with regulated densities of charge and disulfide cross-linking directed to enhance gene expression. J Am Chem Soc 126:2355–2361
79. Kwok KY, McKenzie DL, Evers DL, Rice KG (1999) Formulation of highly soluble poly (ethylene glycol)-peptide DNA condensates. J Pharm Sci 88:996–1003
80. Kanayama N, Fukushima S, Nishiyama N, Itaka K, Jang WD, Miyata K, Yamasaki Y, Chung UI, Kataoka K (2006) A PEG-based biocompatible block catiomer with high buffering capacity for the construction of polyplex micelles showing efficient gene transfer toward primary cells. ChemMedChem 2006:439–444
81. Nomoto T, Matsumoto Y, Miyata K, Oba M, Fukushima S, Nishiyama N, Yamasoba T, Kataoka K (2011) In situ quantitative monitoring of polyplexes and polyplex micelles in the blood circulation using intravital real-time confocal laser scanning microscopy. J Control Release 151:104–109
82. Ge Z, Chen Q, Osada K, Liu X, Tockary TA, Uchida S, Dirisala A, Ishii T, Nomoto T, Toh K, Matsumoto Y, Oba M, Kano MR, Itaka K, Kataoka K (2014) Targeted gene delivery by polyplex micelles with crowded PEG palisade and cRGD moiety for systemic treatment of pancreatic tumors. Biomaterials 35:3416–3426
83. Smith SB, Cui Y, Bustamante C (1996) Overstretching B-DNA: the elastic response of individual double-stranded and single-stranded DNA molecules. Science 271:795–799

Atomic Force Microscopy: Imaging and Rheology of Living Cells

15

Takaharu Okajima

Abstract

Atomic force microscopy (AFM) has been widely used for characterizing physical properties of adherent living cells because it provides high-resolution images and accurate measurements of mechanical properties without modifications to the cells. In this chapter, we review recent advances in AFM single-cell imaging and rheology. Techniques for AFM imaging and mechanical measurements of living cells are first reviewed. We then discuss how rheological properties of cells, which are described as power-law rheology model, are quantified for single-cell diagnostics. In addition to micro- and nano-measurements of cell moduli, we introduce an AFM method combined with a micro-fabricated substrate as a force sensor for investigating how forces propagate inside cells through the cytoskeleton, which is deeply associated with various cell functions. Finally, we reviewed scanning ion conductance microscopy, which allows us to obtain noncontact image of cell membrane topography and to quantify cell membrane fluctuations that are inaccessible to AFM.

Keywords

Atomic force microscopy • Scanning ion conductance microscopy • Cell imaging • Cell rheology • Single-cell diagnostics • Micro-fabricated substrates

T. Okajima (✉)
Graduate School of Information Science & Technology, Hokkaido University, Kita-ku N14 W9, Sapporo 060-0814, Japan
e-mail: okajima@ist.hokudai.ac.jp

R. Kita, T. Dobashi (eds.), *Nano/Micro Science and Technology in Biorheology*, DOI 10.1007/978-4-431-54886-7_15

15.1 Principles of AFM

15.1.1 Imaging

Atomic force microscopy (AFM) [1] is a scanning probe technique that has been adapted for high-resolution imaging, characterization of mechanical properties, and manipulation of living cells. AFM uses a cantilever as a sensor (Fig. 15.1) of forces between a sample surface and a probe tip attached at the end of the cantilever. The forces are detected by either cantilever deflection, damped vibration amplitude, or frequency and/or phase shifts [2, 3]. During image acquisition, the cantilever is raster scanned laterally (x–y axes) over the sample surface, while its vertical position (z-axis) is regulated with a feedback circuit that maintains a constant force between the probe and sample. For high-resolution imaging, a sharp probe tip (<20 nm radius) is required.

In contact mode AFM operation, the force between the tip and the surface is kept constant by maintaining a static deflection of the cantilever (Fig. 15.2a). Thus, the z-position corresponds to the topography of the cell sample. In the case of soft samples such as cells, the contact force may cause local deformation (Figs. 15.2c and 15.3a). To minimize deformation, the contact force should be <1 nN [4], and a low-stiffness (~0.1 N/m) cantilever should be used.

The deflection of the cantilever would be unchanged if the feedback was perfectly regulated. During imaging, however, the deflection signal changes slightly with abrupt changes in the cell surface morphology. The corresponding

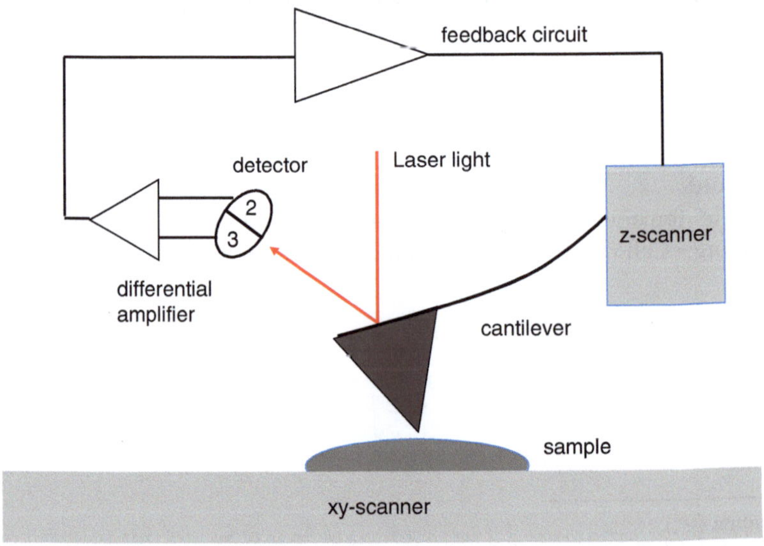

Fig. 15.1 Principles of AFM. The force between the cantilever tip and the sample is detected by deflection of the cantilever and an optical lever coupled with a position-sensitive photodetector (PSD). The PSD signal is used to regulate the force

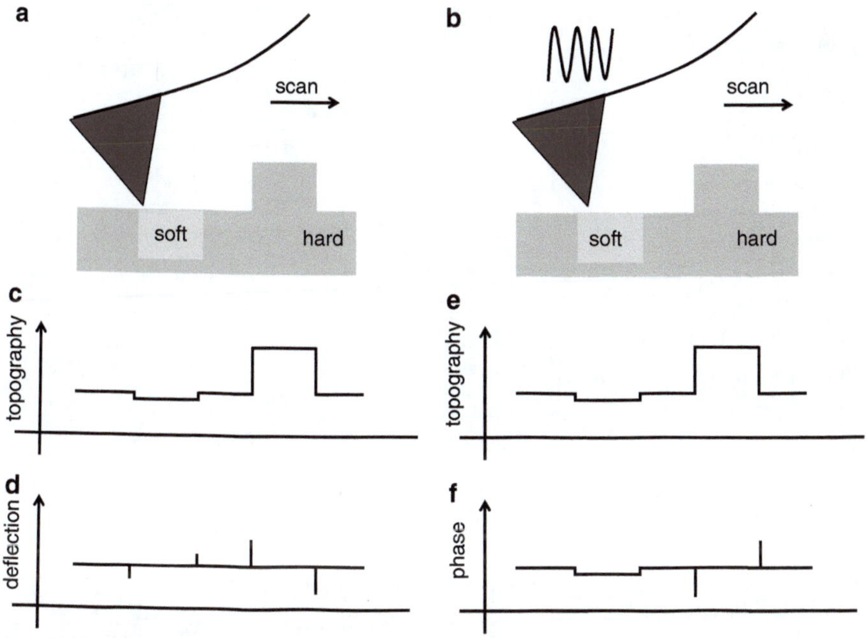

Fig. 15.2 Imaging modes of AFM: (**a**) contact mode and (**b**) AM mode. The sample is a hard elastic material with a soft viscoelastic region. (**c**) Topography and (**d**) deflection images in contact mode. (**e**) Topography and (**f**) phase-shift images in AM mode

error signal can be used to construct images in which changes in topography are sharply identified (Fig. 15.2d). In Fig. 15.3, a deflection error image reveals underlying cytoskeletal structures more clearly than the corresponding topography (z-piezo) image [5].

In contact mode, the scanning tip may generate lateral forces that could drag the cell surface. To reduce lateral forces, amplitude modulation (AM) mode or "tapping" mode AFM was developed. For intermittent contact, the cantilever is continuously oscillated in the z-direction (Fig. 15.2d). The small amplitude of the oscillation near the cantilever's fundamental resonance frequency f_0 is used as the feedback signal for maintaining intermittent contact during topographic image acquisition. As in contact mode, the intermittent contact force and lateral forces could also deform a soft sample surface (Fig. 15.2e). The phase difference between the oscillation drive signal and the oscillating cantilever is sensitive to the change in the topography and reflects energy dissipation in the sample (Fig. 15.2f).

AM mode is used for high-speed AFM (HS-AFM) [6, 7], at video rates for high-resolution observation of dynamic processes. Recently, HS-AFM combined with a wide-area scanner was used to acquire video images of endocytosis on a living cell surface (Fig. 15.4) [8].

When the cantilever is oscillated at frequency f_0, the motion of the cantilever also includes higher harmonic modes ($2f_0$, $3f_0$, etc.) because of nonlinear

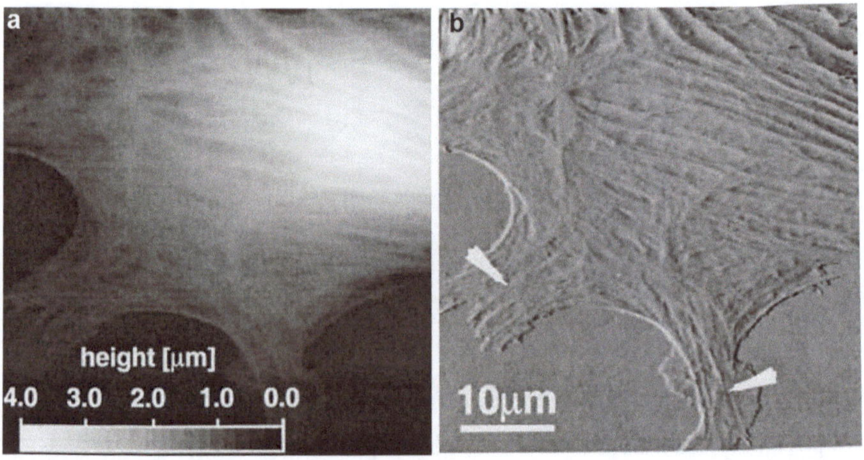

Fig. 15.3 Contact mode height (**a**) and deflection (**b**) images of a living cell showing cytoskeletal structures (Reprinted with permission from [5])

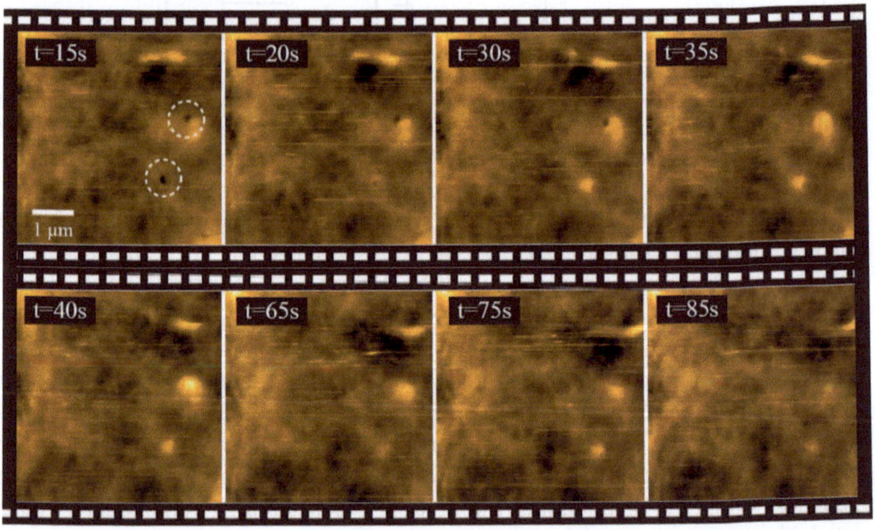

Fig. 15.4 HS-AFM images showing the dynamics of endocytosis (*dotted circles*) in HeLa cells. Scan range and imaging rate are 5 × 5 μm and 5 s/frame over 200 × 200 pixels (Reprinted with permission from [8])

interactions between the AFM tip and the sample [9]. The interaction of the harmonic components with the cell surface can be employed to characterize local stiffness, stiffness gradients, and viscoelastic dissipation at high resolution [10].

In a technique called scanning near-field ultrasonic holography (SNFUH), the sample and the cantilever are simultaneously excited at different ultrasonic frequencies (MHz) f_S and f_C, respectively (Fig. 15.5). The ultrasonic vibration of

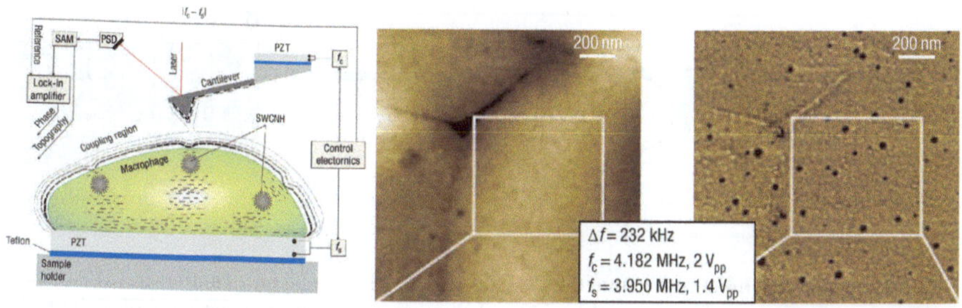

Fig. 15.5 (*Left*) Schematic of scanning near-field ultrasonic holography (SNFUH). The cantilever and the sample are vibrated at fc and fs ultrasonic frequencies, respectively. The AFM cantilever is locked in at the frequency difference |fc-fs|, providing information about local intracellular nano-mechanical structures. (*Right*) Images of nanoparticles in red blood cells (Reprinted with permission from [11])

the cantilever allows us to generate images of subsurface structures in cells because the mechanical waves propagate through the cell and are perturbed by internal structures [11, 12]. The amplitude and the phase shift reflect the local mechanical properties of the subsurface structures. In Fig. 15.5, SNFUH images of nanoparticles embedded in red blood cells are clearly shown [11, 12]. The imaging depth strongly depends on the material properties of the sample [13], but the mechanism of SNFUH imaging is not fully understood.

15.1.2 AFM Probe

The shape of the AFM probe in the region that interacts directly with the sample surface affects imaging resolution as well as mechanical properties. Sharp tips required for a high-resolution imaging are more likely to damage soft cells with fragile structures. Moreover, the exact profile of a sharp tip is hard to determine precisely, precluding quantitative mechanical measurements. For those reasons, a silica or polystyrene colloidal bead [14] with a well-defined spherical shape is widely used for force measurements on cells. It can be attached to a cantilever in various ways (Fig. 15.6). When the bead of the colloidal probe cantilever contacts the cell surface in a liquid environment, the liquid between the cantilever and the cell surface is highly confined and squeezed. This enhances viscous damping of the cantilever and affects the rheological observations. Because the squeezing effect can be reduced by increasing the distance between the cantilever and the surface [17], attaching the bead to the apex of the probe (Fig. 15.6c, d) is suitable for single-cell rheology.

Adhesion between the colloidal bead and the cell surface should be minimized for single-cell rheology because the Hertz model [2, 3, 18, 19], which is the standard model for estimating contact mechanics, assumes that there is no adhesion between contacting materials. Hydrophobic perfluorodecyltrichlorosilane-coated

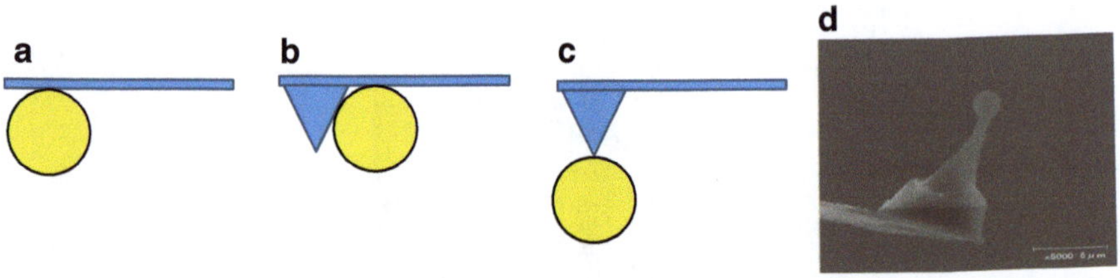

Fig. 15.6 Different mounting geometries for a colloidal bead probe on a cantilever. (**a**) Attached to a tip-less cantilever. (**b**) Attached beside a sharp tip. (**c**) Attached at the apex of a sharp tip. Reprinted with permission from [15]. (**d**) Electron microscope image of a probe attached as in (**c**) (Reprinted with permission from [16])

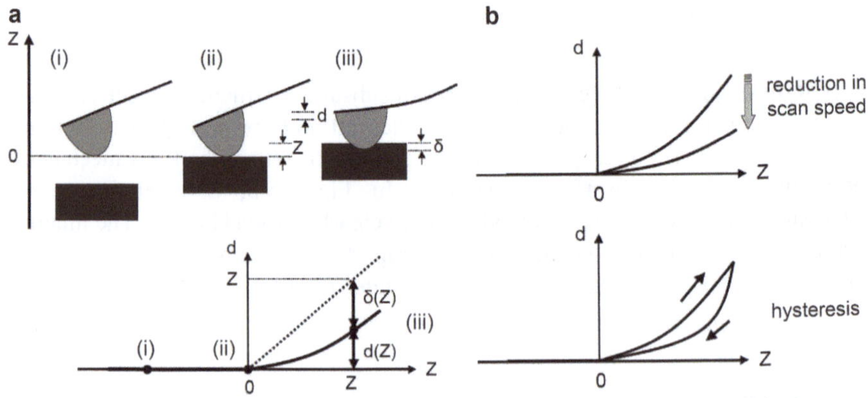

Fig. 15.7 (**a**) Force curve measurements. (i) The cantilever is separated from the sample surface and no deflection occurs. (ii) The AFM cantilever probe contacts the cell surface at the position $Z=0$. (iii) The AFM probe indents the cell sample a Z from the position in (ii). The deflection of the cantilever is d, while the indentation is δ, where $\delta = Z - d$ (Reprinted with permission from [15]). (**b**) Characteristic features of force–distance curves measured in viscoelastic materials

colloidal beads [20] work well to prevent adhesion to the cell surface during force measurements.

15.1.3 Force Measurements

Elastic properties, i.e., reversible deformation of cells, are estimated from the relationship between the loading force F and the normal displacement Z of the AFM probe as it deforms the cell surface (Fig. 15.7a). The indentation δ in the cell surface is determined by subtracting the cantilever deflection d from the displacement Z. The loading force is estimated from Hook's law ($F = kd$), where the spring constant k of the cantilever can be determined by thermal fluctuations [21].

$F(\delta)$ depends on the shape of the AFM probe. According to the Hertz model for a spherical probe with radius R [2, 3, 18, 19], F is given by

$$F = \frac{4}{3}\frac{ER^{1/2}}{1-\nu^2}\delta^{3/2}, \tag{15.1}$$

where ν is Poisson's ratio, which is assumed to be 0.3–0.5 for cells [2, 3], and E is Young's modulus. E has been measured in various animal cells [2, 3], and its spatial heterogeneity has been resolved [5, 22, 23]. In one case, local values of E were attributed to actin filaments rather than microtubules and intermediate filaments [5].

Cells are not completely elastic, but behave more like a compliant viscoelastic material. Therefore, because of energy dissipation in the cells, force–distance curves tend to exhibit hysteresis between approach and retraction (Fig. 15.7b) [24]. Thus, E estimated by force–distance curve measurements may depend on the speed of approach or retraction. Because it increases with increasing speed (Fig. 15.7b), E from force curve measurements is an "apparent" Young's modulus. Thus, frequency and/or time domain AFM measurements are indispensable for quantifying intrinsic mechanical properties of cells.

15.1.4 Frequency Domain AFM

In the force modulation mode [25–27], the dynamic response due to an external periodic strain is measured (Fig. 15.8a). The strain is due to a cantilever that is sinusoidally oscillated with fixed amplitude (usually 10–50 nm) at several frequencies during indentation. The amplitude and phase shift of the cantilever displacement are measured with a lock-in amplifier.

Using the Hertz model from Eq. 15.1, the complex loading force F^* with a small complex amplitude indentation oscillation δ_1^* around an operating indentation δ_0 is approximately expressed [25–30] by a first-order Taylor expansion:

$$F^* = \frac{4R^{1/2}}{3(1-\nu^2)}\left(E_0\delta_0^{3/2} + \frac{3}{2}E_1^*\delta_0^{1/2}\delta_1^*\right), \tag{15.2}$$

where E_0 is Young's modulus at zero frequency and E_1^* is the frequency-dependent Young's modulus, given by $2(1+\nu)G^*$ [19]. Since the oscillating probe experiences hydrodynamic drag forces F_d^* [17], G^* is given by

$$G^* = G' + iG'' = \frac{1-\nu}{4(R\delta_0)^{1/2}}\left[\frac{F_1^*}{\delta_1^*} - ib(0)f\right], \tag{15.3}$$

where $F_1^* = 2(R\delta_0)^{1/2}E_1^*\delta_1^*/(1-\nu^2)$ and b is the drag factor [17]. F_d^* at a separation distance h between the sample surface and the probe with δ_1^* is defined as $F_d^*/\delta_1^* = ib(h)f$. The value $b(0)$ can be determined by the extrapolation of $b(h)$ measured at an oscillating frequency [17]. The phase shift and amplitude of the

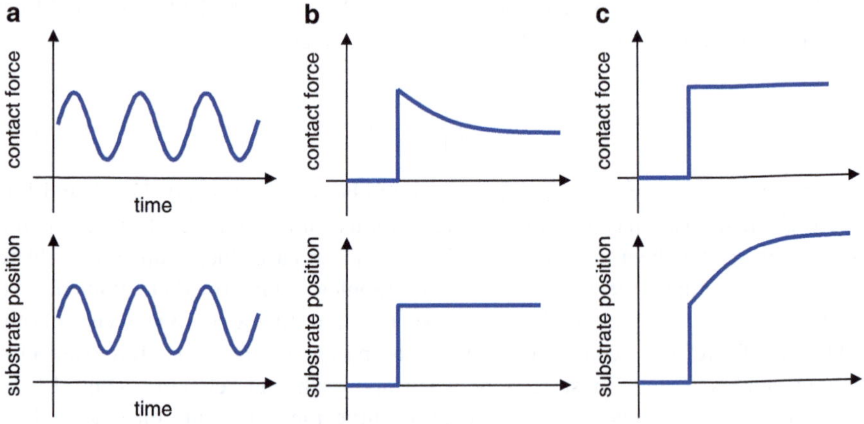

Fig. 15.8 Schematics of AFM rheology measurements: (**a**) force modulation mode, (**b**) stress relaxation, and (**c**) creep relaxation (Reprinted with permission from [15])

AFM instrument at different frequencies can be calibrated with a stiff cantilever in contact with a clean glass substrate in air [17, 28]. By using this method, G^* for several cell types has been measured in detail.

15.1.5 Time Domain AFM

In AFM stress relaxation [16, 20, 24, 31–33], Z is kept constant at the position where the initial force is applied, and F is measured as a function of time t (Fig. 15.8b). The cantilever defection d and the corresponding indentation δ both change during stress relaxation. This situation is not like conventional stress relaxation measurements where the strain is kept a constant value and the stress is measured as a function of t. In cell experiments, the change in d is typically about 1 % (10 nm) of that for δ (1 μm). Therefore, it is assumed that d is approximately constant relative to δ. According to the Hertz model, in which the contact radius a is dependent only on δ with a fixed probe radius R, the average stress is $F/(\pi a^2)$.

Since F, δ, and E are time dependent, and based on the Hertz model of Eq. 15.1, F is given by

$$F(t) = \frac{4R^{1/2}E(t)}{3(1-\nu^2)}\delta^{3/2}(t), \tag{15.4}$$

where $E(t)$ is the relaxation modulus at t.

In the case of stress relaxation with a constant indentation δ_0, $F(t)$ is proportional to $E(t)H(t)$ by Eq. 15.4, using the Heaviside step function $H(t)$ [32, 34]. As shown below, $G(f)$ for cells follows a single power law of frequency f^α at low frequencies. Since the relaxation modulus in the Laplace domain $E(f)$ is proportional to f^α from the relation $E(f) = 2(1+\nu)G(f)$ (in the case where ν is independent of t) [19],

$F(f)$ is proportional to $f^{\alpha-1}$. Therefore, the inverse Laplace transform of $F(f)$ yields the functional form of the loading force for stress relaxation: $F(t) \propto t^{-\alpha}$. Since cells are generally soft, F decreases significantly over long time periods and may approach zero. Therefore, stress relaxation, when compared to creep relaxation, is insensitive for long-term measurements of cell rheology. Furthermore, large initial loading forces required to enhance the signal-to-noise ratio for stress relaxation curves over long times cause large deformations in cells. This may also induce the cells to actively escape from the stress.

Creep relaxation of single cells can also be performed by AFM [20, 35]. In this case, the probe contacts the cell surface at a constant F under feedback, and Z is monitored as a function of t (Fig. 15.8c). Because the contact radius a changes during creep relaxation, the stress applied to the cell is not constant and decreases with t. The relationship between $a(t)$ and the creep compliance $J(t)$ becomes $a^3(t) \propto J(t)$ [18]. If $G(f)$ follows a single power law of the form f^{α}, $J(t)$ is proportional to t^{α}. Moreover, by using $a^2(t) = R\delta$ [18, 19], the indentation for creep relaxation is given by $\delta(t) \propto t^{2\alpha/3}$. For soft cells during creep measurements, δ significantly increases over long time periods, allowing us to easily monitor the relaxation. Conversely, the observed relaxation curve may reflect the highly heterogeneous cell structure with depth. Fluctuation and active movement of the cell also occur because of large δ.

15.2 Single-Cell Rheology

15.2.1 High-Throughput Measurements

Among single cells of the same source and type, rheological properties exhibit spatial, temporal, and intrinsic variations. High-throughput techniques, based on magnetic or optical trapping with micron-sized beads [36–42], micro-fluidic systems [43, 44], and AFM with micro-fabricated substrates [45–48], have been developed to characterize large numbers of cells.

Magnetic twisting cytometry (MTC) is one of the most common methods for investigating rheology statistics of adherent cells. A micron-sized magnetic bead is attached to a cell surface via binding proteins, and the cell modulus is estimated from the displacement of the beads under a periodic, external magnetic force (Fig. 15.9a). Lateral [36–42] or vertical [49] displacement of a large number of microbeads can be simultaneously monitored with optical microscopy. The disadvantages of MTC are that the contact geometry and the degree of binding between the microbeads and the cell surface are not well known, and the positions on the cell surfaces are not precisely controlled. Thus, it is difficult to assess cell-to-cell variations from the experimental data. Furthermore, focal adhesion complexes form at the microbead binding sites; thus, local reorganization of the cytoskeleton may alter the rheology [28].

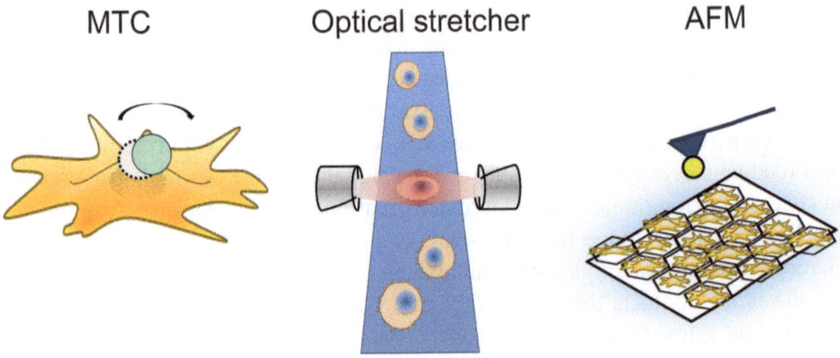

Fig. 15.9 Depiction of single-cell rheology for a large number of cells: magnetic twisting cytometry (MTC), micro-fluidic optical stretcher, and AFM on a micro-fabricated substrate

Micro-fluidic techniques can provide very high-throughput measurements of single-cell rheology. Suspended cells flowing in micro-channels can be deformed by optical pressure (Fig. 15.9b) [43] or hydrodynamic forces [44], and the deformability of whole cells floating in a micro-fluidic chamber is estimated. One disadvantage is that adherent cells have to be detached from their substrate, which may perturb intracellular structures that, in turn, affect cell mechanics.

With micro-fabricated substrates, one can use AFM to characterize the rheology of a large number of single cells rapidly (Fig. 15.9c). It has the advantage of measuring mechanical properties of single adherent cells at any region on the surface without cell surface modification [2, 3]. Thus, AFM is a less-invasive technique for measuring intrinsic mechanical features of single cells.

15.2.2 Power-Law Rheology Model

Because cells have internal organelles, their spatial–temporal rheological properties will vary from cell to cell. In spite of the structural complexities, the rheology of cells has been widely explained in terms of linear viscoelastic [34] or structural dampening models [36, 37, 50, 51].

In linear viscoelastic models, the cell is simulated with linear springs and linear viscous dashpots, and inertia effects are neglected. Therefore, creep and stress relaxations are sums of single-exponential functions in the time domain [34].

Power-law behaviors as a function of f have been observed for cell rheology with MTC and AFM. G' exhibits one single-power-law behavior in the range of 10^0–10^2 Hz, whereas, for other frequency ranges, other power-law models have been proposed: single (Fig. 15.10a) [36, 37] and multiple (Fig. 15.10b) [38, 52].

Fabry et al. reported that G' followed a single-power-law function over 10^{-2}–10^3 Hz, where the exponent α depended on the cytoskeletal architecture, regardless of modifications by chemical drugs, and appeared to cross at $G' = g_0$ at a high

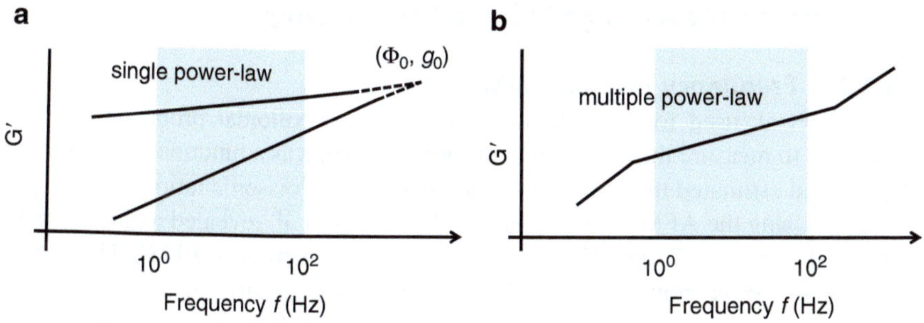

Fig. 15.10 Power-law models of G' as a function of f. (**a**) Single-power-law model for single cells under different conditions and (**b**) multiple-power-law model

frequency $f = \Phi_0$ [36, 37] (Fig. 15.10a). In this model, the complex shear modulus G^* is given by the structural damping model [36, 37, 50, 51]:

$$G^* = G_0(1 + i\eta)\left(\frac{f}{f_0}\right)^{\alpha}\Gamma(1-\alpha)\cos(\pi\alpha/2) + i\mu f, \qquad (15.5)$$

where η is the hysteresivity, which is expressed by $\tan(\pi\alpha/2)$, G_0 is a modulus scale factor at a frequency scale factor of f_0, and Γ denotes the gamma function. The α-value was 0.1–0.4 depending on cell type, where $\alpha = 0$ is solid-like and $\alpha = 1$ is fluidlike. The Newtonian viscous term μf is small, except at high frequencies. Single power laws have been discussed in detail in terms of soft glassy rheology (SGR) [50, 53, 54].

In contrast, two power-law exponents in the frequency domain have been observed; they cross over at around 10^0 Hz or 10^2 Hz (Fig. 15.10b). The exponent for the lower frequencies was 0.5, because of noncovalent protein–protein bond rupture during near-equilibrium loading [52]. Meanwhile, the exponent for the higher frequencies was about 0.75, because of entropic fluctuations of semi-flexible filaments and soft-glass-like dynamics [38].

Multiple-power-law cell rheology has also been observed in time domain experiments. Overby et al. reported that $\alpha = 0.18$ for pulling a single cell in a creep experiment over several seconds and $\alpha = 0.5$ for longer time scales [55]. Using magnetic microbeads, Stamenovic et al. reported that in creep experiments of single cells over a wide range of time scales, there were two power-law regimes with an intervening plateau over 10 s [56]. Desprat et al. employed a uniaxial stretching rheometer to observe that the creep function of pulling a whole cell follows a power-law exponent of 0.24 for periods <200 s, while for periods >200 s, the exponent increased to ≈ 0.5 [57]. These studies commonly showed that in the intermediate frequency range of 10^0–10^2 Hz, the single power law is an intrinsic feature of cell mechanics and is valid at size scales from a few tens of nanometers to the entire cell. However, it is not elucidated whether passive and active cell behaviors are involved in the mechanics over longer time scales.

15.2.3 Ensemble Averaged Single-Cell Rheology

15.2.3.1 Frequency Domain AFM

Mahaffy et al. used force modulation mode with a colloidal probe on an AFM cantilever to measure the mechanical properties of cells as a function of indentation depth δ and estimated the viscoelastic parameters and Poisson's ratio quantitatively [29, 30]. Using the AFM model in Eq. 15.3, Alcaraz et al. revealed a characteristic feature of averaged G' and G'' for single cells as a function of f [28]. The typical behavior of G^* is shown in Fig. 15.11. G' increased linearly in a log–log scale, exhibiting a weak power-law dependence on oscillation frequency. Conversely, G'' displayed a similar frequency dependence at values <10 Hz, and the frequency dependence was more pronounced at higher frequencies. The results fit the structural damping model shown in Eq. 15.5. This power-law frequency dependence has been observed with AFM in different cell types [45, 58, 59]. However, the absolute values of G' and G'' were different between cell types.

15.2.3.2 Time Domain AFM

Darling et al. measured the stress relaxation of single cells for $\approx$60 s with a colloidal probe cantilever [32, 33]. They observed that the stress relaxation was a single-exponential function, obeying a linear viscoelastic model. Moreno-Flores et al. reported that heterogeneities in single-cell rheology could be imaged with stress relaxation AFM [60].

Wu et al. investigated the relationship between viscoelastic properties and the cytoskeletal architecture of cells by using creep relaxation AFM. They demonstrated that creep relaxation for 60 s could be fit with a standard linear solid model consisting of two springs and one dashpot [35]. The creep relaxation of cells treated with various chemical drugs affecting the cytoskeleton was

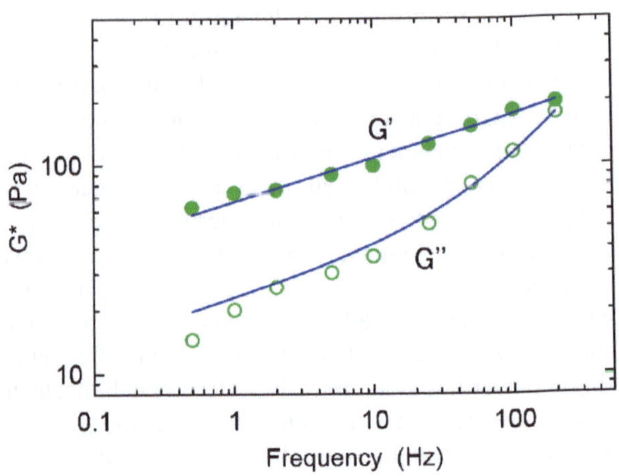

Fig. 15.11 Storage modulus G' (*left*) and loss modulus G'' (*right*) of adherent mouse fibroblast cells

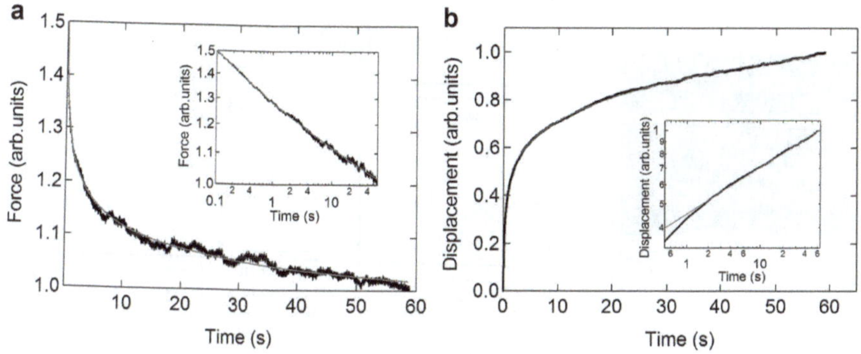

Fig. 15.12 Linear plots of the averaged AFM stress (**a**) and creep (**b**) relaxation curves for NIH3T3 cells on a microarray. The insets show the corresponding relaxations on logarithmic axes. *Solid lines* represent the fit of the power-law functions described in the main text (Reprinted with permission from [20])

examined. Cytochalasin D (cytoD), which depolymerizes actin filaments, reduced both elasticity and viscosity, whereas nocodazole or colcemid, which depolymerizes microtubules, exhibited a marked increase in elasticity and a slight increase in viscosity. Thus, changes in cytoskeletal structure can be detected by using AFM in the time domain.

The results from stress and creep relaxation experiments are inconsistent with those obtained with force modulation mode, where power-law behavior in the frequency domain has been widely observed. In this context, the relaxation behavior of individual cells placed and cultured in microarray wells was characterized with AFM by averaging several relaxation curves (Fig. 15.12) [20]. Tails in both stress and creep relaxation curves at long times follow single power laws over 60 s. Also, $\alpha = 0.1$–0.4, which varies between cells and has an average value in good agreement with that estimated from the force modulation mode [45].

15.2.4 Cell-to-Cell Variability

15.2.4.1 Statistics of Single-Cell Rheology

The statistics of single-cell rheology is crucial to reveal a universal behavior of single cells and to conduct single-cell diagnostics automatically. Hoffman et al. found that the distribution of G^* measured by MTC followed a lognormal and that the amplitude of the rocking motion or mean-square displacement of beads in cells varied dramatically for different methods [40]. Moreover, Massiera et al. showed that by using MTC and laser tracking micro-rheology, the magnitude of G^* at a low frequency exhibited a lognormal distribution, whereas the single-power-law exponent was a normal Gaussian [41]. Using optical trapping and uniaxial stretching of single cells, Balland et al. also showed that α is distributed normally over a cell population and that the prefactors of G^* and J follow a lognormal distribution [42].

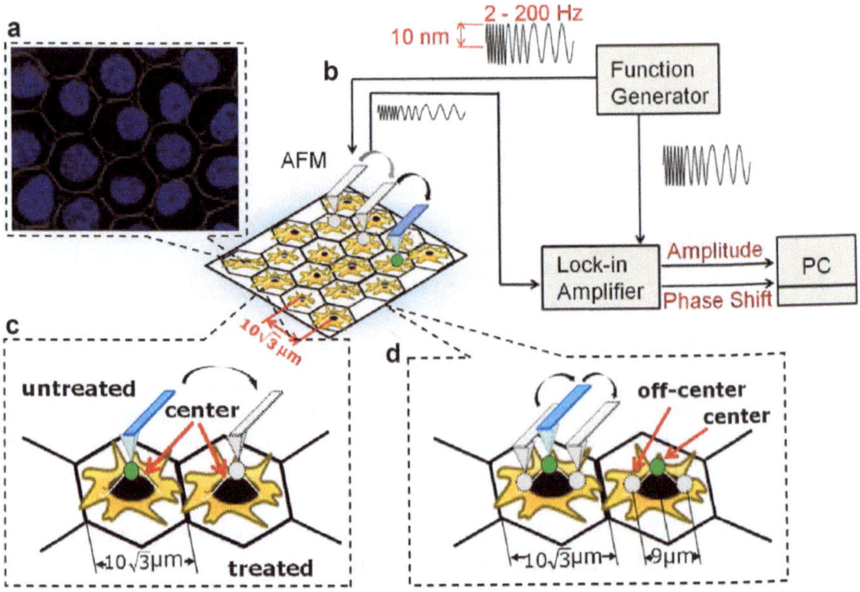

Fig. 15.13 Schematic of AFM for microarrays of cells. Force modulation mode measurements are automatically examined at the centers of the wells after a specific cell is chosen with optical microscopy (Reprinted with permission from [47])

The statistical properties of single-cell rheology were investigated with AFM on a microarray of wells [45] (Fig. 15.13). Experimental variation is minimized because the cell shape is highly controlled in each well and the measurement position of cell is well defined. Force measurements are automatically performed at the centers of each well without confirming the cell positions. Figure 15.14 shows the distributions of single cells cultured in the wells [45]. We observed four characteristic features from the number distributions of mouse fibroblast cells [47]. First, G^* consistently exhibited a lognormal distribution. Second, the geometric mean of G^* ($\overline{G}'$ and $\overline{G}''$) shifted to higher values with increasing f. Third, the distribution of G' became narrower with f, and the distributions of G'' were narrower than those of G'. Fourth, the distribution of G^* for the cytoD-treated cells was narrower than that of the untreated cells.

$\overline{G}'$ and $\overline{G}''$ increased with f and closely followed the structural damping in Eq. 15.5 (Fig. 15.15a, b). The depolymerization of actin filaments resulted in a decrease in $\overline{G}_0$ and an increase in the arithmetic mean of α, $\langle\alpha\rangle$, which were similar to those characteristics measured with MTC [36, 37, 61, 62]. The standard deviation of the complex modulus $\sigma_{\ln G^*}$ was reduced in the treated cells (Fig. 15.15c, d), indicating a strong coupling between cell-to-cell variation and the cytoskeleton, where σ_X represents the standard deviation of X. The $\sigma_{\ln G^*}$ in the untreated and treated cells crossed at the point where the extrapolated lines of $\overline{G}'$ for the treated and untreated cells intersect; this was defined as $\overline{G}' = \overline{g}_0$ at $f = \overline{\Phi}_0$ (Fig. 15.15) [36, 37].

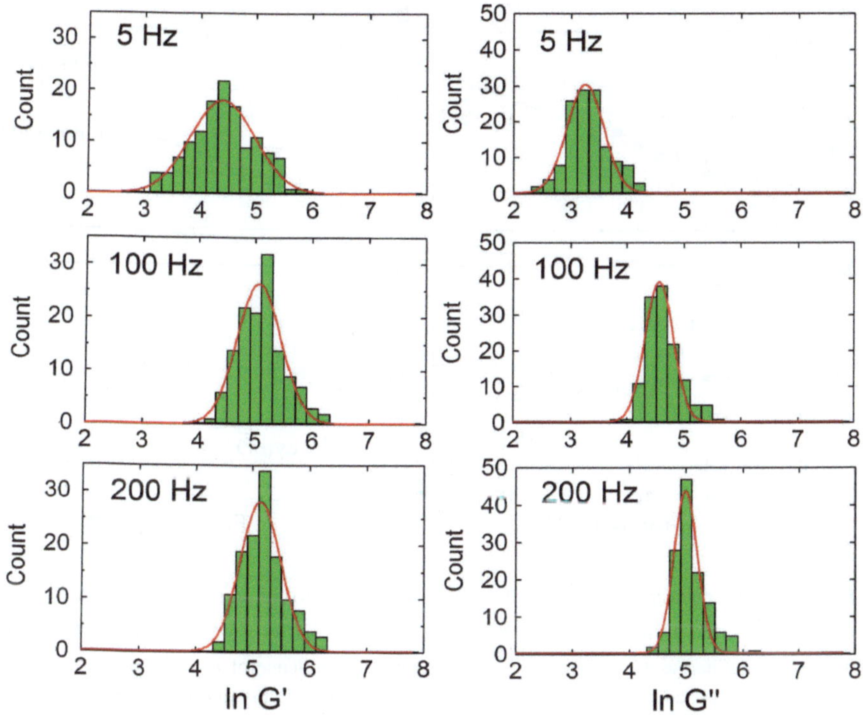

Fig. 15.14 Distributions of the storage G' (*left*) and loss G'' (*right*) moduli of untreated fibroblast cells in microarray wells at different frequencies: (*a*) 5, (*b*) 100, and (*c*) 200 Hz. The *solid line* represents the fitted result using a lognormal distribution function

Regarding the parameters of the single-power-law rheology in Eq. 15.5, G_0 was lognormal with a narrower distribution after cytoD treatment. The power-law exponent α exhibited a Gaussian distribution that also became narrower after cytoD treatment, whereas μ had a lognormal distribution, and its mean value did not change significantly after treatment (Fig. 15.16).

15.2.4.2 Standard Deviation of Cell Storage Modulus

In the single-power-law rheology model, G' for each cell is expressed as

$$G' = g_0\left(\frac{f}{\Phi_0}\right)^{\alpha}, \tag{15.6}$$

where Φ_0 can be estimated by extrapolating G' *vs.* f curves acquired under various conditions. Data for single cells specified by (g_0, Φ_0) varies considerably [47], indicating that cells exhibit mechanical variability that corresponds to the variation in potential energy that a cytoskeletal element must overcome to escape the glass transition, according to the SGR model [50, 53, 54].

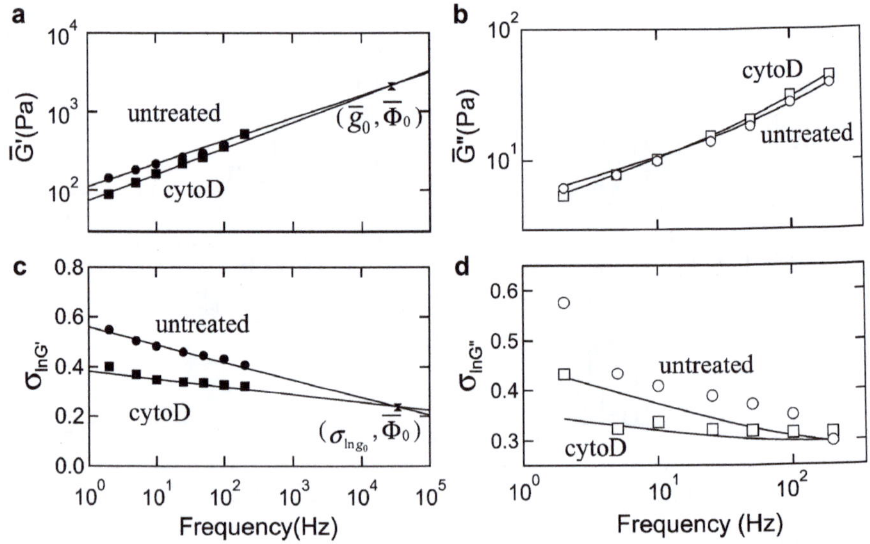

Fig. 15.15 Frequency dependences of $\overline{G}^*$ ($\overline{G}'$ (**a**) and $\overline{G}''$ (**b**)) of untreated (*circle*) and treated (*square*) cells. Solid lines in (**a**) and (**b**) are fits to Eq. 15.3. The point where the curves of $\overline{G}'$ intersect is defined as $\overline{G}' = \overline{g}_0$ at $f = \overline{\Phi}_0$. Frequency dependence of $\sigma_{\ln G'}$ (**c**) and $\sigma_{\ln G'}$ (**d**) of untreated (*circle*) and treated (*square*) cells. *Solid lines* in (**c**) are fits to Eq. 15.8 (Reprinted with permission from [47])

Since $\ln g(\alpha)$ is approximately linear with respect to α [47, 63], the linear relation between $\ln G_0$ and α for each cell in Eq. 15.6 is given by

$$\ln G_0 = \ln \overline{g}_0 - \left[\ln\left(\frac{\overline{\Phi}_0}{f_0}\right) + \frac{d\ln g(\alpha)}{d\alpha}\right]\alpha, \tag{15.7}$$

where $\overline{g}_0$ and $\overline{\Phi}_0$ are the geometric mean of g_0 and Φ_0, respectively.

From Eq. 15.7, we obtain

$$\sigma_{\ln G'} = \sigma_{\ln g_0} + \left(\ln \overline{\Phi}_0 - \ln f\right)\sigma_\alpha, \tag{15.8}$$

which shows that $\sigma_{\ln G'}$ is proportional to $\ln f$ with a slope of $-\sigma_\alpha$ at $f < \overline{\Phi}_0$, which is the frequency-dependent component in terms of the SGR model. The variation, from all sources, in the mechanical responses is characterized by $\sigma_{\ln g_0}$ at $f = \overline{\Phi}_0$, which is the purely elastic component in the SGR model. It was found experimentally that $\widetilde{\sigma}_{\ln G'}$, which is defined as $\sigma_{\ln G'} - \sigma_{\ln g_0}$ and is a function of f, was highly invariant for different cell samples cultured in different dishes (Figs. 15.17 and 15.18).

In Fig. 15.18, $\widetilde{\sigma}_{\ln G'}$ *vs.* f shows that (1) $\widetilde{\sigma}_{\ln G'}$ for cells treated with cytoD was largely reduced relative to that for control cells, and (2) $\widetilde{\sigma}_{\ln G'}$ away from the center of the wells, but still within the nuclear boundary, was smaller than the

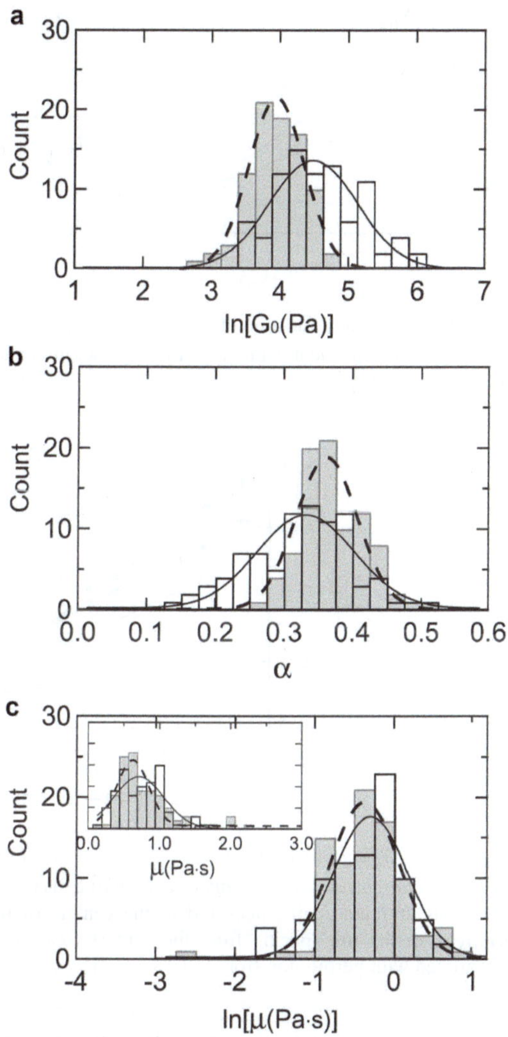

Fig. 15.16 Distributions of (**a**) G_0 on a logarithmic scale, (**b**) α on a linear scale, and (**c**) μ on a logarithmic scale of untreated (*white*) and treated (*gray*) cells. Inset in (**c**) shows the distribution of μ on a linear scale. *Solid* and *dashed lines* represent the fitted results of untreated and treated cells, respectively, using a lognormal distribution function (**a** and **c**) and to a normal distribution function (**b** and inset in **c**) (Reprinted with permission from [47])

corresponding value at the center. The results reveal that the frequency dependence of $\widetilde{\sigma}_{\ln G'}$ varies with the integrity of the actin network and that the cell-to-cell mechanical variation exhibits a spatial dependence. The cell-to-cell variation of G' in the frequency domain is depicted schematically in Fig. 15.19.

In SGR, the power-law exponent of G' is related to the probability of transitions between the potential wells, where the transition rate decreases with a decreasing value of the exponent [50, 53, 54]. SGR elements and energy wells can be identified with myosin motors and the binding energies between myosin and actin, respectively [50], suggesting that the depolymerization of actin filaments by cytoD reduces actin–myosin interactions and enhances the spatial homogeneity of the interactions.

Cells interact mechanically with neighboring cells. The microarray wells were not completely separated, so that neighboring cells were in partial contact. There have

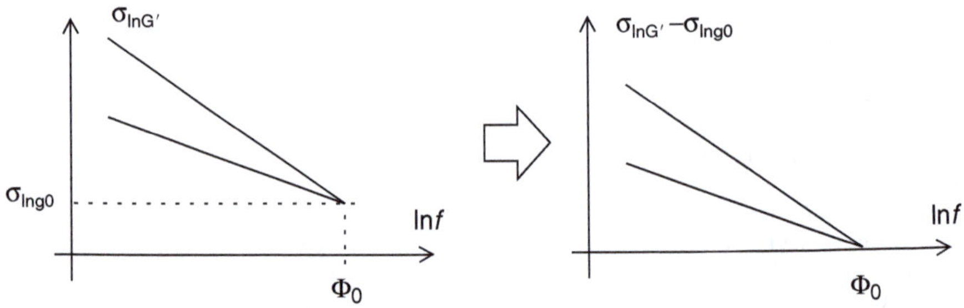

Fig. 15.17 $\sigma_{\ln G'}$ as a function of f as shown in Eq. 15.8. $\tilde{\sigma}_{\ln G'}$, which is defined as $\sigma_{\ln G'} - \sigma_{\ln g_0}$, is the frequency-dependent component that is experimentally invariant among different cell samples of the same cell type

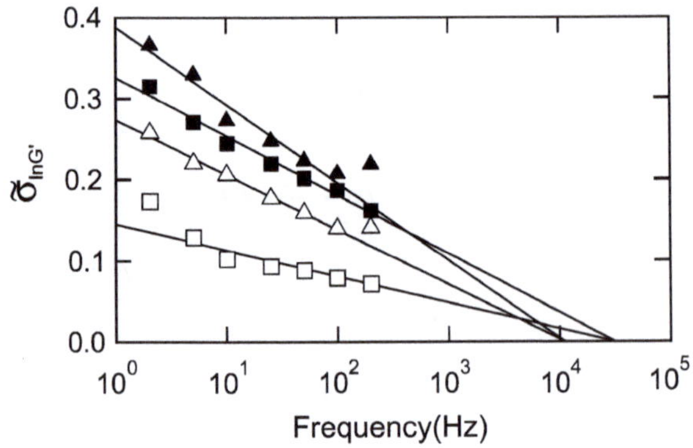

Fig. 15.18 $\tilde{\sigma}_{\ln G'}$, which represents $\sigma_{\ln G'} - \sigma_{\ln g_0}$ as a function of $\ln f$. One array has untreated (*closed rectangle*) and treated (*open rectangle*) cells measured at the center of wells, whereas the other has untreated cells measured at the centers of microarray wells having 20 μm (*closed triangle*) or 4.5 μm spacing from the centers (*open triangle*). *Solid lines* are fits to Eq. 15.8 (Reprinted with permission from [47])

been no reported data regarding the influence of cell-to-cell contact on mechanical measurements [52]. Recently, an AFM study in which NIH3T3 cell migration was highly inhibited on micro-patterned substrates revealed that the power law of G' is not significantly influenced by cell-to-cell contact [64]. Micro-patterned substrates have been widely used for investigating the relationship between cell mechanics and intracellular cytoskeletal structures. Thus, AFM of cells on micro-patterned substrates should be a good way to characterize single-cell mechanical variations.

15.2.4.3 Cancer Cell Detections

The mechanics of living cells are extremely important for understanding motility, division, and adhesion [1–4]. Furthermore, mechanical properties may also be used to distinguish between normal and abnormal cells. Deformability is widely used to

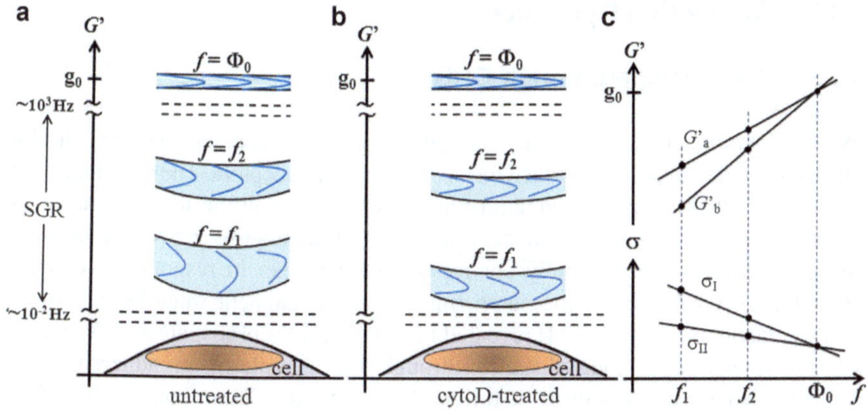

Fig. 15.19 Schematic of G' of untreated cells (**a**) and cytoD-treated cells (**b**) at different frequencies. The cell-to-cell variation of G' depends on intracellular locations: the distribution narrows when changing from cell center to cell nucleus boundaries. The spatial component of the cell-to-cell variation of G' between untreated and treated cells decreases with increasing f, and consequently both cells become spatially homogeneous at $f = \overline{\Phi}_0$ beyond the SGR region (see Eq. 15.8), but the cell-to-cell variation still exists at $f = \overline{\Phi}_0$. The spatial variation of G' for the untreated cells in the SGR region is larger than that for treated cells. One experimental condition is that G'_a (σ_{II}) and G'_b (σ_I) represent values measured at off-center and center locations, respectively, while the other G'_a (σ_I) and G'_b (σ_{II}) are those of the untreated and treated cells, respectively (**c**) (Reprinter with permission from [47])

identify cancer cells [43, 44]. Optical stretchers [43] and deformability cytometry [44] both employ micro-fluidic chambers, in which suspended cells are deformed by either optical pressure (optical stretching) or hydrodynamic forces (deformability cytometer), and the deformations are monitored with a high-speed camera system [43, 44].

To diagnose unmodified single cells attached on substrate, Lekka et al. [65] used AFM to show that Young's modulus E of normal human epithelial cells was one order of magnitude higher than cancer cells. Cross et al. [66] also reported that an ex vivo AFM mechanical analyses of patient cancer cells correlated well with conventional immune-histochemical testing. As mentioned before, the G_0 value, which is related to E, is dependent on the measurement position. Therefore, AFM cancer cell detection is expected to be more precise on micro-fabricated substrates.

AFM has been used to characterize normal and cancerous tissues to understand how the transformation from health to malignancy alters the mechanical properties within the tumor microenvironment [67]. The spatial distributions of E on normal and benign tissues had a single distinct peak, indicating uniform stiffness. In contrast, malignant breast tissues had a broad distribution because of tissue heterogeneity, with a prominent low-stiffness peak representative of cancer cells. The results suggest that AFM provides quantitative indicators at the tissue level for clinical diagnostics of breast cancer with translational significance.

15.3 Single-Cell Dynamics

15.3.1 Force Propagation in Cells

When adherent cells contact a substrate, they form focal contacts that are adhesion sites at the cell–substrate interface formed by integrin receptors. The cells then form stress fibers that are coalesced actin filaments to anchor at the focal contacts and increase stiffness in response to stress applied to the integrins. CSK filaments and nuclear scaffolds are discretely connected to each other in response to external static forces [68, 69]. Force measurements using both magnetic microbeads [70–73] and an elastic micro-pillar [74] revealed that static forces propagate across discrete CSK elements over long distance through the cytoplasm in adherent cells, indicating that pre-stress in the actin bundles is the key determinant of how far a force can propagate. This is known as "action at a distance" behavior [70–73].

AFM has been used to investigate how mechanical perturbations propagate in cells. For example, Rosenbluth and Crow et al. measured the magnitude and timing of intracellular stress propagation, using AFM and fluorescent particle tracking, and showed that AFM deformation of the cell surface exhibited distance dependence that could be eliminated by disruption of the actin cytoskeleton [75]. Silberberg et al. reported that mitochondria displacements, which are markers of microtubule displacements and deformation, were much less sensitive to AFM loading forces at apical surfaces, suggesting that filamentous structures other than actin filaments propagated less mechanical force from apical to basal cell surfaces [76].

A new AFM technique combined with a micro-post substrate [77] was used to characterize the mechanical response of CSK filaments at focal adhesions. Apical and basal cell surfaces were pre-coated with an adhesive protein (fibronectin) and then bound to both a colloidal bead on an AFM cantilever and to polydimethylsiloxane (PDMS) micro-posts [77, 78] (Fig. 15.20). The cantilever was oscillated

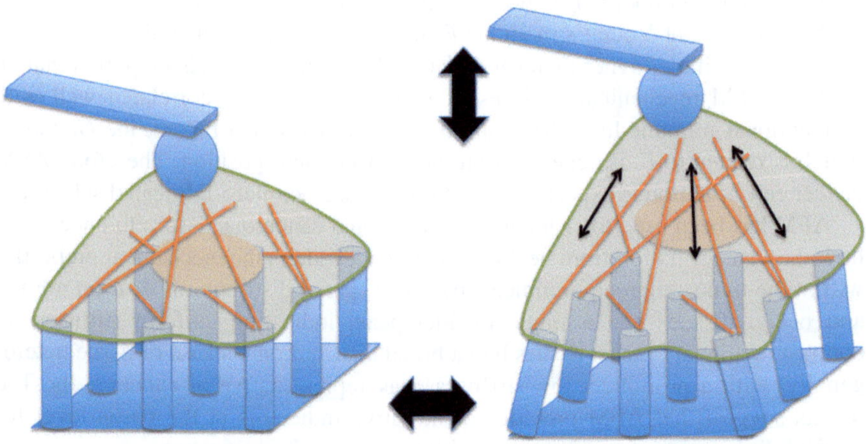

Fig. 15.20 Schematic of AFM of cells atop a micro-post substrate, for measuring force propagation from apical to basal surfaces through the cytoskeleton

normal to the substrate surface with amplitude for a period T at frequency f, while a time series of images of cells on the micro-posts were acquired.

Cells exhibited a power-law rheology with $\alpha \sim 0.2$ at the apical surface, but had no apparent out-of-phase response at the basal surface, thus indicating that the cytoskeletal filaments behave in an elastic manner. As shown in Fig. 15.21, a periodic change in force was observed at these frequencies, but no apparent phase shift in the force magnitude was observed at these frequencies (Fig. 15.21a). The response to the force at the basal cell surface appeared in phase, even at the higher frequency (0.5 Hz) at which the rheological properties of cells were clear for apical surfaces (Fig. 15.21b). At the lower frequency, the response to the force was no longer periodic but more complex. Moreover, force profiles that exhibited periodic responses (Fig. 15.21) were asymmetric, and some (depicted by arrows in Fig. 15.21) showed plateau responses to low forces. These results indicate that the lateral force at the basal cell surface readily propagated when a strong force was applied normal to the apical surface.

The direction of the propagated force was correlated with pre-stress, indicating that the lateral force applied to the micro-posts at the basal surface is directly

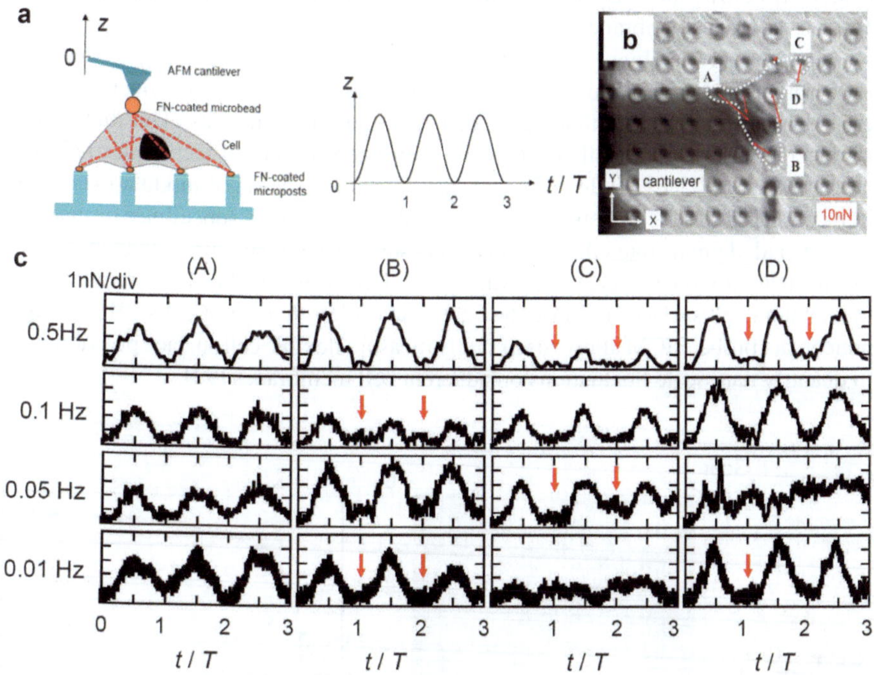

Fig. 15.21 (**a**) AFM data from cells cultured on PDMS micro-posts that were pre-coated with fibronectin. A fibronectin-coated colloidal bead attached at the apex of an AFM tip was bound to the apical cell surface with an initial loading force less than 500 pN for 30 min. An oscillatory pulling force was applied in the frequency range of 0.01–0.5 Hz for three periods ($0 < t < 3T$), while the deflection of the micro-post was measured by phase-contrast or bright-field microscopy. (**b**) Optical microscopic image of micro-posts. An objective lens with a long working distance was focused on the tops of the micro-posts. (**c**) *Red arrows* indicate cell pre-stress, estimated from the deflection of the micro-posts (letters A–D). Time series of lateral force magnitude applied to micro-posts during external modulation at different frequencies by AFM. Letters correspond to those in (**b**). Reprinted with permission from [79]

associated with forces propagated through CSK [79]. The heterogeneities of long-distance force propagation are most likely associated with the deformation of the nucleus [68, 69], remodeling of actin filaments in local regions [80], and entanglement of CSK filaments [81].

15.3.2 Cell Membrane Fluctuations

Scanning ion conductance microscopy (SICM) [82] is used to detect a cell surface at a local position in the noncontact region through an ion current I that flows through the small bore of a pipette (typically 100 nm in diameter) [83–86] (Fig. 15.22a, b). The basic principle of SICM is that I monotonically decreases as the tip approaches the sample surface (Fig. 15.22c); thus, I is used to regulate the tip–surface distance D without tip–surface contact. High-resolution images of microvilli formation and assembly [84] on apical epithelial cell surfaces and fragile neuron cells [85] have been acquired without substantial cell deformation. As described below, the I–D curve provides information about the dynamics of cell surfaces such as membrane fluctuations.

Because cell membranes are flexible and undergo morphological changes with different biological functions, the characterization of cell surface fluctuations is crucial for a better understanding of cell function and dynamics. The thicknesses of adherent mammalian cells have been determined by optical techniques [87–89]. However, the refractive index of a cell may fluctuate because of modulations in the intracellular cytoskeletal network and/or the displacement of subcellular organelles. Pelling et al. demonstrated that AFM can measure local temperature-dependent nanomechanical motion in yeast cell walls [90]. However, an AFM tip may perturb mammalian cell surfaces because they are much softer than yeast cells [91]. Being a noncontact probe, SICM allows us to safely measure flexible cell surface positions and to quantify nanoscale fluctuations on adherent cell membranes [92].

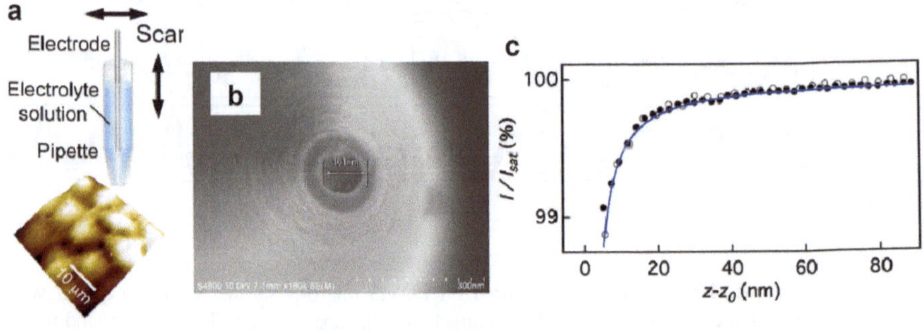

Fig. 15.22 (**a**) Schematic of SICM instrument. (**b**) Scanning electron micrograph of a nano-pipette tip used in this study. The inner and outer diameters were ≈ 100 nm and ≈ 200 nm, respectively. (c) I–D curves measured on a silicon substrate (*filled circle*) and a flat PDMS substrate (*open circle*). The *solid lines* represent fits to Eq. 15.9 (Reprinted with permission from [92])

Figure 15.23 illustrates the method for estimating cell surface fluctuations from *I–D* curves. The cell surface position $z_s(x, t)$ at time t at a normalized lateral position x is $z_s(x, t) = z_0(x) + \delta z_s(x, t)$, where $z_0(x)$ is the average z-position of the apical cell surface and $\delta z_s(x, t)$ is the fluctuation around $z_0(x)$, i.e., $\langle \delta z_s(x) \rangle = 0$, where $<X>$ is the X ensemble or time average. It is assumed that the cell surfaces fluctuate with a Gaussian stochastic distribution P about the root-mean-square (RMS) displacement of surface fluctuations $\langle \delta z_s^2 \rangle^{1/2}$, as given by

$$P\left(z_s - z_0, \langle \delta z_s^2 \rangle\right) = \frac{1}{\sqrt{2\pi \langle \delta z_s^2 \rangle}} \exp\left\{-(z_s - z_0)^2 / \left(2\langle \delta z_s^2 \rangle\right)\right\}. \tag{15.9}$$

The *I–D* curve for a solid substrate is modeled as [93]

$$I_0(z - z_0) = I_{sat}\left[1 + \frac{\zeta}{z - z_0}\right]^{-1}, \tag{15.10}$$

where z is the tip position, z_0 is sample surface position with no fluctuations, I_{sat} is the ion current when the pipette is far from the surface, and ζ is a function of the inner radius of the tip opening, the inner radius of the tip base, the tip length, and the conductivity of the electrolyte in solution in the pipette [93]. ζ is experimentally determined for I_{sat}. Thus, when cell fluctuations obey Eq. 15.9, the average $\langle I \rangle$, measured at z on cells at z_0 with $\langle \delta z_s^2 \rangle^{1/2}$, is given by

$$\left\langle I\left(z - z_0 \langle \delta z_s^2 \rangle\right)\right\rangle = \int_{-\infty}^{\infty} I_0(z - z_s) P\left(z_s - z_0, \langle \delta z_s^2 \rangle\right) dz_s, \tag{15.11}$$

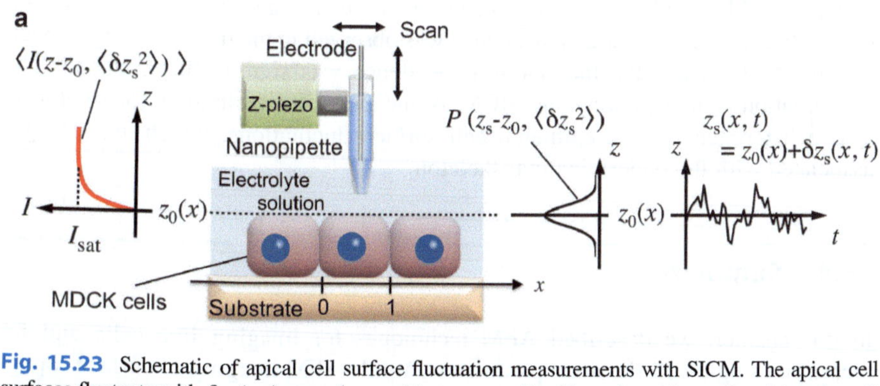

Fig. 15.23 Schematic of apical cell surface fluctuation measurements with SICM. The apical cell surfaces fluctuate with δz_s (x, t) around $z = z_0(x)$ at a normalized lateral position x (0 and 1 at the cell edge and 1/2 at the cell center) at time t. The apical cell surface position is statistically expressed as a Gaussian stochastic distribution P of the RMS displacement of surface fluctuations, which is the apparent amplitude of the cell surface fluctuation. In this case, $<I>$ follows Eq. 15.11 (Reprinted with permission from [92])

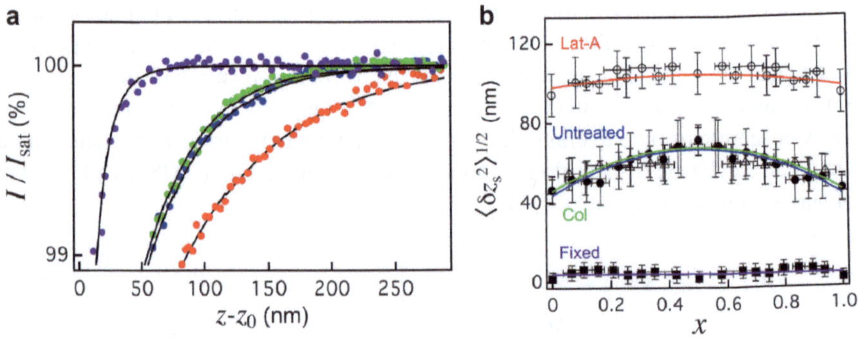

Fig. 15.24 (**a**) *I–D* curves of fixed (*purple*), untreated cells (*blue*), latA-treated cells (*red*), and col-treated cells (*green*). The *solid lines* are fits to Eq. 15.11. (**b**) RMS displacements of surface fluctuations (defined in Eq. 15.9) of fixed cells (*purple line*), untreated cells (*blue line*), col-treated cells (*green line*), and latA-treated cells (*red line*) measured in lines crossing over the center of the cells (the data were fitted as parabolas) (Reprinted with permission from [92])

where z_0 and $\langle \delta z_s^2 \rangle$ are fitting parameters. The measured dynamic range of the cell surface fluctuations is limited by the bandwidth of the SICM preamplifier.

Figure 15.24a shows *I–D* curves for epithelial Madin–Darby canine kidney (MDCK) cell sheets that were almost fully confluent where translational cell migration were highly restricted. For the fixed cells, the RMS displacement of surface fluctuations was 6.5 nm, which is consistent with that determined by optical techniques for fixed red blood cells [94]. The displacements significantly increased to 60 nm in untreated and microtubule-depolymerized, colchicine (col)-treated cells and to 105 nm in actin-depolymerized, latrunculinA (latA)-treated cells.

Figure 15.24b plots the RMS displacement of surface fluctuations estimated at different positions on the cell surface. In the untreated cells, the fluctuations exhibited a clear spatial dependence that increased toward the cell center. Specifically, the RMS displacement was 46 nm at the cell edge and 71 nm at the cell center. In contrast, no spatial dependence was observed in the fixed cells. Moreover, in the latA-treated cells, the spatial dependence was significantly reduced. The investigation demonstrated that SICM is the powerful technique for exploring spatial heterogeneities in epithelial cell surface fluctuations, which are strongly associated with the underlying cytoskeleton.

15.4 Summary

In this chapter, we described AFM techniques for imaging live cells and for measuring single-cell rheology parameters. While AFM imaging of surface topography is well established, imaging of inner structures of cells has great potential. Regarding single-cell rheology, AFM of cells attached to micro-fabricated substrates allows us to characterize frequency- and time-dependent mechanical properties and to quantify cell-to-cell variability, i.e., single-cell diagnostics. We emphasize that AFM biological techniques are still rapidly developing and will be applied more broadly to explore rheological properties of cells, to characterize tissues and organs for bioengineering and for medical applications.

References

1. Binnig G, Quate CF (1986) Atomic force microscope. Phys Rev Lett 56(9):930–933
2. Jena BP, Horber JKH (2002) Atomic force microscopy in cell biology. In: Jena BP, Horber JKH (eds) Methods in cell biology, vol 68. Academic, San Diego
3. Morris VJ, Kirby AR, Gunning AP (2009) Atomic force microscopy for biologists, 2nd edn. Imperial College Press, London
4. Radmacher M (2002) Measuring the elastic properties of living cells by the atomic force microscope. In: Atomic force microscopy in cell biology. Academic Press, San Diego, pp 67–90
5. Rotsch C, Radmacher M (2000) Drug-induced changes of cytoskeletal structure and mechanics in fibroblasts: an atomic force microscopy study. Biophys J 78:520
6. Ando T et al (2001) A high-speed atomic force microscope for studying biological macromolecules. Proc Natl Acad Sci 98(22):12468–12472
7. Ando T, Uchihashi T, Kodera N (2013) High-speed AFM and applications to biomolecular systems. Annu Rev Biophys 42:393–414
8. Watanabe H et al (2013) Wide-area scanner for high-speed atomic force microscopy. Rev Sci Instrum 84(5):053702
9. Garcia R, Herruzo ET (2012) The emergence of multifrequency force microscopy. Nat Nanotechnol 7(4):217–226
10. Raman A et al (2011) Mapping nanomechanical properties of live cells using multi-harmonic atomic force microscopy. Nat Nanotechnol 6(12):809–814
11. Tetard L et al (2008) Imaging nanoparticles in cells by nanomechanical holography. Nat Nanotechnol 3(8):501–505
12. Shekhawat GS, Dravid VP (2005) Nanoscale imaging of buried structures via scanning near-field ultrasound holography. Science 310(5745):89–92
13. Kimura K et al (2013) Imaging of Au nanoparticles deeply buried in polymer matrix by various atomic force microscopy techniques. Ultramicroscopy 133:41–49
14. Ducker WA, Senden TJ, Pashley RM (1991) Direct measurement of colloidal forces using an atomic force microscope. Nature 353:239–241
15. Okajima T (2012) Atomic force microscopy for the examination of single cell rheology. Curr Pharm Biotechnol 13:2623–2631
16. Okajima T et al (2007) Stress relaxation measurement of fibroblast cells with atomic force microscopy. Jpn J Appl Phys 46:5552–5555
17. Alcaraz J et al (2002) Correction of microrheological measurements of soft samples with atomic force microscopy for the hydrodynamic drag on the cantilever. Langmuir 18:716–721
18. Johnson KL (1987) Contact mechanics. Cambridge University Press, Cambridge
19. Landau LD, Lifshiz EM (1986) Theory of elasticity, vol 3, 3rd edn. Pergamon Press, Oxford
20. Hiratsuka S et al (2009) Power-law stress and creep relaxations of single cells measured by colloidal probe atomic force microscopy. Jpn J Appl Phys 48(8):08JB17
21. Hutter JL, Bechhoefer J (1993) Calibration of atomic-force microscope tips. Rev Sci Instrum 64:1868–1873
22. A-Hassan E et al (1998) Relative microelastic mapping of living cells by atomic force microscopy. Biophys J 74:1564–1578
23. Haga H et al (2000) Elasticity mapping of living fibroblasts by AFM and immunofluorescence observation of the cytoskeleton. Ultramicroscopy 82:253–258

24. Okajima T et al (2007) Stress relaxation of HepG2 cells measured by atomic force microscopy. Nanotechnology 18:084010
25. Radmacher M et al (1992) From molecules to cells – imaging soft samples with the atomic force microscope. Science 257(5078):1900–1905
26. Radmacher M, Tilmann RW, Gaub HE (1993) Imaging viscoelasticity by force modulation with the atomic force microscope. Biophys J 64(3):735–742
27. Radmacher M et al (1996) Measuring the viscoelastic properties of human platelets with the atomic force microscope. Biophys J 70:556
28. Alcaraz J et al (2003) Microrheology of human lung epithelial cells measured by atomic force microscopy. Biophys J 84:2071–2079
29. Mahaffy RE et al (2000) Scanning probe-based frequency-dependent microrheology of polymer gels and biological cells. Phys Rev Lett 85(4):880–883
30. Mahaffy RE et al (2004) Quantitative analysis of the viscoelastic properties of thin regions of fibroblasts using atomic force microscopy. Biophys J 86:1777–1793
31. Charras GT, Horton MA (2002) Single cell mechanotransduction and its modulation analyzed by atomic force microscope indentation. Biophys J 82:2970–2981
32. Darling EM, Zauscher S, Guilak F (2006) Viscoelastic properties of zonal articular chondrocytes measured by atomic force microscopy. Osteoarthr Cartil 14(6):571–579
33. Darling EM et al (2007) A thin-layer model for viscoelastic, stress-relaxation testing of cells using atomic force microscopy: do cell properties reflect metastatic potential? Biophys J 92 (5):1784–1791
34. Findley WN, Lai JS, Onaran K (1989) Creep and relaxation of nonlinear viscoelastic materials with an introduction to linear viscoelasticity. Dover Publications Inc., New York
35. Wu HW, Kuhn T, Moy VT (1998) Mechanical properties of L929 cells measured by atomic force microscopy: effects of anticytoskeletal drugs and membrane crosslinking. Scanning 20:389–397
36. Fabry B et al (2001) Scaling the microrheology of living cells. Phys Rev Lett 87(14):148102
37. Fabry B et al (2003) Time scale and other invariants of integrative mechanical behavior in living cells. Phys Rev E 68:041914
38. Deng LH et al (2006) Fast and slow dynamics of the cytoskeleton. Nat Mater 5(8):636–640
39. Van Citters KM et al (2006) The role of F-actin and myosin in epithelial cell rheology. Biophys J 91(10):3946–3956
40. Hoffman BD et al (2006) The consensus mechanics of cultured mammalian cells. Proc Natl Acad Sci U S A 103(27):10259–10264
41. Massiera G et al (2007) Mechanics of single cells: rheology, time dependence, and fluctuations. Biophys J 93(10):3703–3713
42. Balland M et al (2006) Power laws in microrheology experiments on living cells: comparative analysis and modeling. Phys Rev E 74(2):021911
43. Guck J et al (2005) Optical deformability as an inherent cell marker for testing malignant transformation and metastatic competence. Biophys J 88(5):3689–3698
44. Gossett DR et al (2012) Hydrodynamic stretching of single cells for large population mechanical phenotyping. Proc Natl Acad Sci U S A 109(20):7630–7635
45. Hiratsuka S et al (2009) The number distribution of complex shear modulus of single cells measured by atomic force microscopy. Ultramicroscopy 109:937–941
46. Mizutani Y et al (2008) Elasticity of living cells on a microarray during the early stages of adhesion measured by atomic force microscopy. Jpn J Appl Phys 47:6177–6180
47. Cai P et al (2013) Quantifying cell-to-cell variation in power-law rheology. Biophys J 105 (5):1093–1102
48. Miyaoka A et al (2011) Rheological properties of growth-arrested fibroblast cells under serum starvation measured by atomic force microscopy. Jpn J Appl Phys 50(8):08LB16
49. Reed J et al (2008) High throughput cell nanomechanics with mechanical imaging interferometry. Nanotechnology 19(23):235101

50. Kollmannsberger P, Fabry B (2011) Linear and nonlinear rheology of living cells. Annu Rev Mater Res 41:75–97
51. Trepat X et al (2007) Universal physical responses to stretch in the living cell. Nature 447 (7144):592–596
52. Chowdhury F et al (2008) Is cell rheology governed by nonequilibrium-to-equilibrium transition of noncovalent bonds? Biophys J 95(12):5719–5727
53. Kollmannsberger P, Fabry B (2009) Active soft glassy rheology of adherent cells. Soft Matter 5(9):1771–1774
54. Trepat X, Lenormand G, Fredberg JJ (2008) Universality in cell mechanics. Soft Matter 4(9):1750–1759
55. Overby DR et al (2005) Novel dynamic rheological behavior of individual focal adhesions measured within single cells using electromagnetic pulling cytometry. Acta Biomater 1(3):295–303
56. Stamenovic D et al (2007) Rheological behavior of living cells is timescale-dependent. Biophys J 93(8):L39–L41
57. Desprat N et al (2005) Creep function of a single living cell. Biophys J 88(3):2224–2233
58. Smith BA et al (2005) Probing the viscoelastic behavior of cultured airway smooth muscle cells with atomic force microscopy: stiffening induced by contractile agonist. Biophys J 88 (4):2994–3007
59. Roca-Cusachs P et al (2006) Rheology of passive and adhesion-activated neutrophils probed by atomic force microscopy. Biophys J 91(9):3508–3518
60. Moreno-Flores S et al (2010) Stress relaxation microscopy: imaging local stress in cells. J Biomech 43(2):349–354
61. Puig-de-Morales M et al (2004) Cytoskeletal mechanics in adherent human airway smooth muscle cells: probe specificity and scaling of protein-protein dynamics. Am J Physiol Cell Physiol 287(3):C643–C654
62. Laudadio RE et al (2005) Rat airway smooth muscle cell during actin modulation: rheology and glassy dynamics. Am J Physiol Cell Physiol 289(6):C1388–C1395
63. Maloney JM, Van Vliet KJ (2011) On the origin and extent of mechanical variation among cells. arXiv:1104.0702v2
64. Takahashi R et al (2014) Atomic force microscopy measurements of mechanical properties of single cells patterned by microcontact printing. Adv Robot 28:449–455
65. Lekka M, Laidler P, Gil D, Lekki J, Stachura Z, Hrynkiewicz AZ (1999) Elasticity of normal and cancerous human bladder cells studied by scanning force microscopy. Eur Biophys J 28:312–316
66. Cross SE et al (2007) Nanomechanical analysis of cells from cancer patients. Nat Nanotechnol 2(12):780–783
67. Plodinec M et al (2012) The nanomechanical signature of breast cancer. Nat Nanotechnol 7(11):757–765
68. Wang N, Tytell JD, Ingber DE (2009) Mechanotransduction at a distance: mechanically coupling the extracellular matrix with the nucleus. Nat Rev Mol Cell Biol 10:75–82
69. Cai YF, Sheetz MP (2009) Force propagation across cells: mechanical coherence of dynamic cytoskeletons. Curr Opin Cell Biol 21(1):47–50
70. Hu SH et al (2003) Intracellular stress tomography reveals stress focusing and structural anisotropy in cytoskeleton of living cells. Am J Physiol Cell Physiol 285(5):C1082–C1090
71. Hu SH et al (2004) Mechanical anisotropy of adherent cells probed by a three-dimensional magnetic twisting device. Am J Physiol Cell Physiol 287(5):C1184–C1191
72. Hu SH et al (2005) Prestress mediates force propagation into the nucleus. Biochem Biophys Res Commun 329(2):423–428
73. Wang N, Hu SH, Butler JP (2007) Imaging stress propagation in the cytoplasm of a living cell. Methods Cell Biol. 83:179–198
74. Paul R et al (2008) Propagation of mechanical stress through the actin cytoskeleton toward focal adhesions: Model and experiment. Biophys J 94(4):1470–1482

75. Rosenbluth MJ et al (2008) Slow stress propagation in adherent cells. Biophys J 95(12):6052–6059
76. Silberberg YR et al (2008) Mitochondrial displacements in response to nanomechanical forces. J Mol Recognit 21(1):30–36
77. Tan JL et al (2003) Cells lying on a bed of microneedles: an approach to isolate mechanical force. Proc Natl Acad Sci U S A 100(4):1484–1489
78. Gray DS, Tien J, Chen CS (2003) Repositioning of cells by mechanotaxis on surfaces with micropatterned Young's modulus. J Biomed Mater Res A 66A(3):605–614
79. Okada A et al (2011) Direct observation of dynamic force propagation between focal adhesions of cells on microposts by atomic force microscopy. Appl Phys Lett 99(26):263703
80. Chaudhuri O et al (2009) Combined atomic force microscopy and side-view optical imaging for mechanical studies of cells. Nat Methods 6(5):383–388
81. Xu J, Tseng Y, Wirtz D (2000) Strain hardening of actin filament networks. Regulation by the dynamic cross-linking protein alpha-actinin. J Biol Chem 275(46):35886–35892
82. Hansma PK et al (1989) The scanning ion-conductance microscope. Science 243 (4891):641–643
83. Korchev YE et al (1997) Scanning ion conductance microscopy of living cells. Biophys J 73(2):653–658
84. Gorelik J et al (2003) Dynamic assembly of surface structures in living cells. Proc Natl Acad Sci U S A 100(10):5819–5822
85. Novak P et al (2009) Nanoscale live-cell imaging using hopping probe ion conductance microscopy. Nat Methods 6(4):279–281
86. Rheinlaender J et al (2011) Comparison of scanning Ion conductance microscopy with atomic force microscopy for cell imaging. Langmuir 27(2):697–704
87. Reed J et al (2008) Live cell interferometry reveals cellular dynamism during force propagation. ACS Nano 2(5):841
88. Reed J et al (2011) Rapid, massively parallel single-cell drug response measurements via live cell interferometry. Biophys J 101(5):1025
89. Yamauchi T, Iwai H, Yamashita Y (2011) Label-free imaging of intracellular motility by low-coherent quantitative phase microscopy. Opt Express 19:5536–5550
90. Pelling AE et al (2004) Local nanomechanical motion of the cell wall of Saccharomyces cerevisiae. Science 305(5687):1147–1150
91. Pelling AE et al (2007) Mapping correlated membrane pulsations and fluctuations in human cells. J Mol Recognit 20(6):467–475
92. Mizutani Y et al (2013) Nanoscale fluctuations on epithelial cell surfaces investigated by scanning ion conductance microscopy. Appl Phys Lett 102(17):173703
93. Nitz H, Kamp J, Fuchs H (1998) A combined scanning ion-conductance and shear-force microscope. Probe Micrsoc 1:187–200
94. Rappaz B et al (2009) Spatial analysis of erythrocyte membrane fluctuations by digital holographic microscopy. Blood Cell Mol Dis 42(3):228–232

Supplement 16

Rio Kita and Toshiaki Dobashi

16.1 Supplement I. Structure and Properties of Molecules in Nano-/Micro-Space and Related Time Scale

Biological systems are well-known open nonequilibrium systems. They consist of biomolecules, and they assemble in mostly aqueous media to form hierarchical structures. In the systems, various types of weak forces such as hydrogen bonds and van der Waals force play an important role in yielding their unique functions. Here, we summarize the structure and motion of the molecules, and their complexes are considered in this book.

16.1.1 Supplement I. Water and Essential Biomolecules

Water is a transparent and tasteless liquid composed of two hydrogens and one oxygen (H_2O) that freezes at 0 °C (32 °F) and boils at 100 °C (212 °F) at atmospheric pressure. The structure of water is not linear in its molecular coordination, and the oxygen atom has a higher electronegativity than the hydrogen atoms as illustrated in Fig. 16.1. Water is a polar molecule with an electrical dipole moment μ ($\mu = 6.11 \times 10^{-30}$ Cm); thus, water can form intermolecular interactions called hydrogen bonds. Water is a versatile solvent having a great impact on

R. Kita (✉)
Department of Physics, Tokai University, 4-1-1 Kitakaname,
Hiratsuka, Kanagawa 259-1292, Japan
e-mail: rkita@keyaki.cc.u-tokai.ac.jp

T. Dobashi
Division of Molecular Science, Faculty of Science and Technology,
Gunma University, 1-5-1 Tenjin, 376-8515 Kiryu, Gunma, Japan
e-mail: dobashi@gunma-u.ac.jp

R. Kita, T. Dobashi (eds.), *Nano/Micro Science and Technology in Biorheology*,
DOI 10.1007/978-4-431-54886-7_16

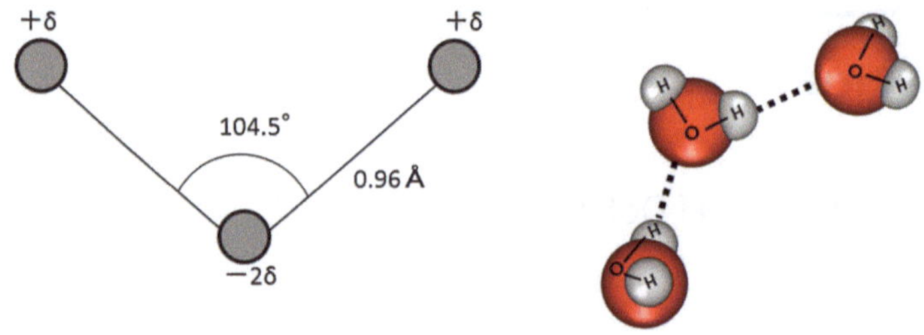

Fig. 16.1 Structural representations of water

biological molecules and other molecules. The viscosity of water is $\eta = 0.89 \times 10^{-3}$ [Pa·s] at 25 °C, and the relaxation time of water has been obtained to be $\tau = 8 \times 10^{-9}$ [s] at 25 °C by dielectric relaxation spectroscopy [1].

Carbohydrates are composed of oxygen, hydrogen, and carbon. There are various types of carbohydrates, the most important of which in biological systems is glucose ($C_6H_{12}O_6$) (Fig. 16.2). Even among the monosaccharides, there are several isomers such as fructose, galactose, mannose, and so forth. Oligosaccharides and polysaccharides are composed of monosaccharides connected by glycosidic bonds. Saccharides have versatile roles in living systems and are a huge research topic that incorporates energy storage and molecular recognition in biological processes. Nucleotides are composed of ribose or 2-deoxyribose, a nucleobase (purine or pyrimidine base) (Fig. 16.3), and phosphate groups. Adenosine diphosphate (ADP) and adenosine triphosphate (ATP) are the key molecules in energy transduction in biological systems. The energy evolved in the dissociation reaction of ATP into ADP is used in a wide range of biochemical reactions. There are 20 amino acids that are the subunits of proteins, with peptide linkages connecting the amino acids. It is of great importance to classify the amino acids into those with hydrophilic and hydrophobic groups (Fig. 16.4), more rigorously, into aliphatic amino acids (gly, ala, val, leu, ile), hydroxyl or sulfuric amino acids (ser, cys, thr, met), aromatic amino acids (phe, tyr, trp), cyclic amino acid (pro), basic amino acids (lys, his, arg), and acidic amino acids (asp, glu, asn, gln). The variation of the content of each amino acid and the order of the connections give different proteins; moreover, the solvent properties (pH and ionic strength) as well as the environment (temperature and pressure) affect the molecular structure and functions of proteins.

16.1.2 Supplement I. Polymer

Polymers are huge molecules consisting of a large number of monomers (subunits). The simplest polymer is linear saturated hydrocarbons containing C–C covalent bonds. The distance between carbon atoms and the angle between successive bonds are 0.154 nm and $\theta = 70.5°$, respectively, because of the tetrahedral structure of the

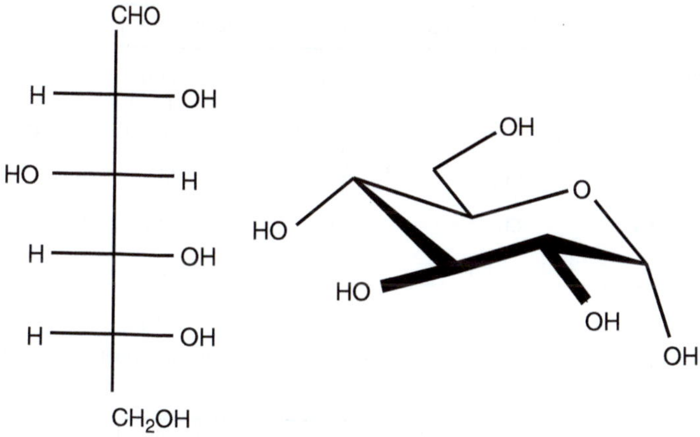

Fig. 16.2 D-Glucose in Fischer projection and in cyclic form (α-D-glucose)

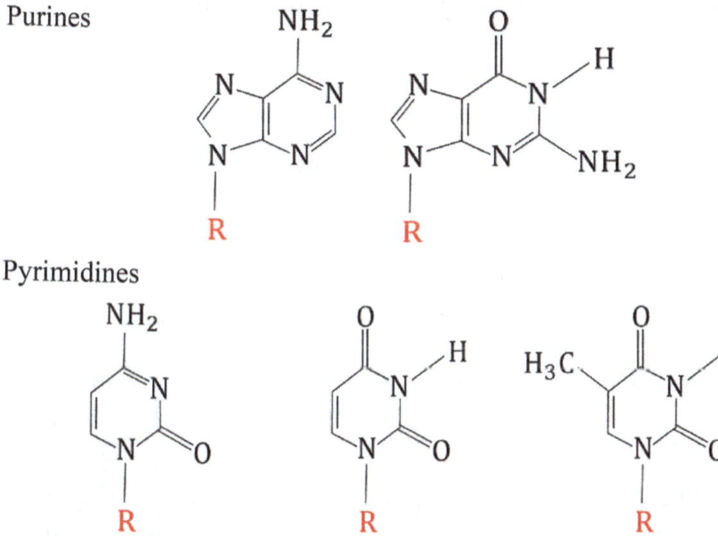

Fig. 16.3 Nucleobase (adenine [A], guanine [G], cytosine [C], uracil [U], and thymine [T]). R- refers to the position connecting to ribose or deoxyribose in RNA or DNA

carbon atoms (Fig. 16.5). The bonds have a freedom of rotational motion around the C–C linkages (internal rotation of the polymer chain), although the bond length and the angle are fixed. The angle of rotation ϕ is defined by the plane formed by the first three carbon atoms and is measured in the clockwise direction from the plane. When a carbon atom is located at the angle $\phi = 0°$, i.e., on the reference plane, it is called the trans position (t). For the case of polyethylene, the minimum potential

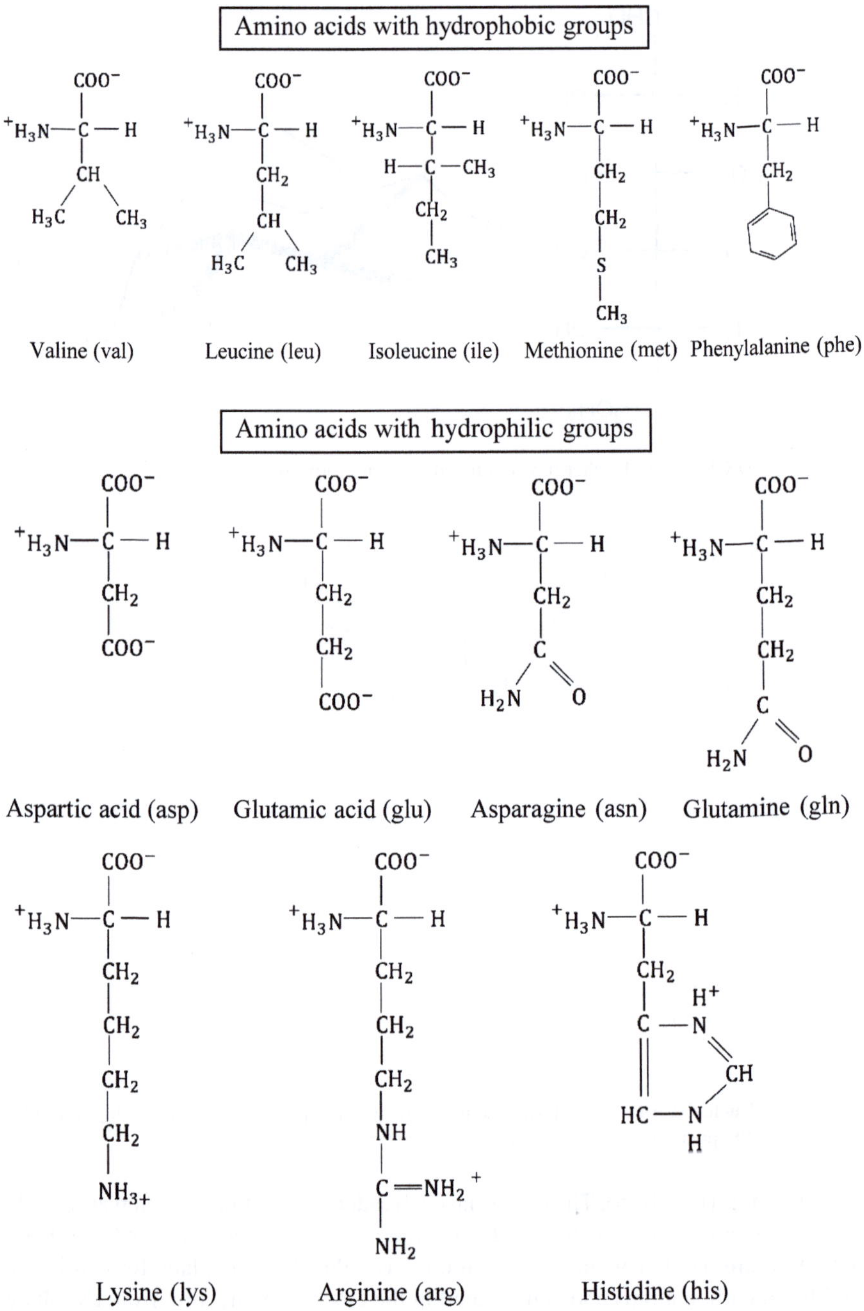

Fig. 16.4 Chemical structure of 20 amino acids

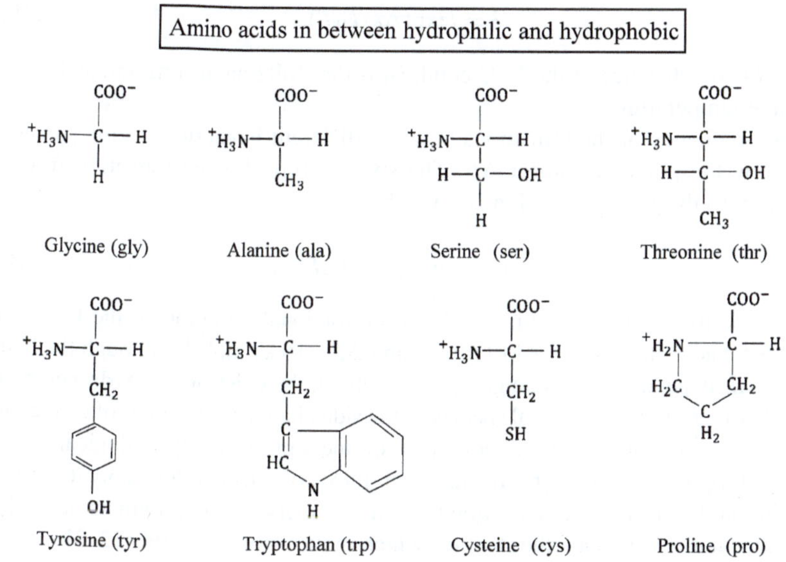

Fig. 16.4 (continued)

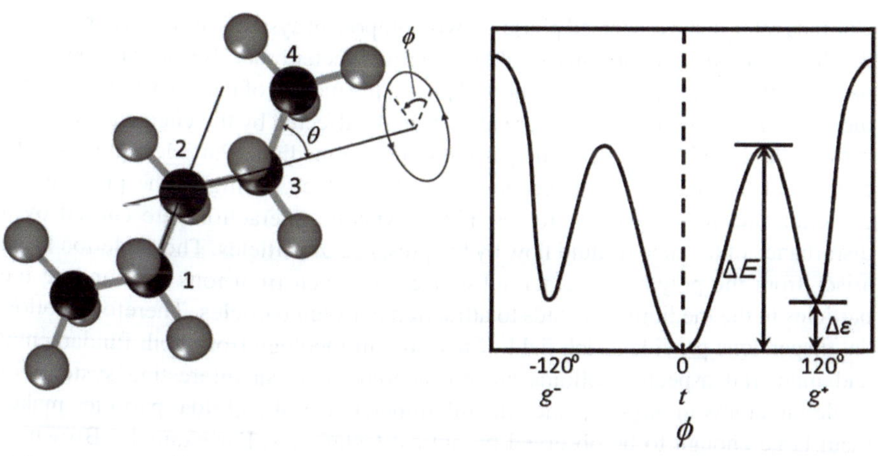

Fig. 16.5 Illustration of the internal rotation ϕ in a polymer chain and the potential energy as a function of rotation angle

energy is obtained at the angle $\phi = \pm 120°$. This is called the gauche position (g^+, g^-). There is an energy difference $\Delta\varepsilon$ between the t position and the g^+ (or g^-) position, which determines the average population of carbon atoms in thermal equilibrium. This is related to the flexibility of the polymer chain. The average length λ of a continuous trans sequence is expressed as

$$\lambda = l\exp(\Delta\varepsilon/k_{\mathrm{B}}T), \tag{16.1}$$

where l is the distance of the C–C bond, k_{B} is the Boltzmann constant, and T is the absolute temperature.

The frequency of the transitions between different isomeric structures is determined by the potential barrier ΔE. The specific time for the transition from the t position to the g^+ (g^-) position is given by

$$\tau = \tau_0\exp(\Delta E/k_{\mathrm{B}}T). \tag{16.2}$$

Here, τ_0 is the microscopic time scale of the torsional vibration of the C–C bond. Polymers take values of $\Delta\varepsilon \sim 4$ kJ mol^{-1} and $\Delta E \sim 16$ kJ mol^{-1}, although the barrier energy ΔE depends on the side group of the chain, the existence of hydrogen bonds, and other factors. The chemical species of the side chain and the order of connections of subunits vary enormously. Among the polymers produced by biological activity, several biopolymers have been classified into polysaccharides, polynucleotides, proteins, and so forth. The relaxation time of polymers and biopolymers depends on the molecular weight and other factors, which induce slow dynamics [2, 3] .

16.1.3 Supplement I. Colloid

Colloids refer to the dispersed phase of two-component systems in which particles in the dispersed phase are usually smaller than micrometer order. Because of the small particle (or droplet) size in the dispersed phase, the motion of particles is affected by thermal forces. Their properties are also strongly affected by the chemical environment at the interface between the particles and the medium. Particles interact with each other throughout the dispersion or at the surface, owing to the presence of depletion and hydrodynamic forces. Hydrodynamic interactions are caused by a disturbance of the field of fluid flow by the presence of particles. The depletion force arises from the polymer or other substances and even from ions surrounding the particles in the medium and leads to attraction between particles. Therefore, colloidal dispersions provide a rich field of research in rheology from both fundamental and industrial aspects. Colloids are considered to be an interesting system for modeling atoms in physics, since the micrometer size of colloidal particles makes them large enough to be observed by optical techniques. For example, Brownian motion and the related studies by Einstein and Smoluchowski were employed to prove the existence of atoms. Furthermore, colloids have been utilized as model systems to study diffusion, aggregation, sedimentation, and other phenomena.

16.1.4 Supplement I. Micelle, Lipid Bilayer, Emulsion, and Microemulsion

Molecules interact with each other in condensed matter to make various types of stable and quasi-stable assemblies or aggregated states in biological systems. When

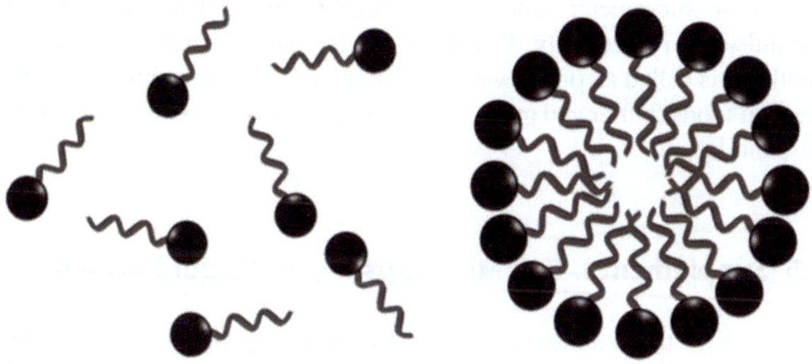

Fig. 16.6 Schematic illustration of surfactant and micelle structure

Fig. 16.7 Schematic illustration of lipid bilayer

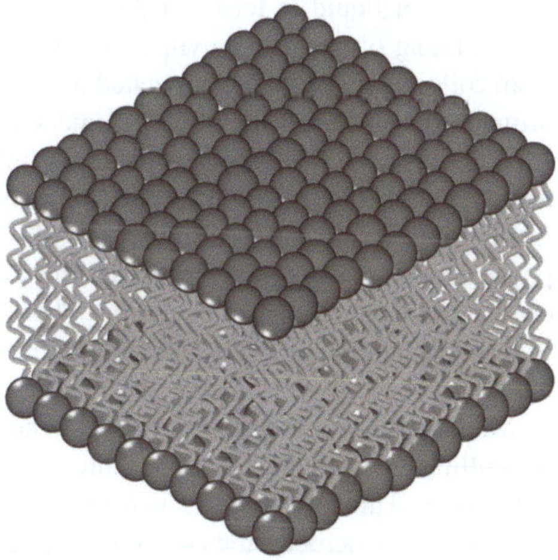

the process yielding these complexes is based on the principle of the second law of thermodynamics, i.e., the principle of the lowest free energy, the process is called self-organization or self-assembly. Micelles are formed by amphiphilic molecules such as surface active agents and block copolymers.

The self-assembly of amphiphilic molecules such as surfactants, lipids, and copolymers provides various types of ordered structures in water. Micellar structures develop spontaneously from surfactants in aqueous solutions above the critical micelle concentration (CMC) (Fig. 16.6). The biological membranes in cells are formed by phospholipids, and cellular membranes are known to have a lipid bilayer structure (Fig. 16.7). Mixtures of two immiscible liquids often form emulsions owing to the effect of surface active agents called emulsifiers (or surfactants). Normally, emulsions are not formed spontaneously because they are unstable energetically. Thus, some energy must be input to form emulsions, such as by stirring or homogenizing, so that the droplets of a liquid can be dispersed in the continuous phase of another liquid. When the size of the droplets is of nanometer or micrometer orders, the emulsions are called nanoemulsions and microemulsions, respectively. When oil droplets are dispersed in a water phase, the emulsion is called an oil-in-water emulsion and vice versa. Emulsions are often observed in foods (butter and mayonnaise), cosmetics, and pharmaceuticals (waxes, creams, and lotions).

16.1.5 Supplement I. Gel, Microcapsule, and Liquid Crystal

Gels are a solid three-dimensional network consisting of polymers and colloids, which include solvent molecules inside the network. The connections in the network are made of functional groups and associative groups, which are called cross-links. Usually, gels show solid-like mechanical properties, although they are mainly composed of liquid in terms of their volume and mass. Networks formed by the cross-linking of primary polymers are called the polymeric gels, while gels made from colloidal particles or aggregated particles are called particulate gels. When the primary molecules consist of biopolymers, they are called biopolymer gels. Chemical gels have covalent bonds at the cross-links; thus, the connections cannot easily be broken, and the cross-links have a permanent structure throughout their lifetime. In contrast, in physical gels, the cross-links comprise physical interactions. In physical gels, even in equilibrium, the connected and disconnected states of cross-links have an average lifetime. Therefore, the time window in experiments is an important factor when studying the mechanical and rheological properties of gels (Fig. 16.8) [4].

Microcapsules are spherical entities whose diameter is typically micrometer or submillimeter that have a core (internal phase) with a thin surrounding wall membrane. Thus, the core is isolated or partially isolated from its surrounding environment. Microcapsules are utilized to control the rate of release of the internal phase through the wall membrane and are thus used as drug delivery systems. The wall is normally composed of polymers, the cross-linking of which is proceeded in

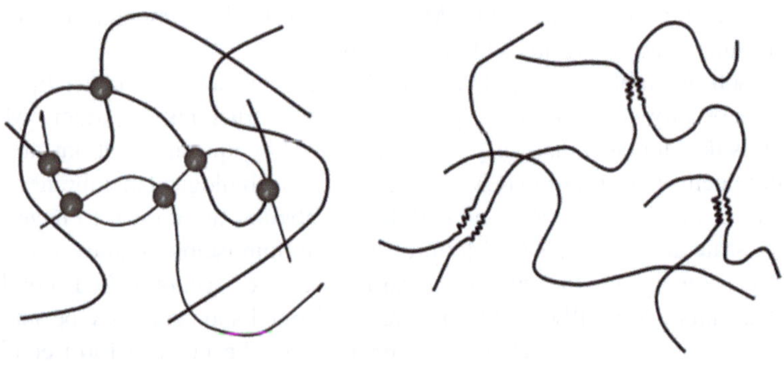

Fig. 16.8 Schematic illustrations of gel network made of covalent bonds and physical association as cross-linker

the preparation process. Therefore, the polymer network is more or less swollen with the solvent. This means that the thin wall membrane of microcapsules is mostly in the gel state.

Liquid crystals are also in an intermediate state between liquids and solids. The molecules of a liquid crystal are oriented in an ordered structure similarly to those in a crystal, while they flow similarly to a liquid. There are various types of liquid crystals consisted of biological and synthetic materials. Electronic displays are widely used as an application of liquid crystals. From a biological viewpoint, lipids, carbohydrates, nucleic acids, and proteins often form a liquid crystal and change their state according to changes in the thermodynamic environment. For example, spider silk has low viscosity, allowing it to flow in the liquid crystalline phase, although it becomes a fiber to form a web after extrusion.

16.1.6 Supplement I. Tissue and Organ

Cells are the building blocks of living organisms. Most cells are visible under an optical microscope, have micrometer-order size, and consist of a protoplasm enclosed in a plasma membrane. An ensemble of similar cells from a tissue has a specific function. Organs are constructed from multiple tissues. Organs have a specific and complicated structure in which numerous molecules aggregate to form cells, tissues, and organs.

16.2 Supplement II. Brief Review of the Classical Theory in Relation to Rheological Measurements

Rheology at the nano- and microscale is associated with the transport, diffusion, aggregation, and relaxation behavior of molecules. The rheological behavior of each system is unique when it undergoes a phase transition or glass transition. The methodologies for studying rheological properties applied to nano-/micro-science and technology need to be understood in relation to the force and molecular motion in systems together with the principles of statistical physics and equilibrium/nonequilibrium thermodynamics. In this Sect. 16.2, a general survey of classical theory together with experimental principles and related phenomena is described.

16.2.1 Supplement II. Force

The structural formation and biological functions of molecules are induced by various forces in biological systems. In most experiments, a stimulus is given to a system to obtain physical quantities as a response. The forces used in experiments include mechanic, electronic, magnetic, thermal force, and so forth.

A fundamental definition of the van der Waals interaction between two molecules is the attractive force formulated in quantum mechanics. In practice, the potential $V_{12}(r)$, which describes the dispersive force between molecules 1 and 2, is a function of the distance r between two molecules,

$$V_{12}(r) = -\frac{A_{12}}{r^6}. \tag{16.3}$$

Here, A_{12} is the Hamaker constant. This implies that there is a characteristic length scale that decays as $1/r^6$ relative to the potential with the constant of proportionality, A_{12}. The Hamaker constant A_{12} is a system-dependent quantity. Typically, the strength of van der Waals forces is on the order of 1 kJ mol^{-1}, which is relatively weak compared with other forces. The Lennard-Jones potential is often used for molecular dynamics simulations as the energy of the interaction,

$$U(r) = 4\varepsilon\left[\left(\frac{\sigma}{r}\right)^{12} - \left(\frac{\sigma}{r}\right)^{6}\right], \tag{16.4}$$

where r is the distance between the particles and ε and σ are characteristic constants.

Hydrogen bonds are stronger than van der Waals force, and their energy is in the range of 10–40 kJ mol^{-1}, although they are still weaker than ionic interactions and covalent bonds by an order of magnitude. Hydrogen bonds are formed between the proton-donating groups in molecules and the proton-accepting atoms. The donor group is a polar group such as O–H, N–H, or S–H, whereas the acceptor is an electronegative atom such as oxygen, nitrogen, or sulfur. Hydrogen bonding plays a key role in the molecular dynamics of aqueous systems such as micelles, membrane formation in biological cells, and in the higher-order conformation of proteins and polynucleotides.

Coulomb's law is the fundamental formula used to consider the electrostatic interactions in systems. The interaction energy E between two point charges at a distance r is

$$E(r) = \frac{q_1 q_2}{4\pi\varepsilon_r\varepsilon_0 r}, \tag{16.5}$$

where q_1 and q_2 are the magnitudes of the point charges and ε_r and ε_0 are the relative dielectric permittivity of materials and the dielectric permittivity in vacuum, respectively. The electrostatic force propagates over a long distance and is often referred to as a long-range interaction, and the strength of ionic bonds between molecules is typically on the order of 100 kJ mol^{-1}.

The screening effect of the Coulombic force is also an important factor for aqueous solutions. An electric double layer forms around a charged group, which screens the Coulombic interactions of electrolytes in aqueous solutions. The screening effect is considered using the characteristic length scale (κ^{-1}) defined by the Debye screening length,

$$\kappa^2 = \frac{(ze)^2 \rho}{2\varepsilon_r \varepsilon_0 k_B T}. \tag{16.6}$$

Here, z is the valence of the counterion, e is the electronic charge, and ρ is the charge on the surface.

16.2.2 Supplement II. Molecular Motion and Diffusion

The Arrhenius equation, which is equivalent to the Eyring equation, is widely accepted as an empirical relationship holding in various physical and chemical phenomena such as the temperature variation of the relaxation of molecular motion, the diffusion, mechanical relaxation of materials, common chemical reactions, and other thermally induced processes/reactions. It gives a fundamental means of describing the relationship among the rate constant k of various processes/reactions, the activation energy E_a, and the absolute temperature T,

$$k = Ae^{-E_a/k_B T}. \tag{16.7}$$

Here, A is the prefactor, and k_B is the Boltzmann constant. The rate constant k indicates the number of collisions resulting in a process or reaction per second, A is the total number of collisions per second, and the term $e^{-E_a/k_B T}$ gives the probability of observing collisions in the reaction or process.

Fick's first equation gives the flux J_x of particles diffusing in accordance with the gradient of the particle concentration $\partial c/\partial x$ as

$$J_x = -D\frac{\partial c}{\partial x}. \tag{16.8}$$

The constant of proportionality is the diffusion coefficient D, which describes the translational diffusion of particles. When the concentration c of particles is in mole/m^3, the diffusion coefficient has the unit m^{2}s^{-1}.

Fick's second law is written as

$$\frac{\partial c}{\partial t} = D\frac{\partial c^2}{\partial x^2}. \tag{16.9}$$

When describing in the three-dimensional form, the Fick's first and second laws can be written as

$$\mathbf{J} = -D\nabla c, \tag{16.10}$$

$$\frac{\partial c}{\partial t} = D\nabla^2 c. \tag{16.11}$$

Here, ∇ is the differential operator $\nabla \equiv \left(\frac{\partial}{\partial x}, \frac{\partial}{\partial y}, \frac{\partial}{\partial z}\right)$ and $\nabla^2 = \frac{\partial^2}{\partial x^2} + \frac{\partial^2}{\partial y^2} + \frac{\partial^2}{\partial z^2}$ is the Laplace operator.

The diffusion coefficient D is related to the friction of the particle in the solvent ς, and the formula shown below is called the Einstein relation:

$$D = \frac{k_B T}{\varsigma}. \tag{16.12}$$

When the diffusing particle is a rigid sphere with radius r migrating in a solvent, the solvent viscosity η and friction coefficient are related by Stokes' law as

$$\varsigma = 6\pi\eta r. \tag{16.13}$$

Combining these two equations gives the Stokes–Einstein equation

$$D = \frac{k_\mathrm{B} T}{6\pi\eta r}. \tag{16.14}$$

In the case of polymer solutions, the radius r of the sphere is replaced by the radius of the hydrodynamically equivalent one R_H, which is called the hydrodynamic radius.

16.2.3 Supplement II. Phenomenological Equation in Nonequilibrium Thermodynamics

Complex natural processes are irreversible and are often described by nonequilibrium thermodynamics. Physical and chemical events considered in nano-/micro-science and technology in biorheology are in principle inhomogeneous, for which the external forces and reaction rates should be considered. In an empirical treatment, the irreversible flow is expressed as a linear function of thermodynamic forces. Fick's first law in Eq. (16.8) is based on this concept. As a more general description of the flux in phenomenological irreversible processes, the linear relation is written as

$$J_i = \sum_{j=1}^{k} L_{ij} X_i \quad (i = 1, 2, \cdots, k), \tag{16.15}$$

where X_i is a thermodynamic force and L_{ij} is called a phenomenological coefficient. The entropy production σ in the system has the form $\sigma = \sum_i J_i X_i$. Assuming an ideal binary mixture having a concentration gradient for simplicity, the flux of component 1 induced by the gradient of the chemical potential $\nabla\mu_1$ as the driving force is given by

$$J_1 = L_1 \cdot \nabla\mu_1 \\ = c_1 v_1 , \tag{16.16}$$

where c_1 is the concentration and v_1 is the velocity of migration of component 1. For an ideal case, the gradient of the chemical potential is expressed in terms of the concentration gradient and is related to the gas constant R and temperature T as

$$\nabla\mu_1 = \left(\frac{\partial\mu_1}{\partial c_1}\right)_{p,T,c_2} \nabla c_1 = \frac{RT}{c_1}\nabla c_1. \tag{16.17}$$

The velocity of component 1, which is inversely proportional to the friction ς and driven by the gradient of the chemical potential, is expressed by combining Eq. (16.17) to give

$$v_1 = -\frac{1}{\varsigma}\nabla\mu_1 = -\frac{RT}{\varsigma c_1}\nabla c_1. \tag{16.18}$$

Comparing Eq. (16.10) or (16.16) with Eq. (16.18), we obtain the relations $D = RT/\varsigma$ and $L_1 = -c_1/\varsigma$.

16.2.4 Supplement II. Phase Transition and Glass Transition

The term "phase" is used to describe solid, liquid, gas, and plasma thermodynamic systems. A phase transition is the transformation of a system from one phase to another. Figure 16.9 shows a phase diagram of water, where S, L, and G refer to the solid, liquid, and gas phases, respectively. TP is the triple point where all three phases coexist and CP is the critical point of the liquid–gas phase transition. The plasma phase is usually achieved in a high-energy environment (for instance, at a high temperature) where the molecules are in the state of an ionized gas. Although the scope of plasma science has broadened to include applications such as the modification of materials including biological substances, in this textbook, we focus on the solid, liquid, and gas phases.

At the phase transition point (for instance, the boiling point or condensation point), the liquid and gas phases have identical free energies. When phases I and II coexist in thermodynamic equilibrium, the chemical potentials μ^{I} and μ^{II} of the phases satisfy

$$\mu^{\mathrm{I}}(p,T) = \mu^{\mathrm{II}}(p,T). \tag{16.19}$$

Differentiating the equation with respect to T gives

$$\frac{\partial\mu^{\mathrm{I}}}{\partial T} + \frac{\partial\mu^{\mathrm{I}}}{\partial p}\frac{dp}{dT} = \frac{\partial\mu^{\mathrm{II}}}{\partial T} + \frac{\partial\mu^{\mathrm{II}}}{\partial p}\frac{dp}{dT}, \tag{16.20}$$

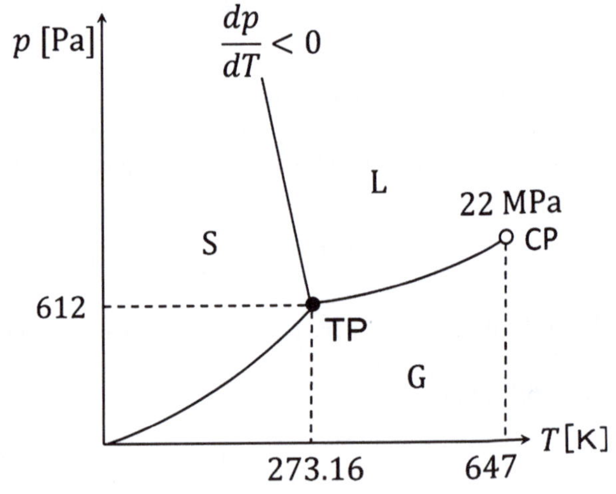

Fig. 16.9 Phase diagram of water

which can be rewritten as

$$-S^{\mathrm{I}} + V^{\mathrm{I}}\frac{dp}{dT} = -S^{\mathrm{II}} + V^{\mathrm{II}}\frac{dp}{dT}, \tag{16.21}$$

where the equations $(\partial\mu/\partial T)_p = -S$ and $(\partial\mu/\partial p)_T = V$ and the Gibbs–Duhem relation $d\mu = -SdT + Vdp$ are used. Here, S and V are the entropy and volume of the system, respectively. From this equation, the slope dp/dT of the liquid–gas boundary can be obtained as

$$\frac{dp}{dT} = \frac{S^{\mathrm{II}} - S^{\mathrm{I}}}{V^{\mathrm{II}} - V^{\mathrm{I}}} = \frac{\Delta H}{T\Delta V}. \tag{16.22}$$

This is called the Clausius–Clapeyron relation. ΔH is the change in enthalpy at the phase transition, expressed as $\Delta H = T(S^{\mathrm{II}} - S^{\mathrm{I}})$. Correspondingly, $\Delta V = V^{\mathrm{II}} - V^{\mathrm{I}}$ is defined. Because $S^{\mathrm{II}} > S^{\mathrm{I}}$ holds to entropy, the enthalpy change should be $\Delta H > 0$, which implies (1) $\Delta V > 0$ when $dp/dT > 0$, and (2) $\Delta V < 0$ when $dp/dT < 0$. In the p–T phase diagram of water, the slope at the solid–liquid boundary is negative ($dp/dT < 0$), which means that the volume of water decreases ($\Delta V < 0$) when ice is melted by heating. This is rather unusual and is due to the effect of the hydrogen bonds in water [2].

Many materials normally exhibit freezing behavior, observed as a first-order phase transition, upon cooling, although rapid cooling often prevents the transition of liquids to solids, resulting in a glass transition. Many polymers are known to exist in a glassy state with an amorphous structure at room temperature and atmospheric pressure. The liquid to glass transition is not a transition between states described by thermodynamics. The glass state is considered to be a nonequilibrium state which exists in a kinetically pinned state of components; thus, the glass transition is primarily a dynamic phenomenon depending on the thermal history.

When considering the rotational motion around C–C covalent bonds in polymer chains, the frequency of transitions between different isomeric structures is characterized by the characteristic time, as described by Eq. (16.2). According to the equation, lowering the temperature leads to the characteristic time τ of motion of different isomers becoming larger than the observation time window. In this case, the polymer chain appears to be frozen. Typically, the glass transition temperature T_g for polymers is defined as corresponding of a relaxation time of $\tau = 100–1{,}000$ s by dielectric relaxation spectroscopy. The glass transition temperature T_g determined by dielectric relaxation spectroscopy is coincident with that determined by the thermal analytical technique of differential scanning calorimetry (DSC).

Recently, a jamming transition of granular media, colloidal suspensions, and other complex materials has attracted the interest of physicists, especially concerning whether the jamming transition is a new type of phase transition or an analog to the glass transition. The jamming transition occurs when the density of a system increases which exhibiting nonequilibrium transitions from a fluidlike to a solid-like state. This topic is of interest in nano-/micro-science and technology since it is associated with measurements of changes in viscoelasticity, which are directly related to the properties of particles at nano-/microscale.

In addition, there are many types of transition even among the structural transition, for instance, the sol to gel transition and the liquid crystal phase to gel phase transition in lipid bilayers. Micelles and liquid crystals have various structural phases depending on their concentration, temperature, and the presence of additives. Biological systems appear to be able to effectively use the structural phase transitions of DNA and proteins to form compact structures in volume in living cells. These transitions are often reversible, similarly to those in thermodynamically treated systems, although their molecular mechanisms are still interesting topics.

16.3 Supplement III. Principle of Experimental Method and Related Physical Property

16.3.1 Supplement III. Viscoelasticity

In the case of the small deformation of isotropic materials, there is a simple relationship between the stress σ (=F/A; the force F divided by the area A) and the strain γ (= $\Delta l/l$; the change in length Δl is divided by the original length l) (Fig. 16.10). For linear materials, the stress–strain relationship has the form

$$\sigma = G\gamma, \tag{16.23}$$

where the constant of proportionality is the elastic modulus G. This relationship is called Hooke's law, and solids that obey Hooke's law are called Hookean solids. G is often called the shear modulus and the rigidity, when considering the deformation of isotropic solids and the shear strain.

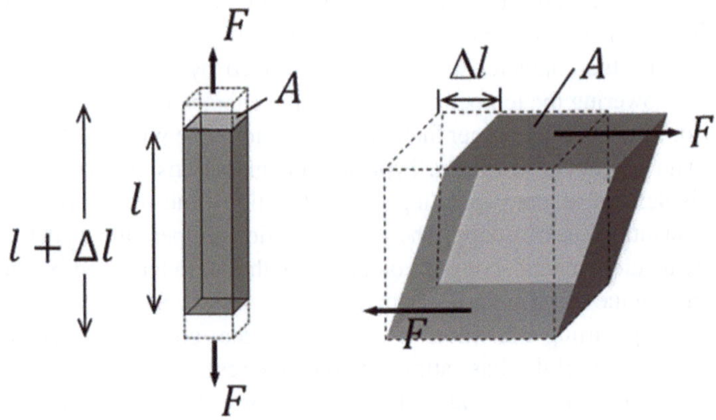

Fig. 16.10 Geometries to define the stress–strain relationship

In the case of longitudinal strain, the constant of proportionality is called Young's modulus E,

$$\sigma = E\gamma. \tag{16.24}$$

The velocity of a flowing liquid is proportional to the stress applied to the liquid. This relationship is described by the stress σ and shear rate $\dot{\gamma} = d\gamma/dt$ as

$$\sigma = \eta \frac{d\gamma}{dt}. \tag{16.25}$$

The constant of proportionality η is called the viscosity. This is known as Newton's law of viscosity, and liquids that obey Newton's law are called Newtonian fluids. In practice, most simple liquids of low-molecular-weight materials are considered to be Newtonian fluids. In this case, the viscosity η is a material constant, although it depends on the temperature and other thermodynamic factors. On the other hand, polymer solutions, colloidal suspensions, and most liquids of biological sources are non-Newtonian fluids. Thus, the viscosity η of a non-Newtonian fluid depends on the shear stress and shear rate.

A Hookean body or Newtonian fluid is an ideal material. In practice, materials considered in biorheology have the properties of both an elastic body and a viscous fluid owing to their complex molecular structures. Thus, biological materials are considered to be viscoelastic materials. The force applied to a Hookean body is stored as elastic energy in materials, whereas the energy (the force applied to a Newtonian fluid) disperses as thermal energy to the environment. To understand the properties of viscoelastic materials, models relating the applied force (stress) to the resultant deformation (strain) have been developed. The Maxwell model and the Voigt (or Kelvin) model are the simplest models for viscoelastic materials and can be described using a dashpot and a spring (Fig. 16.11). An ideal elastic body can be

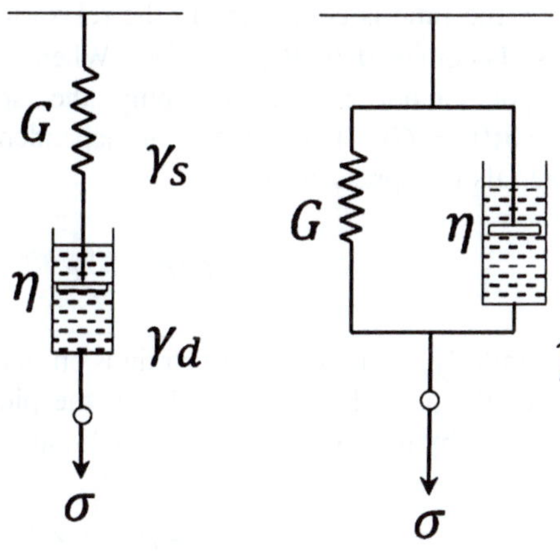

Fig. 16.11 Schematic drawing of Maxwell model (*left*) and Voigt model (*right*)

described using a spring that follows Hooke's law, and an ideal viscous fluid can be treated with Newton's law.

The Maxwell model is constructed by connecting a dashpot and a spring in series. The strain rates of the dashpot and the spring can be summed linearly, resulting in the total strain being the sum of each component, which is described as

$$\dot{\gamma} = \frac{\dot{\sigma}}{G} + \frac{\sigma}{\eta} \tag{16.26}$$

or

$$\gamma = \gamma_s + \gamma_d, \tag{16.27}$$

The Voigt model is described by the parallel connection of a dashpot and a spring. The total stress of the Voigt model is the sum of the stresses of the dashpot and the spring, with the strains being coincident for both components, as

$$\sigma = G\gamma + \eta\dot{\gamma} \tag{16.28}$$

or

$$\gamma = \gamma_s = \gamma_d. \tag{16.29}$$

The relaxation of stress with time after a rapid increase in strain can be calculated using the Maxwell model from the solution of Eq. (16.26) as

$$\sigma = \sigma(t) = \sigma_0 e^{-t/\tau}, \tag{16.30}$$

where the time constant τ is the relaxation time ($\tau = \eta/G$) and σ_0 is the initial stress applied. The relaxation time characterizes the relaxation behavior of materials.

When the time is equivalent to the relaxation time ($t = \tau$), the magnitude of the stress becomes $0.3679\sigma_0 = e^{-1}\sigma_0$. When a rapid stepwise strain is applied, the dashpot cannot move, and only the spring responds to the strain, i.e., $\sigma_0 = \sigma(0) = G\gamma$. Thus, the time-dependent relaxation modulus for Maxwell materials is expressed as

$$G(t) = \frac{\sigma_0}{\gamma} e^{-t/\tau} = Ge^{-t/\tau}. \tag{16.31}$$

Similarly, the relaxation of strain is observed when a constant stress σ_0 is applied in the Voigt model. This results in the phenomenon known as creep, which is described by the solution of Eq. (16.28) as

$$\gamma = \gamma(t) = \gamma_0\left(1 - e^{-t/\tau}\right). \tag{16.32}$$

Here, γ_0 is the constant $\gamma_0 = \gamma(\infty) = \sigma_0/G$ since the rate of strain $\dot{\gamma}$ becomes zero at infinite time. Thus, the creep compliance J, defined as $J(t) = \gamma(t)/\sigma$, is expressed as

$$J(t) = J\left(1 - e^{-t/\tau}\right). \tag{16.33}$$

In the case of the creep compliance, the relaxation time τ is often called the retardation time.

For more general cases, appropriate models are necessary to describe the experimental observations. It is also necessary to consider the suitable model for each material under observation. This is because the model affects the final results, where the relationship between the molecular structure and viscoelasticity is one of the most important subjects of nano-/micro-science and technology in biorheology yet to be elucidated.

In measurements by mechanical spectroscopy to study linear viscoelasticity of materials, a mechanical perturbation (a stress or strain) is applied to a material, and its response is measured as a function of time or frequency. When a sinusoidal perturbation with angular frequency ω ($\omega = 2\pi f$) is applied to a material,

$$\gamma(t) = \gamma_0 \cos(\omega t), \tag{16.34}$$

the resulting strain is obtained in the range of linear viscoelasticity:

$$\sigma(t) = \sigma_0 \cos(\omega t + \delta). \tag{16.35}$$

Imaginary expressions of Eqs. (16.34) and (16.35) are

$$\gamma(t) = \gamma_0 e^{i\omega t}, \tag{16.36}$$

$$\sigma(t) = \sigma_0 e^{i(\omega t + \delta)}, \tag{16.37}$$

where i is the imaginary unit $(i^2 = -1)$ and δ is the phase angle. In this case, the dynamic modulus is expressed in the complex form G^* as

$$G^* = \frac{\sigma(t)}{\gamma(t)} = \frac{\sigma_0}{\gamma_0} e^{i\delta} = G' + iG''. \tag{16.38}$$

The real part G' and imaginary part G'' of the complex modulus G^* are often called the storage modulus and loss modulus, respectively. Here, the phase angle of G'' is rotated 90° from the direction of the applied perturbation and corresponds to the phase angle of viscosity element in the stress-rate of strain relation $(\sigma = \eta\dot{\gamma})$. Introducing the complex viscosity η^* for the sinusoidal perturbation gives

$$\sigma = \eta^* \dot{\gamma} = i\omega\eta^* \gamma. \tag{16.39}$$

Then, by comparing with Eq. (16.38), we obtain

$$\frac{\sigma}{\gamma} = G^* = i\omega\eta^*. \tag{16.40}$$

For the Maxwell model, the complex modulus G^* is derived as

$$G^* = G \frac{i\omega\tau}{1 + i\omega\tau} \tag{16.41}$$

with

$$G' = G \frac{\omega^2\tau^2}{1 + \omega^2\tau^2}, \quad G'' = G \frac{\omega\tau}{1 + \omega^2\tau^2}. \tag{16.42}$$

16.3.2 Supplement III. Electromagnetic Wave and Scattering Method

The methods used for the structural determination of biological molecules are classified into real- and inverse-space observations. Microscopy (with light or electrons as the source) gives real-space images. Scattering methods (with neutron, X-ray, or visible light as the source) are utilized to determine the structure of molecules and their aggregates. Furthermore, modern methods have intensively been developed for use in nano-/micro-science and technology in biorheology, for instance, AFM, NMR imaging, and fluorescence experiments.

In particular, the momentum transfer q for an elastic scattering process or energy-conserving scattering is expressed as $|\mathbf{q}| = q = \frac{4\pi}{\lambda} \sin\left(\frac{\theta}{2}\right)$, where λ is the scattering wavelength and θ is the angle between the incident radiation and the scattered light (Fig. 16.12). The magnitude of the wave vector q has the dimension

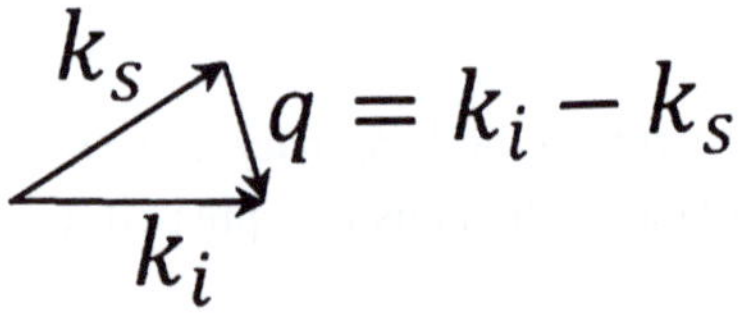

Fig. 16.12 Geometry to describe the relationship of the vector q, incident wave vector k_i, and scattered radiation vector k_s

of the inverse of the length [m^{-1}], instead of the angle θ, which is useful for comparing different sources of radiation (neutron, X-ray, visible light, and forth).

In addition to the structural determination of biological molecules, scattering methods can be used to study the dynamics of biological molecules without perturbing the sample morphology. The dynamics of molecules can be studied by observing the time decay behavior of the detected light owing to the molecular motion of particles or the association of molecules with the radiation energy [4–6].

16.3.3 Supplement III. SPM

Atomic force microscopy (AFM) is a method based on the scanning probe microscopy (SPM) that has entered widespread use in materials science. An atomic force microscope employs a cantilever, with a sharp tip at its end to scan the specimen surface. When the tip is located on the specimen surface, the force between the tip and the specimen depends on the surface properties and surface morphology. Forces that are measured by AFM include van der Waals forces, electrostatic forces, solvation forces, and mechanical contact forces. The deflection of a laser beam is used to analyze the movement of the cantilever. AFM has an imaging resolution typically on the order of ~10 nm. Because the principle of AFM depends on the interactions between the surface components and the tip, various measurement modes are available such as the contact mode, noncontact mode, tapping mode, force curve measurement mode, and so forth.

16.3.4 Supplement III. Magnetic Property and NMR Spectroscopy

Nuclear magnetic resonance (NMR) is a powerful tool for studying the physical and chemical properties of molecules to obtain information about their structure, dynamics, and physicochemical environment. Nuclei (such as ^1H or ^{13}C) absorb and reemit electromagnetic radiation in a magnetic field. The energy of the absorbance depends on the strength of the magnetic field and the magnetic properties of the isotope of the atoms. NMR is also used as a medical imaging technique, where it is known as magnetic resonance imaging (MRI). For the study of molecular dynamics, the relaxation times of spin–lattice interactions (T_1) and spin–spin interactions (T_2) are determined, which are the physical processes responsible for the relaxation of components.

The nuclear spin magnetization vector M is considered to be parallel or perpendicular to the external magnetic field B_0, which is assumed to be oriented along the z-axis. The spin–lattice (longitudinal) relaxation time T_1 provides the relaxation constant of the z-component of the nuclear spin magnetization M_z when it relaxes to its thermal equilibrium state as

$$M_z(t) = M_{zeq} - \left[M_{zeq} - M_z(0)\right] e^{-t/T_1}. \tag{16.43}$$

Here, M_{zeq} is the value of M_z at thermal equilibrium. The spin–spin (transverse) relaxation time T_2 indicates the decay behavior of the component M_{xy} perpendicular to B_0, which is described as

$$M_{xy}(t) = M_{xy}(0)e^{-t/T_2}. \tag{16.44}$$

16.3.5 Dielectric Property and Relaxation Spectroscopy

Dielectric relaxation spectroscopy is used to measure the response function of the electric dipole moment of materials by applying an electromagnetic field to the materials. The interpretation of the resultant permittivity spectra is related to the molecular dynamics of materials. Dielectric relaxation spectroscopy now covers a frequency range of about 18 decades, in frequency 10^{-7}–10^{11} Hz, which gives information on the molecular motion of glass-forming materials, the micro-Brownian motion and segmental motion of polymers, the rotational motion of low-molecular-weight materials, and so forth.

The permittivity ε of a dielectric material is related to its electric displacement D and the electric field E. When an electric field E is applied to a dielectric material, the induced electric displacement D in the material is

$$D = \varepsilon_0 E + P \tag{16.45}$$

where ε_0 is the dielectric constant in vacuum and P is the depolarization in the material ($P = \mu N$ with magnitude of dipole moment μ and number of dipole moments N).

When an alternating electric field is applied, the permittivity is derived in the form of the complex permittivity $\varepsilon*$

$$\varepsilon^* = \varepsilon' - i\varepsilon'', \tag{16.46}$$

where the real and the imaginary parts of permittivity are described as

$$\varepsilon' - \varepsilon_\infty = \frac{\varepsilon_0 - \varepsilon_\infty}{1 + \omega^2\tau^2}, \quad \varepsilon'' = \frac{(\varepsilon_0 - \varepsilon_\infty)\omega\tau}{1 + \omega^2\tau^2}. \tag{16.47}$$

Here, ω and τ are the angular frequency and dielectric relaxation time, respectively. ε_0 and ε_∞ are the low- and high-frequency limit permittivity, respectively. This is called the Debye equation.

References

1. Waigh TA (2007) Applied biophysics; a molecular approach for physical scientists. Wiley, Chichester
2. Tanaka F (2011) Polymer physics; application to molecular association and thermoreversible gelation. Cambridge University Press, Cambridge
3. Kamide K, Dobashi T (2000) Physical chemistry of polymer solutions. Elsevier, Amsterdam
4. Djabourov M, Nishinari K, Ross-Murphys SB (2013) Physical gels from biological and synthetic polymers. Cambridge University Press, Cambridge
5. Macosko CW (1994) Rheology; principles, measurements, and applications. Wiley-VCH, New York
6. Shaw MT (2012) Introduction to polymer rheology. Wiley, Hoboken

Index

R. Kita, T. Dobashi (eds.), *Nano/Micro Science and Technology in Biorheology*, DOI 10.1007/978-4-431-54886-7

D

MIX
Papier aus verantwortungsvollen Quellen
Paper from responsible sources
FSC® C105338

FSC
www.fsc.org

If you have any concerns about our products,
you can contact us on
ProductSafety@springernature.com

In case Publisher is established outside the EU,
the EU authorized representative is:
Springer Nature Customer Service Center GmbH
Europaplatz 3, 69115 Heidelberg, Germany

Printed by Libri Plureos GmbH
in Hamburg, Germany